AF333802

Oceanography of the East Sea (Japan Sea)

Kyung-Il Chang · Chang-Ik Zhang · Chul Park
Dong-Jin Kang · Se-Jong Ju · Sang-Hoon Lee
Mark Wimbush
Editors

Oceanography of the East Sea (Japan Sea)

 Springer

Editors
Kyung-Il Chang
Research Institute of Oceanography
School of Earth and Environmental Sciences
Seoul National University
Seoul
Republic of Korea

Chang-Ik Zhang
Division of Marine Production
  System Management
Pukyong National University
Busan
Republic of Korea

Chul Park
Department of Oceanography and Ocean
  Environmental Sciences
Chungnam National University
Daejeon
Republic of Korea

Dong-Jin Kang
Marine Chemistry and Geochemistry Division
Korea Institute of Ocean Science
  and Technology
Ansan
Republic of Korea

Se-Jong Ju
Deep-Sea and Seabed Resources
  Research Center
Korea Institute of Ocean Science
  and Technology
Ansan
Republic of Korea

Sang-Hoon Lee
Korean Seas Geosystem Research Unit
Korea Institute of Ocean Science
  and Technology
Ansan
Republic of Korea

Mark Wimbush
Graduate School of Oceanography
University of Rhode Island
Rhode Island
USA

ISBN 978-3-319-22719-1   ISBN 978-3-319-22720-7   (eBook)
DOI 10.1007/978-3-319-22720-7

Library of Congress Control Number: 2015946090

Springer Cham Heidelberg New York Dordrecht London
© Springer International Publishing Switzerland 2016

Cover Image: The image shows a distribution of chlorophyll *a* concentration over the East Sea derived from Geostationary Ocean Color Imager (GOCI) observations taken in September, 2011. It reveals various ocean surface features such as fronts, plumes, filaments and eddies. Natural color composite is shown on the land.
GOCI, the first ocean color instrument operated on geostationary orbit, is collecting ocean color radiometry data since July, 2010. GOCI has an unprecedented capability to provide eight images a day with a 500 m resolution for the North East Asian seas around Korean peninsula.

Printed on acid-free paper

Springer International Publishing AG Switzerland is part of Springer Science+Business Media
(www.springer.com)

Preface

Because its characteristics resemble those of global oceans, the East Sea (Japan Sea) has long been studied and explored. Marine expeditions in the East Sea date back to the mid-1800s, and investigations of some important aspects of its oceanographic characteristics, especially water masses and surface circulation, were begun in the early 1900s. With the advent of computer resources as a powerful tool for numerical model simulations, together with modern methods of investigation such as satellite altimetry, Argos drifters, ARGO floats, and moored current measurements, the study of the East Sea has accelerated during the past two decades. In particular, major scientific achievements have flourished since the late 1990s. Six special issues of major international journals were published from 1999 to 2009 containing results from international programs and workshops dedicated to the East Sea. Two important events contributed to this scientific advancement: CREAMS and PAMS. CREAMS (Circulation Research of the East Asian Marginal Seas) began in 1993 as a comprehensive observational program organized by scientists from Japan, Korea, and Russia, the three countries bordering the East Sea. Many joint international cruises have been carried out over the entire East Sea, thus accelerating our understanding of the Sea. Joint cruises are still ongoing, and CREAMS is now an official program of the North Pacific Marine Science Organization (PICES). The special issues mentioned above were publications of results from CREAMS that were presented in related workshops and symposia. PAMS is an acronym for Pacific-Asian Marginal Seas and refers to a biennial workshop that provides a venue for scientists to present knowledge and share ideas about PAMS, and promote international cooperation in PAMS. The first workshop convened in 1981 with the initial name of JECSS (Japan and East China Seas Study), and its first 10 years PAMS exclusively focused on the physical and chemical oceanography of the East Sea and the East China Sea. The region of interest eventually broadened to include other northwestern Pacific marginal seas, and its name was changed to PAMS/JECSS in 1993 and finally to PAMS in 2009. Regular PAMS workshops have subsequently led to the publication of papers in 11 special issues, including books.

One of the major scientific topics relating to the East Sea oceanography is its own thermohaline circulation, similar to that occurring in the North Atlantic. Sill depths of the straits connecting the East Sea with adjacent seas and the North Pacific are shallower than 200 m, and subsurface waters below about 300 m constitute the Proper Water of the East Sea, first named by the famous Japanese oceanographer, Prof. Michitaka Uda in 1934. The Proper Water and intermediate water masses occupy more than 90 % of the East Sea water volume. They are formed in the northern East Sea, discharged to the south, and modified within the East Sea. Another unique property of the East Sea is its high biological productivity, especially in its southwestern part, the Ulleung Basin, where the nutrient-depleted Kuroshio branch, the Tsushima Warm Current, prevails. The primary productivity in the Ulleung Basin is as high as that of the ocean's major upwelling regions. Coastal upwelling, large- and meso-scale circulation are thought to play a role in maintaining this high productivity. Thus the East Sea is an ideal place to address the calibration of a wide range of proxies for ocean ventilation and productivity based on present oceanic conditions, as well as down-core records of the past. The evidence for rapid changes of physical and biogeochemical properties in the East Sea is compelling. Despite its long history of observations and studies, a comprehensive understanding of the ongoing changes and future projections of the East Sea is yet to be provided.

This book was written as a monograph summarizing current knowledge in the various field of oceanography of the East Sea, with the editors' hope that it will provide a useful compilation of previous important studies on each topic, and thus serve as a reference for anyone interested in the East Sea as well as providing motivation for more in-depth, future studies.

The book consists of 18 chapters, covering physical oceanography in Chaps. 2–5, chemical oceanography in Chaps. 6–9, biological and fisheries oceanography in Chaps. 10–15, and geological oceanography in Chaps. 16–18, together with a general introduction and details of the CREAMS program in Chap. 1. Each chapter serves as a stand-alone article addressing a specific topic in the form of a single scientific paper, including its own list of references. Interdisciplinary discussions of processes, such as physical-biological coupling, are somewhat scattered throughout various chapters. Editors and authors of each chapter have made a special effort to include prior publications exhaustively, though not completely, especially non-English papers. A list of abbreviations and a subject index will help readers to understand terminology specific to the East Sea and to use this book as a reference handbook. Readers are also recommended to refer to Chaps. 1 and 4 for topographic features and names, and basin- and meso-scale upper circulation features of the East Sea, which are often mentioned in other chapters.

We gratefully acknowledge financial support provided by the Korean Society of Oceanography. Our appreciation is also extended to external reviewers who read manuscripts critically and suggested valuable comments. Last but not least, we wish to express our sincere gratitude to Dr. Jong Yup Han at the Korea Institute of Ocean Science and Technology for his assistance with the production of this book.

Kyung-Il Chang
Chang-Ik Zhang
Chul Park
Dong-Jin Kang
Se-Jong Ju
Sang-Hoon Lee
Mark Wimbush

Contents

Contributors

Soonmo An Department of Oceanography, Pusan National University, Busan, Republic of Korea

Jang Jun Bahk Petroleum and Marine Research Division, Korea Institute of Geoscience and Mineral Resources, Daejeon, Republic of Korea

Hyun Ju Cha Research Institute of Marine Sciences, Chungnam National University, Daejeon, Republic of Korea

Kyung-Il Chang School of Earth and Environmental Sciences, Seoul National University, Seoul, Republic of Korea

Dong Han Choi Biological Oceanography and Marine Biology Division, Korea Institute of Ocean Science and Technology, Ansan, Republic of Korea

Jin-Woo Choi South Sea Research Institute, Korea Institute of Ocean Science and Technology, Geoje, Republic of Korea

Joong Ki Choi Department of Oceanography, Inha University, Incheon, Republic of Korea

Jeomshik Hwang School of Earth and Environmental Sciences, Seoul National University, Seoul, Republic of Korea

Jung-Ho Hyun Department of Marine Sciences and Convergent Technology, Hanyang University, Ansan, Republic of Korea

Se-Jong Ju Deep-Sea and Seabed Resources Research Center, Korea Institute of Ocean Science and Technology, Ansan, Republic of Korea

Uk-Jae Jung School of Earth and Environmental Sciences, Seoul National University, Seoul, Republic of Korea

Dong-Jin Kang Marine Chemistry and Gochemistry Division, Korea Institute of Ocean Science and Technology, Ansan, Republic of Korea

Young-Shil Kang Korea Fisheries Resources Agency, Busan, Republic of Korea

Guebuem Kim School of Earth and Environmental Sciences, Seoul National University, Seoul, Republic of Korea

Han-Joon Kim Korean Seas Geosystem Research Unit, Korea Institute of Ocean Science and Technology, Ansan, Republic of Korea

Il-Nam Kim Department of Marine Science, Incheon National University, Incheon, Republic of Korea

Jeonghyun Kim School of Earth and Environmental Sciences, Seoul National University, Seoul, Republic of Korea

Kyung-Ryul Kim GIST College, Gwangju Institute of Science and Technology, Gwangju, Republic of Korea

Seong-Pil Kim Korea Institute of Geoscience and Mineral Resources, Pohang Branch, Pohang, Republic of Korea

Suam Kim Department of Marine Biology, Pukyong National University, Busan, Republic of Korea

Tae-Hoon Kim Department of Earth and Marine Sciences, Jeju National University, Jeju, Republic of Korea

Young-Gyu Kim Naval Systems R&D Institute, Agency for Defense Development, Jinhae, Republic of Korea

Young Ho Kim Physical Oceanography Division, Korea Institute of Ocean Science and Technology, Ansan, Republic of Korea

Yun-Bae Kim Ulleungdo-Dokdo Ocean Science Station, Korea Institute of Ocean Science and Technology, Ulleung-Gun, Republic of Korea

Dong-Kyu Lee Department of Oceanography, Pusan National University, Busan, Republic of Korea

Gwang Hoon Lee Department of Energy Resources Engineering, Pukyong National University, Busan, Republic of Korea

Kitack Lee School of Environmental Science and Engineering, Pohang University of Science and Technology, Pohang, Republic of Korea

Sang Heon Lee Department of Oceanography, Pusan National University, Busan, Republic of Korea

Sang Hoon Lee Korean Seas Geosystem Research Unit, Korea Institute of Ocean Science and Technology, Ansan, Republic of Korea

Tongsup Lee Department of Oceanography, Pusan National University, Busan, Republic of Korea

Hanna Na Faculty of Science, Hokkaido University, Sapporo, Hokkaido, Japan

SungHyun Nam School of Earth and Environmental Sciences, Seoul National University, Seoul, Republic of Korea

Jae Hoon Noh Biological Oceanography and Marine Biology Division, Korea Institute of Ocean Science and Technology, Ansan, Republic of Korea

Tatiana Orlova Zhirmunsky Institute of Marine Biology, Far East Division, Russian Academy of Sciences, Vladivostok, Russia

Chul Park Department of Oceanography and Ocean Environmental Sciences, Chungnam National University, Daejeon, Republic of Korea

Jae-Hun Park Department of Ocean Sciences, Inha University, Incheon, Republic of Korea

Jong Jin Park School of Earth System Sciences, Kyungpook National University, Daegu, Republic of Korea

Jun-Yong Park Korean Seas Geosystem Research Unit, Korea Institute of Ocean Science and Technology, Ansan, Republic of Korea

Kyum Joon Park Cetacean Research Institute, National Fisheries Research and Development Institute, Ulsan, Republic of Korea

Kyung-Ae Park Department of Earth Science Education, Seoul National University, Seoul, Republic of Korea

Mi-Ok Park Department of Oceanography, Pukyong National University, Busan, Republic of Korea

Young-Je Park Korea Ocean Satellite Center, Korea Insitute of Ocean Science and Technology, Ansan, Republic of Korea

TaeKeun Rho Ocean Observation and Information Section, Korea Institute of Ocean Science and Technology, Ansan, Republic of Korea

Young Ho Seung Department of Oceanography, Inha University, Incheon, Republic of Korea

Chang-Woong Shin Physical Oceanography Division, Korea Institute of Ocean Science and Technology, Ansan, Republic of Korea

Hong-Ryeol Shin Department of Atmospheric Science, Kongju National University, Kongju, Republic of Korea

Seunghyun Son Cooperative Institute for Research in the Atmosphere, Colorado State University, Fort Collins, CO, USA

Inna Stonik Zhirmunsky Institute of Marine Biology, Far East Division, Russian Academy of Sciences, Vladivostok, Russia

Hae-Lip Suh Department of Oceanography, Chonnam National University, Gwangju, Republic of Korea

Young-Sang Suh Fishery and Ocean Information Division, National Fisheries Research and Development Institute, Busan, Republic of Korea

Eun-Jin Yang Korea Polar Research Institute, KIOST, Incheon, Republic of Korea

Seok Hoon Yoon Department of Earth and Marine Sciences, Jeju National University, Jeju, Republic of Korea

Jae-Yul Yun Research Institute of Oceanography, Seoul National University, Seoul, Republic of Korea

Chang-Ik Zhang Division of Marine Production System Management, Pukyong National University, Busan, Republic of Korea

Abbreviations

AO	Arctic Oscillation
AOI	Arctic Oscillation Index
ARGO	Array for Real-time Geostrophic Oceanography
CDW	Changjiang Diluted Water
CREAMS	Circulation Research of the East Asian Marginal Seas
DAC	Dokdo Abyssal Current
DCE	Dok Cold Eddy
EAST-I	East Asian Seas Time Series-I
EKWC	East Korea Warm Current
ENSO	El Niño-Southern Oscillation
ESBW	East Sea Bottom Water
ESCW	East Sea Central Water
ESDW	East Sea Deep Water
ESIW	East Sea Intermediate Water
GRACE	Gravity Recovery And Climate Experiment
HSIW	High Salinity Intermediate Water
HSTWW	High Salinity Tsushima Warm Water
IGBP	International Geosphere-Biosphere Programme
IPCC	Intergovernmental Panel on Climate Change
JB	Japan Basin
JGOFS	Joint Global Ocean Flux Studies
JSPW	Japan Sea Proper Water
KSBCW	Korea Strait Bottom Cold Water
LC	Liman Current
LOICZ	Land-Ocean Interaction in the Coastal Zone
LSTWW	Low Salinity Tsushima Warm Water
LTMER	Long-Term Ecological Research
MetOp-A	Europe's Meteorological Operational Satellite Program—A
MLD	Mixed layer depth
MOI	Monsoon Index

NB	Nearshore Branch
NKCC	North Korea Cold Current
NKCW	North Korea Cold Water
NPI	North Pacific Index
NUIS	North Ulleung Interplain Seamount
OB	Offshore Branch
PDO	Pacific Decadal Oscillation
PGB	Peter the Great Bay
SVP	Surface Velocity Program
TOPEX	The Ocean Topography Experiment
TWC	Tsushima Warm Current
TWW	Tsushima Warm Water
UB	Ulleung Basin
UIG	Ulleung Interplain Gap
UPJSPW	Upper Portion of the Japan Sea Proper Water
UWE	Ulleung Warm Eddy
YB	Yamato Basin

Chapter 1
General Introduction

Kyung-Ryul Kim, Sang Hoon Lee, Kyung-Ae Park, Jong Jin Park, Young-Sang Suh, Dong-Kyu Lee, Dong-Jin Kang and Kyung-Il Chang

Abstract The East Sea (Japan Sea) is a semi-enclosed marginal sea surrounded by the East Asian continent and Japanese Islands in the northwestern Pacific. It is topographically isolated from the North Pacific which allows only a small portion of the Kuroshio penetrating into the sea. In spite of its small basin size of about 10^6 km^2 and semi-isolated topography, many unique open ocean processes have been observed and identified, especially the existence of its own thermohaline circulation. Early oceanographic survey in the East Sea dates back to mid-19th century followed by numerous basin- or subbasin-scale regular and/or process-oriented observations conducted by surrounding countries, independently or cooperatively, and also by overseas countries. The advent of satellite-tracked drifters and floats and the availability of various satellite-derived products have greatly

K.-R. Kim
GIST College, Gwangju Institute of Science and Technology, Gwangju 500-712,
Republic of Korea
e-mail: krkim@snu.ac.kr

S.H. Lee
Korean Seas Geosystem Research Unit,
Korea Institute of Ocean Science and Technology, Ansan 426-744, Republic of Korea
e-mail: sanglee@kiost.ac.kr

K.-A. Park
Department of Earth Science Education, Seoul National University,
Seoul 151-748, Republic of Korea
e-mail: kapark@snu.ac.kr

J.J. Park
School of Earth System Sciences, Kyungpook National University,
Daegu 702-701, Republic of Korea
e-mail: jjpark@knu.ac.kr

Y.-S. Suh
Fishery and Ocean Information Division,
National Fisheries Research and Development Institute, Busan 619-705, Republic of Korea
e-mail: yssuhkorea@korea.kr

© Springer International Publishing Switzerland 2016
K.-I. Chang et al. (eds.), *Oceanography of the East Sea (Japan Sea)*,
DOI 10.1007/978-3-319-22720-7_1

contributed to understanding and finding important oceanographic processes. Mooring technology used in the East Sea has made progress for establishing some long-term local and global time-series stations.

Keywords History of survey · Topography · Observations · CREAMS · East Sea (Japan Sea)

1.1 Geography and Topography

The East Sea is a semi-enclosed marginal sea surrounded by the East Asian continent and the Japanese Islands (Fig. 1.1; Chough et al. 2000). The East Sea is connected with the Pacific Ocean through four shallow straits, i.e. the Korea (140 m deep), Tsugaru (130 m deep), Soya (55 m deep), and Tatarsky (12 m deep) straits. The northern and eastern margins of the sea are characterized by slopes which are relatively straight and steep, compared to those of the southern and western margins. The southern and western margins include several submarine topographic highs (plateau, ridge, and bank) (Fig. 1.1). According to ETOPO1 dataset (http://www.ngdc.noaa.gov/mgg/global/global.html), average depth of the East Sea is about 1650 m and the maximum depth is about 3800 m. There are three deep basins (the Ulleung, Japan, and Yamato basins) separated by submarine topographic highs (the Korea Plateau, Oki Bank, and Yamato Rise) that rise to within about 500 m of the sea surface (Fig. 1.1).

The Ulleung Basin in the southwestern part of the sea is a deep, bowl-shaped depression (2000–2300 m deep) delimited by continental slopes of the Korean Peninsula and the southwestern Japanese Islands on the west and south, respectively, and by the Korea Plateau and the Oki Bank on the north and east, respectively (Fig. 1.2). Its basin floor deepens to the northeast and is connected to the Japan Basin through the Ulleung Interplain Gap (UIG) between the South Korea Plateau and Oki Bank. The Japan Basin in the northern part of the sea is about 700 km long and 200–300 km wide, trending NE–SW. The basin floor is 3500–3800 m in depth and the deepest area is located between Russia and the

D.-K. Lee
Department of Oceanography, Pusan National University, Busan 609-735, Republic of Korea
e-mail: dglee@pnu.edu

D.-J. Kang
Marine Chemistry and Gochemistry Division, Korea Institute of Ocean Science
and Technology, Ansan 426-744, Republic of Korea
e-mail: djocean@kiost.ac.kr

K.-I. Chang (⊠)
School of Earth and Environmental Sciences, Seoul National University,
Seoul 151-742, Republic of Korea
e-mail: kichang@snu.ac.kr

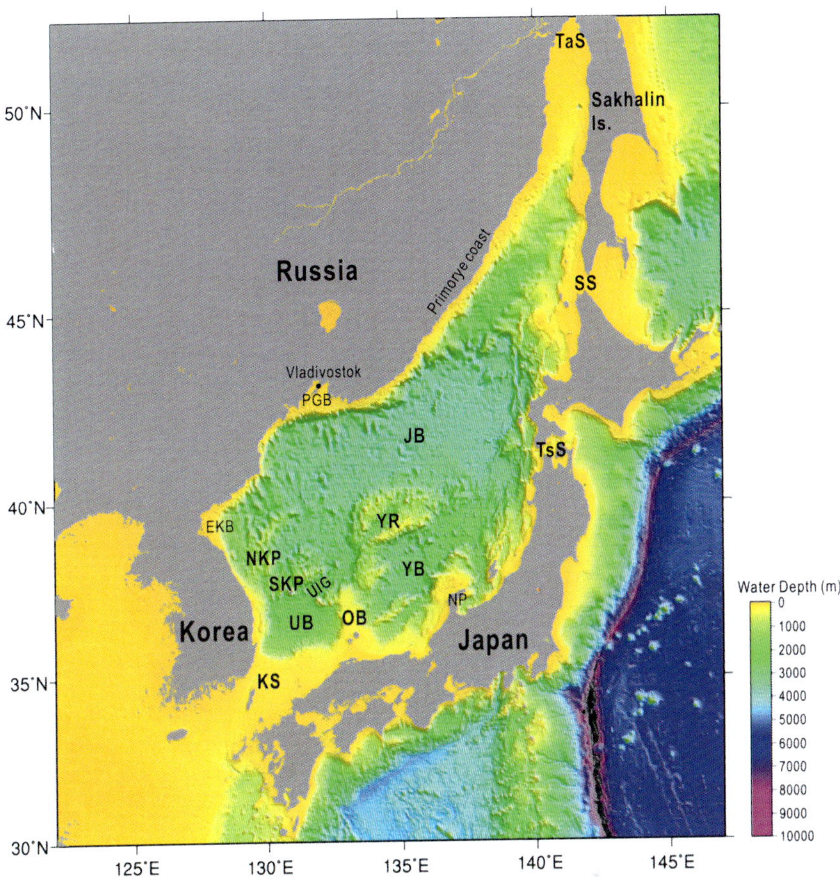

Fig. 1.1 Major physiographic features of the East Sea. *EKB* East Korea Bay; *JB* Japan Basin; *KS* Korea Strait; *NKP* North Korea Plateau; *NP* Noto Peninsula; *OB* Oki Bank; *PGB* Peter the Great Bay; *SKP* South Korea Plateau; *SS* Soya Strait; *TaS* Tatarsky Strait; *TsS* Tsugaru Strait; *UB* Ulleung Basin; *UIG* Ulleung Interplain Gap; *YB* Yamato Basin; *YR* Yamato Rise

southwestern part of Hokkaido. In the southeastern part of the sea, the Yamato Basin trending NE–SW has an average water depth of 2500–2700 m. The basin floors of the Japan and Yamato basins are rather smooth and flat, except for several submarine seamounts and hills that rise up to about 2000 m above the surrounding seafloor (Fig. 1.1).

In the western part of the sea, the Korea Plateau is a topographically rugged feature with 1000–1500 m of relief (Fig. 1.1). It is divided into two components (i.e. the South Korea Plateau and North Korea Plateau) by the indenting Japan Basin (Chough et al. 2000). The South Korea Plateau consists of numerous ridges, seamount chains, and intervening troughs/basins (Lee et al. 2002). It is divided into a western part (called the Gangwon Plateau) and an eastern part (called the Ulleung Plateau) by the Usan Trough (Fig. 1.2). In the central part of the sea, a

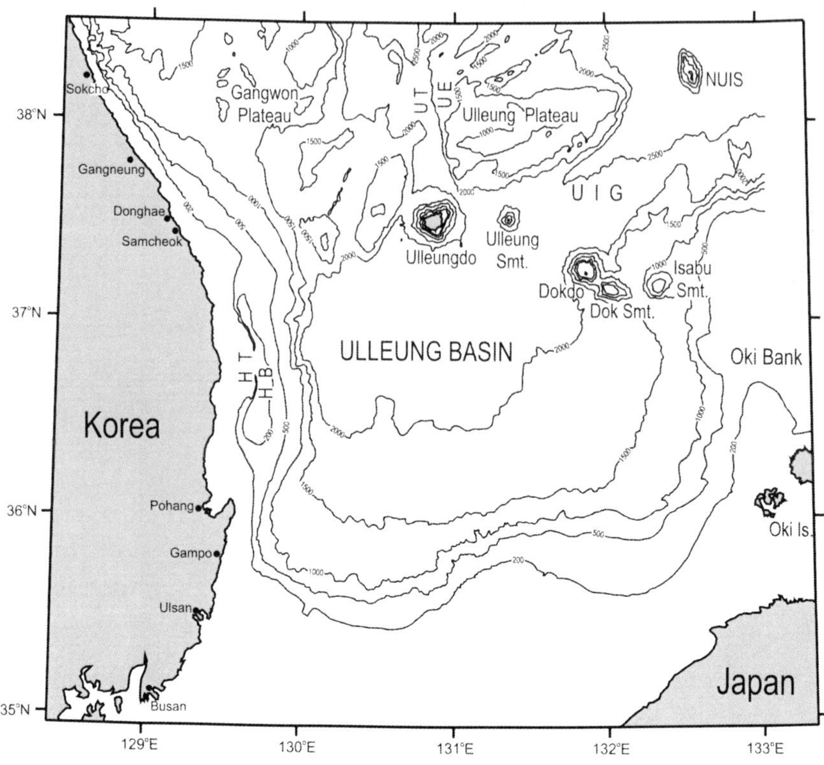

Fig. 1.2 Bathymetric features of the southwestern part of the East Sea. Bathymetry in meters. *HB* Hupo Bank; *HT* Hupo Trough; *NUIS* North Ulleung Interplain Seamount; *Smt.* Seamount; *UE* Usan Escarpment; *UIG* Ulleung Interplain Gap; UT = Usan Trough

volcanic topographic high (Yamato Rise) is parallel to the long axis of the Yamato Basin (Fig. 1.1). The Yamato Rise consists of a few topographic highs, separated from each other by transverse depressions. The prominent Kita-Yamato Trough, a NE–SW-trending longitudinal depression, divides the Yamato Rise into a northwestern component (the Kita-Yamato Bank) and a southeastern component (the Yamato and Takuyo banks). In the southeastern margin of the sea, the Toyama Deep-Sea Channel, a prominent submarine channel (about 750 km long), runs to the abyssal Yamato and Japan basins (Fig. 1.1; Nakajima et al. 1998). Detailed physiographic features of the East Sea are described in Chap. 16.

1.2 History of Hydrographic Surveys

Ocean science in the East Sea was initiated by its surrounding countries in the mid-19th century. In 1859, the Russian Navy conducted the first systematic hydrographic surveys including water temperature, density, and currents in the East

Sea. After World War II, with increased interest in this area, the Soviet Academy of Sciences began intensive investigations of the East Sea in 1950s and the Far Eastern Regional Hydrometeorological Research Institute has carried out large-scale surveys since its establishment in 1950 (Danchenkov et al. 2006).

Japan carried out systematic ocean and fisheries investigations in the East Sea from the 1920s until 1945. In 1932 and 1933, Uda conducted the first simultaneous, basin-scale, oceanographic surveys in the East Sea, including deep water investigations, with two research vessels and more than 50 fisheries vessels (Danchenkov et al. 2006), and he suggested a schematic diagram of the current system and a water mass classification (Uda 1934). Regular ocean surveys by Japanese governmental agencies or local prefectures are still continuing at present. The Japan Meteorological Agency (JMA) has been conducting seasonal ocean observations around Japan to monitor ocean variation and climate change. The most prominent observation line is the PM-line maintained by JMA's Maizuru Marine Observatory. Physical, chemical and biological oceanographic factors, such as water temperature, salinity, dissolved oxygen, nutrients, chlorophyll *a* and zooplankton, have been observed in every season since 1972. The Japan Coast Guard also monitors ocean variation in the East Sea, focusing on tides, currents, and ocean pollution, and operates the Japan Oceanographic Data Center which collects and provides oceanographic data and information. Much effort on ocean observation has been made by the Fisheries Research Agency as well as prefectural fisheries experimental stations in order to carry out research on ocean ecosystems and manage fishery resources.

Korean hydrographic surveys began with the establishment of the National Fisheries Research and Development Institute (NFRDI). NFRDI started as the Fisheries Experimental Station in 1921, and ever since has carried out NFRDI Serial Oceanographic Observations (NSO). NSO is the only integrated ocean observation program that regularly monitors the physical, chemical, and biological states of the Korean Seas. NSO was initially conducted for the purposes of collecting information on fishing grounds and monitoring the ocean state. It began in the 1920s with 6 observation lines that were surveyed occasionally from 2 to 6 times a year; in 1935, 14 observation lines covered the entire sea regions around Korea and extended up to 100 miles from the coast (Fig. 1.3).

In 1961, to get a better understanding of the complicated ocean structure and variability of ocean properties around the Korean Peninsula, NFRDI reorganized the NSO and increased the number of observation lines from 14 to 22 with 175 stations. After adding an additional 3 lines in the 1990s, the NSO is now composed of 25 observations lines with 207 stations, and 8 of these lines with 69 stations are situated in the East Sea (Fig. 1.4). Through long-term observations, the NSO now gives users much important information on climate change in the East Sea. It also supports operational oceanography for prediction of oceanic variability and prevention of natural disasters, and provides the basis for assessing the status of ecosystems in this area. The NSO data are released through the Korea Oceanographic Data Center website (http://kodc.nfrdi.re.kr), which is operated by NFRDI.

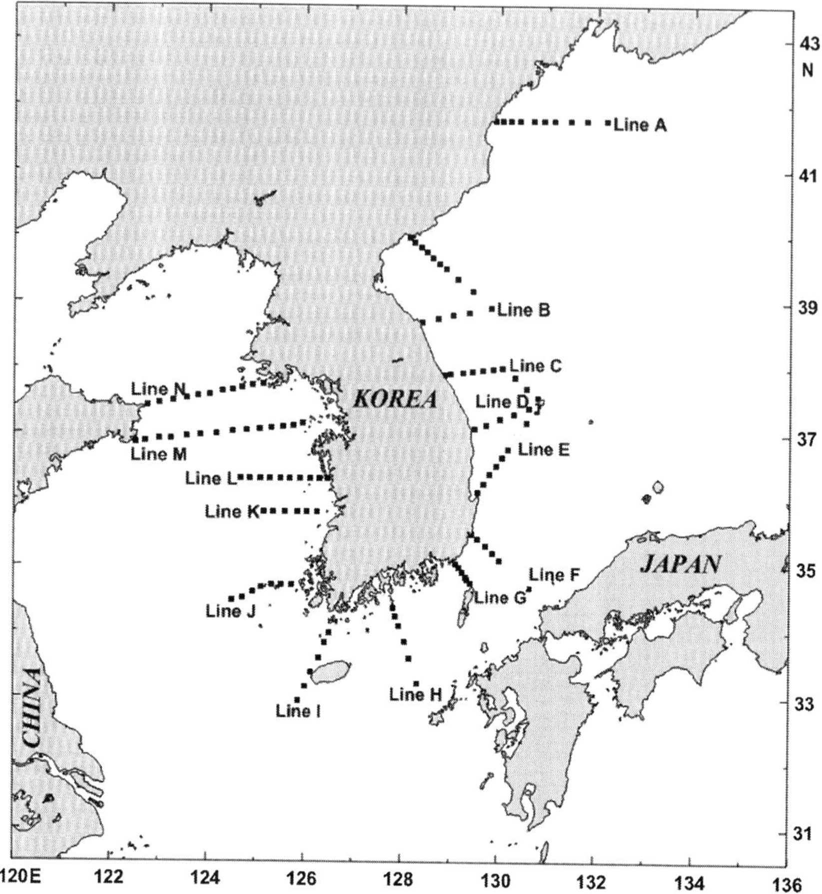

Fig. 1.3 Stations for NSO (1921–1960)

1.3 CREAMS Program

CREAMS, an acronym for "Circulation Research of the East Asian Marginal Seas", began in 1993 as a Japan-Korea-Russia international research program to understand water mass structure and circulation in the East Sea. CREAMS comprised analyses of historical data, field observations, numerical modeling, and laboratory experiments by scientists from Japan, Korea, and Russia in joint collaboration. From its inception, Prof. M. Takematsu, and Prof. Jong-Hwan Yoon at the Research Institute for Applied Mechanics, Kyushu University led the CREAMS program and were enthusiastically joined by Dr. Yuri Volkov at the Far Eastern Regional Hydrometeorological Research Institute, Russia, and Prof. Kuh Kim at

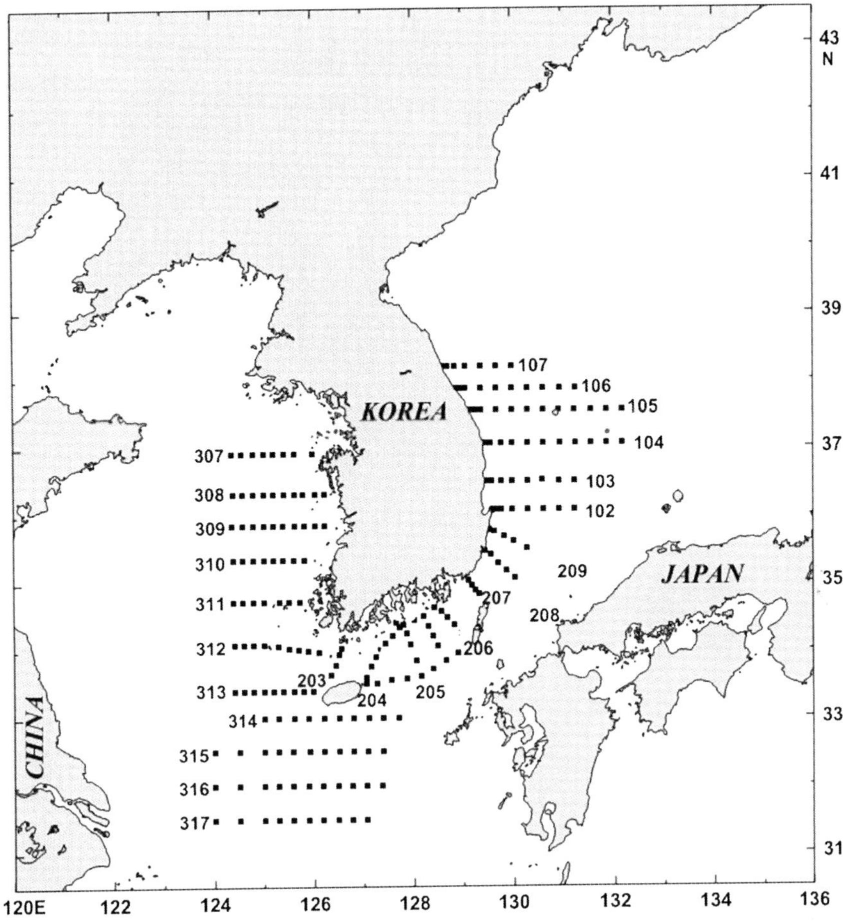

Fig. 1.4 Stations for NSO (1961–present)

Seoul National University (SNU), Korea who played essential roles in CREAMS's successful operation. The initiation of CREAMS marked one of the most important turning points in East Sea studies.

1.3.1 Important Findings Before CREAMS

Prior to the CREAMS era, most of our knowledge of East Sea hydrography mainly originated from the famous report of Uda (1934), who analyzed physical and chemical data acquired in the entire East Sea, even though there had also been

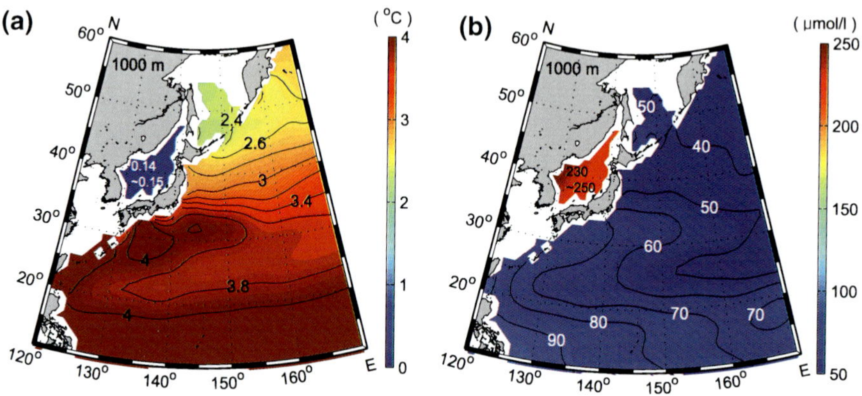

Fig. 1.5 Horizontal distributions of temperature (*left*) and dissolved oxygen (*right*) at 1000 m depth in the East Sea and the Pacific Ocean (data from Locarnini et al. 2010; Antonov et al. 2010; Garcia et al. 2010). The presence of cold and high-oxygen waters in the East Sea is clearly seen. **a** Potential temperature, **b** dissolved oxygen

some reports by Russian scientists in the early 20th century. In the early 1930s, Uda, mobilizing 53 vessels, explored the seas around Korea and found that the entire East Sea was filled with waters of relatively uniform physicochemical characteristics. Deep water in the East Sea being completely isolated from the North Pacific has lower temperature and higher dissolved oxygen content than the same depth levels in the North Pacific (Fig. 1.5). Below about 300 m, water has a nearly constant temperature of 1.0 °C. Uda (1934) called this abundant cold water mass "Japan Sea Proper Water" (JSPW), which makes up about 86 % of the East Sea (Yasui et al. 1967). The high oxygen indicates that the East Sea is a place of very rapid ventilation at depth.

Uda (1934) also observed that these cold waters are very high in dissolved oxygen (DO), with concentrations over ~250 μm. The Russian Vytiaz expedition, carried out in the East Sea during the early 1950s, further confirmed Uda's earlier observation (Fig. 1.6), strongly implying the existence of an active ventilation process in the East Sea, although few details were known about where and how such cold waters are formed in winter and under what conditions. This was later quantified by several tracer studies (e.g. Tsunogai et al. 1993) showing that turnover time in the East Sea is of the order of 100 years, which is at least one order of magnitude smaller than that of the oceanic conveyor-belt system (Broecker and Peng 1982).

However, based on the first Conductivity-Temperature-Depth (CTD) observations, Gamo and Horibe (1983) showed that there are structures in temperature and salinity within the JSPW, even though the range of variations is very small, strongly indicating that the JSPW is not a single water mass, as previously thought, but consists of several water masses such as Deep Water and adiabatic Bottom Water (Fig. 1.7).

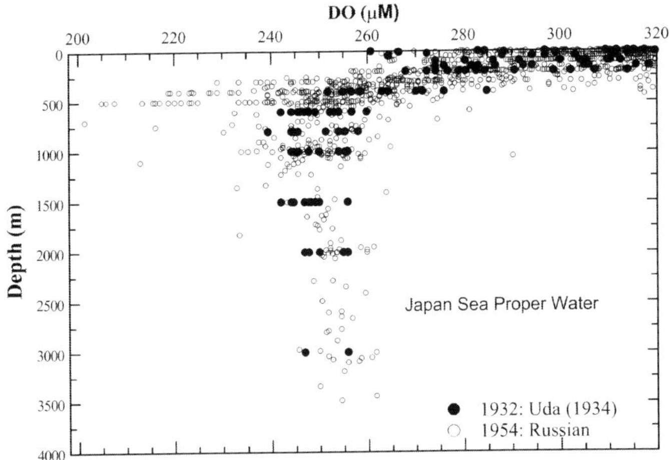

Fig. 1.6 Profiles of dissolved oxygen obtained during Uda's study and the Russian Vityaz study

Fig. 1.7 The first CTD profile taken in the East Sea, clearly showing several structures such as deep water and bottom water within so-called JSPW (redrawn from Gamo and Horibe 1983)

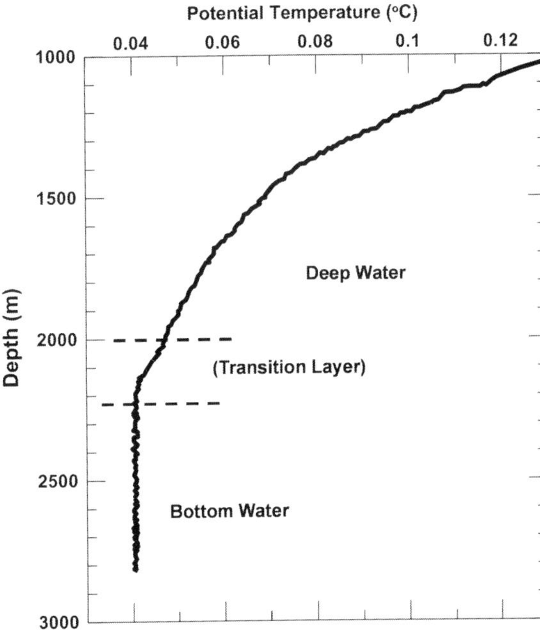

1.3.2 CREAMS Studies

Figure 1.8 shows CTD and chemistry stations together with locations of moorings, deployment positions of satellite drifters and towed Acoustic Doppler Current Profiler (ADCP) lines, carried out in the summer of 1993. Although the locations

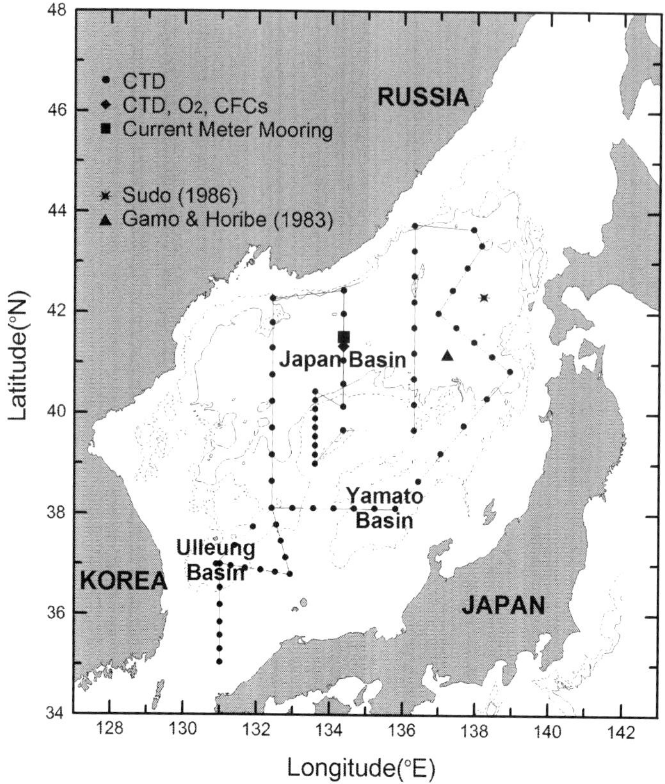

Fig. 1.8 Cruise map of the 1993 CREAMS expedition with stations of previous studies

of hydrographic stations and moorings have changed slightly each year due to some restrictions, it should be recognized that CREAMS studies have provided a rare opportunity to carry out precise measurements of salinity, temperature, and chemical tracers as well as long-term current meter mooring deployment in an extensive manner, in all major basins of the East Sea, for the first time in more than 60 years since Uda's investigation. Previously, as part of the Cooperative Study of the Kuroshio, hydrographic surveys, organized by the Intergovernmental Oceanographic Commission, were conducted mainly in the southern warm region of the East Sea in 1965–1970. However, for the purposes of CREAMS the northern cold region has been the area of main interest. In particular, deep stations and long-term current data became available for the first time in this region.

1.3.2.1 Important Discovery I: Oceanic Structures

The first important discovery during CREAMS studies was confirmation of mid-depth salinity and dissolved oxygen minima, reminiscent of typical oceanic structures, at deep stations in the Japan Basin, as shown in Fig. 1.9. However,

Fig. 1.9 Vertical profiles
of potential temperature (θ),
salinity (S), and dissolved
oxygen (O₂) at the northern
Japan Basin. According to the
θ-S diagram (inset), central
water (CW), deep water
(DW), and bottom water
(BW) could be defined

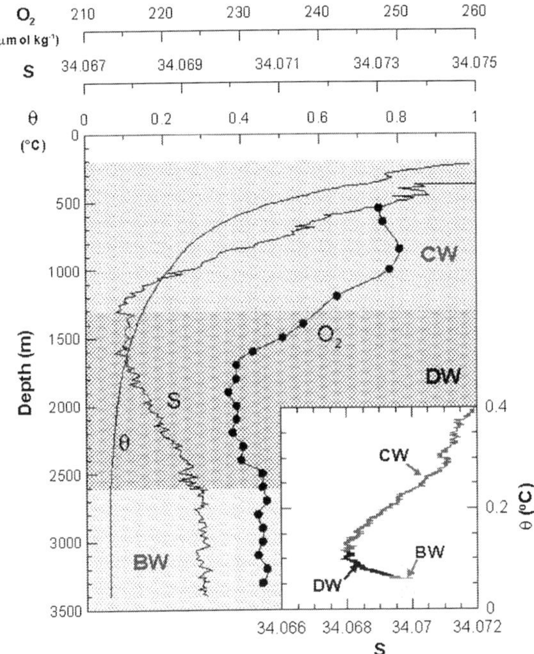

it is remarkable to note that the ranges of variation, when compared with those
observed in the open ocean (Östlund et al. 1987), are so small in the East Sea,
being one order of magnitude smaller in temperature, and even two orders of
magnitude smaller in salinity, with a salinity minimum represented by a difference of only 0.002 psu. Based on the θ-S (potential temperature-salinity) analysis
as shown in Fig. 1.9, new names such as Central Water, Deep Water, and Bottom
Water were proposed to identify these different water masses within the so-called
JSPW (Kim et al. 1996).

It was also fascinating to examine the first long-term time series of currents
taken in the Japan Basin, as shown in Fig. 1.10 (Takematsu et al. 1999). At
nominal depths of 2000 and 3000 m, low-passed currents were very weak until
November, but suddenly strong currents of 15–20 cm s⁻¹ began to fluctuate with
a time scale of a few weeks to months, lasting through the following spring. It is
surprising that even with such extremely weak vertical stratification with frequent
very strong horizontal currents up to 20 cm s⁻¹ in deep waters, the vertical structures shown in Fig. 1.9 have been maintained.

1.3.2.2 Important Discovery II: Dramatic Structural
Changes in the East Sea

Secondly, and more importantly, CREAMS studies confirmed dramatic structural
changes in the East Sea, as reflected in the profiles of properties such as temperature, salinity, and dissolved oxygen shown in Fig. 1.11. The profiles show

that temperature has gone up as much as 0.1–0.5 °C in the upper 1000 m and by 0.01 °C below 2000 m during the last 30 years, increasing the heat content below 500 m at a rate of 0.54 W m^{-2}.

During this warming period, DO went through a drastic change in such a way that the oxygen minimum depths deepened from a few hundred meters in the late 1960s to below 1500 m in the mid-1990s. This deepening of oxygen minimum depths was accompanied by decreases in DO concentrations of up to 20 μm in deep waters, as noted earlier by Gamo et al. (1986). Nevertheless, it is very

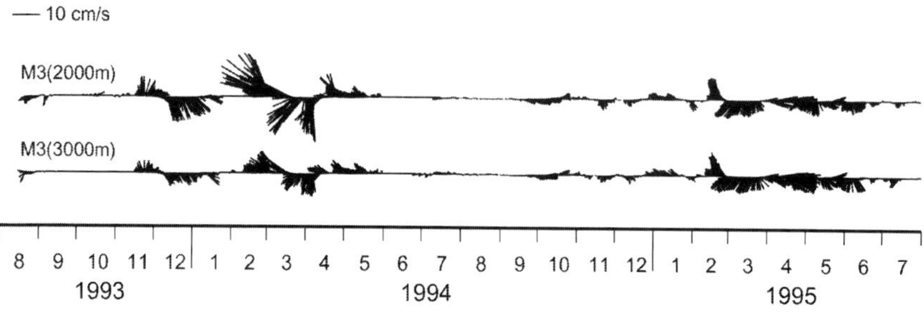

Fig. 1.10 Time series of currents at nominal depths of 2000 m and 3000 m taken in the central part of the Japan Basin (41° 29.7′ N, 134° 21.4′ E) at a depth of 3500 m from August 1993 to July 1995 (Redrawn from Takematsu et al. 1999)

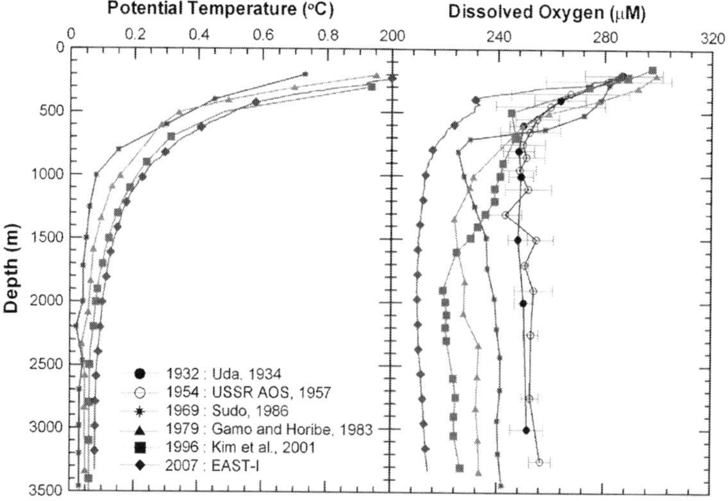

Fig. 1.11 Changes in profiles of potential temperature and dissolved oxygen (DO) in the Japan Basin during the last 30 years. Average profiles of dissolved oxygen in the 1930s and early 1950s are also shown. Property changes such as warming of deep waters and a drastic change of DO structures in the East Sea, since sometime during the 1950s, are clear. Some data are from earlier studies in the area (Redrawn from Uda 1934; USSR AOS 1957; Gamo and Horibe 1983; Sudo 1986)

important to note that DO concentrations instead increased during the same period at intermediate depths above the oxygen minimum. The first continuous salinity profile in the East Sea, obtained in the late 1970s with a CTD (Gamo and Horibe 1983), did show a salinity minimum depth shallower than 800 m, so this depth has essentially doubled since then. Average profiles of DO in the 1930s and 1950s, also shown in the figure, clearly show similarity in profiles in those two decades, strongly implying that the changes started sometime during the 1950s.

It has been shown (Kim and Kim 1996; Kim et al. 1999) that changes in temperature, salinity, and DO profiles in the East Sea are due to a change in its ventilation system. The application of an advection-diffusion model (Craig 1969) to oxygen profiles, in particular, resulted in an apparent 'net production' of oxygen despite 'actual consumption' in the Central Water, strongly supporting an active injection of fresh surface waters into intermediate-water depths occurring at the present time. This has been further supported by chlorofluorocarbon (CFC) measurements, as shown in Fig. 1.12, showing CFC ages no younger than 20 years for the deep waters. According to a moving-boundary box model (MBBM), developed

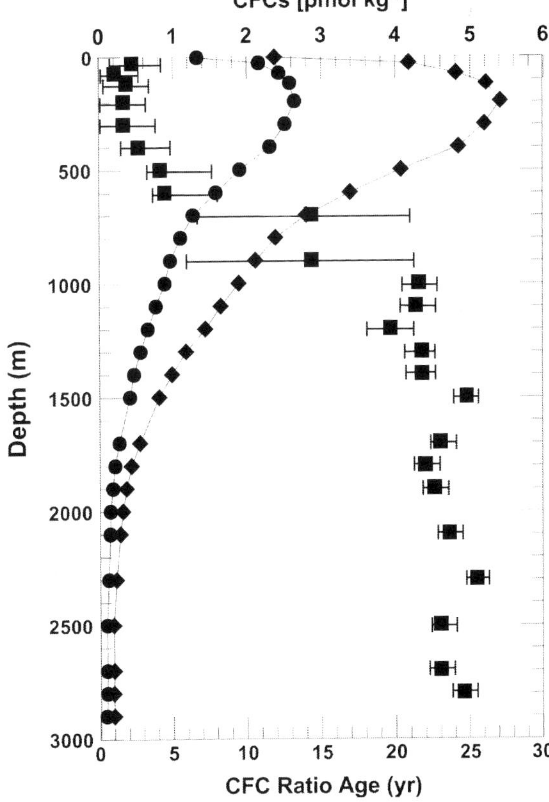

Fig. 1.12 Profiles of CFC-11 (diamond) and CFC-12 (circle) at a station in the central Japan Basin. The CFC ratio age is also shown in the figure. The CFC ratio ages (square) of the waters deeper than 1000 m are no younger than 20 years (Redrawn from Kim et al. 2001)

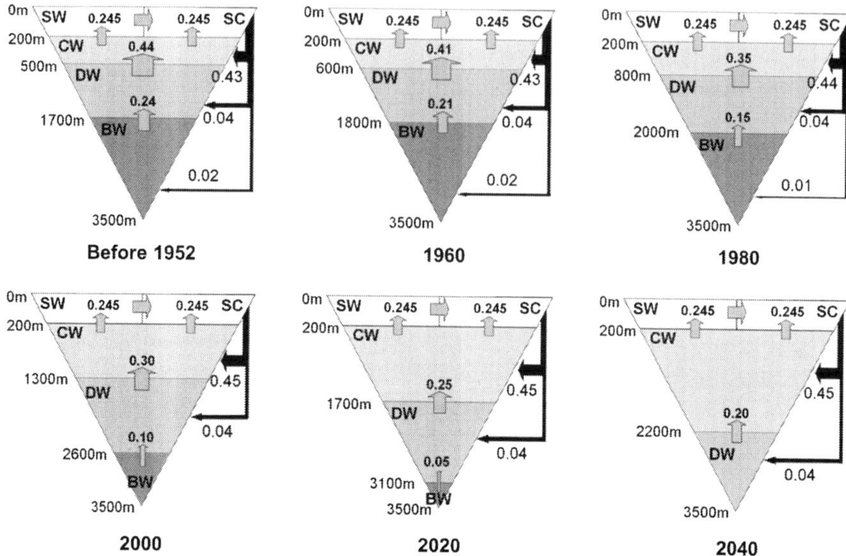

Fig. 1.13 Structural changes with time estimated from a moving-boundary box model. The model predicts that bottom water will disappear completely in 2040 (Redrawn from Kang et al. 2003)

to describe these changes and calibrated with CFCs and tritium results (Kang et al. 2003), bottom water formation appears to have ceased since the mid 1980s (Fig. 1.13).

There have also been some claims that the conveyor-belt in the East Sea has slowed down recently (e.g. Gamo 1999). The cessation of bottom water formation, however, is only a part of the whole story occurring in the East Sea. The results from the MBBM clearly show that the overall ventilation system in the East Sea is still very active, and that the decrease of bottom water formation is counterbalanced by the enhancement of intermediate water formation. The immediate consequence of this shift is a fast expansion of oxygen-rich Central Water in mid-depths in recent years (Kim et al. 1999). The historic temperature data in the Japan Basin also show that the depth of the 0.15 °C isotherm, the boundary between Central Water and Deep Water at the present time, started to deepen sometime in the early 1950s (Kim et al. 2002a) as shown in Fig. 1.14.

1.3.3 Globalization of East Sea Studies

1.3.3.1 The Birth of CREAMS-II

At the fourth CREAMS workshop held in Vladivostok, February 12–13, 1996, it was recommended that CREAMS be expanded to a CREAMS-II phase beginning

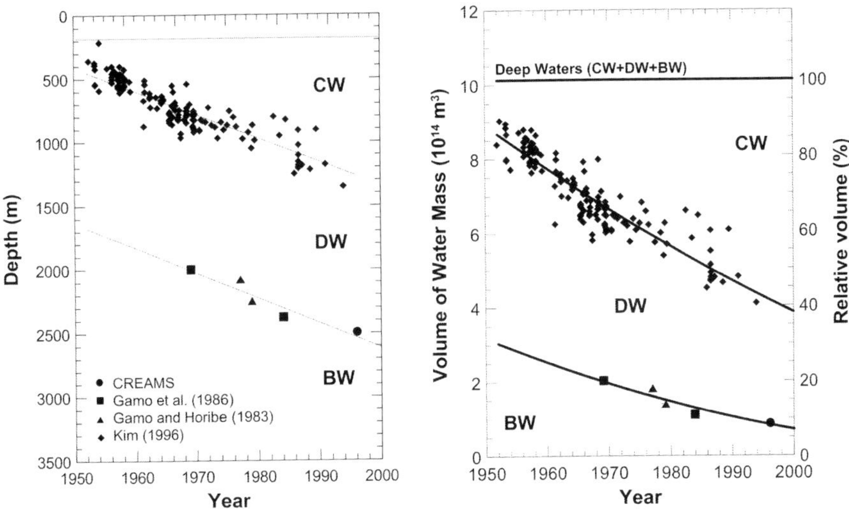

Fig. 1.14 Changes in the depths of the boundaries between *CW* and *DW*, and *DW* and *BW* over time, estimated from the historical data. Volume changes of water masses over time, calculated from the linear relationship between depth and area of the East Sea (Redrawn from Kim et al. 2002a)

in 1998, inviting new research programs and scientists from abroad including participants from the US, and a new name for the East Sea "JES (Japan/East Sea)" was also introduced in the field mainly by American scientists. A new program in the US was initiated that supported approximately 15 US investigators to make detailed observations and develop models of the East Sea circulation during the period 1999–2003.

The development of future research programs in the East Sea and the collaboration between scientists in the countries studying the East Sea were the topics of two days of discussion at the 5th CREAMS Workshop, held in Seoul on April 28–29, 1998. Nearly 50 scientists from Korea, Japan, Russia, and the US attended the meeting, and a number of important discussions took place that were centered on short-term and long-term plans for the exploration of the East Sea and how this exciting work could best be carried out. In addition, since all of these countries are also members of PICES (North Pacific Marine Science Organization), there were several important discussions of the relationship of PICES to this expanded research effort in the East Sea.

1.3.3.2 CREAMS/PICES EAST (East Asian Seas Time Series)-I Program

As the countries involved in East Sea research are all members of PICES, the possibility that PICES might be able to help in some aspects of coordinating research in the East Sea was discussed and it was generally agreed that PICES might be able to provide important help to East Sea researchers, especially in the area of obtaining permission for scientific work to be carried out across the various EEZ (Exclusive Economic Zone) boundaries of the East Sea.

As a result of discussions that took place at the PICES 6th Annual Meeting in Busan in October of 1997, the PICES Science Board formally recognized the scientific value of the work in CREAMS and approved the development of stronger ties between PICES and CREAMS. The final important step in accomplishing this endeavor took place during the First CREAMS/PICES workshop on EAST (East Asian Seas Time Series)-I, held at SNU, in April, 2005. Later that year, an official CREAMS/PICES study of the East Sea was approved under the PICES umbrella and an International Advisory Panel was officially formed, strongly implying that the East Sea has the potential to become one of the world's most active regions of international cooperation in oceanographic research. Under the scope of the CREAMS/PICES umbrella, the Korean EAST-I project, funded by the Ministry of Oceans and Fisheries, was officially launched in 2006, and has been continuing for the last 7 years as two consecutive 5-year programs.

1.3.3.3 IPCC 4th Report and Nobel Peace Prize, 2007

The global nature and importance of the East Sea studies were unequivocally confirmed during the publication of the IPCC (Intergovernmental Panel on Climate Change) 4th report (IPCC 2007). In Chap. 5 of this prestigious report on climate change, titled "Oceanic Climate Change and Sea Level", the dramatic physico-chemical changes in the East Sea observed during CREAMS studies were presented as a representative example occurring in the Pacific Ocean summarized as follows:

> The marginal seas of the Pacific Ocean are also subject to climate variability and change. Like the Mediterranean in the North Atlantic, the Japan (or East) Sea is nearly completely isolated from the adjacent ocean basin, and forms all of its own waters beneath the shallow pycnocline.
>
> Because of this sea's limited size, it responds quickly through its entire depth to surface forcing changes. The warming evident through the global ocean is clearly apparent in this isolated basin, which warmed by 0.1°C at 1000 m and 0.05°C below 2500 m since the 1960s. Salinity at these depths also changed, by 0.06 psu per century for depths of 300 to 1,000 m and by -0.02 psu per century below 1,500 m (Kwon et al. 2004).
>
> These changes have been attributed to reduced surface heat loss and increased surface salinity, which have changed the mode of ventilation (Kim et al. 2004a). Deep water production in the Japan (East) Sea slowed for many decades, with a marked decrease in dissolved oxygen from the 1930s to 2000 at a rate of about 0.8 μmol kg^{-1} year^{-1} (Gamo et al. 1986; Minami et al. 1998). However, possibly because of weakened vertical stratification at mid-depths associated with the decades-long warming, deep-water production reappeared after the 2000-2001 severe winter (e.g., Kim et al. 2002b; Senjyu et al. 2002; Talley et al. 2003).Nevertheless, the overall trend has continued with lower deep-water production in subsequent years.

Furthermore, the approximately 1000 scientists involved in preparing this report were awarded half of the Nobel Peace Prize in 2007. Prof. Kuh Kim as a leading author for the East Sea was one of the winners, who shared the Peace Prize.

1.3.4 Concluding Remarks

The East Sea is a unique site: most of the physical, chemical, and biological processes that occur anywhere in any ocean occur also in the East Sea. The southern portion of the East Sea is subtropical most of the year, and the northern sector behaves as a localized subpolar sea with ice-cover in winter. Strong frontal mixing along a subtropical-subpolar frontal boundary and deep convection are two processes that do not take place in close proximity to each other in most of the world's oceans, but are active at the same time within a few degrees of latitude and longitude in the East Sea. The occurrence of a large variety of physicochemical processes within a relatively small geographic region makes the East Sea an ideal laboratory for improving our general oceanographic knowledge.

The most important legacy of CREAMS studies is that many important "Big Questions in Oceanography," especially those associated with recent climate change, can be addressed, clearly, demonstrating the value of the East Sea as a miniature ocean despite its regional character. Continuing this legacy should be a big task for the coming generation involved in East Sea research.

1.4 Recent Observational Programs

1.4.1 Surface Drifters

World ocean currents have been estimated using ship drift data logged by sailors. But since few merchant ships crossed the East Sea, basin-wide measurements of surface current in the East Sea were rare and usually estimated from shipborne hydrographic surveys until the development of satellite tracked drifters (Niiler et al. 1987, 1995). The Surface Velocity Program (SVP) drifter has a spherical surface buoy and a holey-sock drogue centered at 15 m depth. Satellites determine locations of the drifters and transmit them along with sea surface temperature. The first satellite tracked drifter entering the East Sea was deployed by the Woods Hole Oceanographic Institution in the East China Sea in 1988 (Beardsley et al. 1992). In 1990 a few more drifters were deployed in the East Sea by the Korea Institute of Ocean Science and Technology (KIOST) to study coastal currents and eddies in the Ulleung Basin (Lie et al. 1995). SNU deployed two drifters in the northern East Sea in 1994 (Yang 1996), and air deployments of drifters with a very small drogue (Compact Meteorological and Oceanographic Drifter) were carried out in 1992–1996 by the US Navy in the subpolar frontal area.

Small numbers of SVP drifters were released regularly by various investigators from Pusan National University, KIOST, SNU, NFRDI, and the Japan Hydrographic Office beginning in 1996. SVP drifters and mini-meteorological drifters were extensively deployed bimonthly during 1998–2001 in the southern East Sea through the JES Program sponsored by the Office of Naval Research (Lee et al. 2000; Lee and Niiler 2005). From 2003 to 2006, the Korea Hydrographic and Oceanographic Administration (KHOA) deployed SVP drifters bimonthly in the Korea Strait to study the East Korea Warm Current (Lee and Niiler 2010). A total of 325 drifters have been deployed in the East Sea and all data are archived in the Global Drifter Center (ftp.aoml.noaa.gov).

1.4.2 Argo (Array for Real-Time Geostrophic Oceanography)

The primary goal of the international Argo program is to create an observation network in the global ocean using more than 3000 free-drifting profiling floats that measure the temperature and salinity of the upper ocean (http://www.argo.net). The observation network is dedicated to detecting climate variability of large-scale temperature and salinity, as well as currents, on seasonal to decadal time-scales, to delivering information required for calibration of satellite measurements, and to providing input data for initialization and constraint of climate models (Freeland et al. 2010). The Argo network is clearly distinguished from previous subsurface measurements by its global coverage and especially the effort made to sample both the northern and southern hemispheres without bias. As of 2013, with more than 3600 floats active, over 10,000 profiles per month are being delivered by the Argo float array. This unprecedented amount of data is having a large impact on ocean state estimation and on sea surface temperature forecasting skill (e.g. Balmaseda et al. 2007; Fujii et al. 2008).

The predecessor of the Argo float was originally developed to measure deep currents in a Lagrangian way (Davis et al. 1992; Davis and Zenk 2001); it is known as the Autonomous Lagrangian Circulation Explorer (ALACE). In 1996, Prof. Riser at the University of Washington (UW) tested the first float equipped with a temperature sensor in the East Sea; it was named Profiling-ALACE (PALACE). Two PALACE floats deployed at 350 and 1000 m have successfully collected temperature profiles for several years. The East Sea is a confined, bowl-shaped basin with average depth of 1700 m, resembling open oceans in many aspects (Kim et al. 2002b; Talley et al. 2003). Hence the East Sea has become a good test location for newly developed autonomous floats. In 1999, UW deployed 36 profiling floats at 800 m, equipped with CTD sensors, in the northern part of the East Sea. These profiling floats named Autonomous Profiling Explorer floats are nowadays the most widely used floats in the Argo program. The reliability and accuracy of their CTD sensors (the same sensors used in the present Argo floats) have been quantitatively tested using data from shipboard CTD observations (Park and Kim 2007). Furthermore, an Argo float with optical sensors was tested in this marginal sea in spring 2000 to observe the development of a phytoplankton bloom, demonstrating the possibility of a bio-geo-chemical Argo float (Perry and Rudnick 2003).

Under the Korean Argo program initiated in the year 2000, about 10–30 floats have been deployed every year in the East Sea, the Northwestern Pacific, and the Southern Ocean by KIOST and the Korea Meteorological Administration (KMA). It is noted that KIOST had started to deploy profiling floats in 1998. The total number of Argo floats contributed by Korea and registered in the Argo Global Data Assembly Center reached more than 230 in 2012 (personal communication with Drs. Kang and Suk in KIOST). About 14,000 temperature and salinity profiles have been obtained from the Korean Argo floats and UW floats in the East Sea during 1998–2012, comparable to the total number of shipboard CTD profiles taken by Korean institutions in the East Sea during the last 50 years.

Such large amounts of high-quality Argo data have stimulated much oceanographic research aimed at understanding East Sea circulation as well as contributing to the international Argo program. Since the East Sea is a region with both the highest density of Argo data in the world (especially the Ulleung Basin where more than 100 profiles have been obtained within a 0.5° × 0.5° bin) and numerous regularly taken accurate shipboard measurements, the quality of Argo data passing the Argo Delayed-Mode Quality Control (DMQC) procedure could be statistically quantified using the East Sea Argo data for the first time (Park and Kim 2007; Thadathil et al. 2012). According to Park and Kim (2007), after the proper DMQC procedure the one-standard deviation error in salinity is about 0.003 psu, comparable to the manufacturer's stated accuracy. However, they also noted the importance of recent shipboard measurement in order to calibrate Argo data in the East Sea where significant salinity trends exist. Another notable contribution is to develop a method to process Argo trajectory data and thereby estimate deep currents (Park et al. 2005a). Deep currents can be calculated from the successive surface positions of drifting Argo floats. Without a proper correction of unknown drifts for half an hour to a few hours at the sea surface, the estimated deep currents may contain large uncertainty. Park et al. (2005a) provided a way to correct for the unknown surface drifts and to quantify the errors in the deep currents by using abundant Argo float data in the East Sea. Their method has become a standard procedure to treat trajectory data in the international Argo community, especially for floats using the Argos satellite system. However, for floats using the Iridium satellite system, the method is no longer applied because the unknown surface drift during satellite communication is very small.

Argo floats have also been utilized to answer scientific questions in the East Sea. Their trajectory data help reveal comprehensive deep circulation patterns in the East Sea. It has been found there is remarkable inter-basin dependency of the deep currents: weaker mean and eddy kinetic energy in the Ulleung Basin, stronger mean kinetic energy in the eastern Japan Basin, and stronger eddy kinetic energy in the western Japan Basin (Danchenkov and Riser 2001; Choi and Yoon 2010; Mooers et al. 2006; Park and Kim 2013). Specific deep circulation patterns around the Ulleung Basin and possible linkage between the Ulleung and Yamato basins have also been suggested by Argo float trajectory data (Park et al. 2004b, 2010). Details of deep circulation in the East Sea will be described in Chap. 4. Application effectiveness of Argo float data can be maximized by using them in conjunction with historical hydrographic data. Minobe et al. (2004) constructed time series of three dimensional temperature maps through the optimal interpolation method, using historical CTD and Argo float data. Minobe et al. (2004) and Na et al. (2012) showed interannual and decadal variability of upper ocean temperature in the East Sea. Also, Lim et al. (2012b) compiled expendable bathythermograph (XBT), CTD, and Argo float data, reporting climatology of the mixed layer depth in the East Sea. Details about long-term variability of temperature and patterns of mixed layer depth in the East Sea will be described in Chap. 3. Additionally, Argo float data together with historical shipboard measurements and satellite data have been assimilated into the Modular Ocean Model (Ver. 3) by means of the three-dimensional variational method, providing reanalysis of temperature, salinity, and velocity fields in the East Sea (Kim et al. 2009b).

The Korean Argo program has been very successful. As already described, Argo float data in the East Sea are contributing greatly both to improving Argo float data processing and to answering valuable scientific questions regarding East Sea circulation and variability. Moreover, even though the Korean Argo program has deployed many floats in this marginal sea where a lot of fishing activities take place and the continental margin potentially threatens the floats' operation, the average lifetime of Korean Argo floats is about 3.6 years in the East Sea, longer than the global Argo float average of about 3.1 years.

At present, the Korean Argo program is in its second decade. As with the international Argo program, its future evolution will be fuelled in part by its success during its first 10 years. The next phase of the Argo program is heading toward sustaining data coverage and quality as well as employing multidisciplinary sensors. An appropriate sampling strategy covering the entire East Sea and utilizing additional sensors will broaden Argo float applications, including surface layer sampling, ocean mixing, and biogeochemical impacts of climate variability, which will prominently improve our understanding of the East Sea.

1.4.3 Moored Current Observations

The existence of deep convection and associated thermohaline circulation in the East Sea has long been recognized from tracer distributions (e.g. Gamo and Horibe 1983), and circulation of a specific water mass has also been inferred qualitatively from tracer distributions (e.g. Senjyu and Sudo 1993). Nevertheless, quantitative descriptions of deep circulation and deep currents have been largely unavailable until recently, mainly due to the characteristic homogeneity of the East Sea's water properties. Moored current measurements have been conducted since the late 1980s to quantify deep currents and to better understand the ocean dynamics, a major finding being that deep water in the East Sea is not stagnant but has energetic movements (e.g. Takematsu et al. 1999).

Current measurement moorings deployed in the East Sea can be classified into two categories, one, at deep locations, has a taut-wire mooring line equipped with current meters and other sensors for measuring physical properties (e.g. Kim et al. 2009a), and the other has an upward-looking ADCP mounted inside a specifically designed bottom mount at shallow locations to protect the mooring from fishing activities (Perkins et al. 2000). The taut-wire moorings themselves fall into two categories, surface moorings and subsurface moorings. The surface mooring is used for measuring biogeochemical parameters together with surface currents above the shallow (~200 m deep) pycnocline (Son et al. 2014). Most of the moored current measurements in the East Sea aim to measure deep currents below the pycnocline using subsurface moorings. In a few cases, an upward-looking ADCP mounted on a subsurface float at around 100–200 m has also been used to measure upper-layer currents (Chang et al. 2004; Kim et al. 2009a). This section introduces a brief history of moored current measurements in the Korea Strait using bottom-mounted ADCPs and in the East Sea using subsurface moorings. Except for the Korea Strait, coastal current measurements at depths shallower than 500 m are not included in this review.

Since the Korea Strait is the inflow channel of the Tsushima Warm Current, the quantification of mass and material transports through the Korea Strait, together with their temporal variability, has been recognized as an important contributor to the shaping of the oceanographic environment of the East Sea and its changes (e.g. Kim et al. 2013). Direct current measurement efforts in the Korea Strait date back to the 1920s: measuring currents at a number of locations in the Korea Strait for about 25 h by manually lowering and recovering current meters from a ship (Nishida 1927). Attempts at estimating volume transport through the strait have been made using sea level difference (Yi 1966), submarine cable voltage measurement (Kim et al. 2004b), and an ADCP mounted on a passenger ship (Fukudome et al. 2010). Intense fishing and trawling activities had hindered long-term moored current measurements in the Strait until the first usage of trawl-resistant bottom mounts (TRBM) (Perkins et al. 2000). Twelve bottom-mounted ADCPs inside the TRBMs were deployed along two sections across the full-breadth of the Korea Strait for 10 months from May 1999 to March 2000 with a turnaround of moorings in October 1999 (Teague et al. 2002). Although trawl scrapings were found on TRBM surfaces, the TRBMs turned out to be very effective in protecting the instruments from trawl fishing activities. Only one mooring could not be recovered during the mooring turnaround. The failure of the recovery was not due to the fishing activities because the mooring responded to acoustic interrogation, and it is thought to have sunk into the sticky muddy bottom (Perkins et al. 2000). The data successfully recovered from these moorings in the Korea Strait have been exploited in quantifying the volume transport with its high- and low-frequency variability (Jacobs et al. 2001; Teague et al. 2002) and converting the submarine cable voltage to section-averaged volume transport (Kim et al. 2004b).

Moored current observations using subsurface moorings were first made in 1985–1986 for 58–198 days at a number of locations deeper than approximately 1000 m in Toyama Bay and the Yamato Basin (Kitani 1987; Yamada 1999). Three subsurface moorings deployed in the Yamato Basin carried 3–5 current meters on each mooring from 680 m down to about 100 m above the seabed. Maximum speed of the near-bottom currents reached 13.7 cm s^{-1}, and barotropic fluctuations of subsurface currents and inertial motions were noted (Kitani 1987). This was followed by moored current measurements in the Yamato Basin and eastern Japan Basin by various Japanese institutions including the Japan Sea National Fisheries Research Institute and Maritime Safety Agency, collaborating in some cases with Russian and Korean institutions (Senjyu et al. 2005; MSA Reports 1995–2007). From reports and journal papers we can trace the moored current measurement activities in the Yamato Basin only until 2005.

The pioneering deployment of subsurface current meter moorings in the Japan Basin, where deep convection and water mass formation occur, was made about 20 years ago in August 1993 as a CREAMS activity (see Sect. 1.3.2). Current measurements at 7 locations in the Japan Basin exhibited energetic deep currents with barotropic fluctuations and strong seasonality (Takematsu et al. 1999). Since then, moored current measurements have flourished in investigations of deep Japan Basin currents and circulation. After the pioneering mooring work in the Japan Basin, moored current measurements in it were followed by those of various Japanese institutions in collaboration with Russian institutions, until about 2003 (Senjyu et al. 2005; MSA Reports 1995–2007). As far as we know, no mooring work in the Japan Basin has been documented since then.

In the Ulleung Basin, the first long-term subsurface current meter mooring was deployed in 1996 as a collaborative work of SNU, KIOST, and the Woods Hole Oceanographic Institution (Chang et al. 2002). The mooring location, designated EC1, was in the UIG, a choke point where deep water exchange occurs between the Japan and Ulleung basins. The mooring at EC1 has been in operation since 1996, hence it is the longest-duration deep current monitoring station in the East Sea. It usually carries current meters at three depth levels, between 300 and 500 m, between 1300 and 1500 m, and below 1900 m. At times, additional current meters, including ADCPs, have been installed to measure the full-depth currents (Kim et al. 2009a). It became one of the OceanSITES stations in 2012 (http://www.oc eansites.org). Initial results from the mooring at EC1 revealed a mean southward deep flow of about 2.0 cm s^{-1} (Chang et al. 2002), indicating deep water discharge from the northern Japan Basin to the Ulleung Basin. Later studies suggested a possible outflow on the other side of the UIG (Park et al. 2004b; Teague et al. 2005). An array of 5 moorings was deployed in the UIG from November 2002 to April 2004 to investigate the circulation in the UIG, and the resulting measurements revealed an asymmetric two-way circulation consisting of a broad and weak inflow to the Ulleung Basin in the western UIG and a narrow and strong outflow called the Dokdo Abyssal Current at the eastern boundary of the UIG (Chang et al. 2009). The net deep water transport below 1800 m was almost negligible.

For two years between 1999 and 2001, an unprecedented set of mooring observations was accomplished as part of the US JES program, in collaboration with research projects of Korea and Japan. These observations, from an array of 16 current meters and 23 pressure-gauge equipped inverted echo sounders (PIES) moored near the seabed at depths greater than 1000 m, revealed upper and deep circulation features of the Ulleung Basin (Chang et al. 2004; Mitchell et al. 2005; Teague et al. 2005). Red snow crab fishing activities by Korean fishermen are widespread in the Ulleung Basin at about 40 locations with depths ranging from 500 to 2000 m. Before deploying the instruments, Korean and US scientists involved in the mooring work had a meeting with the local crab fishermen to adjust mooring locations to avoid conflict with the crab fishing activities. Prearrangement of mooring locations in consultation with local fishermen turned out to be very effective in enabling successful recovery of the moored instruments. Among the 39 instruments only 2 PIESs were lost (Mitchell et al. 2005). One of them was later found adrift at the surface and recovered (Dr. Jae-Hun Park, personal communication), indicating that the one loss was not due to crab fishing.

Based on the compilation of mooring data at 55 locations acquired prior to 2000, Senjyu et al. (2005) proposed a schematic abyssal circulation map, which was later corroborated by mean current distributions based on multi-year trajectories of subsurface floats (Choi and Yoon 2010; Park and Kim 2013). Moored current measurements have certainly contributed to better understanding of the deep circulation and other currents, together with their dynamics (e.g. Yoshikawa 2012). But more observational efforts are required to answer some unresolved issues, such as the quantification of deep water transport in the Japan Basin, the interaction of the East Korea Warm Current and the North Korea Cold Current, the coupling of upper and deep currents and their response to air-sea interaction in the Japan Basin. Moored current measurements will remain useful for advancing our understanding of ocean dynamics in the East Sea. Finally it should be emphasized that the success of mooring operations relies heavily on participation of well-trained personnel and experienced technical experts.

1.4.4 Satellite Oceanography

Satellites have amazing capabilities to observe diverse oceanic features in the East Sea from space. Since the launch of Sea Satellite (SeaSAT) in 1978, a variety of satellite data have been used in oceanographic studies of the East Sea. From satellites we have been able to obtain multiple oceanic variables, such as sea surface temperature (SST) from infrared sensors on near-polar-orbiting and geostationary satellites, sea level and significant wave height from altimeters, near-surface winds from scatterometers and passive microwave sensors, chlorophyll-a concentration and suspended particulates from ocean color sensors, and high-resolution Synthetic Aperture Radar (SAR)-observed variables such as near-surface winds, oil spills, waves, and internal waves for diverse scientific applications.

Of the many oceanic parameters measured from satellites, SST is the most fundamentally important for East Sea research. Prior to the quantitative use of SST, there were a few studies using infrared images qualitatively to distinguish oceanic fronts and to confirm existence of the East Korea Warm Current (Huh 1982; Kim and Legeckis 1986). Substantial use of satellite-derived SST data in the East Sea was initiated with the installation of High Resolution Picture Transmission ground stations for the acquisition of near-polar-orbiting National Oceanic and Atmospheric Administration (NOAA) satellite data. In Korea, the first ground station for oceanography was installed at SNU's Research Institute of Oceanography and the second at NFRDI in November 1989. A similar ground station was established at KIOST in 1998. A portable antenna system was also installed on a ship of KHOA in December 2002. Although KMA has operated the ground station for NOAA satellites, the Advanced Very High Resolution Radiometer (AVHRR) data had been utilized as auxiliary imagery data to support weather forecasts till 1990s.

Until now, a number of studies have been performed to investigate spatial and temporal variability of SST, the subpolar front, sea ice in the Tatarsky Strait, and the annual cycle of SST and its relation to bathymetry and wind forcings, using the copious NOAA-series data from NOAA-9 to NOAA-19 (e.g. Park et al. 2004a, 2005b, 2006, 2007). Accuracies of SSTs from NOAA-series satellites were determined by comparison with in situ measurements from surface drifters (e.g. Park et al. 1994, 1999). To overcome significant bias errors like underestimation (overestimation) of satellite SST in winter (summer) and skin-bulk temperature differences, regional optimization of SST for the East Sea has been accomplished by deriving new coefficients in the SST algorithm for each satellite (e.g. Park et al. 1999, 2008a, 2008b).

The Next Generation Sea Surface Temperature (NGSST) product from Tohoku University has facilitated the use of satellite SST data for scientific research, input data for numerical models, and many other applications in the East Sea. It is composed of SSTs from NOAA/AVHRR, Meteorological Satellite (MTSAT)-series geostationary satellites, and Advanced Microwave Scanning Radiometer-Earth Observation System (AMSR-E) combined through optimal interpolation, and was distributed to the scientific community for the Global Ocean Data Assimilation Experiment project until October 2011 (Guan and Kawamura 2004). Now a variety of SST data sets for studies of the East Sea can be obtained from overseas institutions such as the National Aeronautics and Space Administration (NASA)/ Jet Propulsion Laboratory, NOAA's Satellite and Information Service, and so on. However, such datasets have long been based on low-resolution Global Area Coverage data for the global ocean. Thus, SSTs with more accuracy and higher-resolution (about 1.1 km) should be continuously and consistently produced to support the scientific community conducting the East Sea research.

There have been scatterometers such as SeaSAT in 1978, the European Remote Sensing (ERS)-1 in 1992, and ERS-2 in 1995, but substantial use of wind data for studies of the East Sea began in September 1996 with the NASA Scatterometer (NSCAT). NSCAT winds were compared with calculated winds based on atmospheric pressure maps (Na et al. 1997), and utilized to understand Ekman transport and wind stress curl in the East Sea (Lee 1998). During a recent decade, satellite scatterometer winds have been used to investigate spatial and temporal variability of the near-surface wind field and wind stresses over the East Sea, their relations to changes in SSTs and contribution to heat flux, and as input data representing one of the surface forcings in numerous numerical models (e.g. Nam et al. 2005; Park et al. 2005b). Subsequent launches of the Quick Scatterometer (QuikSCAT) in 1999 and the Advanced Scatterometer (ASCAT) in 2006 have given us further opportunities to continue research projects using wind vectors over the East Sea at high spatial and temporal resolutions of about 25 km and 1–2 days, respectively (e.g. Wentz and Smith 1999). However, wind vectors from scatterometers have been found to have many errors associated with a wind direction ambiguity problem, uncertainty at low wind speeds, and a saturation problem at high wind speeds. Hence, careful use of the wind data is essential for East Sea studies. Considering the variety of active uses of near-surface wind fields employed hitherto, continuous and consistent launches of satellite scatterometers with high precision should be carried out to obtain longer time series of surface wind vectors.

One of the other remarkable applications of satellites is the determination of sea level or sea surface height by satellite altimetry. There have been many altimeters such as SeaSAT in 1978, Geodetic Satellite (GEOSAT) in 1985, TOPEX/Poseidon in 1992, Jason-1 in 2002, Environmental Satellite (Envisat) in 2002, in 2002, and Jason-2 in 2008. In the East Sea, satellite-observed sea level data have been utilized in studies of the long-term trend of sea level associated with global warming, and its spatiotemporal variability related to mesoscale eddies (e.g. Choi et al. 2004; Kang et al. 2005). Sea level rise in the southern part of the East Sea (6.6 ± 0.4 mm year^{-1}), obtained from 9 years of TOPEX/Poseidon data, was found to be about twice as large as the average rise rate of sea level in the global ocean (3.1 ± 0.4 mm year^{-1}) (Kang et al. 2005). Very recently, KHOA has been partly using the Archiving, Validation and Interpretation of Satellite Oceanographic (AVISO) sea level data to construct monthly maps of sea surface currents based on geostrophic balance (http://www.khoa.go.kr).

SAR backscatter signals have been used to study diverse features of the coastal region in the East Sea since the 2000s. SAR data from ERS-2, RADARSAT-1/2, Envisat Advanced SAR (ASAR), TerraSAR-X, and Advanced Land Observation Satellite (ALOS)/Phased Array type L-band SAR (PALSAR) have been used to monitor features such as vector winds, internal waves, surface waves, and wave height and to understand their physical processes in the East Sea. Near-surface wind speeds in the coastal region have been estimated from backscattering cross sections using C-band Model (CMOD)-4, 5, CMOD-IFR2 of Institut Français de Recherche pour l'exploitation de la mer (IFREMER), and L-band wind models, and their errors with their potential causes have been discussed (Kim and Moon 2002; Yoon 2008; Kim et al. 2010; Shimada et al. 2010; Kim and Park 2011; Kim et al. 2012). The signatures of eddies, fronts, filaments, currents, and sea ice on SAR images in the East Sea have been described (e.g. Yoon et al. 2007; Mitnik et al. 2009). Internal waves near the east coast of Korea have been studied using both in situ oceanic measurements and SAR images (Kim et al. 2005a, b; Gan et al. 2008).

Along with physical parameters measured from satellites, satellite ocean color data have been utilized to study the impact of changes in the physical environment on biological processes in the East Sea. Many research projects have been carried out to understand physio-biogeochemical processes by considering the variability of mixed layer depth and wind forcing using diverse satellite data from the Coastal Zone Color Scanner (CZCS), Sea-viewing Wide Field-of-view Sensor (SeaWiFS), Moderate Resolution Imaging Spectroradiometer (MODIS), and Medium Resolution Imaging Spectrometer (MERIS) of Envisat (Kim et al. 2000; Yamada et al. 2004, 2005; Yoo and Kim 2004; Yamada and Ishizaka 2006; Kim et al. 2007; Jo et al. 2007; Yoo and Park 2009; Park et al. 2013). Recently, data from the Geostationary Ocean Color Imager (GOCI) satellite, the world's first geostationary color-imaging satellite, has opened a new world of opportunities to study short-term variability of oceanic eddies and dynamic features in the upper ocean of the East Sea since 2010 (e.g. Lim et al. 2012a; Park et al. 2012).

The main advantage of satellite data derives from the simultaneous, extensively broad, and repetitive observations over the ocean. Considering the previous difficulties and limitations of in situ ocean measurements in the East Sea, satellite remote sensing is expected to support the scientific community through continuous and consistent monitoring of diverse oceanic phenomena with the help of increasing numbers of satellites and new types of sensors with higher precision than ever before.

References

Antonov JI, Seidov D, Boyer TP et al (2010) World ocean atlas 2009, volume 2: salinity. In: Levitus S (ed) NOAA Atlas NESDIS 68. U.S. Government Printing Office, Washington DC, 184 pp

Balmaseda MA, Anderson D, Vidard A (2007) Impact of Argo on analyses of the global ocean. Geophys Res Lett 34:L16605. doi:10.1029/2007GL030452

Beardsley RC, Limeburner R, Kim K et al (1992) Lagrangian flow observations in the East China, Yellow and Japan Seas. La Mer 30:297–314

Broecker WS, Peng T (1982) Tracers in the sea. Columbia University, Columbia

Chang KI, Hogg NG, Suk MS et al (2002) Mean flow and variability in the southwestern East Sea. Deep-Sea Res Pt I 49:2261–2279

Chang KI, Teague WJ, Lyu SJ et al (2004) Circulation and currents in the southwestern East/Japan Sea: overview and review. Prog Oceanogr 61:105–156

Chang KI, Kim K, Kim YB et al (2009) Deep flow and transport through the Ulleung Interplain Gap in the southwestern East/Japan Sea. Deep-Sea Res Pt I 56:61–72

Choi YJ, Yoon JH (2010) Structure and seasonal variability of the deep mean circulation of the East Sea (Sea of Japan). J Oceanogr 66:349–361

Choi BJ, Haidvogel D, Cho YK (2004) Nonseasonal sea level variation in the Japan/East Sea from satellite altimeter data. J Geophys Res 109:C12028. doi:10.1029/2004JC002387

Chough SK, Lee IIJ, Yoon SH (2000) Marine geology of Korean Seas, 2nd edn. Elsevier, Amsterdam

Craig H (1969) Abyssal carbon and radiocarbon in the Pacific. J Geophy Res 74:5491–5506

Danchenkov MA, Riser SC (2001) Observations of currents, temperature and salinity in the Japan Sea in 1999–2000 by PALACE floats. Oceanography of the Japan Sea. In: Proceedings of CREAMS' 2000 international symposium, Far Eastern regional hydrometeorological research institute, pp 33–40

Danchenkov MA, Lobanov VB, Riser SC et al (2006) A history of physical oceanographic research in the Japan/East Sea. Oceanography 19:18–31

Davis RE, Zenk W (2001) Subsurface Lagrangian observation during the 1990s. In: Siedler G, Church J, Gould J (eds) Ocean circulation and climate: observing and modelling the global ocean. Academic Press, London, pp 123–139

Davis RE, Webb DC, Regier LA et al (1992) The autonomous lagrangian circulation explorer (ALACE). J Atmos Oceanic Technol 9:264–285

Freeland, H, Roemmich D, Garzoli S et al (2010) Argo—a decade of progress. In: Hall J, Harrison DE, Stammer D (eds) Proceedings of OceanObs'09: sustained ocean observations and information for society, vol 2. ESA Publication WPP-306, Venice, Italy, 21–25 Sept 2009

Fujii Y, Yasuda T, Matsumoto S et al (2008) Observing system evaluation (OSE) using the El Niño forecasting system in Japan Meteorological Agency. In: Proceedings of the Oceanographic Society of Japan fall meeting (in Japanese)

Fukudome KI, Yoon JH, Ostrovskii A et al (2010) Seasonal volume transport variation in the Tsushima Warm Current through the Tsushima Straits from 10 years of ADCP observations. J Oceanogr 66:539–551

Gamo T (1999) Global warming may have showed down the deep conveyor belt of a marginal sea of the northwestern Pacific: Japan Sea. Geophys Res Lett 26:3137–3140

Gamo T, Horibe Y (1983) Abyssal circulation in the Japan Sea. J Oceanogr Soc Japan 39:220–230

Gamo T, Nozaki Y, Sakai H et al (1986) Spacial and temporal variations of water characteristics in the Japan Sea bottom layer. J Mar Res 44:781–793

Gan XL, Huang WG, Li XF et al (2008) Coastally trapped atmospheric gravity waves on SAR, AVHRR and MODIS images. Int J Remote Sens 29:1621–1634

Garcia HE, Locarnini RA, Boyer TP et al (2010) World Ocean Atlas 2009, volume 3: dissolved oxygen, apparent oxygen utilization, and oxygen saturation. In: Levitus S (ed) NOAA Atlas NESDIS 68, U.S. Government Printing Office, Washington, 344 pp

Guan L, Kawamura H (2004) Merging satellite infrared and microwave SSTs: methodology and evaluation of the new SST. J Oceanogr 60:905–912

Huh OK (1982) Spring season flow of the Tsushima Current and its separation from the Kuroshio: satellite evidence. J Geophys Res 37:9687–9693

IPCC (2007) Contribution of Working Group I to the fourth assessment report of the intergovernmental panel on climate change. In: Solomon S, Qin D, Manning M et al (eds) IPCC fourth assessment report: climate change 2007. Cambridge University Press, Cambridge, p 996

Jacobs GA, Book JW, Perkins HT et al (2001) Inertial oscillations in the Korea Strait. J Geophys Res 106:26943–26957

Jo CO, Lee JY, Park KA et al (2007) Asian dust initiated early spring bloom in the northern East/Japan Sea. Geophys Res Lett 34:L05602. doi:10.1029/2006GL027395

Kang DJ, Park S, Kim YG et al (2003) A moving-boundary box model for oceans in change: an application to the East/Japan Sea. Geophys Res Lett 30:1299. doi:10.1029/2002GL016486

Kang SK, Cherniawsky JY, Foreman MGG et al (2005) Patterns of recent sea level rise in the East/Japan Sea from satellite altimetry and in situ data. J Geophys Res 110. doi:10.1029/2004jc002565

Kim KR, Kim K (1996) What is happening in the East Sea (Japan Sea)? Recent chemical observations during CREAMS 93–96. J Korean Soc Oceanogr 31:164–172

Kim K, Legeckis R (1986) Branching of the Tsushima Current in 1981–83. Prog Oceanogr 17:265–276

Kim DJ, Moon WM (2002) Estimation of sea surface wind vector using RADARSAT data. Remote Sens Environ 80:55–64

Kim TS, Park KA (2011) Estimation of polarization ratio for sea surface wind retrieval from SIR-C SAR data. Korean J Remote Sens 27:729–741

Kim K, Kim KR, Chung JY et al (1996) New findings from CREAMS observations: water masses and eddies in the East Sea. J Korean Soc Oceanogr 31:155–163

Kim KR, Kim K, Kang DJ et al (1999) The East Sea (Japan Sea) in change: a story of dissolved oxygen. Mar Technol Soc J 33:15–22

Kim SW, Saitoh SI, Ishizaka J et al (2000) Temporal and spatial variability of phytoplankton pigment concentrations in the Japan Sea derived from CZCS images. J Oceanogr 56:527–538

Kim K, Kim KR, Min DH et al (2001) Warming and structural changes in the East (Japan) Sea: a clue to future changes in global oceans? Geophys Res Lett 28:3293–3296

Kim KR, Kim G, Kim K et al (2002a) A sudden bottom water formation during the severe winter 2000–2001: the case of the East/Japan Sea. Geophys Res Lett 29. doi:10.1029/2001GL014498

Kim KR, Kim K, Kang DJ et al (2002b) The changes in the East/Japan Sea found by CREAMS. Oceanogr Jpn 11:419–429 (in Japanese)

Kim K, Kim KR, Kim YG et al (2004a) Water masses and decadal variability in the East Sea (Sea of Japan). Prog Oceanogr 61:157–174

Kim K, Lyu SJ, Kim YG et al (2004b) Monitoring volume transport through measurement of cable voltage across the Korea Strait. J Atmos Ocean Tech 21:671–682

Kim DJ, Nam SH, Kim HR et al (2005a) Can near-inertial internal waves in the East Sea be observed by synthetic aperture radar? Geophys Res Lett 32:L02606. doi:10.1029/2004GL021532

Kim HR, Nam SH, Kim D-J et al (2005b) Reply to comment by Q. Zheng on "can near-inertial internal waves in the East Sea be observed by synthetic aperture radar?". Geophys Res Lett 32:L20607. doi:10.1029/2005GL024351

Kim HC, Yoo S, Oh IS (2007) Relationship between phytoplankton bloom and wind stress in the sub-polar frontal area of the Japan/East Sea. J Mar Syst 67:205–216

Kim YB, Chang KI, Kim K et al (2009a) Vertical structure of low frequency currents in the southwestern East Sea (Sea of Japan). J Oceanogr 65:259–271

Kim YH, Chang KI, Park JJ et al (2009b) Comparison between a reanalyzed product by the 3D-Var assimilation technique and observations in the Ulleung Basin of the East/Japan Sea. J Mar Sys 78:249–264

Kim TS, Park KA, Moon WI (2010) Wind vector retrieval from SIR-C SAR data off the east coast of Korea. J Korean Earth Sci Soc 31:475–487

Kim TS, Park KA, Choi WM et al (2012) L-band SAR-derived sea surface wind retrieval off the east coast of Korea and error characteristics. Korean J Remote Sens 28:477–487

Kim SK, Chang KI, Kim B et al (2013) Contribution of currents to the N abundance in the northwestern Pacific marginal seas. Geophys Res Lett 40:143–148

Kitani K (1987) Direct current measurements of the Japan Sea Proper Water. Rep Japan Sea Nat Fish Res Inst 341:1–6 (in Japanese)

Kwon YO, Kim K, Kim YG et al (2004) Diagnosing long-term trends of the water mass properties in the East Sea (Sea of Japan). Geophys Res Lett 31:L20306. doi:10.1029/2004GL020881

Lee DK (1998) Ocean surface winds over the seas around Korea measured by the NSCAT (NASA scatterometer). J Korean Soc Remote Sens 14:37–52 (in Korean)

Lee DK, Niiler P (2005) The energetic surface circulation of the Japan/East Sea. Deep-Sea Res Pt II 52:1547–1563

Lee DK, Niiler P (2010) Surface circulation in the Southwestern Japan/East Sea from drifters and sea surface height. Deep-Sea Res Pt II 57:1222–1232

Lee DK, Niiler P, Lee SR et al (2000) Energetics of the surface circulation of the Japan/East Sea. J Geophy Res 105:19561–19573

Lee SH, Chough SK, Back GG et al (2002) Chirp (2–7 kHz) echo characters of the South Korea Plateau: styles of mass movement and sediment gravity flow. Mar Geol 184:227–247

Lie HJ, Byun SK, Bang IK et al (1995) Physical structure of eddies in the southern East Sea. J Korean Soc Oceanogr 30:170–183

Lim JH, Son S, Park JW et al (2012a) Enhanced biological activity by an anticyclonic warm eddy during early spring in the East Sea (Japan Sea) detected by the geostationary ocean color satellite. Ocean Sci J 47:377–385

Lim SS, Jang CJ, Oh IS et al (2012b) Mixed layer climatology for the East Sea (Japan Sea). J Mar Sys 96–97:1–14

Locarnini RA, Mishonov AV, Antonov JI et al (2010) World Ocean Atlas 2009, volume 1: temperature. In: Levitus S (ed) NOAA Atlas NESDIS 68, U.S. Government Printing Office, Washington, 184 pp

Minami H, Kano Y, Ogawa K (1998) Long-term variations of potential temperature and dissolved oxygen of the Japan Sea water. J Oceanogr 55:197–205

Minobe S, Sako A, Nakamura M (2004) Interannual to interdecadal variability in the Japan Sea based on a new gridded upper water temperature dataset. J Phys Oceanogr 34:2382–2397

Mitchell DA, Watts DR, Wimbush M et al (2005) Upper circulation patterns in the Ulleung Basin. Deep-Sea Res Pt II 52:1617–1638

Mitnik L, Dubina V, Konstantinov O et al (2009) Remote sensing of surface films as a tool for the study of oceanic dynamic processes. Ocean Polar Res 31:111–119

Mooers CNK, Kang H, Bang I et al (2006) Some lessons learned from comparisons of numerical simulations and observations of the Japan/East Sea circulation. Oceanography 19:86–95

MSA Reports (1995–2007) Report of marine pollution surveys. Hydrographic Department, Maritime Safety Agency, Japan (in Japanese)

Na JY, Han SK, Seo JW et al (1997) Empirical orthogonal function analysis of surface pressure, sea surface temperature and winds over the East Sea of the Korea (Japan Sea). J Korean Fish Soc 30:188–202 (in Korean)

Na H, Kim KY, Chang KI et al (2012) Decadal variability of the upper ocean heat content in the East/Japan Sea and its relationship with the northwestern Pacific variability. J Geophys Res 117:C02017. doi:10.1029/2011JC007369

Nakajima T, Satoh M, Okamura Y (1998) Channel-levee complexes, terminal deep-sea fan and sediment wave fields associated with the Toyama Deep-Sea channel system in the Japan Sea. Mar Geol 147:25–41

Nam SH, Kim YH, Park KA et al (2005) Spatio-temporal variability in sea surface wind stress near and off the east coast of Korea. Acta Oceanol Sin 24:107–114

Niiler PP, Davis RE, White HJ (1987) Water-following characteristics of a mixed layer drifter. Deep-Sea Res Pt I 34:1867–1881

Niiler PP, Sybrandy AS, Bi KN et al (1995) Measurements of the water-following capability of holey-sock and TRISTAR drifters. Deep-Sea Res Pt I 42:1951–1955

Nishida K (1927) Report of the oceanographical investigation, No. 2, report of the current observations. In: The first report—results of the current measurements in the adjacent seas of Korea, 1923–1926. Gov Fish Exp Sta, pp 1–68 (in Japanese)

Östlund HG, Craig H, Broecker WS et al (1987) Shorebased data and graphics. In: Etal HG (ed) GEOSECS Atlantic, Pacific, and Indian Ocean expeditions, vol 7. Government Printing Office, Washington

Park JJ, Kim K (2007) Evaluation of calibrated salinity from profiling floats with high resolution conductivity-temperature-depth data in the East/Japan Sea. J Geophys Res 112:C05049. doi :10.1029/2006JC003869

Park JJ, Kim K (2013) Deep currents obtained from ARGO float trajectories in the Japan/East Sea. Deep-Sea Res Pt II 85:169–181

Park KA, Chung JY, Kim K et al (1994) A study on comparison of satellite drifter temperature with satellite derived sea surface temperature of NOAA/NESDIS. J Korean Soc Remote Sens 11:83–107

Park KA, Chung JY, Kim K et al (1999) Sea surface temperature retrievals optimized to the East Sea (Sea of Japan) using NOAA/AVHRR data. Mar Technol Soc J 33:23–35

Park KA, Chung JY, Kim K (2004a) Sea surface temperature fronts in the East (Japan) Sea and temporal variations. Geophys Res Lett 31:L07304. doi:10.1029/2004gl019424

Park YG, Oh KH, Chang KI et al (2004b) Intermediate level circulation of the southwestern part of the East/Japan Sea estimated from autonomous isobaric profiling floats. Geophys Res Lett 31:L13213. doi:10.1029/2004GL020424

Park JJ, Kim K, King BA, Riser SC (2005a) An advanced method to estimate deep currents from profiling floats. J Atmos Oceanic Tech 22:1294–1304

Park KA, Chung JY, Kim K et al (2005b) Wind and bathymetric forcing of the annual sea surface temperature signal in the East (Japan) Sea. Geophys Res Lett 32:L05610. doi:10.1029/200 4gl022197

Park KA, Cornillon PC, Codiga DL (2006) Modification of surface wind near ocean fronts: effects of Gulf Stream rings on scatterometer (QuikSCAT, NSCAT) wind observations. J Geophys Res 111:C03021. doi:10.1029/2005JC003016

Park KA, Ullman DS, Kim K et al (2007) Spatial and temporal variability of satellite-observed subpolar front in the East/Japan Sea. Deep-Sea Res Pt I 54:453–470

Park KA, Sakaida F, Kawamura H (2008a) Error characteristics of satellite-observed sea surface temperatures in the northeast Asian Sea. J Korean Soc Earth Sci 29:280–289

Park KA, Sakaida F, Kawamura H (2008b) Oceanic skin-bulk temperature difference through the comparison of satellite-observed sea surface temperature and in-situ measurements. Korean J Remote Sens 24:1–15

Park YG, Choi A, Kim YH et al (2010) Direct flows from the Ulleung Basin into the Yamato Basin in the East/Japan Sea. Deep-Sea Res Pt I 57:731–738

Park KA, Woo HJ, Ryu JH (2012) Spatial scales of mesoscale eddies from GOCI chlorophyll-*a* concentration images in the East/Japan Sea. Ocean Sci J 47:347–358

Park KA, Chae HJ, Park JE (2013) Characteristics of satellite chlorophyll-*a* concentration speckles and are moval method in a composite process in the East/Japan Sea. Int J Remote Sens 34:4610–4635

Perkins HT, Teague WJ, Jacobs GA et al (2000) Currents in Korea-Tsushima Strait during summer 1999. Geophys Res Lett 27:3033–3036

Perry MJ, Rudnick DL (2003) Observing the ocean with autonomous and lagrangian platforms and sensors (ALPS): the role of ALPS in sustained ocean observing systems. Oceanography 16:31–36

Senjyu T, Sudo H (1993) Water characteristics and circulation of the upper portion of the Japan Sea Proper Water. J Mar Sys 4:349–362

Senjyu T, Aramaki T, Otosaka S et al (2002) Renewal of the bottom water after the winter 2000–2001 may spin-up the thermohaline circulation in the Japan Sea. Geophys Res Lett 29:1149. doi:10.1029/2001GL014093

Senjyu T, Shin HR, Yoon JH et al (2005) Deep flow field in the Japan/East Sea as deduced from direct current measurements. Deep-Sea Res Pt II 52:1726–1741

Shimada T, Sawada M, Sha W et al (2010) Low-level easterly winds blowing through the Tsugaru Strait, Japan. Part I: case study and statistical characteristics based on observations. Mon Wea Rev 138:3806–3821

Son YT, Chang KI, Yoon ST et al (2014) A newly observed physical cause of the onset of the subsurface spring phytoplankton bloom in the southwestern East/Japan Sea. Biogeosciences 11:1319–1329

Sudo H (1986) A note on the Japan Sea Proper Water. Prog Oceanogr 17:313–336

Takematsu M, Nagano Z, Ostrovskii AG et al (1999) Direct measurements of deep currents in the northern Japan Sea. J Oceanogr 55:207–216

Talley LD, Lobanov V, Ponomarev V et al (2003) Deep convection and brine rejection in the Japan Sea. Geophys Res Lett 30:1159. doi:10.1029/2002GL016451

Teague WJ, Jacobs GA, Perkins HT et al (2002) Low-frequency current observations in the Korea/Tsushima Strait. J Phys Oceanogr 32:1621–1641

Teague WJ, Tracey KL, Watts DR et al (2005) Observed deep circulation in the Ulleung Basin. Deep-Sea Res Pt II 52:1802–1826

Thadathil P, Bajish CC, Behera S et al (2012) Drift in salinity data from Argo profiling floats in the Sea of Japan. J Atmos Oceanic Technol 29:129–138

Tsunogai S, Watanabe YW, Harada K et al (1993) Dynamics of the Japan Sea Deep Water studied with chemical and radiochemical tracers. In: Teramoto T (ed) Deep Ocean Circulation, Physical and Chemical aspects. Elsevier, Amsterdam, pp 105–120

Uda M (1934) The results of simultaneous oceanographical investigations in the Japan Sea and its adjacent waters in May and June 1932. J Imp Fish Exp Sta 5:57–190 (in Japanese)

USSR Academy of Science (1957) Reports of multi-disciplinary oceanography expeditions on R/V Vityaz. vol 6. unpublished

Wentz FJ, Smith DK (1999) A model function for the ocean-normalized radar cross-section at 14 GHz derived from NSCAT observations. J Geophys Res 104:11449–11514

Yamada H (1999) Movement of the Japan Sea Proper Water from the results of moored direct current measurements. Umi to Sora (Sea and Sky) 74:168–171 (in Japanese)

Yamada K, Ishizaka J (2006) Estimation of interdecadal change of spring bloom timing, in the case of the Japan Sea. Geophys Res Lett 33:L02608. doi:10.1029/2005GL024792

Yamada K, Ishizaka J, Yoo S et al (2004) Seasonal and interannual variability of sea surface chlorophyll-*a* concentration in the Japan/East Sea (JES). Prog Oceanogr 61:193–211

Yamada K, Ishizaka J, Nagata H (2005) Spatial and temporal variability of satellite primary production in the Japan Sea from 1998 to 2002. J Oceanogr 61:857–869

Yang JY (1996) Current structure derived from satellite tracked drifter in the northern part of the East Sea. Dissertation, Seoul National University

Yasui M, Yasuoka T, Shiota O (1967) Oceanographical studies of the Japan Sea (1) water characteristics. Oceanogr Mag 19:177–192

Yi SU (1966) Seasonal and secular variations of the water volume transport across the Korea Strait. J Oceanol Soc Korea 1:7–13

Yoo S, Kim HC (2004) Suppression and enhancement of the spring bloom in the southwestern East Sea/Japan Sea. Deep-Sea Res Pt II 51:1093–1111

Yoo S, Park J (2009) Why is the southwest the most productive region of the East Sea/Sea of Japan? J Mar Syst 78:301–315

Yoon HJ (2008) Analyses on the sea surface wind field data by satellite remote sensing. J Korea Inst Marit Infor Commun Sci 12:149–157

Yoon HJ, Mitnik LM, Kang HS et al (2007) ERS SAR observations of the Korean coastal waters. Korean J Remote Sens 23:65–69

Yoshikawa Y (2012) An eddy-driven abyssal circulation in a bowl-shaped basin due to deep water formation. J Oceanogr 68:971–983

Chapter 2
Forcings

Kyung-Ae Park, Kyung-Il Chang, Hanna Na and Uk-Jae Jung

Abstract The East Sea (Japan Sea) is strongly influenced by the Asian monsoon with prevalent northerly and southerly winds in winter and summer, respectively. It gains heat from April to August and loses heat in other seasons with annual net heat loss ranging from 25 to 108 W m^{-2}. Extremely strong winds of severe Siberian cold-air outbreaks typify the winter season from December to February, which occasionally results in cold bottom water formation. A spatially distinct pattern of wind stress curl during the outbreak periods appears south of Vladivostok and near East Korea Bay. The cold-air outbreaks result in significant surface heat losses through sensible and latent heat fluxes exceeding 500 W m^{-2}, which was estimated to be even larger, as high as 1000 W m^{-2}, during a deep convection period in winter. Wintertime surface heat loss is also an important factor in deep penetration of frontal subduction, resulting in the formation of East Sea Intermediate Water found south of the subpolar front. Net heat loss at the sea surface in the East Sea is compensated by its warm inflow-outflow system. Inflow of the Tsushima Warm Current through the Korea Strait (2.6 Sv), which is balanced with outflows through the Tsugaru Strait and the Soya Strait, exhibits large

K.-A. Park (✉)
Department of Earth Science Education, Seoul National University, Seoul 151-748, Republic of Korea
e-mail: kapark@snu.ac.kr

K.-A. Park · K.-I. Chang
Research Institute of Oceanography, Seoul National University, Seoul 151-742, Republic of Korea

K.-I. Chang · U.-J. Jung
School of Earth and Environmental Sciences, Seoul National University, Seoul 151-742, Republic of Korea
e-mail: ujjung@curl.snu.ac.kr

H. Na
Faculty of Science, Hokkaido University, Sapporo, Hokkaido 060-0810, Japan
e-mail: hna@mail.sci.hokudai.ac.jp

© Springer International Publishing Switzerland 2016
K.-I. Chang et al. (eds.), *Oceanography of the East Sea (Japan Sea)*,
DOI 10.1007/978-3-319-22720-7_2

volume transports in summer to autumn and small in winter to spring with a range of about 1 Sv (1 Sv $= 10^6$ m^3 s^{-1}). While the mean transport is larger in the Tsugaru Strait compared to that in the Soya Strait, the seasonal variability is larger in the Soya Strait.

Keywords Sea surface wind · Surface heat flux · Volume transport · Tsushima Warm Current · Korea Strait · Tsugaru Strait · Soya Strait · East Sea (Japan Sea)

2.1 Introduction

The East Sea has many characteristics similar to those of the global ocean and may be considered its miniature. There are remarkable atmospheric and oceanic forcings such as wind forcing, surface heat flux at the air-sea interface, and boundary flux through straits, which have strongly affected the basin-wide surface-to-deepwater circulations, formation and distribution of water masses, and many other properties of the East Sea. It is almost impossible to discuss substantial changes in water masses, currents, and circulations without a priori knowledge of the spatial and temporal variability of the fundamental forcing fields over the East Sea. Thus, in the following sections, we have attempted to review the previous literature and recent results on wind forcing (Sect. 2.2), surface heat flux (Sect. 2.3), and boundary flux (Sect. 2.4). Section 2.5 is a summary, and a discussion of remaining issues.

2.2 Surface Wind

2.2.1 Accuracy of Satellite Scatterometer Wind Vectors

Satellite scatterometers have measured near-surface wind velocity vectors under all weather conditions with accuracy of less than 2 m s^{-1} in speed and 20° in direction and spatial resolution of 25 km (Jet Propulsion Laboratory 1998). Scatterometers such as NASA Scatterometer (NSCAT) of the ADvanced Earth Observing Satellite (ADEOS), SeaWinds of the Quick Scatterometer (QuikSCAT), and the Advanced Scatterometer (ASCAT) of Europe's Meteorological Operational Satellite Program—A have been successfully operated with relatively high spatial and temporal coverage since September 1996.

Over the decade prior to the launch of satellite scatterometers in the 1990s, wind fields for the period 1978–1997 produced by Na et al. (1997), called 'Na wind' hereafter, was the most renowned in the scientific community and extensively utilized for diverse studies in the seas adjacent to Korea (e.g. Kim and Yoon 1996). Unlike automatic weather station (AWS) measurements, it is not based on direct measurements of wind vectors but on those indirectly calculated from the distribution of atmospheric pressure on daily weather maps using the Cardone model (Cardone 1969). The Na wind field has both similarity and dissimilarity to the scatterometer wind field. Comparing the two, the spatial structures are mostly

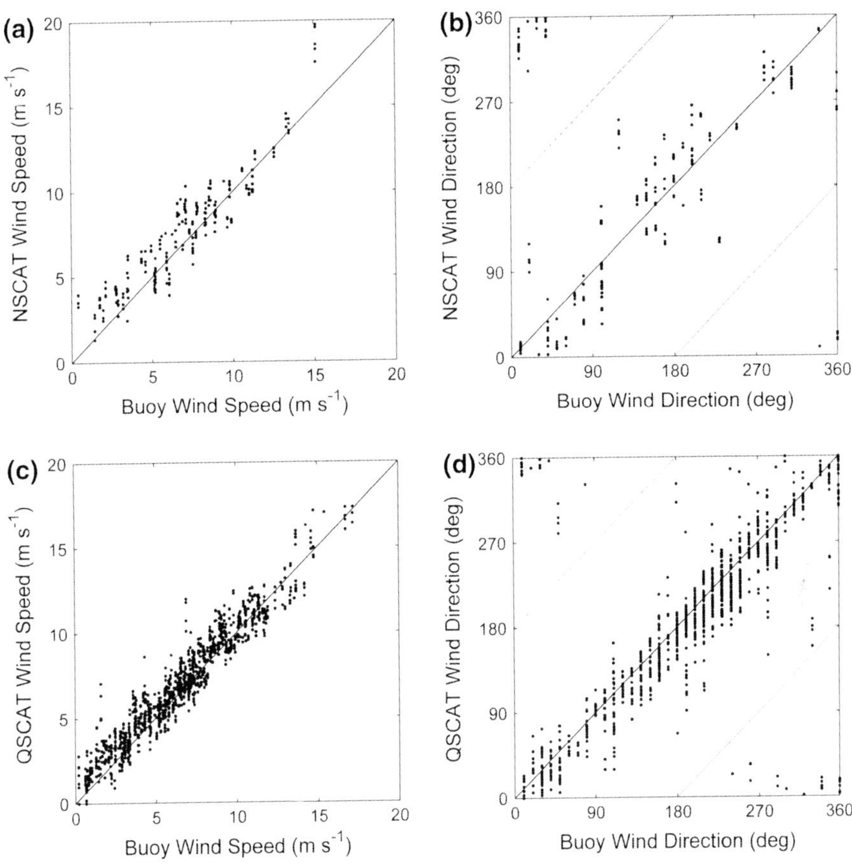

Fig. 2.1 Comparison of satellite-derived winds with winds from three Japan Meteorological Agency (JMA) buoys in the seas adjacent to Korea: **a** NSCAT wind speed, **b** NSCAT wind direction, **c** QuikSCAT wind speed, and **d** QuikSCAT wind direction (from Kim et al. 2005)

coherent on large basin-wide scales, but significant differences and energy loss are detected in the Na wind at spatial scales less than 100 km (Park et al. 2003).

Comparison of satellite scatterometer-derived wind speeds from NSCAT and QuikSCAT/SeaWinds with buoy-measured wind speeds reveals relatively small root-mean-squared (rms) errors with a range of 0.9–1.3 m s^{-1} in the East Sea; this meets the scatterometer mission error limit requirement of 2 m s^{-1} as shown in Fig. 2.1 (Ebuchi 1997; Lee 1998; Park et al. 2003; Kim et al. 2005). However, scatterometer wind directions show large rms errors of 15°–40° due to ambiguity problems, random and systematic errors, and the uncertainty of geophysical model functions (Freilich 1997; Park et al. 2003). Regional biases of wind vectors in the East Sea are large at low wind speeds in particular; these biases depend on wind speed and atmospheric stability in the marine-atmospheric boundary layer (MABL) (Park and Cornillon 2002; Chelton et al. 2004; Park et al. 2006a). In addition, scatterometers have saturation problems at high wind speeds causing an underestimation of wind speeds when compared with buoy measurements of

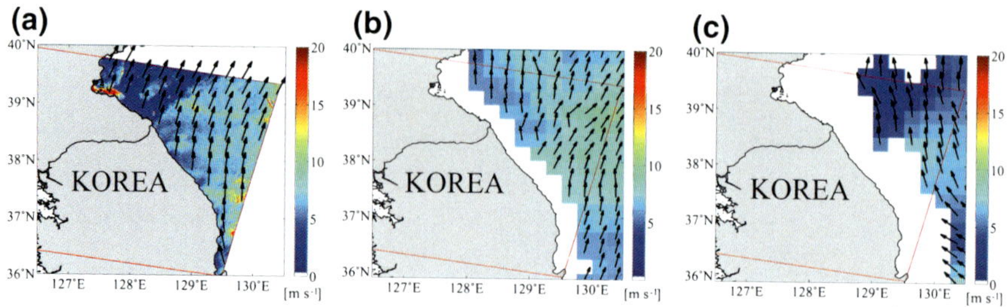

Fig. 2.2 Distribution of wind speed (*color scale* in m s^{-1}) and direction (*arrow*) off the east coast of Korea from **a** ALOS PALSAR, **b** QuikSCAT, and **c** ASCAT, on the same date (11 August 2007), where the *red box* indicates the boundary of ALOS PALSAR observations (redrawn from Kim et al. 2012)

winds (Liu and Xie 2006). Since the oceanic near-surface wind field also plays an important role in the estimation of latent heat flux and sensible heat flux, it should be continuously produced with high quality to reduce uncertainty in the estimation of net heat flux in the East Sea.

Since a satellite scatterometer observes winds at relatively low resolution of about 25 km, scatterometer data are not readily applicable in coastal areas. By contrast, Synthetic Aperture Radar (SAR) can provide a high-resolution wind field even within 25 km from the coast (Fig. 2.2). In the East Sea, some studies have been performed to validate the SAR-observed winds by comparison with winds from moored buoys, AWS near the coast, and numerical models (Isoguchi and Kawamura 2007). Overall, the rms errors of wind speeds from C-band and L-band SARs meet the error limit of 2 m s^{-1} in the East Sea (Kim and Moon 2002; Kim et al. 2010, 2012; Kim and Park 2011). The high-resolution coastal winds of the East Sea are anticipated to promote deep understanding of physical and biological processes at the coastal areas.

2.2.2 Spatial and Temporal Variability of Near-Surface Winds

Figure 2.3 shows the distribution of monthly-mean wind vectors calculated using QuikSCAT data from 2000 through 2008. The scatterometer wind fields reveal detailed spatial structures with large spectral energy even at high wavenumbers (Lee 1998; Park et al. 2003). The monthly-mean speeds range from 3 to 12 m s^{-1} over the basin. Overall, the speeds are larger during the wintertime, with values greater than 8 m s^{-1} beginning in November and reaching a peak in January (>10 m s^{-1}), particularly south of Vladivostok and along the Primorye coast. Northwesterly cold-air outbreaks through Vladivostok in winter cause the formation of cold bottom water in the East Sea and drive a cyclonic circulation in the Japan Basin (e.g. Kawamura and Wu 1998; Chen et al. 2001; Chu et al. 2001; Kim et al. 2002; Dorman et al. 2004). Strong winds also appear over East Korea Bay

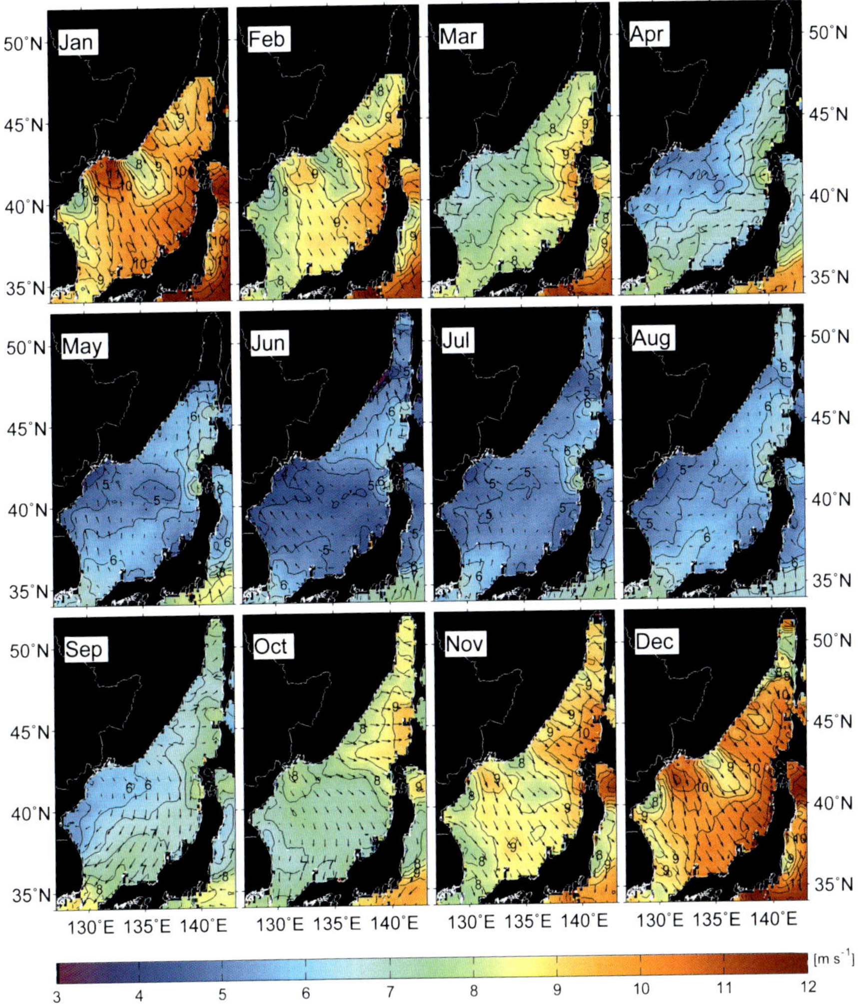

Fig. 2.3 Spatial distribution of monthly-mean wind vectors from 9 years of QuikSCAT data for the period from 2000 to 2008 over the East Sea

(see Fig. 1.1) (Park et al. 2003; Nam et al. 2005). Southerly winds begin to blow in April, with lowest speeds of less than 5 m s^{-1} in June, and last into August (Fig. 2.3) when the wind direction changes from southerly to northerly. Northerly winds continue through fall and winter until March of the next year.

Northerly winds passing through the orographic gap near Vladivostok begin in October, become stronger at ~10 m s^{-1} in December, achieve greatest strength of about 12 m s^{-1} as cold-air outbreaks in January, and last, at ~8 m s^{-1}, until February (Fig. 2.3). During such cold outbreaks, the 0 °C surface air isotherm extends well southward of 40°N, the surface heat losses in the center of the East

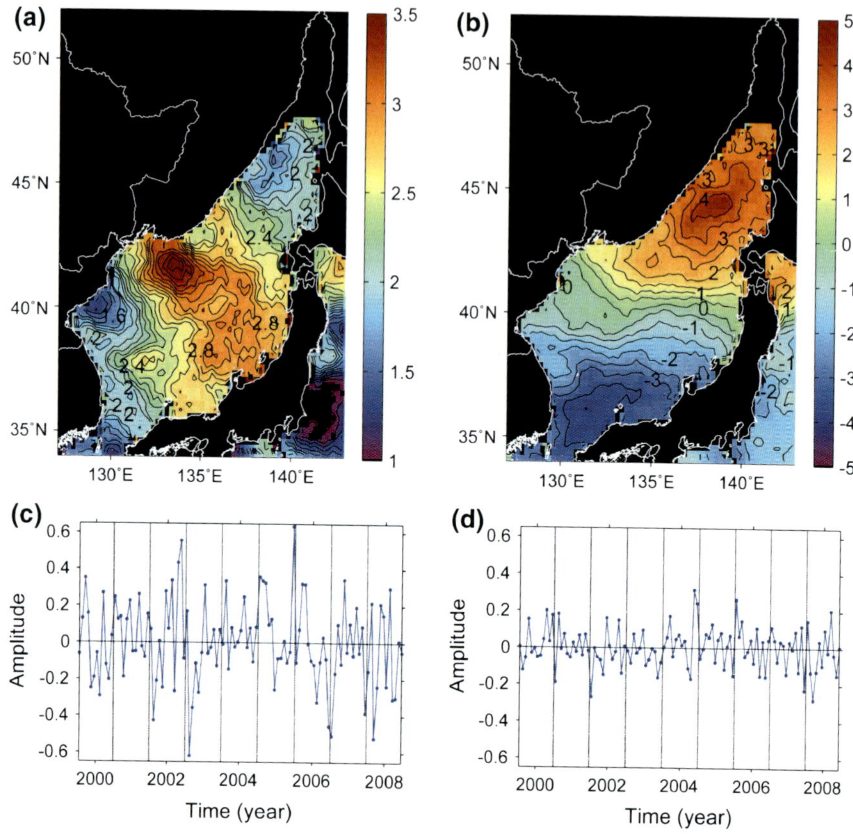

Fig. 2.4 a First and **b** second EOF modes of monthly-mean QuikSCAT wind speeds over the East Sea from 2000 to 2008, and **c–d** their time-varying amplitudes

Sea can exceed 600 W m^{-2}, and cloud streets with individual roll clouds cover most of the East Sea, extending from the Russian coast to Honshu (Dorman et al. 2004). The average number of strong cold-air outbreaks per winter (November–March) season is about 13, occupying 39 % of the winter period and contributing 43 % of the net heat loss from the East Sea (Dorman et al. 2004). More details about the surface buoyancy flux are in Sect. 2.3.

Figure 2.4a shows the first EOF mode of 9-year QuikSCAT wind speed anomalies over the East Sea from 2000 through 2008, accounting for 44.3 % of the total variance. Overall, it represents the strong winds from the continental side to the offshore region during cold-air outbreaks in winter. The winds are spatially shielded by mountainous land masses between Vladivostok and the northeastern part off the Primorye coast, producing low speeds in that region (Park et al. 2003). The time-varying amplitudes usually reach a peak in January each year (Fig. 2.4c). Large values over the southeastern region near the Japanese coast are related to relatively high wind speeds due to strengthening of unstable condition within the MABL (Park and Cornillon 2002; Chelton et al. 2004; Park et al. 2006a; Shimada and Kawamura 2006). The second EOF mode (Fig. 2.4b), explaining 10.1 % of the QuikSCAT wind variance, is dominant in the region between the Russian coast and Hokkaido from 40°N to 48°N in winter. The impact of such wind forcing on oceanic circulation has not been studied vigorously and should be studied in the future.

2.2.3 Wind Stress and Its Curl

Nam et al. (2005) analyzed near-shore and off-shore wind stress and showed typical amplitudes of wind stress off the east coast of Korea in the seasonal (>90 days), intra-seasonal (20–90 days), and synoptic (2–20 days) bands are 0.20, 0.03, and 0.04–0.18 N m^{-2}, respectively. The synoptic-band wind stress shows seasonal modulation by becoming stronger in winter (0.5–0.6 N m^{-2}) than in summer. The wind stress curl causes a pair of positive and negative values on the left and right sides of the downwind air flow, owing to the intensification of winds through the orographic gap near Vladivostok (Kawamura and Wu 1998; Park et al. 2003; Dorman et al. 2004; Nam et al. 2005; Shimada and Kawamura 2006). Another paired structure of positive/negative curl is found off Wonsan over East Korea Bay, for which the mechanism is similar to that south of Vladivostok, being driven by winds through an orographic gap between mountains (Park et al. 2003; Nam et al. 2005). Wind stress curl fields result also from accelerations and decelerations of wind as it blows over subpolar fronts in the central part of the East Sea (Shimada and Kawamura 2006).

The positive wind stress curl of the dipole structure occupying most of the East Sea in winter generates a large basin-wide cyclonic circulation (Yoon et al. 2005). The cyclonic wind stress curl in East Korea Bay may play an important role in the separation of the East Korea Warm Current (EKWC) (Kim and Yoon 1996; Yoon et al. 2005). Trusenkova et al. (2009) indicated that wind stress curl can be an additional factor forcing the Tsushima Warm Current (TWC) to branch off at the Korea Strait into the EKWC and the Offshore Branch. Sensitivity of the circulation to the different wind forcings has proved to be significant, which means the wind forcing, along with surface buoyancy forcing, is substantial in determining the different patterns and magnitudes of the basin-wide circulation (Hogan and Hulbert 2000).

2.3 Surface Heat Flux

2.3.1 Comparison of Heat Flux Estimates

Heat flux estimates in the East Sea are available from several studies (Table 2.1). The surface heat fluxes are usually computed by means of bulk formulas using atmospheric and marine surface data at the air-sea interface. The net heat flux Q_{net}

Table 2.1 Comparisons of summertime (Sum.; June–August), wintertime (Win.; December–February), and annual mean (Ann.) latent, sensible, and net heat fluxes (W m^{-2}) in the East Sea

Sources	Periods	Q_e			Q_h			Q_{net}		
		Sum.	Win.	Ann.	Sum.	Win.	Ann.	Sum.	Win.	Ann.
Kato and Asai (1983)	78–79	−29	−169	−102	5	−132	−47			−35
Kondo et al. (1994)	65–90	−26	−124	−78	5	−88	−31			−25
Park et al. (1995)	61–90	−17	−161	−90	2	−97	−32			−51
Hirose et al. (1996)	60–90	−35	−143	−93	1	−108	−40	113.3	−251	−53
Ahn et al. (1997)	80–94			−141			−41			−90
Na et al. (1999)	78–95	−42[a]	−172[a]	−114[a]	−4[a]	−144[a]	−60[a]	92[a]	−332[a]	−108[a]
Hong et al. (2005)	88–00	−48	−144	−108	0.5	−79	−32			−70
Dorman et al. (2004, 2005)	Summer 5/19/99–6/02/00 6/24/99–7/18/99 7/19/99–8/11/99 Winter 1/17/00–2/04/00 3/02/00–3/16/00	−27/−28/5	−160/−69		4/−3/7	−142/−41		137/−138	−295/−77	
OAFlux	89–09	−28.6	−163.1	−97.7	2.2	−116.9	−43.9	130.7	−276	−59
ECMWF	91–01	−28.4	−145.2	−95.1	0	−106.3	−52.8	154.4	−263.7	−47.6

Flux estimates from the OAFlux products and ECMWF model results are also shown

[a]Values estimated from figures, by eye

is the sum of four components, $Q_{net} = Q_s + Q_b + Q_h + Q_e$, where Q_s is the net shortwave radiation flux, Q_b the net long wave radiation flux, Q_h the sensible heat flux, and Q_e the latent heat flux. Different parameterizations were used for each of these components. As a sign convention, positive (negative) heat flux components denote oceanic heat gain (loss).

Heat flux estimates in gridded boxes with $0.2°–2°$ resolutions based on different bulk formulas and long-term averaged input variables show considerable differences (Table 2.1, Fig. 2.5). We have also compared those previous estimates with long-term mean heat fluxes from 1991 to 2001 produced by the European Centre for Medium-range Weather Forecast (ECMWF) model (Dorman et al. 2005), and from 1989 to 2009 obtained from the Objectively Analyzed air-sea Fluxes (OAFlux) project at the Woods Hole Oceanographic Institution (Yu and Weller 2007). The annual climatological Q_{net}, winter Q_h, and winter Q_e from those estimates all show negative values, indicating oceanic heat loss, with amplitudes ranging 25–108, 79–144, and 124–172 W m^{-2}, respectively. The long-term mean Q_{net} values from the ECMWF model, OAFlux products, Park et al. (1995), and Hirose et al. (1996) (HKY96 hereafter) all show similar values of about -50 W m^{-2} (corresponding to about 0.05 PW over the entire East Sea), which is about the same as the surface heat loss over the East China Sea (0.04 PW; Liu et al. 2010). Net heat losses in Kato and Asai (1983) and Kondo et al. (1994), however, are smaller ranging from -35 to -25 W m^{-2}. The other studies reported larger-magnitude estimates between -108 and -70 W m^{-2}, mainly due to their larger winter Q_h or Q_e (Ahn et al. 1997; Na et al. 1999 (NSL99 hereafter); Hong et al. 2005). The relatively wide range of the long-term estimates may arise from differences of datasets used in the calculations, parameterizations used in the bulk formulas, temporal resolutions of input data (NSL99), and different time periods reflecting interannual variation of the fluxes. The range of interannual variation is from about -35 to -22 W m^{-2} with a peak-to-peak difference of -13 W m^{-2}, according to a time series of annual mean Q_{net} from the OAFlux product (not shown); this implies that the flux discrepancies in Table 2.1 are not due to interannual variation of the heat fluxes. Summer Q_h and Q_e magnitudes are only about 1–6 and 11–33 % of their respective winter values, and the large heat loss in winter is caused by the dry and cold Siberian cold air mass sweeping over the warm sea surface.

2.3.2 Temporal Variations

The climatological annual cycles of heat fluxes obtained from the ECMWF model between 1991 and 2001 (Dorman et al. 2005) are shown in Fig. 2.5 together with those from the OAFlux products, HKY96, and NSL99. Also included are estimates from Dorman et al. (2004, 2005) based on highly accurate atmospheric datasets collected during surveys over the East Sea from research vessels. Mean Q_{net} values are positive from April to August with annual maximum heat gain between May and July, and negative from September to March with maximum heat loss

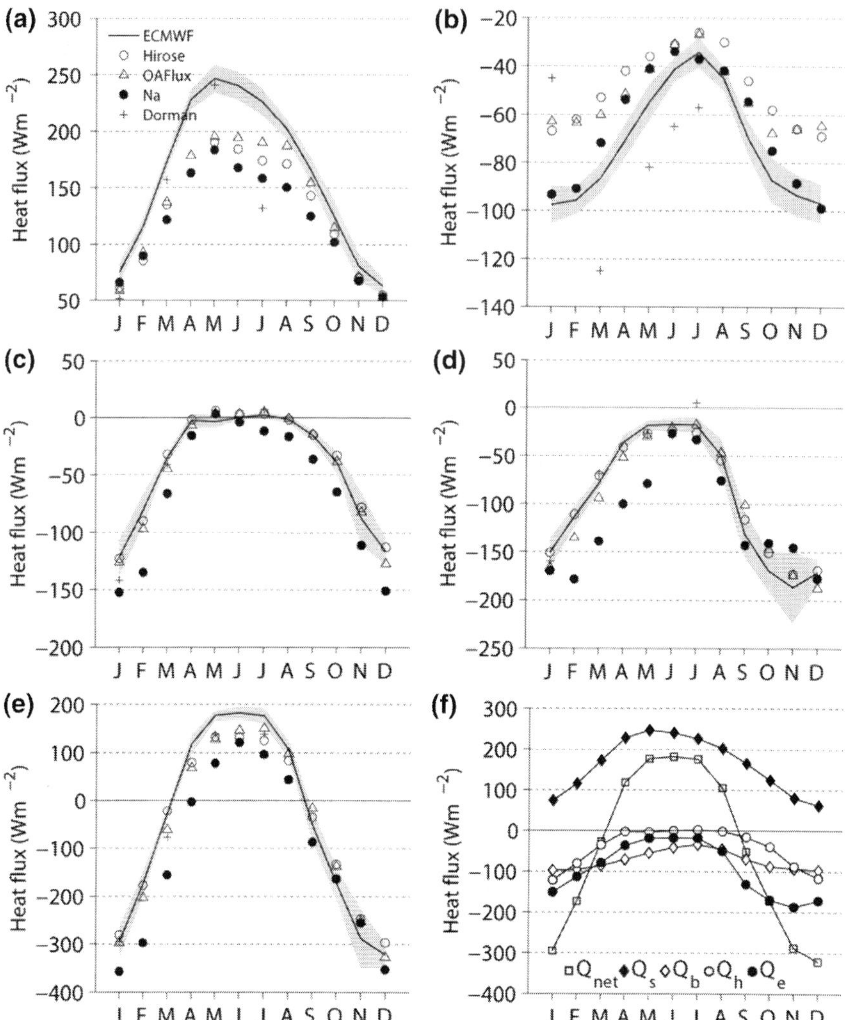

Fig. 2.5 Monthly-averaged heat flux components: **a** net shortwave flux (Q_s), **b** net longwave flux (Q_b), **c** sensible heat flux (Q_h), **d** latent heat flux (Q_e), **e** net heat flux (Q_{net}). Heat flux components based on ECMWF model results between 1991 and 2001 are shown as *solid lines, with shading* to indicate standard deviations. Both the monthly mean values and standard deviations for ECMWF model results are taken from Dorman et al. (2005). Heat flux components estimated by Hirose et al. (1996), Na et al. (1999), Dorman et al. (2004, 2005), and from OAFlux products are indicated by *open circles, closed circles, crosses,* and *triangles,* respectively. **f** Monthly-averaged heat flux components from ECMWF models results

in December and January (Fig. 2.5). In summer, surface heat flux is dominated by shortwave (Q_s) and longwave (Q_b) radiations, while the heat loss components (Q_b, Q_h, Q_e) show their annual minima due to reduced air-sea temperature differences and weak winds (Dorman et al. 2005). Major heat loss occurs in winter, especially through the latent (Q_e) and sensible (Q_h) heat fluxes. Heat losses due to sensible and latent fluxes between January and May estimated by NSL99 are larger than those from other estimates, yielding their larger net negative heat fluxes in those periods (see also Table 2.1).

The radiation components from the ECMWF model show large discrepancies compared with those components from other estimates. The ECMWF model results are characterized by large Q_s in summer and large heat loss through Q_b in winter. The Q_s in May 1999 (May 19–June 2) from Dorman et al. (2005) well matches the climatological Q_s from the ECMWF model. On the other hand, the Q_s value in July 1999 (July 19–August 11) shows a big discrepancy compared to the other climatological Q_s. The largest monthly Q_s values commonly occur in May due to the low cloud cover (HKY96). Compared to other estimates, the climatological monthly Q_b values from the ECMWF model and NSL99 have large-amplitude annual cycles due to large heat losses in winter.

Statistically significant negative Q_{net} trends occurred for 1984–2004 over the regions around the Kuroshio and the Kuroshio Extension, including the East Sea (Li et al. 2011). The negative trend over the East Sea was about -10 W m^{-2}/decade based on the OAFlux (Yu and Weller 2007). The negative Q_{net} trend was ascribed to the negative trend of Q_e due to an increase in the saturated specific humidity at the increasing sea surface temperature (Li et al. 2011). It should be noted that the trends based on a different dataset (National Oceanography Centre Southampton Flux Dataset v2.0, Berry and Kent 2009) show a negative trend in the southeastern East Sea, similar to the results from the OAFlux estimates, while a positive trend occurs in the northwestern East Sea.

The annual climatological Q_{net} over the East Sea is negative, the same as in western boundary current regions including the Kuroshio and the Kuroshio Extension (Han and Kang 2003; Li et al. 2011), meaning that, on average, the East Sea loses heat to the atmosphere. Further it implies that, under steady-state conditions, the East Sea receives heat from the surrounding seas by means of heat transported by the TWC through the Korea Strait (Isoda 1999; Han and Kang 2003). The estimated mean volume transport through the Korea Strait based on the surface heat flux is 2.20 Sv (HKY96), which is consistent with other independent estimates mentioned in Sect. 2.4.

2.3.3 Spatial Distribution

The spatial distribution of heat flux components shows significant spatial inhomogeneity especially in winter (HKY96). The long-term mean Q_{net} is negative over the East Sea except for a small region near the Russian coast, and its

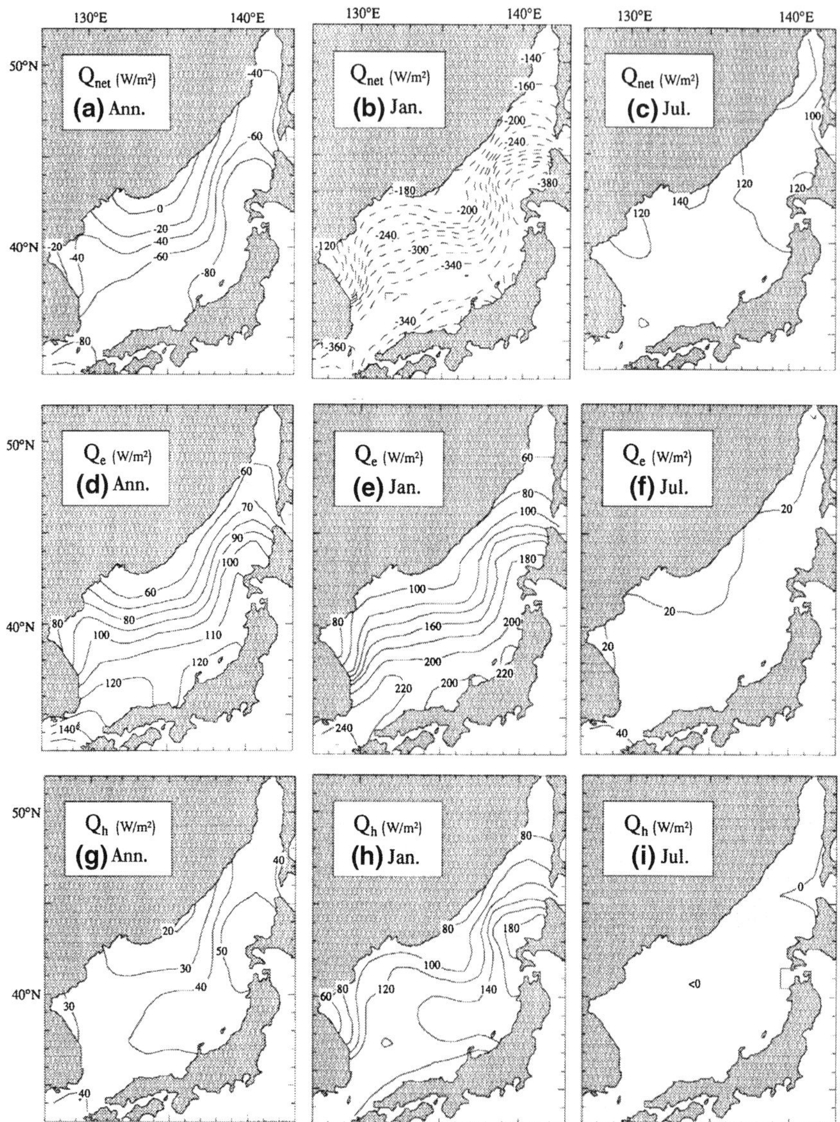

Fig. 2.6 Long-term annual mean heat flux components (*left panel*), and monthly mean heat flux components for January (*middle panel*) and for July (*right panel*). Q_{net}, Q_e, Q_h denote the sensible, latent, and net heat fluxes, respectively (Redrawn from Hirose et al. 1996)

spatial pattern is characterized by an increase in the heat loss toward the southeast (Fig. 2.6a). The large meridional difference in the Q_{net} field in winter (Fig. 2.6b) mainly determines the spatial pattern of the annual mean Q_{net}. And the wintertime heat losses due to the Q_e and Q_h (Fig. 2.6e, h) mainly account for the spatial variation of the Q_{net} field. The spatial distribution of long-term mean Q_b (not shown) is relatively homogeneous. The spatial variation of Q_s is mainly determined by latitude and cloudiness, and the long-term mean Q_s value is large (small) in the southwestern (northeastern) part of the East Sea (not shown).

2.4 Boundary Flux

2.4.1 Korea Strait

The Korea Strait is about 180 km wide and 330 km long with a sill depth of 140 m. It is divided by Tsushima Island into a narrower (about 40 km) but deeper (about 220 m maximum) western channel and a wider (about 140 km) but shallower (about 110 m maximum) eastern channel (Fig. 2.7). A number of continuous observations since the late 1990s (Table 2.2) have revealed mean and temporal variations of the currents and transports in the Korea Strait (Chang et al. 2004 and references therein).

The TWC flows northeastward through the Korea Strait with a maximum speed near the center of the strait in its upstream region (to the south of Tsushima Island), while it splits into two branches in the downstream region (to the north of Tsushima Island, Fig. 2.8; see also Sect. 4.2). The core in the western channel is generally stronger than that in the eastern channel. A southwestward countercurrent is observed in the lee of Tsushima Island and is explained as part of a current-induced island wake (Takikawa et al. 2005; Teague et al. 2005;

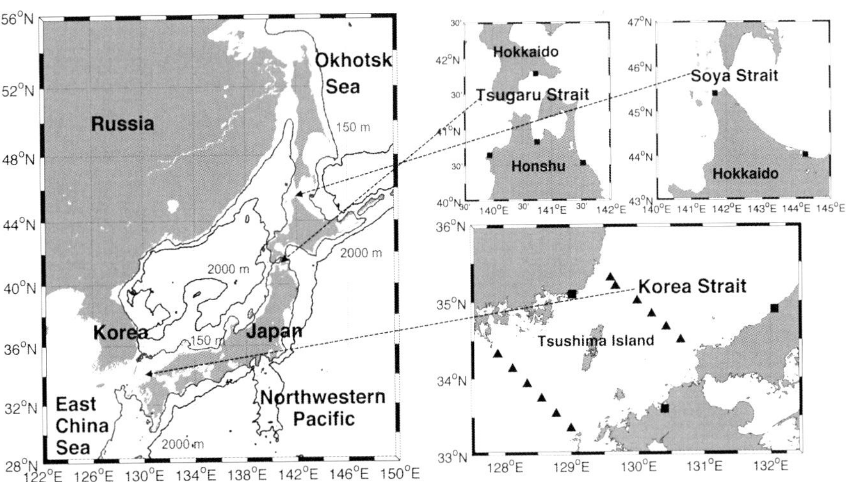

Fig. 2.7 Bottom topography (2000 and 150 m contours) and straits in the East Sea. *Squares* and *triangles* denote major tide gauge stations and bottom-mounted ADCP sites, respectively (from Teague et al. 2006 and references therein)

Table 2.2 Continuous observations in the straits

Straits	Periods	Methods	References
Korea	March 1998–present	Cable voltage measurement across the Strait	Kim et al. (2004)
	February 1997–present	Vessel-mounted ADCP across the Strait (Camellia)	Takikawa et al. (2005)
	May 1999–March 2000	Bottom-mounted ADCP across the Strait (LINKS program)	Teague et al. (2006) and references therein
	February 2002–present	High frequency radar around Tsushima Island	Yoshikawa et al. (2006)
Tsugaru	November 1999–March 2000	Vessel-mounted ADCP across the Strait	Ito et al. (2003)
	April 2000–December 2007	Vessel-mounted ADCP across the Strait	Kuroda et al. (2004)
Soya	August 2003–present	High frequency radar along the coast of Hokkaido	Ebuchi et al. (2006)

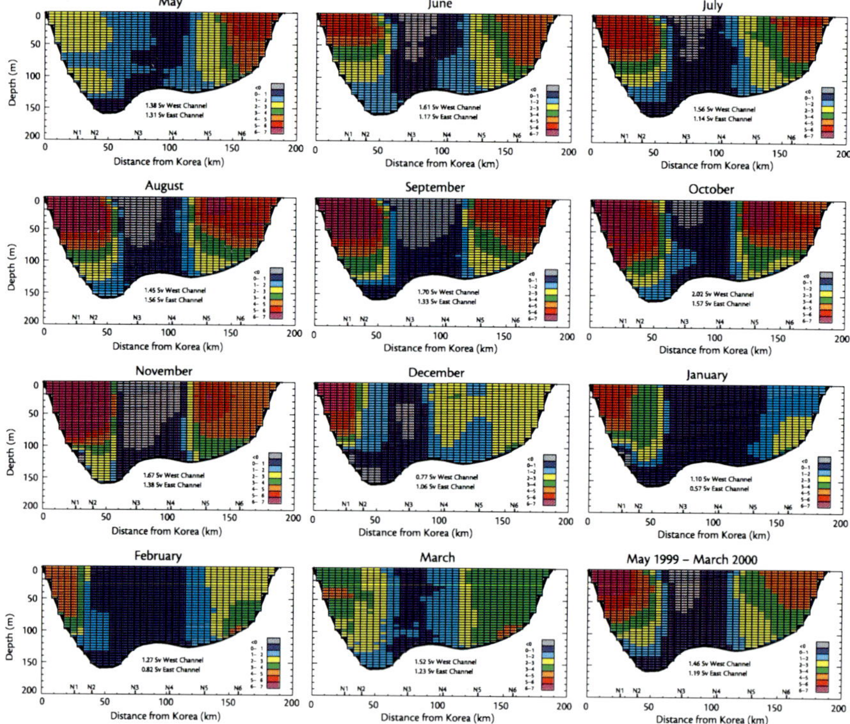

Fig. 2.8 Monthly averaged transports from May 1999 to March 2000, and averaged transport for the same period in the downstream region of the Korea Strait (to the north of Tsushima Island, Fig. 2.7) (from Teague et al. 2006). Units are 1 Sv $= 10^6$ m^3 s^{-1}. Each *grid square* is approximately 5-km across strait by 4 m in depth, except near the bottom where the depth dimension may be less

Yoshikawa et al. 2010). Tidal currents in the Korea Strait, which are stronger than those inside the East Sea, have maximum speed about 50–70 cm s^{-1} (Isobe et al. 1994; Teague et al. 2002; Takikawa et al. 2003). The maximum subtidal current is about 100 cm s^{-1} (Teague et al. 2002; Takikawa et al. 2005).

Mean volume transport through the Korea Strait is estimated to be about 2.6 Sv from various observational data (Isobe 2008 and references therein); this is roughly one-tenth of the mean positive (northeastward) transport of the Kuroshio in the East China Sea (24.0 ± 0.9 Sv; Andres et al. 2008). Subinertial (2–10 days) variability is in the range of 2–3 Sv and shows amplification at the period of Helmholtz resonance (3–5 days) between the East Sea and the Pacific Ocean (Lyu et al. 2002). It is forced by atmospheric pressure disturbances because this subinertial period is too short for sea level inside the East Sea to respond isostatically, due to flow restrictions at the narrow and shallow straits (Nam et al. 2004; Lyu and Kim 2005; Park and Watts 2005; Kim and Fukumori 2008; see also Sect. 5.2). Seasonally, the volume transport is maximum from summer to autumn and minimum from winter to spring, with monthly standard deviation of about 1 Sv (Fig. 2.8; Teague et al. 2002; Lyu and Kim 2003; Takikawa et al. 2005). Interannual variability was reported to be as strong as seasonal variability (Kim et al. 2004).

Large-scale wind forcing over the North Pacific is known to be mainly responsible for the mean (Tsujino et al. 2008) and seasonal to interannual variability (Lyu and Kim 2005; Tsujino et al. 2008; Ma et al. 2012; see also Sect. 4.5). Local monsoonal wind forcing plays an additional role to modify the seasonal cycle of the TWC transport (Moon et al. 2009; Ma et al. 2012; Cho et al. 2013). It is notable that wind stress to the east of Soya Strait (over the Okhotsk Sea) was shown to be also important from the numerical modeling studies (Kim and Fukumori 2008; Tsujino et al. 2008).

The origin of the TWC is known to be the Kuroshio in the East China Sea (Nitani 1972; Lie et al. 1998) or the Taiwan Warm Current from the Taiwan Strait (Beardsley et al. 1985; Fang et al. 1991). Recent studies appear to agree that the contribution of the Kuroshio is larger in winter and that of the Taiwan Warm Current is larger in summer (Isobe 2008 and references therein). Cho et al. (2009) quantitatively showed that about 83 % of the volume transport through the Korea Strait is from the Kuroshio in winter and about 66 % comes from the Taiwan Warm Current in summer. Park et al. (2013) argued, however, that the branch current west of Kyushu could be overestimated in numerical modeling studies due to complex flow patterns, and consequently the Taiwan Warm Current is responsible for more than half of the TWC transport throughout the year.

In delivering warm water from the south, the TWC transports heat through the Korea Strait into the East Sea. Isobe et al. (2002), using observational data, calculated annual mean heat transport through the Korea Strait to be 0.17 PW (1PW = 10^{15} W). A recent numerical modeling study showed that about 0.21 PW of heat should be transported through the Korea Strait, based on calculation of the oceanic heat budget over the East China Sea (Liu et al. 2010). These estimates are about one-tenth of the heat transport by the Kuroshio in the East China Sea (1.69 PW; Zhang et al. 2012).

Fresh water from the East China Sea is also delivered by the TWC to the East Sea through the Korea Strait. Isobe et al. (2002) estimated the freshwater transport as to be 3.3×10^4 m^3 s^{-1} and suggested that at least 70 % of total river discharge around the Yellow and East China Seas flows into the East Sea. The Changjiang discharge accounts for about 90 % of this total river discharge and exhibits large seasonal variations with maximum in July and minimum in January (Chen et al. 1994). Salinity in the Korea Strait (both western and eastern channels) is correlated with the Changjiang discharge at interannual time scales (Senjyu et al. 2006). Wind stress over the East China Sea also affects salinity variation in the Korea Strait, particularly in the western channel, by modifying currents in the Cheju Strait (Senjyu et al. 2009).

The TWC also transports nutrients and other materials that greatly affect biological productivity in the East Sea (Onitsuka et al. 2007; Kim et al. 2013). Direct estimates of material transports through the straits are sparse due to the lack of concentration and transport data. Repeated observations in the eastern channel showed high nutrient transports in summer and autumn (Morimoto et al. 2009) and large interannual variability (Morimoto et al. 2012). For a complete understanding of material transports in the Korea Strait, it is necessary to conduct hydrographic observations simultaneously in both the western and eastern channels. Observations at the western channel are particularly important as the current in the western channel is stronger than that in the eastern channel (Fig. 2.8) and water masses together with materials originating from the Yellow Sea and the East China Sea mainly pass through the western channel (Chung et al. 2000; Teague et al. 2002; Takikawa et al. 2005).

There is a water mass not delivered by the TWC: Korea Strait Bottom Cold Water (KSBCW), a cold ($<$10 °C) water mass found mainly in the bottom layer of the western channel originating in the northern region of the East Sea (see also Sect. 3.3.2). It is observed within 70 km of the Korean coast almost throughout the year (Johnson and Teague 2002; Kim et al. 2006). Using simultaneous temperature and current data, Kim et al. (2006) showed that the bottom temperature of KSBCW decreased when the southwestward bottom current strengthened (maximum 15 cm s^{-1} of subtidal velocity). The KSBCW also exhibits interannual variations (Min et al. 2006; Na et al. 2010) associated with changes of upper water temperature, particularly in the southwestern region of the East Sea (Yun et al. 2004; Na et al. 2010).

2.4.2 Tsugaru Strait and Soya Strait

The Tsugaru Strait is about 30 km wide and 110 km long (narrower and shorter than the Korea Strait) with a sill depth of 130 m (Fig. 2.7). The eastward outflow to the Northwestern Pacific through the Tsugaru Strait is a part of the TWC system in the East Sea and is called the Tsugaru Warm Current. Continuous observation of currents across the strait using a vessel-mounted Acoustic Doppler Current Profiler

(ADCP) (Table 2.2) has improved our understanding of the mean and temporal variations of the Tsugaru Strait throughflow. Tidal currents are well known to be strong with maximum speeds exceeding 100 cm s^{-1} (Onishi et al. 2004), comparable to or greater than those in the Korea Strait. Luu et al. (2011) pointed out that the tidal residual current is not negligible (up to about 30 cm s^{-1}). Maximum subtidal current is about 130 cm s^{-1} near the middle of the strait where bottom depth is deeper than 200 m (Ito et al. 2003). Mean volume transport is estimated to be about 1.5 Sv (Toba et al. 1982; Onishi and Ohtani 1997; Ito et al. 2003; Nishida et al. 2003).

The Soya Strait is about 40 km wide and less than 20 km long with a sill depth of 55 m (narrowest, shortest, and shallowest among the three major straits). The southeastward (along the coast of Hokkaido, Fig. 2.7) outflow to the Okhotsk Sea through the Soya Strait is a part of the TWC system and is called the Soya Warm Current. Tidal currents are not strong compared to those in the Korea and Tsugaru Straits, and diurnal tidal constituents are dominant (Aota and Matsuyama 1987). Maximum subtidal currents are comparable to those in the Korea and Tsugaru Straits, being about 100–120 cm s^{-1} in summer and autumn (Matsuyama et al. 2006; Ebuchi et al. 2006, 2009; Fukamachi et al. 2008, 2010). Mean transport is estimated to be about 1 Sv (Fukamachi et al. 2008) with large interannual variability (Onishi and Ohtani 1997; Fukamachi et al. 2010).

In terms of mean volume transports, outflows through the Tsugaru and Soya Straits are balanced by the inflow through the Korea Strait, thus about 60–70 % of the inflow volume flows out through the Tsugaru Strait and the rest flows out through the Soya Strait (Na et al. 2009 and references therein). However, the annual range of the transport through the Tsugaru Strait (about 0.4 Sv; Nishida et al. 2003) appears to be about half of that through the Soya Strait (about 1 Sv; Fukamachi et al. 2008). Cho et al. (2009) conducted high-resolution numerical modeling in the Northwestern Pacific region and showed that near-steady seasonal transport through the Tsugaru Strait contributes to a seasonally-varying ratio of Tsugaru Strait outflow transport to that of the Soya Strait that is high in winter and low in summer (Fig. 2.9). Seung et al. (2012) applied Godfrey's island rule (Godfrey 1989) with bottom friction and explained that the relatively small seasonal variability of Tsugaru Strait transport is because the latitude of zero-wind-stress curl over the North Pacific is located east of Hokkaido (between the Tsugaru and Soya Straits, see also Sect. 4.5).

The Tatarsky Strait is another strait, located in the northernmost region of the East Sea. Volume transport through this narrow (about 10 km) and shallow (about 10 m) strait is known to be negligible because of the cyclonic recirculation (about 2 Sv) associated with the Liman Current (Riser et al. 1999). The circulation near the Tatarsky Strait, however, could play a significant role in the freshwater budget in the East Sea, due to discharge from the Amur River (Yanagi 2002; Yoon and Kim 2009). Thus, a better knowledge of the circulation in the Tatarsky Strait may contribute to complete understanding of freshwater and salt budgets in the East Sea. Moreover, it could indirectly affect the inflow and outflow of the East Sea by altering surface circulation in the northern region (Park et al. 2006b; Yoon and Kim 2009).

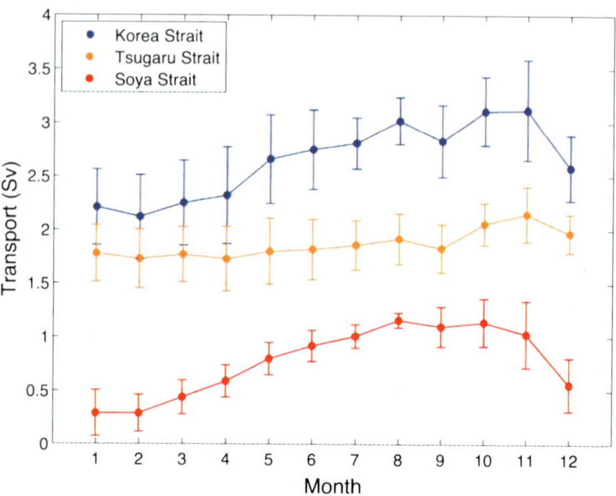

Fig. 2.9 Monthly mean transports and standard deviations in the Korea, Tsugaru, and Soya straits from 1994 to 2003 derived by a primitive-equation ocean circulation model simulated for 10 years (Redrawn from Cho et al. 2009). The *vertical lines* represent standard deviations in each month

2.4.3 Long-Term Variability

Interannual to decadal variability of the boundary transports appears to be as large as seasonal variability (about 1 Sv of volume transport; Kim et al. 2004; Lyu and Kim 2005; Morimoto et al. 2012), but, compared to seasonal variability, it is not well understood. Sea level differences (SLD) from tide gauge data (Fig. 2.7) could provide multi-decadal proxies for volume transport variations and allow simultaneous comparison of the transports at different straits. SLD across the Korea Strait showed a good linear relationship with the Korea Strait volume transport (Lyu and Kim 2003; Takikawa and Yoon 2005). In the case of the Tsugaru Strait, both SLD across the Tsugaru Strait (Nishida et al. 2003) and SLD between inside and outside the East Sea (Ito et al. 2003) agreed with the Tsugaru Strait transport variations. Along-strait SLD could also be used to estimate transport variations in the case of the Soya Strait (Fukamachi et al. 2008, 2010).

Long-term time series estimates can also be used to examine relationships between East Sea variability and larger-scale variability in the Northwestern Pacific (see also Sect. 3.4). The TWC may contribute as a link between upper-ocean heat content variability in the East Sea and that in the Northwestern Pacific on a decadal time scale (Na et al. 2012). A physical-biogeochemical modeling study (Liu and Chai 2009) indicated that nutrient transport in the East Sea is connected to large-scale changes in the Pacific Decadal Oscillation (PDO), which is thought to be related to TWC variability (Gordon and Giulivi 2004), and hence to Kuroshio variability in the East China Sea (Andres et al. 2009). During positive phases of the PDO, Kuroshio transport in the East China Sea tends to decrease and TWC transport to increase, which in turn delivers more buoyant subtropical water to the East Sea. The observed long-term increasing trend of inorganic nitrogen in the East Sea is also considered to have been mainly influenced by influx through the Korea Strait (Kim et al. 2013) rather than local nitrogen flux from atmosphere (Kim et al. 2011).

2.5 Summary and Discussion

We have given an overview of current knowledge of East Sea forcings such as wind field, surface heat flux, and boundary flux through the three straits connecting to the open ocean. One of the dominant features of wind speeds in the East Sea is occasional outbreaks of very cold Siberian air. Such cold-air outbreaks through the orographic gap near Vladivostok have been studied intensively to investigate their impacts on the circulation in the basin. They play an important role in generating cold bottom water formation south of the Peter the Great Bay (see Fig. 1.1) through brine rejection during sea ice formation (Talley et al. 2003). In addition, they have a significant effect on the circulation of the cyclonic gyre in the Japan Basin. Wind stress curl and divergence fields in the path of cold continental air outbreaks are modified by wind-sea surface temperature coupling across the subpolar front due to changes in MABL stability.

In contrast with the cold-air outbreak through Vladivostok, the role of strong winds over East Korea Bay during the outbreak period has not been thoroughly studied. The winds are expected to control relatively small-scale cyclonic and anticyclonic circulation and mesoscale eddies in the local region. In addition, the relationship between wind forcings and subpolar frontal dynamics should be further investigated (Yoshikawa et al. 2012). SST-wind coupling and its feedback mechanisms near the frontal region have not been vigorously studied and need further studies. Year-to-year variation and long-term variability of wind forcing and its relation to climate change should be studied continuously. Issues related to climate change and local changes in wind forcings should be also considered further in the future.

The East Sea gains heat from April to August and loses heat in the other months. The annual mean net heat flux is negative with a large range of 25–108 W m^{-2} according to previous estimates. The East Sea's inflow–outflow system compensates for the net heat loss. The winter (December–February) net heat losses, which are important in deep water formation in the northern East Sea and frontal subduction along the subpolar front, range from 250 to 330 W m^{-2}. Latent and sensible heat fluxes are mainly responsible for net heat losses in winter due to the winter monsoon with intermittent strong cold-air outbreaks. The net loss becomes enhanced when cold and dry air sweeps over the warm sea surface south of the subpolar front; this results in spatial differences in net and latent heat fluxes in winter. In summer, the net heat flux is mainly determined by radiation components, with heat loss components, including long wave cooling, showing their annual minima. The summertime (June–August) net heat flux ranges from 90 to 155 W m^{-2}.

Heat flux estimates presented here share some common features, such as their seasonal variations, but there are also considerable quantitative differences in all heat flux components. Interannual variation and long-term trends of surface heat fluxes are poorly known in the East Sea, although many studies have documented long-term changes and variability of deep water masses that are mainly formed in the northern East Sea during winter. It needs to be determined whether or not long-term variabilities of water properties and deep circulation are associated with varying surface air-sea fluxes. Atmospheric reanalysis products have been widely used to understand spatiotemporal variation of air-sea fluxes; they will also be very useful in the East Sea, but must be used with caution. Intercomparison of different products, and quantitative evaluation of the products based on in situ measurements, are prerequisites in the use of any specific product.

There has been significant progress in understanding boundary flux in the East Sea since continuous monitoring started in the late 1990s or early 2000s, along with extensive studies of currents and circulation in the straits. Outflows through the Tsugaru and Soya Straits are closely linked with inflow through the Korea Strait (2.6 Sv) because of the semi-enclosed nature of the East Sea and negligible flow through the Tatarsky Strait. Seasonally, the inflow and outflows are generally large in summer to autumn and small in winter to spring with the range of about 1 Sv. Seasonal variability in the Korea Strait tends to be reflected more in the Soya Strait rather than in the Tsugaru Strait, although mean outflow through the Tsugaru Strait is larger than that through the Soya Strait (Fig. 2.9). The TWC system in the East Sea is controlled by large-scale and local wind forcing on time scales longer than a month. Interannual variability (comparable to seasonal variability) and decadal variability of the TWC system needs further study along with its relationship to Northwestern Pacific variability.

Material transports in the East Sea have been less studied than volume transports, because of limited observations. Although strongly related to volume transport through the straits, other factors, such as intrusion of different water masses, alter material transport significantly. Moreover, comparison of material transports at different straits has been very limited. Thus, simultaneous observations at the straits including hydrographic and biogeochemical sampling is most needed for future progress in understanding boundary transports in the East Sea.

Acknowledgments Authors would like to thank Eun-Young Lee (Seoul National University) for her editorial work. This study was supported partly by EAST-I project of the Ministry of Oceans and Fisheries of Korea.

References

Ahn JB, Ryu JH, Yoon YH (1997) Comparative analysis and estimates of heat fluxes over the ocean around Korean Peninsula. J Korean Meteorol Soc 33:725–736 (in Korean)

Andres M, Wimbush M, Park JH et al (2008) Observations of Kuroshio flow variations in the East China Sea. J Geophys Res 113:C05013. doi:10.1029/2007JC004200

Andres M, Park JH, Wimbush M et al (2009) Manifestation of the Pacific decadal oscillation in the Kuroshio. Geophys Res Lett 36:l16602. doi:10.1029/2009GL039216

Aota M, Matsuyama M (1987) Tidal current fluctuations in the Soya Current. J Oceanogr Soc Jpn 43:276–282

Beardsley RC, Limeburner R, Yu H et al (1985) Discharge of the Changjiang (Yangtze River) into the East China Sea. Cont Shelf Res 4:57–76

Berry DI, Kent EC (2009) A new air-sea interaction gridded dataset from ICOADS with uncertainty estimates. Bull Am Meteorol Soc 90:645–656

Cardone VJ (1969) Specification of the wind field distribution in the marine boundary layer for wave forecasting. New York University, New York

Chang KI, Teague WJ, Lyu SJ et al (2004) Circulation and currents in the southwestern East/Japan Sea: overview and review. Prog Oceanogr 61:105–156

Chelton DB, Schlax MG, Freilich MH et al (2004) Satellite measurements reveal persistent small-scale features in ocean winds. Science 303:978–983

Chen C, Beardsley RC, Limeburner R et al (1994) Comparison of winter and summer hydrographic observations in the yellow and East China Sea and adjacent Kuroshio during 1986. Cont Shelf Res 14:909–929

Chen SS, Zhao W, Tenerelli JE et al (2001) Impact of the AVHRR sea surface temperature on atmospheric forcing in the Japan/East Sea. Geophys Res Lett 28:4539–4542

Cho YK, Seo GH, Choi BJ et al (2009) Connectivity among straits of the northwest Pacific marginal seas. J Geophys Res 114:C06018. doi:10.1029/2008JC005218

Cho YK, Seo GH, Kim CS et al (2013) Role of wind stress in causing maximum transport through the Korea Strait in autumn. J Mar Syst 115–116:33–39

Chu PC, Lan J, Fan C (2001) Japan Sea thermohaline structure and circulation part I: climatology. J Phys Oceanogr 31:244–271

Chung CS, Hong GH, Kim SH et al (2000) Biogeochemical fluxes through the Cheju Strait. J Korean Soc Oceanogr 5:208–215

Dorman CE, Beardsley RC, Dashko NA et al (2004) Winter marine atmospheric conditions over the Japan Sea. J Geophys Res 109:C12011. doi:10.1029/2001JC001197

Dorman CE, Beardsley RC, Limeburner R et al (2005) Summer atmospheric conditions over the Japan/East Sea. Deep-Sea Res Part II 52:1393–1420

Ebuchi N (1997) Statistical distribution of wind speeds and directions contained in the preliminary NSCAT science data products. J Adv Mar Sci Tech Soc 3(2):141–156

Ebuchi N, Fukamachi Y, Ohshima KI et al (2006) Observation of the Soya warm current using HF ocean radar. J Oceanogr 62:47–61

Ebuchi N, Fukamachi Y, Ohshima KI et al (2009) Subinertial and seasonal variations in the Soya Warm Current revealed by HF ocean radars, coastal tide gauges, and bottom-mounted ADCP. J Oceanogr 65:31–43

Fang G, Zhao B, Zhu Y (1991) Water volume transport through the Taiwan Strait and the continental shelf of the East China Sea measured with current meters. In: Takano K (ed) Oceanography: Asian marginal seas. Elsevier, New York, pp 345–358

Freilich MH (1997) Validation of vector magnitude datasets: effects of random component errors. J Atmos Ocean Technol 14:695–703

Fukamachi Y, Tanaka I, Ohshima KI et al (2008) Volume transport of the Soya Warm Current revealed by bottom-mounted ADCP and ocean-radar measurement. J Oceanogr 64:385–392

Fukamachi Y, Ohshima KI, Ebuchi N et al (2010) Volume transport in the Soya Strait during 2006–2008. J Oceanogr 66:685–696

Godfrey JS (1989) A Sverdrup model of the depth-integrated flow from the world ocean allowing for island circulations. Geophys Astrophys Fluid Dyn 45:89–112

Gordon AL, Giulivi CF (2004) Pacific decadal oscillation and sea level in the Japan/East Sea. Deep-Sea Res Part I 51:653–663

Han IS, Kang YQ (2003) Supply of heat by Tsushima Warm Current in the East Sea (Japan Sea). J Oceanogr 59:317–323

Hirose N, Kim CH, Yoon JH (1996) Heat budget in the Japan Sea. J Oceanogr 52:553–574

Hogan PJ, Hulbert HE (2000) Impact of upper ocean-topographical coupling and isopycnal outcropping in Japan/East Sea models with 1/8° to 1/64° resolution. J Phys Oceanogr 30:2535–2561

Hong GM, Kwon BH, Kim YS (2005) Heat fluxes in the marine atmospheric surface layer around the Korean Peninsula based on satellite data. J Fish Mar Sci Educ 29:210–217 (in Korean)

Isobe A (2008) Recent advances in ocean-circulation research on the Yellow Sea and East China Sea shelves. J Oceanogr 64:569–584

Isobe A, Tawara S, Kaneko A et al (1994) Seasonal variability in the Tsushima Warm Current, Tsushima-Korea Strait. Cont Shelf Res 14:23–35

Isobe A, Ando M, Watanabe T et al (2002) Freshwater and temperature transports through the Tsushima-Korea Straits. J Geophys Res 107:C73065. doi:10.1029/2000JC000702

Isoda Y (1999) Cooling-induced current in the upper ocean of the Japan Sea. J Oceanogr 55:585–596

Isoguchi O, Kawamura H (2007) Coastal wind jets flowing into the Tsuhima Strait and their effect on wind-wave development. J Atmos Sci 64:564–578

Ito T, Togawa O, Ohnishi M et al (2003) Variation of velocity and volume transport of the Tsugaru Warm Current in the winter of 1999–2000. Geophys Res Lett 30(13):1678. doi:10.1029/2003GL017522

Jet Propulsion Laboratory (1998) Science data product (NSCAT-2) user's manual ver 1.2. In: Overview and geophysical data products. Jet Propulsion Laboratory, Pasadena

Johnson DR, Teague WJ (2002) Observations of the Korea Strait bottom cold water. Cont Shelf Res 22:821–831

Kato K, Asai T (1983) Seasonal variations of heat budgets in both the atmosphere and the sea in the Japan Sea area. J Meteorol Soc Japan 61:222–237

Kawamura H, Wu P (1998) Formation mechanism of Japan Sea proper water in the flux center off Vladivostok. J Geophys Res 103(C10):21611–21622

Kim SB, Fukumori I (2008) A near uniform basin-wide sea level fluctuation over the Japan/East Sea: a semienclosed sea with multiple Straits. J Geophys Res 113:C06031. doi:10.1029/2007JC004409

Kim DJ, Moon WM (2002) Estimation of sea surface wind vector using RADARSAT data. Remote Sens Environ 80(1):55–64

Kim TS, Park KA (2011) Estimation of polarization ratio for sea surface wind retrieval from SIR-C SAR data. Korean J Remote Sens 27(6):729–741

Kim CH, Yoon JH (1996) Modeling of the wind-driven circulation in the Japan Sea using a reduced gravity model. J Oceanogr 52(3):359–373

Kim KR, Kim G, Kim K et al (2002) A sudden bottom water formation during the severe winter 2000–2001: the case of the East/Japan Sea. Geophys Res Lett 29(8):1234. doi:10.1029/2001GL014498

Kim K, Lyu SJ, Kim YG et al (2004) Monitoring volume transport through measurement of cable voltage across the Korea Strait. J Atmos Ocean Technol 21:671–682

Kim K, Kim YB, Park JJ et al (2005) Long-term and real-time monitoring system of the East/Japan Sea. Ocean Sci J 40(1):25–44

Kim YH, Kim YB, Kim K et al (2006) Seasonal variation of the Korea Strait bottom cold water and its relation to the bottom current. Geophys Res Lett 33:L24604. doi:10.1029/2006GL027625

Kim TS, Park KA, Moon WI (2010) Wind vector retrieval from SIR-C SAR data off the east coast of Korea. J Korean Earth Sci Soc 31(5):475–487

Kim TW, Lee K, Najjar RG et al (2011) Increasing N abundance in the Northwestern Pacific Ocean due to atmospheric nitrogen deposition. Science 334:504–509

Kim TS, Park KA, Choi WM et al (2012) L-band SAR-derived sea surface wind retrieval off the east coast of Korea and error characteristics. Korean J Remote Sens 28(5):477–487

Kim SK, Chang KI, Kim B et al (2013) Contribution of ocean current to the increase in N abundance in the northwestern Pacific marginal seas. Geophys Res Lett 40:143–148. doi:10.1029/2012GL054545

Kondo T, Ostrovskii A, Umatani S (1994) Climatologies of the heat fluxes over the Japan Sea. In: Proceedings of CREAMS'94 international symposium

Kuroda H, Isoda Y, Ohnishi M et al (2004) Examination of harmonic analysis methods using semi-regular sampling data from an ADCP installed on a regular ferry: evaluation of tidal and residual currents in the eastern mouth of the Tsugaru Strait. Umi no Kenkyu 13:553–564 (in Japanese)

Lee DK (1998) Ocean surface winds over the seas around Korea measured by the NSCAT (NASA Scatterometer). J Korean Soc Remote Sens 14(1):37–52 (in Korean)

Li G, Ren B, Zheng J et al (2011) Net air-sea surface heat flux during 1984–2004 over the North Pacific and North Atlantic oceans (10°N–50°N): annual mean climatology and trend. Theor Appl Climatol 104:387–401

Lie HJ, Cho CH, Lee JH et al (1998) Separation of the Kuroshio water and its penetration onto the continental shelf west of Kyushu. J Geophys Res 103:2963–2976

Liu G, Chai F (2009) Seasonal and interannual variation of physical and biological processes during 1994–2001 in the Sea of Japan/East Sea: a three-dimensional physical-biogeochemical modeling study. J Mar Syst 78:265–277

Liu WT, Xie X (2006) Measuring ocean surface wind from space. In: Gower J (ed) Remote sensing of the marine environment, manual of remote sensing, vol 6, 3rd edn., American Society for photogrammetry and remote sensingBethesda, USA, pp 149–178

Liu N, Eden C, Dietze H et al (2010) Model-based estimate of the heat budget in the East China Sea. J Geophys Res 115:C08026. doi:10.1029/2009JC005869

Luu QH, Ito K, Ishikawa Y et al (2011) Tidal transport through the Tsugaru Strait—part 1: characteristics of the major tidal flow and its residual current. Ocean Sci J 46:273–288

Lyu SJ, Kim K (2003) Absolute transport from sea level difference across the Korea Strait. Geophys Res Lett 30(6):1285. doi:10.1029/2002GL016233

Lyu SJ, Kim K (2005) Subinertial to interannual transport variations in the Korea Strait and their possible mechanisms. J Geophys Res 110:C12016. doi:10.1029/2004JC002651

Lyu SJ, Kim K, Perkins HT (2002) Atmospheric pressure-forced subinertial variations in the transport through the Korea Strait. Geophys Res Lett 29(9):1294. doi:10.1029/2001GL014366

Ma C, Wu D, Lin X et al (2012) On the mechanism of seasonal variation of the Tsushima Warm Current. Cont Shelf Res 48:1–7

Matsuyama M, Wadaka M, Abe T et al (2006) Current structure and volume transport of the Soya Warm Current in summer. J Oceanogr 62:197–205

Min HS, Kim YH et al (2006) Year-to-year variation of cold waters around the Korea Strait. Ocean Sci J 41:227–234

Moon JH, Hirose N, Yoon J-H et al (2009) Effect of the along-strait wind on the volume transport through the Tsushima/Korea Strait in September. J Oceanogr 65:17–29

Morimoto A, Takikawa T, Onitsuka G et al (2009) Seasonal variation of horizontal material transport through the eastern channel of the Tsushima Straits. J Oceanogr 65:61–71

Morimoto A, Watanabe A, Onitsuka G et al (2012) Interannual variations in material transport through the eastern channel of the Tsushima/Korea Straits. Prog Oceanogr 105:38–46

Na JY, Han SK, Seo JW et al (1997) Empirical orthogonal function analysis of surface pressure, sea surface temperature and winds over the East Sea of Korea (Japan Sea). J Korean Fish Soc 30:188–202 (in Korean)

Na J, Seo J, Lie HJ (1999) Annual and seasonal variations of the sea surface heat fluxes in the East Asian marginal seas. J Oceanogr 55:257–270

Na H, Isoda Y, Kim K et al (2009) Recent observations in the straits of the East/Japan Sea: a review of hydrography, currents and volume transports. J Mar Syst 78(2):200–205

Na H, Kim KY, Chang KI et al (2010) Interannual variability of the Korea Strait bottom cold water and its relationship with the upper water temperatures and atmospheric forcing in the Sea of Japan (East Sea). J Geophys Res 115:C09031. doi:10.1029/2010JC006347

Na H, Kim KY, Chang KI et al (2012) Decadal variability of the upper-ocean heat content in the East/Japan Sea and its relationship to Northwestern Pacific variability. J Geophys Res 117:C02017. doi:10.1029/2011JC007369

Nam SH, Lyu SJ, Kim YH et al (2004) Correction on TOPEX/POSEIDON altimeter data for nonisostatic sea level response to atmospheric pressure in the Japan/East Sea. Geophys Res Lett 31:L02304. doi:10.1029/2003GL018487

Nam SH, Kim YH, Park KA et al (2005) Spatio-temporal variability in sea surface wind stress near and off the east coast of Korea. Acta Oceanol Sin 24(1):107–114

Nishida Y, Kanomata I, Tanaka I et al (2003) Seasonal and interannual variations of the volume transport through the Tsugaru Strait. Umi no Kenkyu 12:487–499 (in Japanese)

Nitani H (1972) Beginning of the Kuroshio. In: Yoshida K (ed) Stommel H. Its physical aspects. University of Tokyo press, Kuroshio, pp 129–163

Onishi M, Ohtani K (1997) Volume transport of the Tsushima Warm Current, west of Tsugaru Strait bifurcation area. J Oceanogr 53:27–34

Onishi M, Isoda Y, Kuroda H et al (2004) Winter transport and tidal current in the Tsugaru Strait. Bull Fish Sci Hokkaido Univ 55:105–119

Onitsuka G, Yanagi T, Yoon JH (2007) A numerical study on nutrient sources in the surface layer of the Japan Sea using a coupled physical-ecosystem model. J Geophys Res 112:C05042. doi:10.1029/2006JC003981

Park KA, Cornillon PC (2002) Stability-induced modification of sea surface wind over Gulf Stream rings. Geophys Res Lett 29(24):2211. doi:10.1029/2001GL014236

Park JH, Watts DR (2005) Response of the southwestern Japan/East Sea to atmospheric pressure. Deep-Sea Res Part II 52:1671–1683

Park W, Oh IS, Shim T (1995) Temporal and spatial distributions of heat fluxes in the East Sea (Sea of Japan). J Korean Soc Oceanogr 30:91–115 (in Korean)

Park KA, Kim KR, Chung JY et al (2003) Comparison of the wind speed from an atomospheric pressure map (Na wind) and satellite scatterometer-observed wind speed (NSCAT) over the East (Japan) Sea. J Korean Soc Oceanogr 38(4):173–184

Park KA, Cornillon PC, Codiga DL (2006a) Modification of surface winds near ocean fronts: effects of Gulf Stream rings on scatterometer (QuikSCAT, NSCAT) wind observations. J Geophys Res 111:C03021. doi:10.1029/2005JC003016

Park KA, Kim K, Cornillon PC et al (2006b) Relationship between satellite-observed cold water along the Primorye coast and sea ice in the East Sea (the Sea of Japan). Geophys Res Lett 33:L10602. doi:10.1029/2005GL025611

Park YG, Yeh SW, Hwang JH et al (2013) Origin of the Tsushima Warm Current in a high resolution ocean circulation model. J Coast Res 65:2041–2046

Riser SC, Warner MJ, Yurasov GI (1999) Circulation and mixing of water masses of Tatar Strait and the northwestern boundary region of the Japan Sea. J Oceanogr 55:133–156

Senjyu T, Enomoto H, Matsuno T et al (2006) Interannual salinity variations in the Tsushima Strait and its relation to the Changjiang discharge. J Oceanogr 62:681–692

Senjyu T, Han IS, Matsui S (2009) Connectivity between the interannual salinity variation in the western channel of the Tsushima Strait and hydrographic conditions in the Cheju Strait. J Oceanogr 65:511–524

Seung YH, Han SY, Lim EP (2012) Seasonal variation of volume transport through the straits of the East/Japan Sea viewed from the Island rule. Ocean Polar Res 34:403–411

Shimada T, Kawamura H (2006) Satellite observations of sea surface temperature and sea surface wind coupling in the Japan Sea. J Geophys Res 111:C08010. doi:10.1029/2005JC003345

Takikawa T, Yoon JH (2005) Volume transport through the Tsushima Straits estimated from sea level difference. J Oceanogr 61:699–708

Takikawa T, Yoon JH, Cho KD (2003) Tidal currents in the Tsushima Straits estimated from ADCP data by ferryboat. J Oceanogr 59:37–47

Takikawa T, Yoon JH, Cho KD (2005) The Tsushima Warm Current through Tsushima Straits estimated from ferryboat ADCP data. J Phys Oceanogr 35:1154–1168

Talley LD, Lobanov V, Ponomarev V et al (2003) Deep convection and brine rejection in the Japan Sea. Geophys Res Lett 30(4):1159. doi:10.1029/2002GL016451

Teague WJ, Jacobs GA, Perkins HT et al (2002) Low frequency current observations in the Korea/Tsushima Strait. J Phys Oceanogr 32:1621–1641

Teague WJ, Hwang PA, Jacobs GA et al (2005) Transport variability across the Korea/Tsushima Strait and the Tsushima Island wake. Deep-Sea Res Part II 52:1784–1801

Teague WJ, Ko DS, Jacobs JA et al (2006) Currents through the Korea/Tsushima Strait: a review of LINKS observations. Oceanography 19:50–63

Toba Y, Tomizawa K, Kurasawa Y et al (1982) Seasonal and year-to-year variability of the Tsushima-Tsugaru Warm Current System with its possible cause. La Mer 20:41–51

Trusenkova O, Nikitin A, Lobanov V (2009) Circulation features in the Japan/East Sea related to statistically obtained wind patterns. J Mar Syst 78:214–225

Tsujino H, Nakano H, Motoi T (2008) Mechanism of currents through the straits of the Japan Sea: mean state and seasonal variation. J Oceanogr 64:141–161

Yanagi T (2002) Water, salt, phosphorus and nitrogen budgets of the Japan Sea. J Oceanogr 58:797–804

Yoon JH, Kim YJ (2009) Review on the seasonal variation of the surface circulation in the Japan/East Sea. J Mar Syst 78:226–236

Yoon JH, Abe K, Ogata T et al (2005) The effects of wind-stress curl on the Japan/East Sea circulation. Deep-Sea Res Pt II 52:1827–1844

Yoshikawa Y, Masuda A, Marubayashi K et al (2006) On the accuracy of HF radar measurement in the Tsushima Strait. J Geophys Res 111:C04009. doi:10.1029/2005JC003232

Yoshikawa Y, Masuda A, Marubayashi K et al (2010) Seasonal variations of the surface currents in the Tsushima Strait. J Oceanogr 66:223–232

Yoshikawa Y, Lee CM, Thomas LN (2012) The subpolar front of the Japan/East Sea part III: competing roles of frontal dynamics and atmospheric forcing in driving ageostrophic vertical circulation and subduction. J Phys Oceangr 42:991–1011

Yu L, Weller RA (2007) Objectively analyzed air-sea fluxes for the global ice-free oceans (1981–2005). Bull Am Meteorol Soc 88:527–539

Yun JY, Magaard L, Kim K et al (2004) Spatial and temporal variability of the North Korean cold water leading to the near-bottom cold water intrusion in Korea Strait. Prog Oceanogr 60:99–131

Zhang Q, Hou Y, Yan T (2012) Inter-annual and inter-decadal variability of Kuroshio heat transport in the East China Sea. Int J Climatol 32:481–488

Chapter 3
Water Masses and Their Long-Term Variability

Jong Jin Park, Kyung-Ae Park, Young-Gyu Kim and Jae-Yul Yun

Abstract We here review previous studies of spatiotemporal variability of surface and subsurface water properties in the East Sea (Japan Sea), including (1) sea surface temperature (SST) and mixed layer depth (MLD), (2) water masses, and (3) their long-term variability. Intra-annual, annual, and interannual variability of SSTs and SST-front structures are introduced together with a description of regionally adapted SST calibration methods. The MLD has distinct spatial patterns in its seasonal variability, which is determined by atmospheric forcing as well as advection effects. Water masses in the East Sea are listed in a comprehensive manner, covering not only subsurface water masses (e.g. Intermediate Waters) formed in the northern part of the sea but also near-surface water masses (e.g. Tsushima Warm Water) mostly originating from outside the sea. Where previous studies are available, we attempt to describe the definition, formation, and fate of the various water masses. Also, in terms of climate change in the East Sea, long-term variability found in the water masses or water properties is presented and its possible relationship with interannual to decadal variability of atmospheric forcing and heat content in the North Pacific is also considered.

J.J. Park (✉)
School of Earth System Sciences, Kyungpook National University,
Daegu 702-701, Republic of Korea
e-mail: jjpark@knu.ac.kr

K.-A. Park
Department of Earth Science Education, Seoul National University,
Seoul 151-748, Republic of Korea
e-mail: kapark@snu.ac.kr

Y.-G. Kim
Naval Systems R&D Institute, Agency for Defense Development,
Jinhae 645-600, Republic of Korea
e-mail: youngkim@add.re.kr

J.-Y. Yun
Research Institute of Oceanography, Seoul National University,
Seoul 151-742, Republic of Korea
e-mail: jaeyyun@snu.ac.kr

© Springer International Publishing Switzerland 2016
K.-I. Chang et al. (eds.), *Oceanography of the East Sea (Japan Sea)*,
DOI 10.1007/978-3-319-22720-7_3

Keywords Sea surface temperature · Mixed layer depth · Front · Water masses · Sea level change · Temporal variability · East Sea (Japan Sea)

3.1 Introduction

Measurements of physical properties, such as temperature and salinity, are fundamental components of oceanography; they have a longer history and wider spatial coverage than any other kinds of measurements in the East Sea. These property measurements have been conducted using various instruments, such as shipboard instruments (bottle and Conductivity-Temperature-Depth profiler), moored instruments, satellite imagery, and autonomous floats. Such diverse instruments have enabled us to secure long-term datasets over most of the East Sea, helping to reveal detailed characteristics of water masses, spatial structure of water properties, and even their long-term variability in the East Sea. In this chapter, we introduce spatial, climatological, long-term variation of water properties near the sea surface as well as the underlying layers. Surface properties, including sea surface temperature (SST), frontal structure, and the surface mixed layer, are described in Sect. 3.2. Also, various kinds of water masses in the East Sea are introduced in Sect. 3.3 and climate-scale variability of surface and subsurface water properties is reviewed in Sect. 3.4. Finally summary and remaining questions are provided in Sect. 3.5.

3.2 Sea Surface Temperature and Mixed Layer Depth

SST patterns, like thermal fronts, are a response to atmospheric forcing as well as an important driver of atmospheric circulation. The ocean having a large heat capacity strongly influences climate variability. Hence surface-layer heat content (including the mixed layer), rather than just SST, becomes an important variable in terms of understanding air-sea interaction. The mixed layer is defined as the vertically well-mixed layer near the sea surface that can be formed both by atmospheric buoyancy or frictional forcing and by ocean mixing processes resulting from surface/internal wave breaking, Langmuir circulation, and ocean currents, etc. The mixed layer is one of the important components in understanding the oceanic role in the Earth's climate system, because it is an air–sea exchange conveyor of heat and momentum. Major oceanic biological and chemical processes that play crucial roles in Earth's climate also occur in the mixed layer (Falkowski et al. 1998). Mixed layer variability is fundamentally important information for understanding spatiotemporal variability of biogeochemical materials and for evaluating performance of numerical models (Lim et al. 2012).

3.2.1 Sea Surface Temperature

SST is one of the most important geophysical variables in monitoring, diagnosing, understanding, and predicting diverse oceanic phenomena. SST data have long been used to detect and investigate temporal and spatial variability of mesoscale eddies, thermal fronts, and a variety of features in the East Sea.

Satellite SST images have revealed remarkable spatial and temporal variability of eddies and currents in the regions of the East Korea Warm Current (EKWC) along the east coast of Korea and the Tsushima Warm Current (TWC) in the southern East Sea (see Chap. 4) (Kim and Legeckis 1986; Isoda and Saitoh 1993; Park and Chung 1999). The spatial scale of the mesoscale eddies tends to become smaller as one goes north due to changes in the vertical structure of water properties and the spatial variation of internal Rossby deformation radius (Park and Chung 1999; Park et al. 2012).

SSTs in the East Sea have large annual variations, the difference between summer and winter being approximately 20 °C in the northern region and 15 °C in the southern region (e.g. Nakayama 1951). The annual cycle accounts for 88–96 % of the total SST variance in the East Sea, which is a larger percentage than that of the North Pacific Ocean (Yashayaev and Zveryaev 2001; Park et al. 2005). There are large north-south SST differences of about 16 °C (−1 to 15 °C) in winter and 10 °C (16–26 °C) in summer (Fig. 3.1). After onset of the winter monsoon, the cold SST anomaly first occurs in the north to northwestern boundary from Tatarsky Strait to Peter the Great Bay (PGB) (see Fig. 1.1) in November with a minimum SST value of 6–8 °C (Chu et al. 2001). Large amplitudes of the annual cycle (>11 °C) are found in the PGB, East Korea Bay, and south of the Siberian seamount in the western Japan Basin; these are related to cooling centers and deep convection sites of bottom-water formation (Kawamura and Wu 1998; Kim et al. 2002; Talley et al. 2003; Park et al. 2005). Spatial disparity of SST cooling generates an asymmetry of seasonal SST variations, which creates a semiannual component of SST variability with a relatively large amplitude of about 2 °C, about 20 % of the annual cycle's amplitude, in the northern region of the East Sea (Park and Chung 1999; Park and Lee 2014).

3.2.2 Surface Front

Oceanic fronts are accompanied by intensified spatial distinctions of hydrographic properties such as temperature, salinity, and biogeochemical properties (Fedorov 1986). The East Sea is a miniature version of the global ocean, and there are well-developed surface fronts in the central region of the East Sea where cold northern and warm southern water form an SST thermal boundary called the subpolar front (SPF) (Moriyasu 1972; Kim and Legeckis 1986; Kubota 1990). The main portion

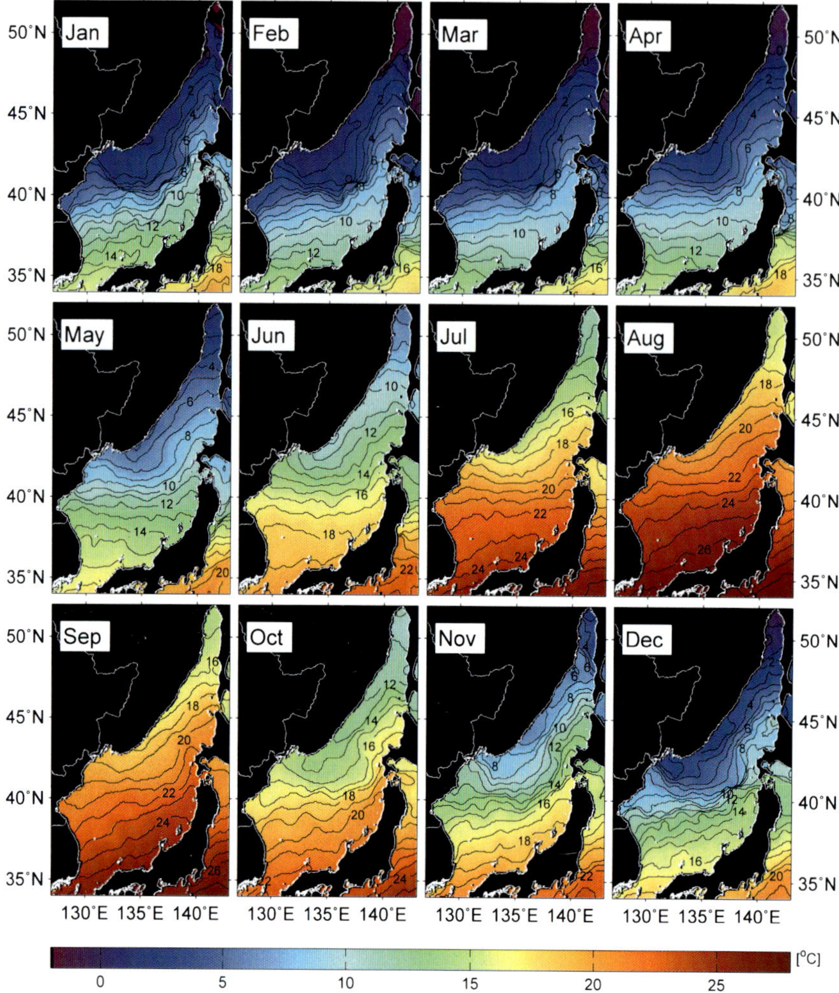

Fig. 3.1 Monthly distribution of SST in the East Sea extracted from SST climatology of the global ocean from NOAA/AVHRR data by NASA/JPL

of the SPF is mostly located approximately along 40°N between 132.3°E and 138°E, and bifurcates at both its eastern and western ends with strong seasonality (Fig. 3.2). Elevated variability of the SPF is attributed mainly to mesoscale eddies at the frontal region (Park et al. 2007). The spatial structure of the SPF shows strong seasonality associated with temporal variations of a warm eddy off the east coast of Korea, surface cooling due to Siberian cold air south of Vladivostok, and bifurcation of the TWC west of the Tsugaru Strait (Belkin and Cornillon 2003; Park et al. 2004, 2007). Spatial variability of wind forcing produces a significant meridional shift of the SPF by generating, in November, a northwestern SPF branch called the North Korean Front. The most conspicuous part of the SPF

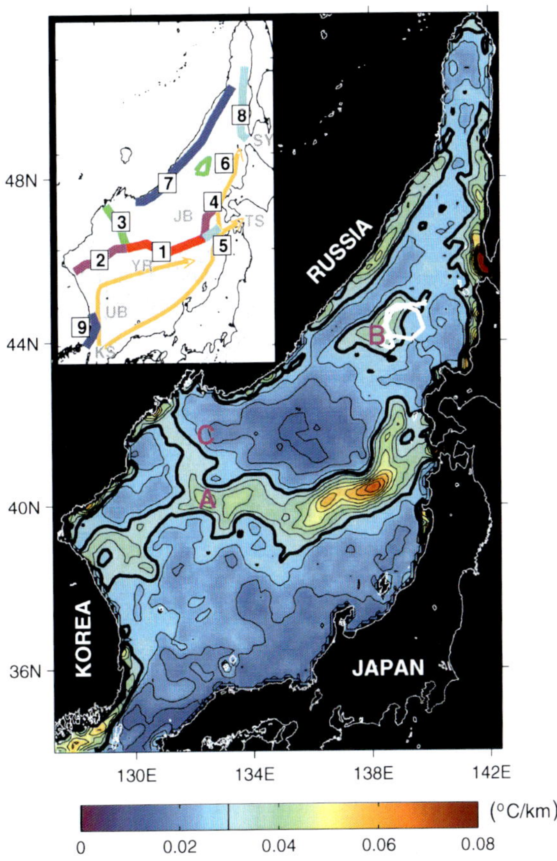

Fig. 3.2 Spatial distribution of magnitudes of SST gradients. The thick lines are contours at 0.03 °C km^{-1} (*inset*). Schematic diagram of the classified SST fronts in the East Sea. Representative fronts are numbered 1 to 9. The *grey lines* are 2000 m contours. Paths of the TWC are superimposed in *brown*. Abbreviations are *KS* Korea Strait; *JB* Japan Basin; *SY* Soya Strait; *TS* Tsugaru Strait; *UB* Ulleung Basin; and *YR* Yamato Rise (from Park et al. 2004)

is topographically trapped at the northern edge of the Yamato Rise, where steep bathymetry stabilizes the fronts (Kubota 1990; Park et al. 2007). Other fronts also develop along the Primorye coast (as a remarkably robust shelf-slope front), the western coast of Sakhalin Island, and the southeast coast of Korea between the EKWC and prevailing upwelling in summer (Kim and Kim 1983; Isoda and Saitoh 1993; Belkin and Cornillon 2003) (Fig. 3.2). Fronts induced by the TWC are clearly distinguishable most of the time, however they are sometimes fuzzy in the mean frontal map due to strong seasonal to interannual variability of the TWC in the southern part of the East Sea.

In addition to seasonality, the SPF in the central part of the East Sea shows significant year-to-year variations in strength. The magnitude of the SPF gradient, with a maximum of about 0.12 °C km^{-1} between 136°E and 138°E, has dominant year-to-year variability in the eastern portion because of a strongly varying seasonal cycle (Park et al. 2007). The SPF modifies the sea surface wind field across the interface between the regions south and north of the frontal zone (Shimada and Kawamura 2006). Year-to-year variations of the SPF in the East Sea may be associated with other indices such as El Niño-Southern Oscillation (ENSO), Pacific Decadal Oscillation (PDO), and other long-term oscillations (Gordon and Giulivi 2004). Decadal variability of SSTs shows large amplitudes trapped near the SPF (Minobe et al. 2004). Further descriptions of long-term SST variability will be provided in Sect. 3.4.

3.2.3 Mixed Layer Depth

Spatiotemporal variability of atmospheric forcing (Dorman et al. 2004, 2005) most likely controls the distribution and variability of the mixed layer depth (MLD) in the East Sea. Jang et al. (1995), on the other hand, found that temporal change of the mixed layer in the Korea Strait in autumn is mainly controlled by advection rather than local air–sea interactions such as wind stress or buoyancy flux. Kim and Isoda (1998) described the MLD focusing on interannual variability over the entire East Sea using the World Ocean Atlas 1994 (Monterey and Levitus 1997) and long-term repeated hydrography along the PM line made since 1972 by the Maizuru Marine Observatory. Moreover there have been some recent studies of the MLD in the East Sea in terms of local water mass formation in specific years (Talley et al. 2003; Kawamura et al. 2007).

Lim et al. (2012) conducted, for the first time, a comprehensive analysis regarding seasonality of basin-wide MLD patterns in the East Sea. MLD climatology obtained from historical hydrographic data revealed that the MLD undergoes strong seasonal variation over the entire East Sea, ranging from 10 to 20 m in summer to more than 400 m in winter (Fig. 3.3). Pronounced seasonality of wind stirring and surface cooling due to monsoons is primarily responsible for this strong seasonality. Spatial distribution of the MLD in winter, showing inhomogeneous structure in the basin, exhibited a shallower MLD near the SPF (~40°N) and deeper MLD in both the southern and northern East Sea as shown in Fig. 3.3. The homogeneity of the summer MLD is clearly due to uniform heating, mostly by large-scale atmospheric radiative heat flux. On the other hand, spatial inhomogeneity of the seasonal variability is likely controlled by the ocean itself, including major current systems such as the Ulleung Warm Eddy, the Nearshore Branch of the TWC, the EKWC, and the North Korea Cold Current (NKCC) (see Chap. 4).

Temperature-based MLDs, mostly less than 20 m, show no significant difference from density-based MLDs for all seasons except winter when the difference becomes significant, especially near the Japanese coast in the southern East Sea and in the deep convection region. Although the temperature-based MLD does not represent all East Sea conditions, it allows us to construct MLD climatology covering nearly the whole East Sea, which is not possible when we use density profiles, because of their poor spatial coverage.

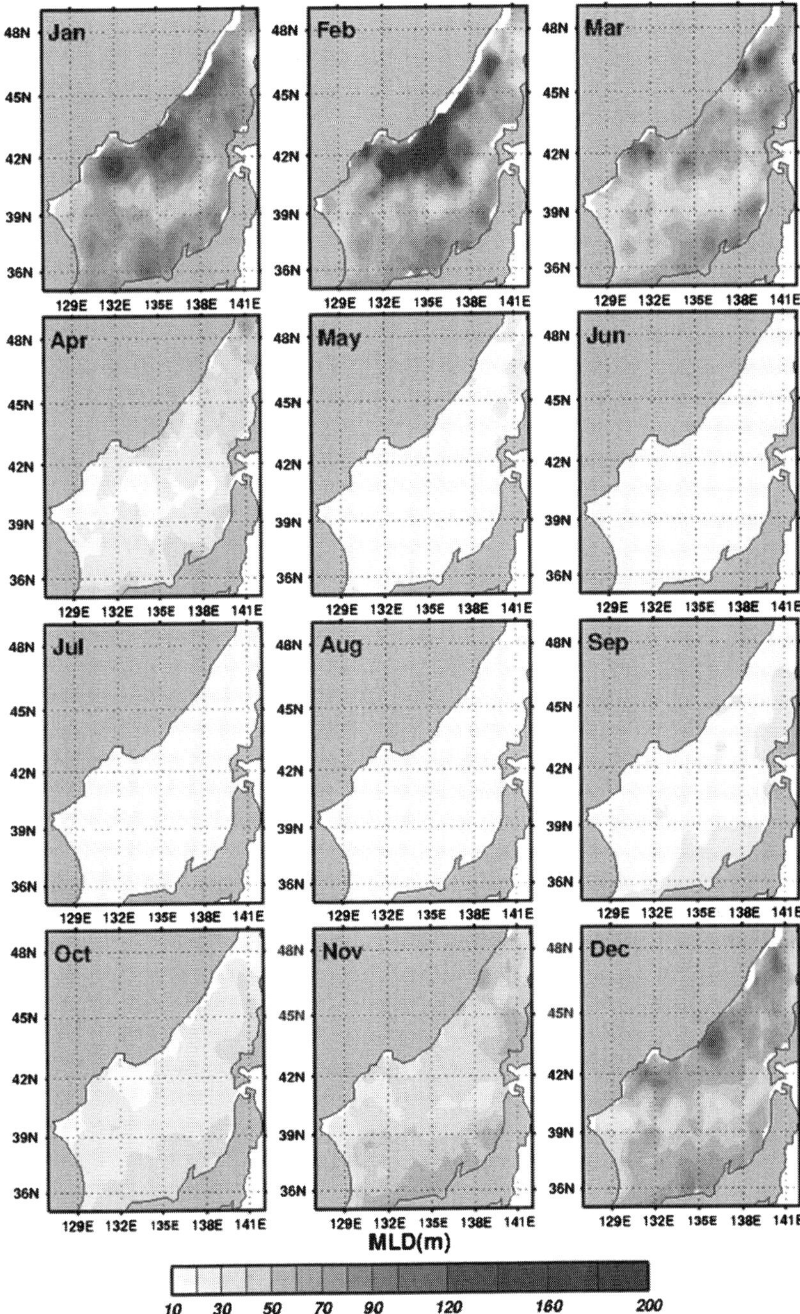

Fig. 3.3 MLD climatology estimated from individual profiles with a temperature difference criterion of $\Delta T = 0.2\ ^\circ C$ from the near-surface value (10 m) (from Lim et al. 2012)

3.3 Water Masses

Water masses have been introduced by identifying their water characteristics and specifying ocean processes responsible for creating large-scale water properties. In open oceans, most water properties are initially determined at the sea surface and subducted into the ocean interior through a process called ventilation. The ventilated water masses maintain their properties because the ocean interior is nearly adiabatic without internal sources of heat or freshwater capable of modifying the water properties. Therefore, water masses with specific properties are helpful for identifying flow paths from the sea surface to the interior as well as subsurface ocean circulation patterns.

More than 90 % of the East Sea is occupied by water colder than 5 °C (Yasui et al. 1967); this water must be formed in the northern part of the East Sea and then spread at depth into the whole East Sea area by a deep circulation which in turn affects the movement of the upper ocean. Water colder than 1 °C, which fills most of the subsurface East Sea, was called "Japan Sea Proper Water" (JSPW) by Uda (1934). JSPW is not a single water mass but consists of various water mass components (e.g. Nitani 1972; Senjyu and Sudo 1993; Kim et al. 2004). Based on CREAMS measurements (see Chap. 1.3), Kim et al. (2004) divided the JSPW into East Sea Central Water (ESCW), East Sea Deep Water (ESDW), and East Sea Bottom Water (ESBW). There are distinct intermediate waters with dissolved oxygen concentration higher than 250 μmol l^{-1} between JSPW and upper ocean water masses, such as East Sea Intermediate Water (ESIW) and High Salinity Intermediate Water (HSIW) (e.g. Kim and Chung 1984; Kim and Kim 1999). The upper ocean is mostly controlled by exterior water masses named Tsushima Warm Waters (TWWs) flowing from outside the East Sea through the Korea Strait. The interior water masses are known to be formed in wintertime, mostly in the northern part of the East Sea. The ventilation processes creating the water masses resemble those in the open oceans (Min and Warner 2005; Riser and Jacobs 2005). That is one of the reasons why the East Sea is called a "mini-ocean" (Ichiye 1984), "miniature ocean" (Kim et al. 2002; Talley et al. 2003) or "natural laboratory" (Kim et al. 2001b).

In the next two sections we will describe upper ocean water masses and their spatial patterns, and introduce various interior water masses in the East Sea such as those described in Table 3.1.

3.3.1 Upper Ocean Water Masses

3.3.1.1 Tsushima Warm Water

In marginal seas like the East Sea, it is also useful to name water masses originating outside the seas, especially when they have large-scale consistency. These

Table 3.1 Definition of water masses in the East/Japan Sea (from Kim et al. 2004)

Water mass	Potential temperature (°C)	Salinity (psu)	Dissolved oxygen (μmol/l)	Remarks
Tsushima Warm Water	>10	>34.3		Primarily south of 41°N
East Sea Intermediate Water	1–5	<34.06	>250	Western Japan Basin, Ulleung basin
High Salinity Intermediate Water	1–5	>34.07	>250	Eastern Japan basin
East Sea Central Water	0.12–0.6	>34.067		Deeper limit at 1500 dbar (DSM)
East Sea Deep Water	<0.12	34.067–34.070		Upper limit at 1500 dbar (DSM)
East Sea Bottom Water	<0.073	34.070		Homogeneous mixed layer

exterior water masses are often distinguished from resident water masses by their physical and chemical characteristics. There are two different kinds of exterior water masses in the East Sea: High Salinity Tsushima Warm Water (HSTWW) and Low Salinity Tsushima Warm Water (LSTWW), which together flow into the East Sea through the Korea Strait (e.g. Miyazaki 1953; Miyazaki and Abe 1960; Moriyasu 1972). These TWWs occupy the upper 100–200 m in the southern part of the East Sea with volume of approximately 8×10^4 km^2 (Kitani 1987), playing a crucial role, together with atmospheric buoyancy forcing, in determining SST spatial patterns.

HSTWW is persistent throughout the year, although its temperature varies seasonally. It usually has temperature higher than 10 °C and salinity higher than 34.3 psu. It is known that HSTWW derives from the Kuroshio in the East China Sea. On the other hand, LSTWW is only found in late summer and early autumn (August–October), occupying the upper 20–50 m above the HSTWW. Using chemical tracer analysis, Kim et al. (2001a) confirmed that LSTWW originates from Changjiang river discharge water. Indeed, the rainy season (June–August) in the East Sea does not match with the time period of LSTWW appearance. Moreover, the amount of local precipitation is not enough to produce the large volume of low salinity water found in the East Sea. Thus, most of the upper low salinity water in summer may come from LSTWW. By examining historical hydrographic data, Senjyu et al. (2006) showed that upper ocean salinity variation in the Korea Strait has a negative correlation with Changjiang river discharge, which tends to confirm the origin of the surface low salinity water in the East Sea.

By vertical convection in winter, TWW trapped in warm eddies (e.g. Ulleung Warm Eddy) of the southern East Sea often turns into a thick thermostad layer with a temperature of around 10 °C (7.8–11 °C) (Kim 1991). Such homogeneous layers are capped by TWW advected from the Korea Strait during the following spring and summer, and are commonly found in hydrographic measurements throughout the year (Fig. 3.4). Due to its resemblance with Mode Waters in the open oceans (e.g. North Pacific Subtropical Mode Water and Eighteen Degree Water in the North Atlantic), we name it Ten Degree Water (TDW) by personal communication with Dr. H-R Kim. However, variability and characteristics of this TDW have been largely unexplored.

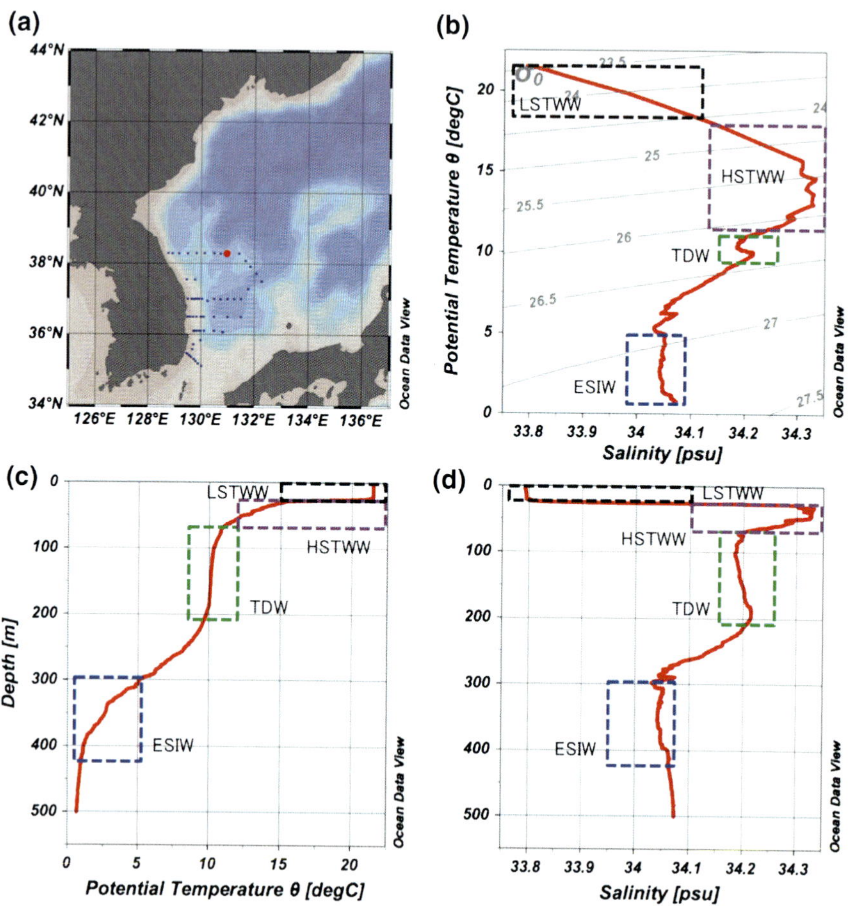

Fig. 3.4 An example of hydrographic data obtained in the southwestern part of the East Sea. **a** The station map of shipboard CTD observation in September 2002. The *red dot* marks the location of an example station, for which **b** T-S diagram, **c** potential temperature profile, and **d** salinity profile are shown. The *colored broken boxes* denote water masses, such as LSTWW (*black*), HSTWW (*purple*), TDW (*green*), and ESIW (*blue*)

3.3.2 Intermediate Waters

3.3.2.1 East Sea Intermediate Water

There are two intermediate waters formed in the East Sea, ESIW and HSIW. They share similar density ranges, but occupy different geographical regions (Fig. 3.5). ESIW is characterized by temperatures ranging from 1 to 5 °C, salinity less than 34.06 psu, and dissolved oxygen greater than 250 μmol l^{-1} as shown in Table 3.1. The subsurface salinity minimum layer was first noticed by Japanese scientists (Miyazaki 1952, 1953; Kajiura et al. 1958; Moriyasu 1972). However, Kim and Chung (1984) gave the name ESIW to this kind of low salinity water mass found in the Ulleung Basin underneath the TWW. They pointed out that the low salinity water with high oxygen concentration must be ventilated at the sea surface in the northern part of the East Sea. Kim and Kim (1999) showed the spatial distribution of ESIW properties (see Fig. 3.5) based on a quantitative definition of the water mass, using high resolution CTD and oxygen data. They revealed that the lowest salinity in the ESIW is located in the western Japan Basin, indicating a potential formation area of the water mass. The region west of the North Korean Front typically has a sea surface temperature of 2–3 °C in winter, and may be an outcropping area of ESIW with similar water properties in its core layer.

The formation process and origin of ESIW in the East Sea have been studied by many researchers (e.g. Senjyu and Sudo 1993; Seung 1997; Kim and Kim 1999; Yoshikawa et al. 1999; Yoon and Kawamura 2002; Kawamura et al. 2007), although shipboard observations restricted to the western Japan Basin were

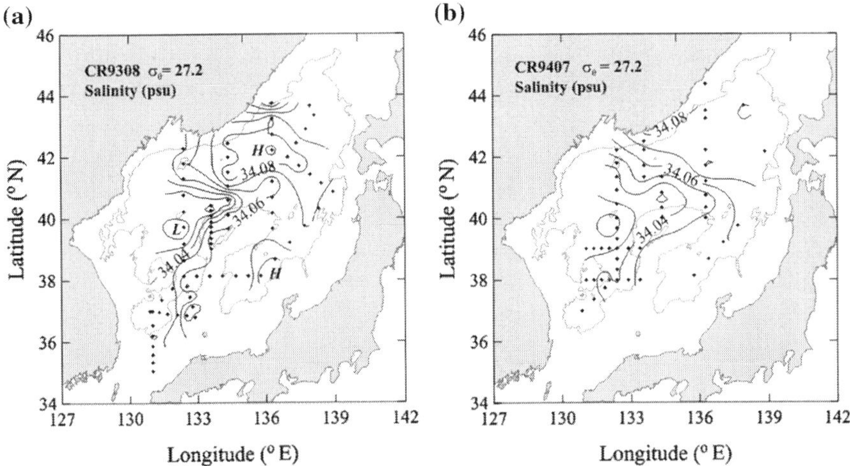

Fig. 3.5 Horizontal distribution of salinity on the surface of $\sigma_\theta = 27.2$ kg m^{-3} in **a** August, 1993 and **b** July, 1994 (Redrawn from Kim and Kim 1999)

insufficient for detailed examination of the ESIW formation area. Previous studies have consistently pointed out that ESIW is presumably formed off Vladivostok in the western Japan Basin. However, among models, ESIW core properties have significant differences, casting a doubt of the simulated formation process and area. Yoshikawa et al. (1999) reported strong subduction of ESIW occurred in their model near the PGB (41–42°N, 131–135°E; see Fig. 1.1), while Yoon and Kawamura (2002) presented a formation region west of 131°E. The most plausible formation process has been hypothesized by Yun et al. (2004), based on analysis of climatological hydrographic data. They noted that the ESIW subduction area in the northwestern part of the East Sea (similar to that of Yoon and Kawamura 2002), where a large extent of surface low salinity resides before convection occurs, is a region of strong negative wind stress curl from December to February due to the orographic effect (Yoon et al. 2005). Thus, traditional subduction controlled by surface cooling and downward Ekman pumping (Stommel 1979; Luyten et al. 1983) or slantwise convection by wind forcing over an oceanic front (Thomas and Lee 2005; Lee et al. 2006) could produce ESIW with core temperature 2–3 °C and salinity less than 34.1 psu during winter.

Furthermore, origin of the ESIW low salinity water is still controversial. One traditional view is that the cold, fresh water flowing southward along the east coast of Russia from the Tartarsky Strait contributes to ESIW formation (e.g., Yoon and Kawamura 2002; Postlethwaite et al. 2005). The other possible source of the low salinity is LSTWW flowing through the Korea Strait in summer and reaching into the western Japan Basin (Yun et al. 2004). Both these low salinity waters are found around the presumed formation area at the sea surface, but it will be challenging to determine which low salinity water contributes most to ESIW formation.

ESIW formed in winter expands from the western Japan Basin southward along the east coast of the Korean Peninsula as well as eastward along the SPF (Fig. 3.5). Such spatial distribution patterns (Cho and Kim 1994; Shin et al. 2013) are roughly consistent with the mid-depth circulations shown in Park and Kim (2013). Yoshikawa et al. (1999) attempted to simulate subduction and circulation of intermediate waters by using a nudging model, suggesting 6–7 years travel time of the simulated ESIW from the formation site to the Ulleung Basin, an order of magnitude larger than the 6 month travel time inferred from Argo float trajectory data (Park 2002; Yanagimoto and Taira 2003). Such a discrepancy may be attributed to unrealistic mid-depth circulations in the model results when compared to the actual ones (Park and Kim 2013).

One can speculate that most of the ESIW may be eventually mixed with the water mass above, such as TWW, potentially affecting the upper ocean heat content. However, the mechanism of its demise remains undetermined. There are three possible mechanisms controlling the ESIW demise: radical transformation of the water mass near the coast and in the Korea Strait (Kim et al. 2006), outflow through the Tsugaru Strait (Park et al. 2008), and gradual mixing with the TWW in the East Sea interior.

3.3.2.2 High Salinity Intermediate Water

Senjyu (1999) noticed the HSIW-like oxygen rich layer from historical hydrographic data taken in 1969, regarding it as a boundary (not a water mass) between overlying and underlying water masses. However, by analyzing high resolution hydrographic data obtained from CREAMS expeditions, Kim and Kim (1999) showed that the high salinity water resided in the eastern Japan Basin as an actual water mass, called HSIW (temperature from 0.6 to 5 °C, salinity higher than 34.075 psu, and dissolved oxygen greater than 250 μmol l^{-1}) and suggested that HSIW is locally formed in winter. The HSIW formation area in the eastern Japan Basin is under the influence of large-scale positive wind-stress curl in winter, where cyclonic circulation is persistent throughout the year (Kim and Kim 1999; Park and Kim 2013). Watanabe et al. (2001) confirmed the existence of HSIW and suggested that its high salinity origin might be the HSTWW presumably outcropped and cooled near the Russian coast after the surface LSTWW was removed. In contrast to the southward subduction of ESIW below the TWW, HSIW is not likely to penetrate south of the SPF, but instead stay within the eastern Japan Basin, north of the SPF. It seems that HSIW formation resembles Warren-type water masses that are locally formed by disappearance of the seasonal thermocline during winter (Warren 1972). Properties of HSIW are highly variable from one year to another (Kim and Kim 1999), yet keeping higher salinity than that of the adjacent ESIW. The large year-to-year variation is presumably related to changing atmospheric and oceanic conditions of water formation (Kim et al. 2004).

3.3.2.3 North Korea Cold Water

The evolution and fate of ESIW have not been fully explored, but kindred water masses, North Korea Cold Water (NKCW) and Korea Strait Bottom Cold Water (KSBCW), presumably advected toward the coastal area from the ESIW pool, have been studied for the past few decades, since both water masses are often found along the east coast of Korea. NKCW is defined as water with salinity less than 34.05 psu and density of 26.9–27.2 kg m^{-3} (Park 1972; Lim 1983; Kim and Kim 1983; Kim and Chung 1984; Min 1994; Cho and Kim 1998; Kim 1999). Cho and Kim (1994) pointed out that ESIW flows into the Ulleung Basin along two different paths: one is along the east coast of Korea and the other is along the isobaths west of the Korea Plateau. The water mass on the former path is called NKCW and is carried by the NKCC originally named by Uda (1934). NKCW is noticeable in April near the Sokcho coastal area (see Fig. 1.2), flowing southward. Such flow is clearly observed over a period of 6–7 months from April to October and is most pronounced in August (Yun et al. 2004). The spatial distribution of NKCW has a distinct pattern of repeated extension and contraction normal to the coastline, with a periodicity of one year. Since the NKCC as a strong deep western boundary current quickly transports the NKCW along the coast, its minimum salinity is usually lower than that of the offshore ESIW.

3.3.2.4 Korea Strait Bottom Cold Water

The resident cold water in the East Sea intrudes into the Korea Strait; it is then called KSBCW, colder than 10 °C, and found near the western trough of the strait. KSBCW was first reported by Nishida (1926) and its origin and variability have been studied for many years (Lim and Chang 1969; Cho and Kim 1998; Isobe 1994; Park et al. 1995; Kim et al. 2006; Min et al. 2006). KSBCW intruded from May to January with two minima in bottom temperature, corresponding to maximum southwestward bottom currents in August/September and December/January, respectively (Kim et al. 2006).

Kim and Kim (1983) and Lim (1983) showed that KSBCW is an extension of NKCW by tracing the salinity-minimum and oxygen maximum layer. KSBCW is usually found on the sloping bottom away from the deepest topography in the Korea Strait. A rotating hydraulic model successfully reproduced such behavior of the KSBCW, suggesting a mechanism of connectivity with NKCW (Park et al. 1995). In addition, it is also plausible that the interior ESIW could spill into the Korea Strait when it shoals sufficiently to extend over the sill (~150 m depth) at the mouth of the Korea Strait (Yun et al. 2004).

3.3.3 Central Water, Deep Water, and Bottom Water

For waters colder than 1 °C, the vertical profile of salinity looks almost homogeneous in traditional measurements (Uda 1934). However, more than 60 years after Uda's observations, CTD measurements with high precision obtained during CREAMS expeditions revealed that this cold water consists of several different water masses, as in the open ocean. The salinity of those water masses varies at the third decimal place in psu. Figure 3.6 shows vertical structure details of temperature, salinity, and oxygen concentration at depths below 200 m. Such expanded views of salinity profiles exhibit an interesting minimum at around 1500 m, called the Deep Salinity Minimum (DSM), which was revealed for the first time in the CREAMS observations. This DSM is found in the entire Japan Basin as well as the Ulleung and Yamato Basins (Kim 1996; Min 2002; Kim et al. 2004). The DSM is regarded not as a certain water mass, but as a boundary between water masses, similarly as in the open ocean. ESCW, which has relatively weak vertical gradients in salinity and oxygen concentration, lies between the intermediate waters and the DSM, and ESDW lies between the DSM and ESBW. ESBW, which was first noted by Japanese scientists (Suda 1932; Nitani 1972), has nearly uniform potential temperature in a 1000 m thick layer above the ocean bottom (below 2500 m in Fig. 3.6). It is arguable whether the uniform temperature is a distinct water mass or just a homogeneous layer produced by adiabatic mixing (Talley et al. 2006). However, after formation of the densest bottom water in the East Sea in 2001 (Kim et al. 2002), the homogeneous layer has been regarded as a distinct water mass.

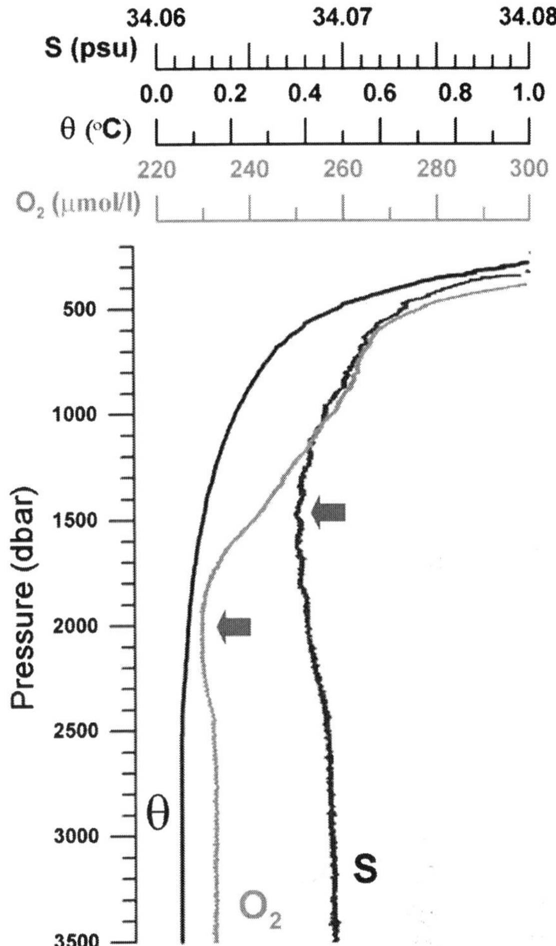

Fig. 3.6 Vertical profiles of potential temperature, salinity, and dissolved oxygen at 42°N, 136°E below 200 dbar taken in the summer of 1995. *Arrows* indicate depths of minima for dissolved oxygen and salinity (from Kim et al. 2004)

It is speculated that ESCW may be formed through open-ocean convection south of Vladivostok (Senjyu and Sudo 1993, 1994; Seung and Yoon 1995; Choi 1996), where cyclonic circulation and a strong heat flux center (Kawamura and Wu 1998) reside in winter. ESBW, on the other hand, is possibly convected at the continental shelf slope near the PGB through the slope convection process accompanied by brine rejection during sea ice formation; this may be a way to create dense water in the East Sea (Ponomarev et al. 1991; Vasiliev and Makashin 1992; Talley et al. 2003).

In Fig. 3.6, the oxygen profile exhibits a minimum at around 2000 m which is different from the DSM depth. Such an oxygen minimum layer was reported by Gamo et al. (1986), but at a shallower depth. A vertical advection-diffusion model with an oxygen consumption process reproduced the oxygen minimum when weakened formation of ESDW was prescribed (Kim and Kim 1995; Kim 1996).

This result supports the idea that deepening of the oxygen minimum results from a response of the East Sea to climate change (e.g. Kim et al. 1996, 2004; Gamo 1999). Details about long-term variation of water masses follow below.

3.4 Long-Term Variability of Water Properties

Understanding the long-term variability of water properties is an essential step for predicting future changes in the East Sea. Furthermore, the East Sea with its resemblances to open oceans has been regarded as a natural laboratory for studying ocean response to global climate change. Since the East Sea has a much shorter response time to a given climate forcing than the open ocean, investigating long-term changes in the East Sea has important implications for understanding global ocean change. During the past several decades, the East Sea has experienced radical changes in water masses and upper ocean structures. In this section, we review previous studies of long-term variability in the East Sea and their possible relationships with North Pacific and atmospheric variability, although detailed mechanisms responsible for the variability still remain undetermined.

3.4.1 Water Masses in Change

A remarkable warming in the East Sea has occurred below 500 dbar since 1969 (e.g. Kim et al. 1996, 2001b; Kim and Kim 1996; Kwon et al. 2004), the rate of which is comparable to rates observed in the world ocean (Levitus et al. 2001). Accompanying the warming trends, dissolved oxygen concentration continued to decrease in the deep and bottom waters, and increase between 700 and 1500 m (Gamo et al. 1986; Kim et al. 1996, 2004; Kim and Kim 1996; Chen et al. 1999; Gamo 1999). Also, salinity has had an increasing trend at 500 and 800 m, and decreasing trend at 2500 m since 1965 (Minami et al. 1999; Park and Kim 2007). These observed changes of temperature and dissolved oxygen concentration have been interpreted as being consistent with a substantial shoaling of the ventilation system, from deep and bottom water ventilation to mid-depth ventilation (Kim and Kim 1996; Riser et al. 1999). Kang et al. (2003) attempted to quantify these changes using a simple box model, suggesting, by a linear extrapolation of volume decrease, the possibility of ESBW disappearance by the year 2050. It should be noted that occasional convection of the ESBW was detected in winter 2000/2001 (Kim et al. 2002; Senjyu et al. 2002).

In the past, a water mass called the Upper Portion of the Japan Sea Proper Water (UPJSPW), which also has a high oxygen concentration, was introduced by Sudo (1986) and Senjyu and Sudo (1993) based on hydrographic data in July 1969. With careful comparisons between the 1969 data and CREAMS data, Min (2002) and Kim et al. (2004) investigated similarities and differences between UPJSPW and ESCW, and concluded they had distinctly different characteristics in terms of their water properties and depth ranges. Such dissimilarities between water masses in the 1960s and in the 1990s may result from modification of the thermohaline circulation by climate change in the East Sea, although it is not certain that UPJSPW in 1969 (Senjyu 1999) has been altered into HSIW and ESCW (Kim et al. 2004; Min and Kim 2006) at present.

Understanding the physical processes responsible for such a water mass transition is far from complete. However, Kwon et al. (2004) diagnosed a change in the vertical mode of ventilation during the past few decades by using a simple inverse model. The inverse model suggested that ESDW and ESBW contractions during the 1990s resulted from warming of their properties as well as changes of the deep circulation, while HSIW and ESCW expansion were mainly due to changes of fresh water fluxes at their surface outcrops. Such a long-term change in water masses contains not only a trend-like variation, presumably associated with global warming (Kim et al. 1999), but also decadal-scale variability. Minami et al. (1999) and Watanabe et al. (2003) noted bi-decadal oscillations (~15 years) of oxygen, phosphate, and water temperature below 2000 m in the Eastern Japan Basin and the Yamato Basin which were synchronized with those in North Pacific Intermediate Water, indicating that forcings (atmospheric buoyancy flux or advective flux by the Kuroshio) which drive such water mass variability in the East Sea and the North Pacific might be closely related.

A numerical experiment performed by Park (2007) suggested that the combined effect of atmospheric cooling and intensification of the upper ocean current could have produced the cold and salty deep convection observed during 2000–2001, hence suggesting that the advection of warm, salty water through the Korea Strait is also an important factor controlling the East Sea thermohaline circulation. This warm and salty TWC is responsible for net heat loss to the atmosphere in the East Sea (Hirose et al. 1996; Han and Kang 2003), and salt carried by the current could play an important role in deep water formation off Vladivostok (Gamo et al. 1986; Sudo 1986; Senjyu and Sudo 1994; Seung and Yoon 1995; Yoshikawa et al. 1999). Also, Takikawa and Yoon (2005) found a ~15 year oscillation in volume transport estimated from sea level difference across the Korea Strait, which has a similar period to the variability of physical and chemical properties in the deep water (Watanabe et al. 2003).

On the other hand, Cui and Senjyu (2010) suggested that interdecadal oscillation, with about 20-year periodicity, of the dissolved oxygen concentration at a depth of 1000 m had a positive correlation with the Arctic Oscillation (AO) index, tightly linked with cold air outbreak activity (Isobe and Beardsley 2007). Thus, a combination of oceanic heat/salt advection and atmospheric forcing is clearly a major controller of long-term water mass variation in the East Sea.

3.4.2 Interannual and Decadal Variation of the Upper Ocean

Long-term variability or trend found in deep waters of the East Sea is inevitably linked with upper ocean variability, because surface property conditions immediately before water mass formations mainly determine the characteristics of the ventilated water masses. Furthermore, spatiotemporal variability of the upper ocean strongly affects the distribution and evolution of marine biota; it also forces the marine atmospheric boundary layer, thereby affecting the climate system. In this section, we reference numerous journal articles reporting on upper ocean variability in the East Sea with timescales longer than a year and discussing possible mechanisms.

3.4.2.1 Upper Ocean Temperature

There have been many research publications on SST variability in the East Sea, on diverse time scales at annual, interannual, and decadal frequencies (Miita and Tawara 1984; Watanabe et al. 1986; Isoda 1994; Gamo 1999; Park and Chung 1999; Park and Oh 2000; Kim et al. 2001b, 2004; Minobe et al. 2004; Takikawa and Yoon 2005). Interannual SST variability is characterized by energetic variability south of the SPF in the western region of the East Sea, which has a high correlation (Pearson's correlation $r > 0.7$) with sea level pressure (SLP) in winter over the western North Pacific (Minobe et al. 2004). In addition, the interannual variability is coherent with adjacent atmospheric systems such as the Aleutian Low and the Siberian High as well as tropical climate systems such as El Niño and La Niña. Park and Chu (2006) showed a positive correlation between SST and surface air temperature, which is probably due to intensive ocean-atmosphere interaction in wintertime. Park and Oh (2000) and Hong et al. (2001) suggested that SST anomaly variations with interannual timescales were closely related to ENSO events such that El Niño (La Niña) events generated 'cold (warm) summer' and 'warm (cold) winter' SST conditions. On the other hand, Minobe et al. (2004) suggested a possible mechanism through which the energetic interannual SST mode in the western part of the East Sea might result from a complex combination of ocean advection and wind forcing: southward propagation of temperature anomalies through alongshore wind driven coastal waves and northward advection of these anomalies by the EKWC. Also, Chang et al. (2004) and Talley et al. (2006) pointed out that variability of temperature in the upper 200 m is mainly governed by the TWW flowing through the Korea Strait.

According to Minobe et al. (2004), the decadal mode of dominant SST variability is likely to be caused by the East Asian winter monsoon due to SLP variability of the Siberian High related to the AO and the North Atlantic Oscillation. Yeh et al. (2010) suggested that decadal fluctuations of SST in the East Sea are mainly associated with the variable advection of cold air in winter and consequent heat loss. Indeed, large-scale atmospheric patterns such as the AO have a significant correlation with cold air outbreak activity over the East Sea in wintertime (Isobe and Beardsley 2007). Nevertheless, the process generating decadal SST variability is not as simple as anomalous oceanic cooling driven by local atmospheric forcing. Na et al. (2012) revealed that spatial patterns of the decadal mode and the warming mode of upper ocean heat content were similar to each other, and their most energetic features were found in the eastern part of the East Sea near the SPF.

The Complex Empirical Orthogonal Function (CEOF) structures of upper ocean temperature are comparable from the sea surface to 300 m, and maximum variances are found at around 100 m depth (Minobe et al. 2004) as shown in Fig. 3.7. Na et al. (2012) also confirmed those patterns by the cyclostationary EOF method.

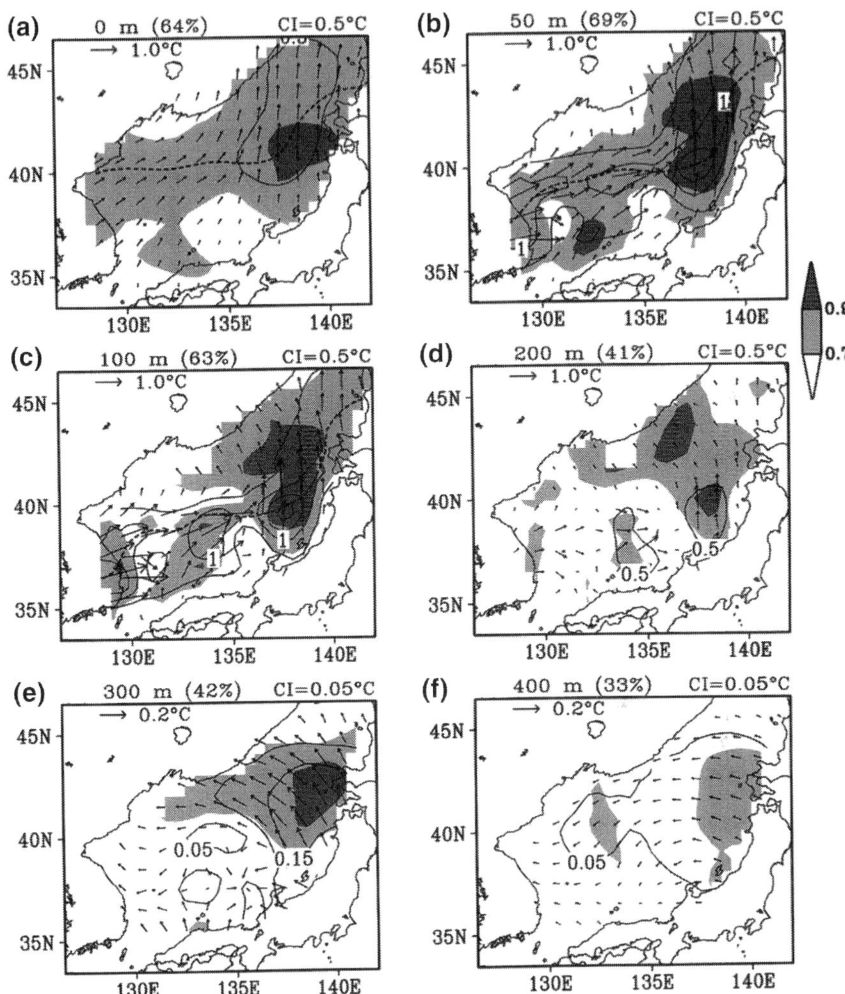

Fig. 3.7 Spatial patterns of the decadal CEOF mode of water temperatures. *Contours* indicate CEOF amplitudes (°C), and the confidence interval (CI) is given at the *top* of each panel. *Dark* (medium) *gray shading* indicates the region where correlation coefficients are higher than 0.7 (0.5) between the temperature and principal component time series. The amplitudes are also indicated by the *length of the arrows* according to the scale in the respective panels. Phases are indicated by the *direction of the arrows*: an arrow pointing to the right indicates 0° phase, an *arrow* pointing to the *top* of the page is 90°, etc. (phase increases in the direction of arrow rotating counterclockwise). The direction of phase increase is the direction of phase propagation (from Minobe et al. 2004)

The spatial inhomogeneity and subsurface variance maximum of the decadal mode imply that decadal migration of upper ocean circulation in the East Sea could be an important factor controlling decadal SST variability, together with the local air-sea heat exchange.

Upper ocean temperature variability in the East Sea is likely part of a profound feedback system affecting the regional weather and climate system. When cold, dry air from the Siberian continent flows over the warm East Sea, an internal marine boundary layer develops in the atmosphere with offshore fetch by extracting large amounts of heat from the ocean surface (Khelif et al. 2005; Shimada and Kawamura 2006, 2008). Hence, the SST condition significantly influences wind direction and speed, turbulent heat flux, and precipitation over the East Sea (Hirose and Fukudome 2006; Hirose et al. 2009). Furthermore, Yamamoto and Hirose (2008, 2009, 2011) demonstrated that mesoscale SST features help improve simulation not only of developing cyclones in the northwestern Pacific but also of the regional winter atmospheric circulations associated with cold air outbreaks over the eastern coast of the Eurasian Continent. Also, Seo et al. (2014) exhibited that the SST variability in the East Sea can have a far-reaching and strongly non-linear impact on large-scale atmospheric circulation over the North Pacific.

3.4.2.2 Sea Surface Height

The accumulation of multi-year satellite altimetry data has allowed investigations of spatial patterns and interannual-decadal variations of upper ocean circulation over the East Sea. Choi (2004) studied non-seasonal variability of sea surface height (SSH) using EOF analysis of merged TOPEX/Poseidon and ERS-1/2 altimeter data and identified dominant patterns of intraseasonal and interannual variations of SSH in the East Sea. Such basin-wide oscillations related to variability of the SPF are correlated with sea level changes near the Korea Strait, implying a close linkage with path variations of the TWC. Satellite SSH data also revealed biennial variability over the Yamato Basin (Hirose and Ostrovskii 2000). Similar oscillations were found in the transport of the TWC (Toba et al. 1982) and in SST variation along 134°E (Isoda 1994), although dynamical relationships among them are unclear. The biennial variability might be an event-like pattern induced by slowly moving eddies or an internal basin mode of the East Sea (Hirose and Ostrovskii 2000; Gordon and Giulivi 2004). From a different point of view, Senjyu et al. (1999) documented interannual (2.8 years) variability of sea levels simultaneously oscillating at the Japanese coast (both inside and outside the East Sea) and, based on a positive correlation with the East Asian winter monsoon index, suggested that the SSH variability can be attributed to anomalous heat loss associated with cold air outbreaks.

Gordon and Giulivi (2004) reported that SSH decadal-scale variability within the East Sea is associated with the PDO, suggesting that negative PDO leads to warmer SST and then higher SSH through the steric effect. Kang et al. (2005) hypothesized that decadal variability of SSH might be attributed to that of heat content anomaly in the upper 300 m of the Pacific Ocean (Levitus et al. 2001) flowing into the East Sea interior, as well as meridional migration of the EKWC/TWC and mesoscale eddies, thus emphasizing the importance of oceanic processes (currents and waves passing through straits) in transferring climate variability in the open ocean into the marginal sea. The long-term sea level trend over the entire East Sea, determined from analyses of satellite altimeter and tide gauge data (Kang et al. 2005, 2008), is 5.4 ± 0.3 mm year^{-1}, much larger than the global rates of 3.1 ± 0.4 mm year^{-1} from Cabanes et al. (2001) and 2.8 ± 0.4 mm year^{-1} from Cazenave and Nerem (2004). Long-term variability of SSH in the East Sea is mainly due to steric height change in the upper layer of the southern region, which is correlated with the heat anomaly pattern in the Pacific Ocean. Kang et al. (2005) hypothesized that the long-term variation of sea level in the Pacific penetrated into the East Sea via the TWC.

3.5 Summary and Remaining Questions

In this chapter, we have introduced spatial and temporal variations of water properties in the East Sea, together with water masses identified in previous studies. SST, SST-front structures, and the MLD in the East Sea all show remarkable seasonality related to ocean circulation and local atmospheric buoyancy forcing. In particular, the spatial structure of the SPF varies due not only to mesoscale eddies near the frontal region, but also to atmospheric wind forcing affecting migration of the EKWC/TWC. The strong subpolar SST front near 40°N hinders winter development of the surface mixed layer, thereby contributing to spatial inhomogeneity of MLD patterns.

The East Sea consists of various water masses, some of which are transferred from outside the marginal sea while others are formed within. The distinct exterior water masses are LSTWW presumably originating from diluted Changjiang river water, and HSTWW derived mainly from Kuroshio water in the Pacific. Those two exterior waters significantly influence water mass formation, especially ESIW and HSIW. The interior water masses are diverse, but most of them are formed in the northern part of the East Sea, TDW being an exception. TDW is a layer with nearly uniform water temperature (around 10 °C), analogous to mode waters in the open ocean; it is often found around the Ulleung Basin in the southwestern part of the East Sea. Among the interior water masses, there are two remarkably different intermediate waters in the East Sea, HSIW and ESIW. Both of these intermediate waters have similar density ranges, but their salinities and geographical locations are different. HSIW with high salinity and high oxygen concentration is trapped within the eastern Japan Basin where it is formed in wintertime. On the other hand, ESIW with a salinity minimum is subducted southward into the permanent thermocline and extends to both the southern part of the East Sea and the eastern part along the SPF front. Some of the ESIW is transported to the Korean coastal region, contributing to NKCW along the east coast of Korea and KSBCW in the Korea Strait.

Below the intermediate waters, there are ESCW, ESDW, and ESBW which have been recently identified in high precision hydrographic data from the CREAMS observations. Valuable datasets obtained from CREAMS revealed not only circulation patterns of deep waters but also long-term changes of those water masses. It has been found that water mass characteristics in the 1990s differ from those in the 1960s.The depth ranges of intermediate water and ESCW have been expanding for several decades, while those of ESDW and ESBW have been continuously shrinking, although occasional formation of ESBW was evident in 2000/2001.

Long-term variability of the East Sea is likely affected by inflows through the Korea Strait as well as local and remote atmospheric forcing. Cold air outbreak activity in winter, which has a significant correlation with larger scale atmospheric patterns like the AO, induces low frequency variability of SST. Although statistical correlations with atmospheric patterns have been reported, a detailed physical process leading to the spatiotemporal variability of the SST with interannual timescales is still uncertain. Also, highly energetic variability near the SPF suggests that upper ocean current meandering, presumably driven by flow instability, local atmospheric forcing and inflow transport, is one of the important factors controlling long-term variability of upper ocean properties. Likewise, analysis of SSH data in the East Sea presents statistically significant correlations with local (monsoon index) and remote (PDO index) atmospheric variability as well as variability in the Korea Strait. Decadal variability of SST, heat content, and SSH in the East Sea is evidently synchronized with variability in the North Pacific. However, it is still unclear whether large-scale atmospheric patterns such as the AO simultaneously force the East Sea and the Pacific or whether variability in the Pacific propagates into the East Sea.

Remaining issues regarding spatiotemporal variability of water properties in the East Sea are mostly related to understanding the dynamical processes involved. The following are relevant questions. How are the intermediate waters formed? Where are they destroyed? Why and how have water masses in the East Sea changed with decadal timescales? What controls the interannual-decadal variability of upper ocean temperature? Although some hypotheses on these questions have been suggested in previous studies, many detailed processes are still uncertain. Answers to these questions will enable us to determine how the East Sea will evolve in the future and how such evolution of physical states affects the biogeochemical environment of the East Sea as well as the regional climate system. From the viewpoint of a miniature ocean, our understanding of East Sea variability should facilitate foreseeing future changes of the global ocean. To achieve these goals, it is obvious that long-term and large-scale comprehensive observations should be maintained in the East Sea.

Acknowledgments The authors would like to thank Dr. Hyong-Rok Kim in the Agency for Defense Development for valuable discussion about 'Ten Degree Water'. We are also grateful to two reviewers for providing constructive comments which have improved the manuscript. J.J. Park is supported by "Research for the Meteorological and Earthquake Observation Technology and Its Application" of National Institute of Meteorological Research.

References

Belkin IM, Cornillon PC (2003) SST fronts of the Pacific coastal and marginal seas. Pac Oceanogr 1(2):90–113

Cabanes C, Cazenave A, Le Provost C (2001) Sea level rise during past 40 years determined from satellite and in situ observations. Science 294:840–842

Cazenave A, Nerem RS (2004) Present-day sea level change: observations and causes. Rev Geophys 42:RG3001. doi:10.1029/2003RG000139

Chang KI, Teague WJ, Lyu SJ et al (2004) Circulation and currents in the southwestern East/Japan Sea: overview and review. Prog Oceanogr 61:105–156. doi:10.1016/j.pocean.2004.06.005

Chen CTA, Bychkov AS, Wang CS et al (1999) An anoxic Sea of Japan by the year 2200? Mar Chem 67:249–265

Cho YK, Kim K (1994) Two modes of the salinity minimum layer in the Ulleung Basin. La Mer 32:271–278

Cho YK, Kim K (1998) Structure of the Korea strait bottom cold water and its seasonal variation in 1991. Cont Shelf Res 18:791–804

Choi YK (1996) Open-ocean convection in the Japan (East) Sea. La Mer 34:259–272

Choi BJ (2004) Nonseasonal sea level variations in the Japan/East Sea from satellite altimeter data. J Geophys Res 109:C12028. doi:10.1029/2004JC002387

Chu PC, Lan J, Fan C (2001) Japan sea thermohaline structure and circulation, part i: climatology. J Phys Oceanogr 31:244–271

Cui Y, Senjyu T (2010) Interdecadal oscillations in the Japan Sea proper water related to the arctic oscillation. J Oceanogr 66:337–348. doi:10.1007/s10872-010-0030-z

Dorman CE, Beardsley RC, Dashko NA et al (2004) Winter marine conditions over the Japan Sea. J Geophys Res 109:C12011. doi:10.1029/2001JC001197

Dorman CE, Beardsley RC, Limeburner R et al (2005) Summer atmospheric conditions over the Japan/East Sea. Deep-Sea Res Pt II 52:1393–1420

Falkowski PG, Barber R, Smetacek V (1998) Biogeochemical controls and feedbacks on ocean primary production. Science 281:200–206

Fedorov KN (1986) The physical nature and structure of oceanic fronts. Lecture notes in coastal estuarine studies, vol 19. AGU, Washington DC. doi:10.1029/LN019

Gamo T (1999) Global warming may have slowed down the deep conveyor belt of a marginal sea of the northwestern Pacific: Japan Sea. Geophys Res Lett 26:3137–3140

Gamo T, Nozaki Y, Sakai H et al (1986) Spatial and temporal variations of water characteristics in the Japan Sea bottom layer. J Mar Res 44:781–793

Gordon AL, Giulivi CF (2004) Pacific decadal oscillation and sea level in the Japan/East Sea. Deep-Sea Res Pt I 51:653–663. doi:10.1016/j.dsr.2004.02.005

Han IS, Kang YQ (2003) Supply of heat by Tsushima current in the East Sea (Japan Sea). J Oceanogr 59:317–323

Hirose N, Fukudome K (2006) Monitoring the Tsushima warm current improves seasonal prediction. Sci Online Lett Atmos 2:61–63. doi:10.2151/sola.2006

Hirose N, Ostrovskii AG (2000) Quasi-biennial variability in the Japan Sea. J Geophys Res 105(C6):14011–14027

Hirose N, Kim CH, Yoon JH (1996) Heat budget in the Japan Sea. J Oceanogr 52:553–569

Hirose N, Nishimura K, Yamamoto M (2009) Observational evidence of a warm ocean current preceding a winter teleconnection pattern in the northwestern Pacific. Geophys Res Lett 36:L09705. doi:10.1029/2009GL037448

Hong CH, Cho KD, Kim HJ (2001) The relationship between ENSO events and sea surface temperature in the East (Japan) Sea. Prog Oceanogr 49:21–40

Ichiye T (1984) Some problems of circulation and hydrography of the Japan Sea and the Tsushima current. In: Ichiye T (ed) Ocean hydrodynamics of the Japan and East China Seas. Elsevier oceanography series vol 39. Elsevier, Amsterdam, pp 15–54

Isobe A (1994) Seasonal variability of the barotropic and baroclinic motion in the Tsushima-Korea strait. J Oceanogr 50:223–238

Isobe A, Beardsley RC (2007) Atmosphere and marginal-sea interaction leading to an interannual variation in cold-air outbreak activity over the Japan Sea. J Climate 20:5707–5714. doi:10.1175/2007JCLI1779.1

Isoda Y (1994) Inter-annual SST variations to the north and south of the polar front in the Japan Sea. La Mer 32:285–294

Isoda Y, Saitoh SI (1993) The northward intruding eddy along the east coast of Korea. J Oceanogr 49:443–458

Jang CJ, Shim TB, Kim K (1995) Short term variation of the mixed layer in the Korea strait in autumn. J Oceanol Soc Korea 30:512–521

Kajiura K, Tsuchiya M, Hidaka K (1958) The analysis of oceanographical condition in the Japan Sea. Rep Dev Fish Resour Tsushima Warm Curr 1:158–170 (in Japanese)

Kang DJ, Park SY, Kim YG et al (2003) A moving-boundary box model (MBBM) for oceans in change: an application to the East/Japan Sea. Geophys Res Lett 30:1299. doi:10.1029/2002GL016486

Kang SK, Cherniawsky JY, Foreman MGG et al (2005) Patterns of recent sea level rise in the East/Japan Sea from satellite altimetry and in situ data. J Geophys Res 110:C07002. doi:10.1029/2004JC002565

Kang SK, Cherniawsky JY, Foreman MGG et al (2008) Spatial variability in annual sea level variations around the Korean peninsula. Geophys Res Lett 35:L03603. doi:10.1029/2007GL032527

Kawamura H, Wu P (1998) Formation mechanism of the Japan Sea proper water in the flux center off Vladivostok. J Geophys Res 103:611–622

Kawamura H, Yoon JH, Ito T (2007) Formation rate of water masses in the Japan Sea. J Oceanogr 63:243–253

Khelif D, Friehe CA, Jonsson H et al (2005) Wintertime boundary-layer structure and air-sea interaction over the Japan/East Sea. Deep-Sea Res Pt II 52:1525–1546. doi:10.1016/j.dsr2.2004.04.005

Kim HR (1991) The vertical structure and temporal variation of the intermediate homogeneous water near Ulleung Island. Dissertation, Seoul National University

Kim YG (1996) A study on the water characteristics and the circulation of the intermediate and deep layer of the East Sea. Dissertation, Seoul National University

Kim YH (1999) Physical characteristics of the North Korean cold water and currents off Jumunjin. Dissertation, Seoul National University

Kim K, Chung JY (1984) On the salinity minimum and dissolved oxygen maximum layer in the East Sea (Sea of Japan). In: Elsevier oceanography series vol 39. Elsevier, Amsterdam, pp 55–65

Kim SW, Isoda Y (1998) Interannual variations of the surface mixed layer in the Tsushima current region. Umi to Sora 74:11–22 (in Japanese)

Kim CH, Kim K (1983) Characteristics and origin of the cold water mass along the east coast of Korea. J Oceanogr Soc Korea 18(1):71–83

Kim K, Kim KR (1995) Changes in deep water characteristics in the East Sea (Sea of Japan): a regional evidence for global warming? In: PICES 4th annual meeting abstracts, pp 31–32

Kim KR, Kim K (1996) What is happening in the East Sea (Japan Sea), recent chemical observation during CREAMS 93-96. J Oceanogr Soc Korea 31:164–172

Kim YG, Kim K (1999) Intermediate waters in the East/Japan Sea. J Oceanogr 55:123–132

Kim K, Legeckis R (1986) Branching of the Tsushima current in 1981–83. Prog Oceanogr 17:265–276

Kim K, Kim KR, Chung JY et al (1996) New findings from CREAMS observations: water masses and eddies in the East Sea. J Korean Soc Oceanogr 31:155–163

Kim KR, Kim K, Kang DJ et al (1999) The East Sea (Japan Sea) in change: a story of dissolved oxygen. Mar Technol Soc J 33:15–22

Kim GB, Yang HS, Kim KR et al (2001a) Radium isotopes in the Pacific-Asian marginal seas as a water-mass tracer. In: The 11th PAMS/JECSS proceeding in Cheju, Korea

Kim K, Kim KR, Min DH et al (2001b) Warming and structural changes in the East (Japan) Sea: a clue to future changes in global oceans? Geophys Res Lett 28:3293–3296

Kim KR, Kim G, Kim K et al (2002) A sudden bottom water formation during the severe winter 2000–2001: the case of the East/Japan Sea. Geophys Res Lett 29(8):1234. doi:10.1029/2001GL014498

Kim K, Kim KR, Kim YG et al (2004) Water masses and decadal variability in the East Sea (Sea of Japan). Prog Oceanogr 61:157–174

Kim YH, Kim YB, Kim K et al (2006) Seasonal variation of the Korea strait bottom cold water and its relation to the bottom current. Geophys Res Lett 33:L24604. doi:10.1029/2006GL027625

Kitani K (1987) Direct current measurements of the Japan Sea proper water. Rep Japan Sea Natl Fish Res Inst 341:1–6 (in Japanese)

Kubota M (1990) Variability of the polar front in the Japan Sea. Sora to Umi 12:35–44 (in Japanese)

Kwon YO, Kim K, Kim YG, Kim KR (2004) Diagnosing long-term trends of the water mass properties in the East Sea (sea of Japan). Geophys Res Lett 31:L20306. doi:10.1029/2004GL020881

Lee C, Thomas L, Yoshikawa Y (2006) Intermediate water formation. Oceanography 19:110–121

Levitus S, Antonov JI, Wan J et al (2001) Anthropogenic warming of earth's climate system. Science 292:267–270

Lim KS (1983) The characteristic and origin of the cold water mass in the southeastern sea off Korea. Dissertation, Seoul National University

Lim DB, Chang SD (1969) On the cold water mass in the Korea strait. J Oceanogr Soc Korea 39:155–162

Lim SS, Jang CJ, Oh IS et al (2012) Mixed layer climatology for the East Sea (Japan Sea). J Mar Syst 96–97:1–14

Luyten JR, Pedlosky J, Stommel H (1983) The ventilated thermocline. J Phys Oceanogr 13:292–309

Miita T, Tawara S (1984) Seasonal and secular variation of water temperature in the East Tsushima strait. J Oceanogr Soc Japan 40:91–97

Min HS (1994) A study on the variation of the bottom water temperature in the Korea strait. Dissertation, Seoul National University

Min HS (2002) Long-term variations of water properties at the inter-mediate depth in the East Sea. Dissertation, Seoul National University

Min HS, Kim CH (2006) Water mass formation variability in the intermediate layer of the East Sea. Ocean Sci J 41:255–260

Min DH, Warner MJ (2005) Basin-wide circulation and ventilation study in the East Sea (sea of Japan) using chlorofluorocarbon tracers. Deep-Sea Res Pt II 52:1580–1616

Min HS, Kim YH, Kim CS (2006) Year-to-year variation of cold waters around the Korea strait. Ocean Sci J 41:227–234

Minami H, Kano Y, Ogawa K (1999) Long-term variations of potential temperature and dissolved oxygen of the Japan Sea proper water. J Oceanogr 55:97–205

Minobe S, Sako A, Nakamura M (2004) Interannual to interdecadal variability in the Japan Sea based on a new gridded upper water temperature dataset. J Phys Oceanogr 34:2382–2397

Miyazaki M (1952) The heat budget of the Japan Sea. Bull Hokkaido Reg Fish Res Lab 4:1–54 (in Japanese)

Miyazaki M (1953) On the water masses of the Japan Sea. Bull Hokkaido Reg Fish Res Lab 7:1–65 (in Japanese)

Miyazaki M, Abe S (1960) On the water masses in the Tsushima current area. J Oceanogr Soc Japan 16:19–28

Monterey GI, Levitus S (1997) Climatological cycle of mixed layer depth in the world ocean. U.S. Government Printing Office, Washington, DC, NOAA NESDIS-5

Moriyasu S (1972) The Tsushima current in Kuroshio-its physical aspects. In: Yoshida K (ed) Stommel H. University of Tokyo Press, Tokyo, pp p353–p369

Na H, Kim KY, Chang KI et al (2012) Decadal variability of the upper ocean heat content in the East/Japan Sea and its possible relationship to northwestern Pacific variability. J Geophys Res 117:1–11. doi:10.1029/2011JC007369

Nakayama I (1951) On the mean surface temperature of seas around Japan. In: Kobe B (ed) Mar Obs 159:1–16 (in Japanese)

Nishida K (1926) Report of the oceanographic investigation, no 1 Gover Fish Exp Stat (in Japanese)

Nitani H (1972) On the deep and bottom waters in the Japan Sea. In: Shoji D (ed) Researches in hydrography and oceanography. Tokyo, pp 145–201

Park CH (1972) Hourly change of temperature and salinity in the Korea strait. J Oceanol Soc Korea 7:15–18 (in Korean)

Park JJ (2002) Deep currents from APEX in the East Sea. Dissertation, Seoul National University

Park YG (2007) The effects of Tsushima warm current on the interdecadal variability of the East/Japan Sea thermohaline circulation. Geophys Res Lett 34:L06609. doi:10.1029/200 6GL029210

Park S, Chu PC (2006) Interannual SST variability in the Japan/East Sea and relationship with environmental variables. J Oceanogr 62:115–132

Park KA, Chung JY (1999) Spatial and temporal scale variations of sea surface temperature in the East Sea using NOAA/AVHRR data. J Oceanogr 55:271–288

Park JJ, Kim K (2007) Evaluation of calibrated salinity from profiling floats with high resolution conductivity-temperature-depth data in the East/Japan Sea. J Geophys Res 112:1–9. doi:10. 1029/2006JC003869

Park JJ, Kim K (2013) Deep currents obtained from Argo float trajectories in the Japan/East Sea. Deep-Sea Res Pt II 85:1–45

Park KA, Lee EY (2014) Semiannual cycle of sea surface temperature in the East/Japan Sea and cooling process. Int J Remote Sens 35:4287–4314. doi:10.1080/01431161.2014.916437

Park WS, Oh IS (2000) Interannual and interdecadal variations of sea surface temperature in the East Asian marginal seas. Prog Oceanogr 47:191–204. doi:10.1016/S0079-6611(00)00036-7

Park YG, Cho YG, Kim K (1995) A hydraulic model of the Korea strait bottom cold current. J Oceanogr 51:713–727

Park KA, Chung JY, Kim K (2004) Sea surface temperature fronts in the East (Japan) sea and temporal variations. Geophys Res Lett 31:L07304. doi:10.1029/2004GL019424

Park KA, Chung JY, Kim K et al (2005) Wind and bathymetric forcing of the annual sea surface temperature signal in the East (Japan) sea. Geophys Res Lett 32:L5610. doi:10.1029/200 4GL022197

Park KA, Ullman DS, Kim K et al (2007) Spatial and temporal variability of satellite-observed subpolar front in the East/Japan Sea. Deep-Sea Res Pt I 54:453–470. doi:10.1016/j.dsr.2006.12.010

Park JJ, Kim K, Yang JY (2008) Aspiration and outflow of the intermediate water in the East/Japan Sea through the Tsugaru strait. Geophys Res Lett 35:1–6. doi:10.1029/200 7GL032981

Park KA, Woo HJ, Ryu JH (2012) Spatial scales of mesoscale eddies from GOCI chlorophyll-a concentration images in the East/Japan Sea. Ocean Sci J 47:347–358

Ponomarev VI, Yurasov GI, Zuenko YI (1991) High salinity bottom water generation on the Japan Sea continental shelf. In: 6th Japan and East China Seas study workshop, Fukuoka, Japan

Postlethwaite CF, Rohling EJ, Jenkins WJ et al (2005) A tracer study of ventilation in the Japan/East Sea. Deep-Sea Res Pt II 52:1684–1704

Riser SC, Jacobs G (2005) The Japan/East Sea: a historical and scientific introduction. Deep-Sea Res Pt II 52:1359–1362

Riser SC, Warner MJ, Yurasov GI (1999) Circulation and mixing of water masses of Tatar strait and the northwestern boundary region of the Japan Sea. J Oceanogr 55:133–156

Senjyu T (1999) The Japan Sea intermediate water; its characteristics and circulation. J Oceanogr 55:111–122

Senjyu T, Sudo H (1993) Water characteristics and circulation of the upper portion of the Japan Sea proper water. J Mar Syst 4:349–362

Senjyu T, Sudo H (1994) The upper portion of the Japan Sea proper water: its source and circulation as deduced from isopycnal analysis. J Oceanogr 50:663–690

Senjyu T, Matsuyama M, Matsubara N (1999) Interannual and decadal sea-level variations along the Japanese Coast. J Oceanogr 55:619–633

Senjyu T, Aramaki T, Otosaka S et al (2002) Renewal of the bottom water after the winter 2000–2001 may spin-up the thermohaline circulation in the Japan Sea. Geophys Res Lett 29(7):1149. doi:10.1029/2001GL014093

Senjyu T, Enomoto H, Masuno T, Matsui S (2006) Interannual salinity variations in the Tsushima strait and its relation to the Changjiang discharge. J Oceanogr 62:681–692

Seo H, Kwon YO, Park JJ (2014) On the effect of the East/Japan Sea SST variability on the North Pacific atmospheric circulation in a regional climate model. J Geophys Res Atmos 119:418–444. doi:10.1002/2013JD020523

Seung YH (1997) Application of ventilation theory to the East Sea. J Korean Soc Oceanogr 32:8–16

Seung YH, Yoon JH (1995) Some features of winter convection in the Japan Sea. J Oceanogr 51:61–73

Shimada T, Kawamura H (2006) Satellite observations of sea surface temperature and sea surface wind coupling in the Japan Sea. J Geophys Res 111:C08010. doi:10.1029/2005JC003345

Shimada T, Kawamura H (2008) Satellite evidence of wintertime atmospheric boundary layer responses to multiple SST fronts in the Japan Sea. Geophys Res Lett 35:L23602. doi:10.1029/2008GL035810

Shin CW, Byun SK, Kim C et al (2013) Seasonal variations in the low-salinity intermediate water in the region south of subpolar front of the East Sea (sea of Japan). Ocean Sci J 48(1):35–47

Stommel H (1979) Determination of water mass properties of water pumped down from the Ekman layer to the geostrophic flow below. Proc Natl Acad Sci US 76:3051–3055

Suda K (1932) On the Japan Sea bottom water. Kaiyo-Jiho 4:221–240 (in Japanese)

Sudo H (1986) A note on the Japan Sea proper water. Prog Oceanogr 17:313–336

Takikawa T, Yoon JH (2005) Volume transport through the Tsushima straits estimated from sea level difference. J Oceanogr 61:699–708

Talley LD, Lobanov V, Ponomarev V et al (2003) Deep convection and brine rejection in the Japan Sea. Geophys Res Lett 30:1159. doi:10.1029/2002GL016451

Talley LD, Min DH, Lobanov VB et al (2006) Japan/East Sea water masses and their relation to the sea's circulation. Oceanography 19:32–49

Thomas LN, Lee CM (2005) Intensification of ocean fronts by down-front winds. J Phys Oceanogr 35:1086–1102

Toba Y, Tomizawa Y, Kurasawa Y et al (1982) Seasonal and year-to-year variability of the Tsushima-Tsugaru warm current system with its possible cause. La Mer 20:41–51

Uda M (1934) The result of simultaneous oceanographical investigations in the Japan Sea and its adjacent water in May and June, 1932. J Imp Fish Exp Sta 5:186–235 (in Japanese)

Vasiliev AS, Makashin VP (1992) Ventilation of the Japan Sea waters in winter. La Mer 30:169–177

Warren BA (1972) Insensitivity of subtropical mode water characteristics to meteorological fluctuations. Deep-Sea Res 19:1–19

Watanabe T, Hanawa K, Toba Y (1986) Analysis of year-to-year variation of water temperature along the coast of the Japan Sea. Prog Oceanogr 17:337–357

Watanabe T, Hirai M, Yamada H (2001) High salinity intermediate water of the Japan Sea in the eastern Japan Basin. J Geophys Res 106:11437–11450

Watanabe YW, Wakita M, Maeda N et al (2003) Synchronous bidecadal periodic changes of oxygen, phosphate and temperature between the Japan Sea deep water and the North Pacific intermediate water. Geophys Res Lett 30:2273. doi:10.1029/2003GL018338

Yamamoto M, Hirose N (2008) Influence of assimilated SST on regional atmospheric simulation: a case of a cold-air outbreak over the Japan Sea. Atmos Sci Lett 9:13–17. doi:10.1002/asl.164

Yamamoto M, Hirose N (2009) Regional atmospheric simulation of monthly precipitation using high-resolution SST obtained from an ocean assimilation model: application to the wintertime Japan Sea. Mon Weather Rev 137:2164–2174. doi:10.1175/2009MWR2488.1

Yamamoto M, Hirose N (2011) Possible modification of atmospheric circulation over the northwestern Pacific induced by a small semi-enclosed ocean. Geophys Res Lett 38:L03804. doi:10.1029/2010GL046214

Yanagimoto D, Taira K (2003) Current measurements of the Japan Sea proper water and the intermediate water by ALACE floats. J Oceanogr 59:359–368

Yashayaev IM, Zveryaev II (2001) Climate of the seasonal cycle in the north Pacific and the north Atlantic oceans. Int J Clim 21:401–417

Yasui M, Yasuoka T, Tanioka K et al (1967) Oceanographic studies of the Japan Sea (1). Oceanogr Mag 18:177–192

Yeh SW, Park YG, Min HS et al (2010) Analysis of characteristics in the sea surface temperature variability in the East/Japan Sea. Prog Oceanogr 85:213–223. doi:10.1016/j.pocean.2010.03.001

Yoon JH, Kawamura H (2002) The Formation and circulation of the intermediate water in the Japan Sea. J Oceanogr 58:197–211

Yoon JH, Abe K, Ogata T et al (2005) The effects of wind-stress curl on the Japan/East Sea circulation. Deep-Sea Res Pt II 52:1827–1844

Yoshikawa Y, Awaji T, Akito K (1999) Formation and circulation processes of intermediate water in the Japan Sea. J Phys Oceanogr 29:1701–1722

Yun JY, Magaard L, Kim K et al (2004) Spatial and temporal variability of the North Korean cold water leading to the near-bottom cold water intrusion in Korea strait. Prog Oceanogr 60(1):99–131

Chapter 4
Circulation

Dong-Kyu Lee, Young Ho Seung, Yun-Bae Kim, Young Ho Kim,
Hong-Ryeol Shin, Chang-Woong Shin and Kyung-Il Chang

Abstract Recent basin-wide direct observations of near-surface circulation in the East Sea (Japan Sea), using satellite technology, reveal new circulation features. The East Korea Warm Current and Eastern Branch, originating from the Tsushima Warm Current, join to form the East Sea Current in the interior of the southern East Sea after leaving the coast. The formation mechanisms of the East Korea Warm Current and the Eastern Branch are diverse and their detailed dynamics are discussed in this chapter. The East Sea Current then exits the East Sea mainly through the Tsugaru Strait and this new current system is well reproduced by numerical models. The

D.-K. Lee (✉)
Department of Oceanography, Pusan National University, Busan 609-735, Republic of Korea
e-mail: dglee@pnu.edu

Y.H. Seung
Department of Oceanography, Inha University, Incheon 402-751, Republic of Korea
e-mail: seung@inha.ac.kr

Y.-B. Kim
Ulleungdo-Dokdo Ocean Science Station, Korea Institute of Ocean Science and Technology,
Ulleung-Gun 799-823, Republic of Korea
e-mail: dokdo512@kiost.ac.kr

Y.H. Kim · C.-W. Shin
Physical Oceanography Division, Korea Institute of Ocean Science and Technology,
Ansan 426-744, Republic of Korea
e-mail: yhkim@kiost.ac.kr

C.-W. Shin
e-mail: cwshin@kiost.ac.kr

H.-R. Shin
Department of Atmospheric Science, Kongju National University, Kongju 314-701,
Republic of Korea
e-mail: hrshin@kongju.ac.kr

K.-I. Chang
School of Earth and Environmental Sciences, Seoul National University,
Seoul 151-742, Republic of Korea
e-mail: kichang@snu.ac.kr

K.-I. Chang et al. (eds.), *Oceanography of the East Sea (Japan Sea)*,
DOI 10.1007/978-3-319-22720-7_4

Nearshore Branch and the North Korea Cold Current are only observed in summer months (May–August) and reversal of the North Korea Cold Current during winter is observed by the drifters. Mesoscale eddies are prevalent in the southern East Sea. They congregate in meridional bands and eddies are born, grow, propagate, and decay in a band determined by the vorticity of the mean flow. Many anticyclonic eddies in the Ulleung Basin (a negative vorticity zone) have life spans longer than one year, being sustained by kinetic energy received from the meandering mean flow. Deep water in the Japan Basin circulates cyclonically, and part of it penetrates into the southern Ulleung and Yamato basins through the channels and gaps. Deep cyclonic circulation in the Japan Basin appears to have strong mean currents of 4–8 cm s^{-1} along the perimeter of the basin and slow flows of 1–3 cm s^{-1} in the basin's interior. In the Ulleung Basin the deep flow is weak and highly variable with alternating cyclonic and anticyclonic sub-basin-scale cells. Deep circulation in the Japan Basin shows a distinct seasonal cycle with maximum speed in March and minimum speed in November. The seasonal variation of deep currents is not strong in the Ulleung and Yamato basins but the dominant variability is found to be in the 5–60-day period band. The deep circulation may be formed by cold-water convection rather than surface wind forcing, but the driving mechanism of the deep circulation is one of the unknown issues in the East Sea. Numerical simulations with realistic topography and forcing, high resolution, and lateral boundary conditions produce pictures of the circulation with many of the features observed in the entire East Sea, and they provide an opportunity to understand the dynamical processes of the East Sea circulation in detail. Numerical models with data assimilation have also been attempted successfully to interpret the near-surface and deep circulations in the East Sea.

Keywords Circulation · Currents · Dynamics · Numerical models · East Sea (Japan Sea)

4.1 Introduction

Maps of ocean near-surface circulation were historically drawn by compiling ship-drift data logged by sailors. Ship-drift data from merchant ships on the East Sea were scant and most of the information on the basin-scale surface circulation was based on hydrographic data. Uda (1934b) first performed basin-wide hydrographic surveys in the East Sea and was able to draw a current map which, until recently, was used extensively. Determination of the absolute current from hydrographic or Acoustic Doppler Current Profiler (ADCP) data has spatial and temporal limitations such as infrequent sampling, insufficient coverage by shipboard observation, and difficulty of defining a reference-level velocity in the shallow coastal sea. But basin-wide absolute current observations have been performed using recently developed satellite remote sensing technology. Sections 4.2 and 4.3 describe mean and temporal variations of East Sea near-surface circulation, including mesoscale eddies, based on 325 satellite-tracked drifters deployed mostly in 1996–2008, and satellite measurements of sea surface temperature (SST) and sea level anomaly (SLA).

The existence of the thermohaline circulation (THC) or the conveyor-belt system has often been referred to as one of the oceanic characteristics of the East Sea (Gamo 1999; Kim et al. 2001). As in the North Atlantic, the THC of the East Sea consists of deep-water formation processes (see Chap. 3 and references therein), interior upwelling processes, surface return flows (Park 2007), and deep circulation. The upwelling processes are largely unknown, however, mass balance considerations cast doubt on significant upwelling occurring across the 1800 m depth level in the central Ulleung Basin (UB) (Chang et al. 2009). Section 4.4 describes deep currents directly observed by Argo or Argo-type floats and current meter moorings in the East Sea as a branch of the THC, followed by a short review of the THC strength. Parking depths of the floats ranged from 700 to 800 m, but they are regarded as representing currents at 800 m since vertical current shear is weak. Water at 800 m depth corresponds to East Sea Central Water according to Kim et al. (2004a). The moored current measurements have been carried out mainly between 700 m depth and near the seabed. The history of float and moored current observations in the East Sea is introduced in Chap. 1. Water mass formation and characteristics of subsurface water masses with their variability are detailed in Chap. 3.

As described in Chap. 3, water masses in the East Sea are usually divided into three layers: the upper, intermediate, and deep layers. It is not clear whether circulation patterns also change from one water mass to another. Observations indicate that coupling between the upper layer occupied by Tsushima Warm Water (TWW) and the underlying layers is very weak, whereas that between the intermediate and deep layers is much stronger (Kim et al. 2009a). In fact, some observed deep currents show a weak bottom-trapped nature extending up to the intermediate layer, and are very different from the overlying currents carrying the TWW (Kim et al. 2013). Weak coupling between the upper and underlying layers can also be inferred dynamically from the fact that currents carrying the TWW are driven by inflow and outflow, whereas the underlying layers, nearly homogeneous in the vertical and much denser than the overlying layer, are isolated within the East Sea basin. The region north of the subpolar front, referred to as the cold-water region, is occupied by water colder than the TWW. In the cold-water region, the coupling between the upper layer and the underlying layers is not clear, but it seems to be stronger than that in the TWW region because of weaker stratification, especially in winter when the cold water-region loses much buoyancy through intense cooling (e.g. Seung and Yoon 1995a). Overall, the circulation in the East Sea may be divided largely into two regimes, the upper-layer circulation carrying the TWW and the underlying deep-layer circulation. It is noted, however, that the coupling between them, though weak, should be further quantified in future studies. Section 4.5 discusses some fundamental dynamical issues relating to upper- and deep-layer circulations in the East Sea, based on recent studies.

Early studies on dynamical processes in the East Sea employed numerical models with idealized and simple topography and coastline configurations. Although the simple models did not simulate realistic circulations and other phenomena in the East Sea, they have contributed to our understanding of the East Sea's circulation dynamics. With increasing computational resources since the 1990s, realistic configurations with real topography, time-varying meteorological forcing, and improved open boundary conditions have appeared in modeling studies of the East Sea circulation. They have simulated realistic surface and sub-surface circulations, thereby expanding our understanding of the East Sea. Furthermore, sensitivity experiments of various numerical configurations have expanded our understanding of the relationship between small-scale phenomena and large-scale circulation in the East Sea. Numerical modeling has also advanced into an operational ocean forecast system. As ocean observational data have increased with advancing observing technology such as Argo floats, satellite remote sensing, and so on, ocean data assimilation techniques have been applied to numerical models in the East Sea. Section 4.6 describes numerical circulation models and data assimilated models which help us to understand observed circulation features of the East Sea.

4.2 Near-Surface Circulation

4.2.1 Mean Surface Current

Based on historical hydrographic data and current data from shipboard ADCPs, earlier studies (e.g. Katoh et al. 1996; Cho and Kim 2000) have suggested that the Tsushima Warm Current (TWC) entering into the East Sea splits into three currents: (1) The East Korea Warm Current (EKWC) flows along the eastern coast of Korea and enters the UB. (2) The Nearshore Branch follows the northern coast of Japan all the way up to the Tsugaru Strait. (3) The Offshore Branch flows into the Yamato Basin (YB) but its path is not as well defined as the other two branches. However, recent deployments of satellite-tracked drifters have revealed different mean flow patterns in the East Sea after the TWC passes through the Korea Strait (Lee and Niiler 2010a). More than 50 % of the time, drifters passing through the western channel of the Korea Strait followed the EKWC path along the Korean coast. The EKWC meanders after entering the UB and organizes itself into a broad East Sea Current after merging with the flow from the eastern channel of the Korea Strait (Eastern Branch) (Fig. 4.1; Lee and Niiler 2005). The EKWC behaves like Arruda et al.'s (2004) simple model in which eddies are generated by a meandering current and the mean current exits through the Tsugaru Strait. Most of the time, the current in the eastern channel of the Korea Strait (Eastern Branch in Fig. 4.1) flows along the coast of Japan either up to the Oki Bank or to the Noto Peninsula and then joins the East Sea Current (Fig. 4.1). In summer only, the Nearshore Branch of the TWC develops along the coast of Japan. The East Sea Current (Lee and Niiler 2005) is the broad current that the EKWC and Eastern Branch form after leaving the respective coasts of Korea and Japan. The naming of this new current is needed to address the observed features: that the EKWC becomes a strong mean current in the middle of the southern East Sea and that the Eastern Branch, most of the time, enters the YB after leaving the coast of Japan. This broad mean current system has also been reproduced well by recent high resolution numerical models (Luneva and Clayson 2006; Clayson et al. 2008).

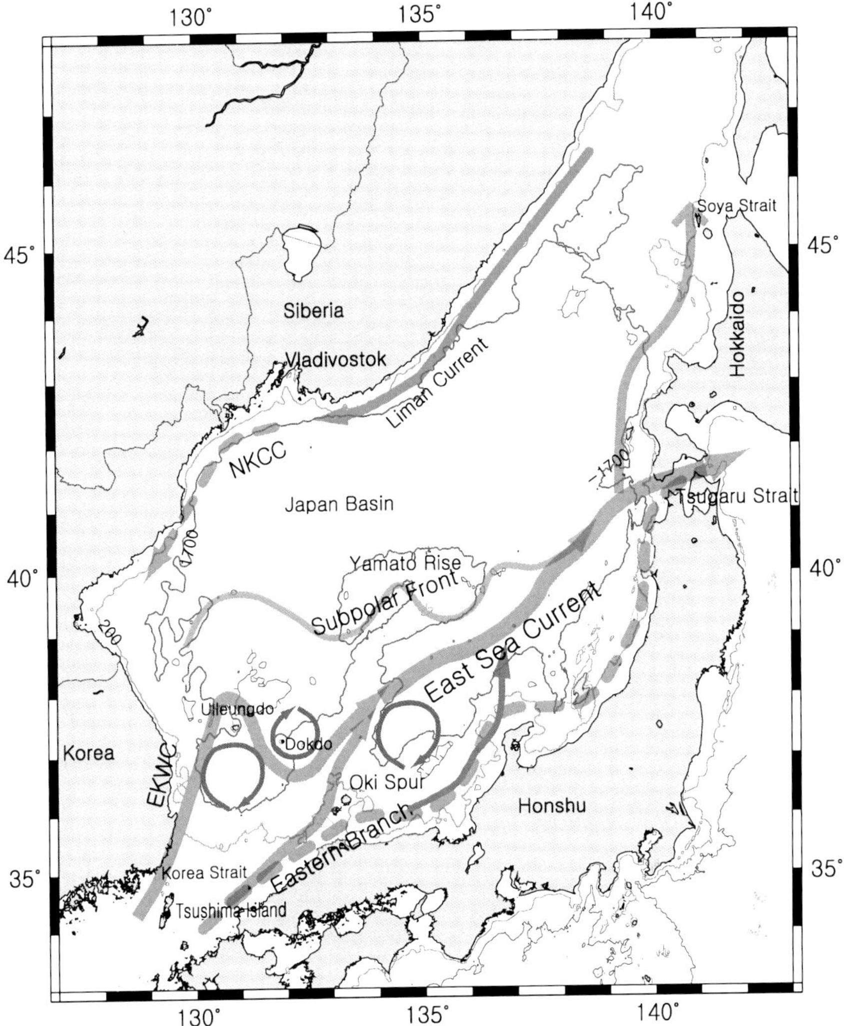

Fig. 4.1 Schematics of the surface circulation of the East Sea based on satellite tracked drifters. *Dotted lines* represent the current observed only in summer months (May–August)

More detailed descriptions of the southern East Sea surface circulation, including an historical view of the branching of the TWC, are found in Sect. 4.5.2.

In the northern East Sea, no distinctive year round mean current is observed except the Liman Current which flows southward along the Siberian coast down to Vladivostok (Fig. 4.1). The Liman Current, first described by Uda (1934a), derives from the freshwater flux related to ice melting in the Tatarsky Strait and from the wind drift (Martin and Kawase 1998; Park et al. 2006). Although the North Korea Cold Current (NKCC) appears, from numerical model simulation, throughout the seasons with seasonal transport variation (Kim and Min 2008), it is only during

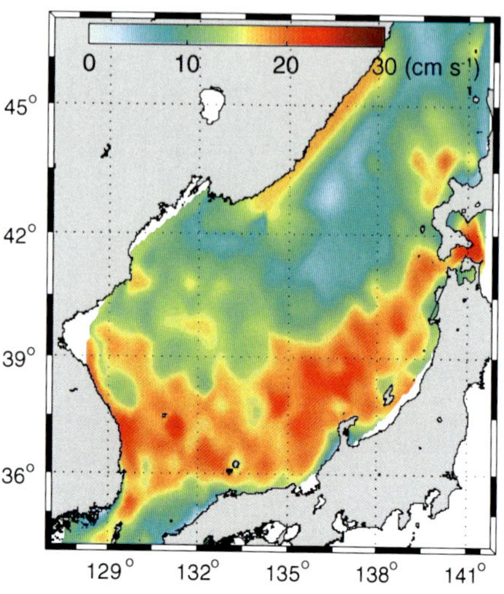

Fig. 4.2 Square root of eddy kinetic energy at the sea surface from surface drifter data (Redrawn from Lee and Niiler 2005)

summer that it has been observed from drifters. From over 325 drifters that were mostly deployed in the southern East Sea, only a few crossed the subpolar front between the warm-energetic southern East Sea and the cold-less-energetic northern East Sea (Fig. 4.2), and thus there were not many drifter observations in the northern East Sea from which a mean circulation can be defined.

4.2.2 Variability of the Surface Current and the Subpolar Front

The subpolar front divides the East Sea into two regions: (1) the energetic southern East Sea with large eddy kinetic energy (EKE) in which a strong mean flow enters the basin and exits through the Tsugaru and Soya straits, and (2) the calm northern East Sea with small EKE in which no distinctive year-round mean flow is found (Fig. 4.2). Historically the NKCC has been regarded as a year round mean current along the coast of North Korea (e.g. Kim and Min 2008). But large seasonal changes including a reversal of the NKCC have been observed (Yang 1996; Lee and Niiler 2005; Yoon et al. 2005). The subpolar front is observed year round nearly along 40°N in the northeastern East Sea (Fig. 4.1), but a separate front with characteristics similar to the subpolar front is found in the middle of the southern East Sea, formed by the reduced transport of warm water through the Korea Strait in spring (Lee and Niiler 2005). The subpolar front inferred from analyzing Lagrangian observations of individual float tracks may differ from the subpolar front described in Chap. 3, but it represents well the boundary between the less energetic northern East Sea and the energetic southern East Sea, formed by warm water inflow from the East China Sea through the Korea Strait.

The EKWC, after leaving the coast, forms various circulation patterns in the UB, depending on eddies and separation location from the coast (Katoh et al. 1996; Mitchell et al. 2005b). Three patterns from drifter tracks are recognized for the flow entering the East Sea through the western channel of the Korea Strait (Lee and Niiler 2010a): (1) Flow with large negative vorticity gains more negative vorticity from the deepening topography and so turns clockwise. It then joins the flow from the eastern channel of the Korea Strait and flows along the coast of Japan up to the Oki Spur (Tsushima Warm Current Pattern). (2) When the flow has mean speed over 55 cm s^{-1} with large positive vorticity, potential vorticity conservation applies and the EKWC follows the isobaths first along the coast and then along a topographic feature north of Ulleungdo (Ulleung Island) (Inertial Boundary Current Pattern). After entering the northern UB, the flow becomes an inertial meandering jet similar to a western boundary current. (3) The EKWC with small negative vorticity follows the southeastern coast of Korea up to 38°N (Ulleung Eddy Pattern) and meanders through the UB after entering it. Branching of the flow after passing through the western channel of the Korea Strait is found to occur only 15 % of the time; thus splitting or branching of the TWC at the entrance of the East Sea reported in hydrographic surveys and in theoretical studies is possibly attributable to the fact that the flow in each channel of the Korea Strait chooses a separate path.

Contrary to previous studies, the Nearshore Branch along the Japan coast and the NKCC along the North Korea coast are observed only in summer months (Lee and Niiler 2005). A northward NKCC along the coast up to the area west of Vladivostok in winter months is a surprising new observation. This reversal of the NKCC was observed from drifters (Yang 1996) and it has also shown up in a numerical model (Yoon et al. 2005). The Liman Current is assumed to be a year round current, but those far northern areas are too remote to aquire enough measurements for studying surface circulation.

4.2.3 Coastal Upwelling

The TWW along the east coast of Korea is carried by the northward flowing EKWC south of the subpolar front of the East Sea. The North Korea Cold Water (NKCW), originating from the western Japan Basin (JB) and carried by the NKCC, occupies the lower layer below the TWW with the permanent thermocline between them. In summer (June–August), the TWW includes the Low Salinity Tsushima Warm Water and High Salinity Tsushima Warm Water (HSTWW) (see Chap. 3), and the water column exhibits a three-layer structure. Due to the northward flowing EKWC and southward flowing NKCC underneath the EKWC (Byun and Seung 1984; Chang et al. 2002), isopycnals (isotherms) slope upward towards the coast. Subsurface isotherms further shoal and outcrop at the sea surface in summer when southerly winds prevail, and the surfacing of subsurface isotherms can occur in a few days (Byun 1989). Hence, coastal upwelling in the east coast

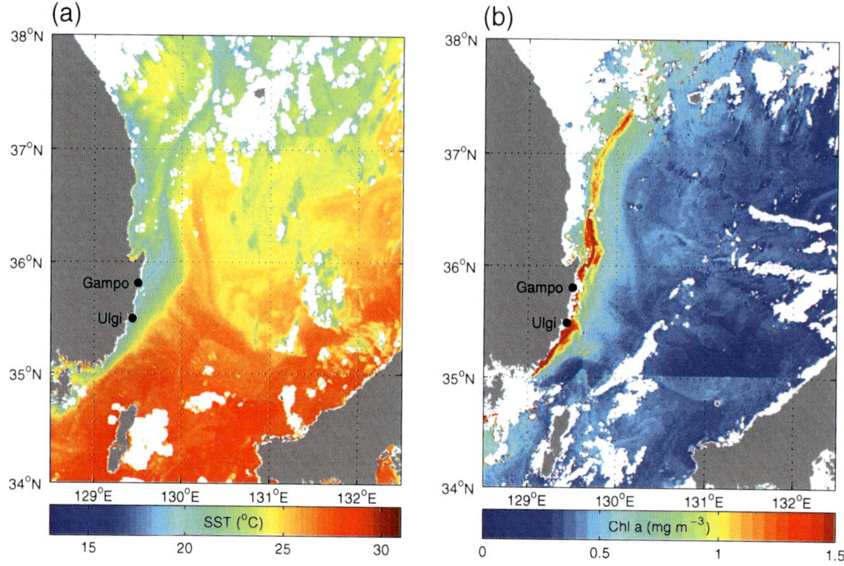

Fig. 4.3 Satellite images of **a** sea surface temperature from NOAA on 20 July 2013 and **b** chlorophyll *a* on 21 July 2013. Surface cold waters along the east coast of Korea including the Gampo-Ulgi area due to coastal upwelling can be seen. Coastal high chlorophyll water off Gampo-Ulgi area can also be seen together with offshore advection of cold and high-chlorophyll coastal water due to the East Korea Warm Current

of Korea occurs due to the combined effects of the background current (EKWC) and upwelling-favorable southerly winds (An 1974; Seung 1974; Lee 1983; Byun and Seung 1984; Lee and Na 1985). The source of cold waters upwelled to the sea surface can be either the HSTWW or the NKCW depending on the intensity of the upwelling (Kim and Kim 1983; Lee and Na 1985; Byun 1989; Lee et al. 1998).

The summertime southerly wind has been shown to be correlated with SST (Lee 1983; Lee et al. 2003b). Surface cold waters, however, have been frequently observed in a localized region, the Gampo-Ulgi area (GU area) (Fig. 4.3), although the southerly wind prevails along the entire east coast of Korea in summer. In the GU area, 80 % of southerly wind events caused the SST to drop more than 1.0 °C (Lee et al. 1998), and the SST decrease resulting from an upwelling-favorable wind occurs in 3–18 h (Lee et al. 2003b). Other factors responsible for localized upwelling in the GU area include steeply sloping isopycnals towards the coast due to an approach of the EKWC, local diverging topography, separation of the EKWC near Gampo, and the degree of stratification (Byun and Seung 1984; Lee and Na 1985; Lee et al. 1998, 2003b). Mean SST drops in the GU area during summer upwelling events are about 3.0–4.0 °C (Lee 1983; Lee et al. 1998) with a maximum SST drop of 12.2 °C in 1978 (Lee 1983) and an even bigger drop of 14.0 °C in 1997 at Gampo (Lee et al. 2003b). When strong winds (3–6 m s^{-1}) persisted for at least 3 days, upwelling events occurred, and the events persisted for 3–33 days (Lee 1983). Byun (1989) applied a 3-layer model of upwelling (Csanady 1982) and showed that wind impulse, the time-integrated wind stress, during upwelling events in 1982 and 1983 ranged from 10 to 40 m^2 s^{-1}, which corresponds to a wind blowing for 5–20 days with speed of 4 m s^{-1}.

In eastern boundary upwelling regions, significant correlations are often found between an upwelling-favorable wind and other variables than SST, such as coastal sea level and alongshore current (e.g. Smith 1974). On the other hand, coastal sea level is poorly correlated with wind in the GU area (Lee et al. 2003b) indicating other factors affecting the subtidal sea level variability (e.g. Lyu et al. 2002; Cho et al. 2014). Subtidal current fluctuations representing EKWC variation were also shown to be only marginally correlated with local wind (Lee et al. 2003b). Coastal upwelling in the GU area also occurred in spring (Byun and Seung 1984; Hyun et al. 2009).

The GU upwelling area often forms a semi-circle of cold water, with a radius of about 10–20 km from the coast to the upwelling front (Byun 1989). Satellite-derived SST images also showed that upwelled cold water in the GU area protrudes to the northeast (Byun 1989; Lee et al. 1998).

Seung (1984), using a simple 2-layer numerical model, suggested several possible upwelling centers, other than the GU area, along the east coast of Korea, due to coastal geometry. The satellite image in Fig. 4.3 clearly shows that coastal cold waters due to upwelling are also found in other coastal areas, which has been poorly documented. A report was given of an observation of the appearance of unusually cold surface water in the mid-east coast of Korea, north of the GU area, caused by the most intense southerly winds for the past 10 years resulting from a large-scale atmospheric condition (Park and Kim 2010).

Coastal upwelling at the east coast of Korea has important local and basin-scale consequences in terms of biogeochemical aspects (see Chap. 10). Locally, the GU upwelling area in summer is characterized by high primary production (Han et al. 1998), high chlorophyll fluorescene (Lee and Kim 2003), high zooplankton abundance (Lee et al. 2004), relatively low detritus percentage, high values of vertically integrated planktonic carbon, high percentage of particulate organic carbon and nitrogen in total suspended matter, and low ratio of particulate organic carbon to chlorophyll *a* (Yang et al. 1998) compared to those in offshore regions and in other seasons. The upwelled water in the area also has high dissolved oxygen content, low pH, and high alkalinity (Lee and Kim 2003). The oxygen rich upwelled water observed in 2001 suggests its origin from the NKCW rather than either the Japan Sea Proper Water or the TWW, at least in 2001 (Lee and Kim 2003).

Coastal waters with high productivity due to coastal upwelling are drawn offshore by large-scale circulation set by the EKWC and/or the mesosale circulation of the Ulleung Warm Eddy (UWE), resulting in enhanced biological productivity in the UB (Onitsuka et al. 2007; Hyun et al. 2009; Yoo and Park 2009) (Fig. 4.3). Without the coastal upwelling, the nutrient-depleted EKWC acted to decrease chlorophyll concentration in the middle of the UB (Son et al. 2014). Hence, the coastal upwelling together with surface circulation would be an important factor in making the UB the most productive basin among the three basins of the East Sea (Yamada et al. 2005; Kwak et al. 2013).

Consecutive hydrographic data support the highly-variable nature of the coastal upwelling (Byun 1989; Lee and Kim 2003). On the other hand, time series data to reveal the multiple scale variability is scarce, partly due to heavy fishing activities in the coastal areas. More observational efforts are required to delineate the spatial and temporal structure of the upwelling and to unravel the physics behind the coastal upwelling. Also biogeochemical and ecosystem responses to coastal upwelling have not yet been fully addressed, although many studies have pointed out the importance of the coastal upwelling in maintaining high primary productivity in the UB. Even a qualitative description of coastal upwelling in coastal regions along the east coast of Korea other than the GU area is lacking.

4.3 Mesoscale Eddies

4.3.1 Characteristics of Eddies

The East Sea is one of the most eddy-rich regions of the world ocean (Ichiye and Takano 1988). In particular, many warm and cold eddies are found near the UB (Lie et al. 1995; Lee and Niiler 2005), and eddies in this region can be classified into three groups (Lee and Niiler 2010b): (1) coastal eddies along the Korean coast, (2) frontal eddies along meanders of the inertial boundary current, (3) UWE and Dok Cold Eddy (DCE). The UWE and DCE are clustered mostly in the negative and positive relative vorticity areas, respectively (Fig. 4.4). The UWE has the largest size (126 × 194 km) and longest mean life (171 days). Although the mean propagation direction of the UWE is southeastward, it sometimes moves northward along the Korean coast due to an oscillating EKWC (or subpolar front) meander and leaves the UB to enter the northern East Sea (Shin et al. 2005). The UWE originates from the EKWC and forms from a southward flow related to anticyclonic circulation around Ulleungdo (Tanioka 1968; Mitchell et al. 2005b). Arruda et al. (2004) argued that the UWE may not be related to instabilities but instead may be due to the effects of the meridional gradient of planetary vorticity (β) and nonlinearities. Temporal evolution of the UWE is presented in Sect. 4.3.2. When the EKWC is weak and linear, there is no eddy formation, in agreement with this observation.

The DCE is located south of Dokdo throughout much of the year (Fig. 4.5). The DCE is generated through an instability process, followed by a retreat of the subpolar front, when a large-amplitude meander pinches off between Ulleungdo and Dokdo (Mitchell et al. 2005a). The DCE is typically about 60 km in diameter and its center was observed between 130.5°E and 132°E from June 1999 to July 2001. After formation from the subpolar front, the DCE stays southwest of Dokdo for 1–6 months before propagating westward toward Korea where it merges with the EKWC, or is engulfed in a southward meander of the subpolar front. When the

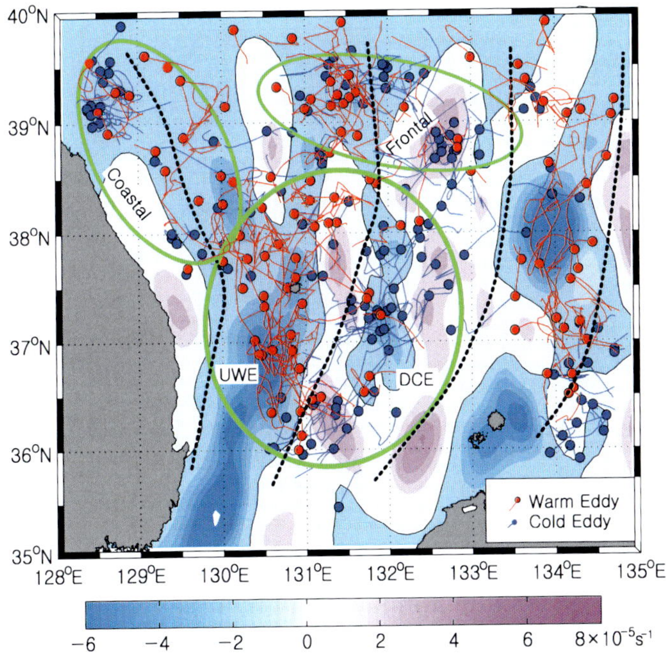

Fig. 4.4 Distribution and propagation of eddies plotted on a colored background of mean relative vorticity (*see color bar*). Four meridional bands of mean relative vorticity and three groups of eddies are demarcated by *dotted lines* and *green lines*, respectively (from Lee and Niiler 2010b)

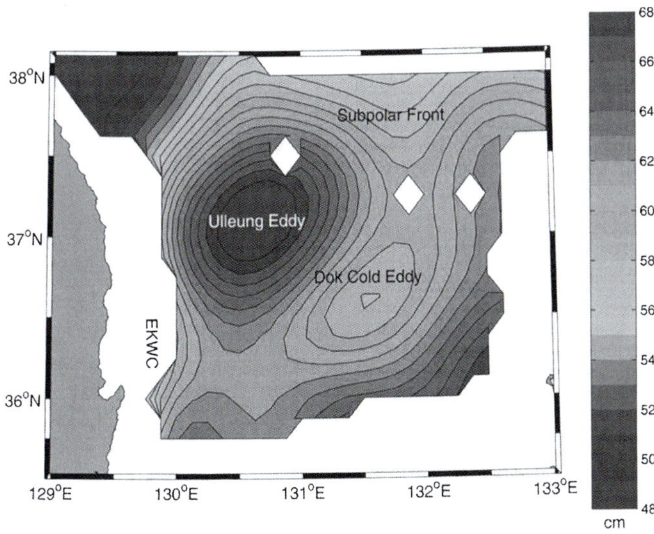

Fig. 4.5 Average dynamic height over a 2-yr deployment in the Ulleung Basin. The mean positions of the EKWC, Ulleung (Warm) Eddy, Subpolar Front, and the Dok Cold Eddy are labeled. The *white diamonds*, from *left* to *right*, are Ulleungdo (Ulleung Island), Dokdo (Dok Island), and a seamount that reaches within 500 m of the surface (from Mitchell et al. 2005a)

DCE approaches the Korean coast, the EKWC diverts eastward around it, suggesting that DCEs play an important role in the path of the EKWC.

Two to four warm eddies always exist in the YB (Isoda 1994; Morimoto et al. 2000) (Fig. 4.6). Warm eddies are generated in spring around the Oki Islands due to the flow over the bottom topography around the Oki Bank. After separating from the Oki Bank, the warm eddies move eastward along the Japanese coast or offshore from the Japanese coast. While they are moving, the warm eddies sometimes coalesce with neighboring warm eddies or split into multiple eddies. The warm eddies move continuously eastward from the Noto Peninsula in winter-spring with mean speeds of 0.5–2 cm s^{-1}, reaching the northwestern coast of Japan in the following winter. In there they typically split into two or three small eddies, and decay within a few months.

The presence of mesoscale eddies and their migration in the upper layer results not only in strong upper layer currents with associated low-frequency variability at a fixed location (Kim et al. 2009a) but also deep current fluctuations below 1000 m (Takematsu et al. 1999b) and excitation of topographic Rossby waves (Kim et al. 2013). Anticyclonic mesoscale eddies, like the UWE in the East Sea, also interact with near-inertial waves funneling near-inertial motions to the deep layer below the eddies (Park and Watts 2005) and reflecting near-inertial waves within the thermostad of the eddy (Byun et al. 2010) (see Chap. 5.4).

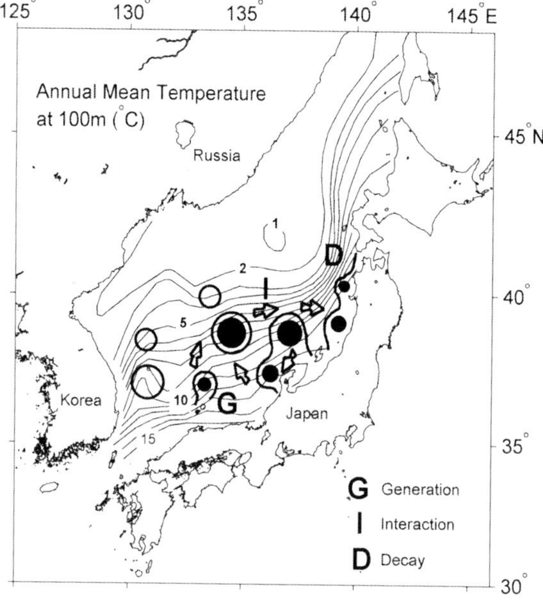

Fig. 4.6 Schematic illustration of the movement process of warm eddies in the southeastern East Sea. *Thin contour lines* indicate the horizontal distribution of annual mean temperature at 100 m depth (Redrawn from Isoda 1994)

4.3.2 Evolution of the Ulleung Warm Eddy

The horizontal distribution of temperature and salinity at a depth of 200 m is well suited for identifying warm eddies in the southwestern East Sea (Shin et al. 2005). In April 1993 the UWE (WE92) had a size of approximately 150 km east-west and 170 km north-south at 200-m depth (Fig. 4.7). The surface homogeneous layer of 10 °C (Ten Degrees Water, see Sect. 3.3.1) and 34.2 psu inside the UWE was produced by vertical convection from sea-surface cooling in winter, and deepened to a maximum of about 250 m in early spring. The homogeneous water displays positive oxygen and negative salinity anomalies relative to the surrounding thermocline water, indicative of formation from winter mixed layer water (Gordon et al. 2002). The deep homogeneous layer is a characteristic of a warm eddy that winters in the East Sea (Shin et al. 1995) and in the Kuroshio Extension (Tomosada 1986; Yasuda et al. 1992). From April to September, WE92 shrunk about 20 km in the east-west and about 80 km in the north-south directions (Fig. 4.7). The thickness of the homogeneous layer was also reduced to depths between 70 and 150 m. When the EKWC and the warm eddy converged around Ulleungdo, the EKWC

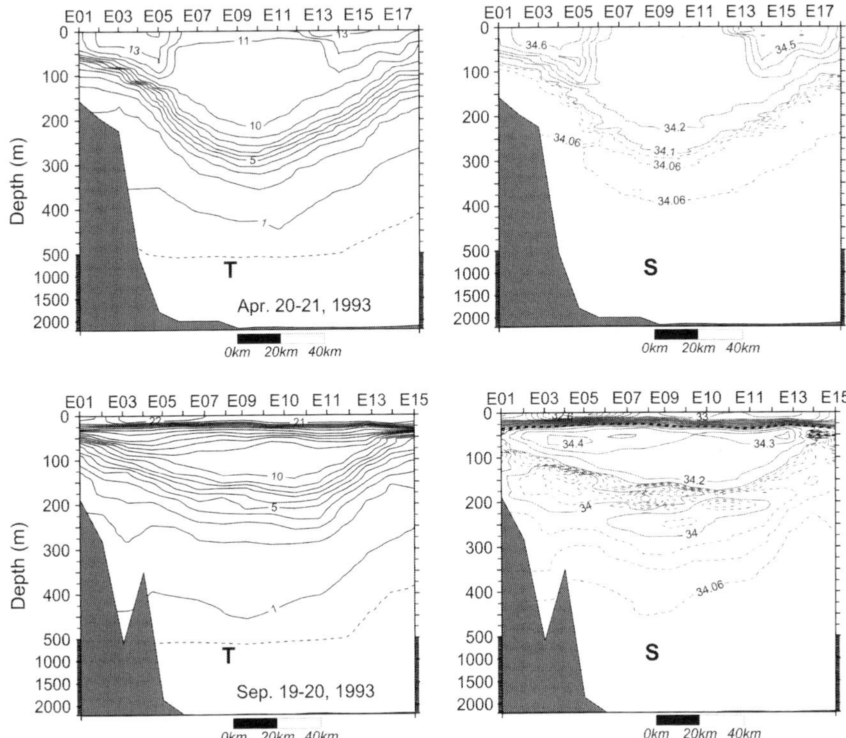

Fig. 4.7 Temperature (°C) and salinity (psu) vertical cross-sections in (*upper*) April and (*lower*) September 1993 (Redrawn from Shin et al. 2005)

supplied water of high temperature and salinity to the UWE; the EKWC flowed around the edge of the UWE in spring (Fig. 4.7) and gradually expanded into the inner portion during summer and fall.

Eddies are important to biological productivity, because of their association with upwelling of subsurface water. One of the mechanisms responsible for nutrient input into the euphotic zone is uplifting of the nutricline by cyclonic eddies, known as "eddy pumping," which results in enhanced primary production inside a cyclonic eddy (Falkowski et al. 1991). High primary productivity in the East Sea during late spring and early fall appears to be sustained by the interaction between eddies and wind, as well as other factors such as coastal upwelling and terrestrial nutrient input during the rainy season (e.g. Hyun et al. 2009).

4.4 Thermohaline Circulation

4.4.1 Rates of Water Mass Formation

The strength of the ocean's THC is often represented by a zonally integrated meridional streamfunction (MOC, meridional overturning circulation). The MOC can be calculated from full-depth hydrographic observations along latitude lines (Talley et al. 2003), numerical model results (Balmaseda et al. 2007), and direct observations (McCarthy et al. 2012). In the East Sea, it is difficult to calculate the MOC using hydrographic data, because the properties of thick deep water masses are almost homogeneous with little spatial density contrast. According to the meridional streamfunction based on numerical model results driven by climatological forcing, there exists a closed streamfunction similar to the Atlantic MOC although the East Sea's inflow-outflow system complicates the upper streamfunction field (Yoshikawa et al. 1999). Model-derived maximum East Sea MOC transport is about 0.6 Sv ($1\ Sv = 10^6\ m^3\ s^{-1}$), and below 1500 m the maximum value is about 0.15 Sv.

The water mass formation rate may set the strength of the THC (Kuhlbrodt et al. 2007). The formation rate of the Upper Portion of the Japan Sea Proper Water (UPJSPW) lying in the 400–1000 m depth range was estimated to be 0.48 and 0.38 Sv, respectively, based on hydrographic data (Senjyu and Sudo 1996) and numerical model results forced by climatological forcing (Kawamura et al. 2007). Box models constrained by observed chemical tracers yielded two different total formation rates of subsurface water masses below 600 m prior to the 1960s: 0.65 Sv (Postlethwaite et al. 2005) and 1.41 Sv (Jenkins 2008). After the 1960s, both estimates decreased to 0.27 Sv, which is about a 60–80 % reduction of the formation rate for those water masses; this was ascribed to the slow-down of the East Sea THC (Gamo 1999). A study using a moving boundary box model (MBBM) also suggests a decline in Bottom Water formation in 2000 compared with that in 1960 (Kang et al. 2003). However, the total formation rate of East Sea Central, Deep, and Bottom Waters in 2000, 0.49 Sv, remained the same as that in 1960 in the MBBM, suggesting a change from bottom water formation to upper water formation without any significant reduction of the total formation rate of sub-surface water masses.

4.4.2 *Deep Currents and Circulation*

One distinctive characteristic of the East Sea is the prevalence of spatially homogeneous deep waters below the pycnocline. Density-based dynamic calculation is thus impractical and would give misleading values for the deep currents. Studies of subsurface circulation of the East Sea have benefited from tracer distributions, numerical modeling, moored current measurements since the late 1980s, and deployments of floats since 1995. Compilations of data from the floats and moored current measurements allow construction of mean circulation maps for the entire East Sea (Senjyu et al. 2005; Choi and Yoon 2010; Park and Kim 2013). Two-year long continuous moored current and bottom pressure measurements between 1999 and 2001 revealed details of the UB deep circulation (Teague et al. 2005). Schematic circulation maps suggested by previous studies are shown in Fig. 4.8.

The subsurface currents below the pycnocline are characterized by weak vertical shear and barotropic fluctuations (Takematsu et al. 1999a; Senjyu et al. 2005; Chang et al. 2009). A limited number of full-depth current measurements from about 40 m to near the seabed shows that while low-passed subsurface current fluctuations below the pycnocline are coherent, the fluctuations of upper currents above 200 m are poorly correlated with those below the pycnocline (Kim et al. 2009a, 2013).

4.4.2.1 Mean Currents

Figure 4.9 shows bin-averaged mean currents at 800 m, together with EKE, with bin size of 0.5° in both longitude and latitude based on a compilation of float data from 1999 to 2010 (Park and Kim 2013). Also Fig. 4.10 shows mean currents at 82 locations measured at the deepest depth level of each mooring (below 1000 m except at 4 locations: U1, U6, Y16, Y17) by current meter moorings between 1986 and 2007. The mooring periods mostly exceed 6 months except at 7 locations (U1, U6, U14, from YB15 to YB18). The deep current map shown in Fig. 4.10 has been updated from previous studies (Chang et al. 2002; Senjyu et al. 2005; Teague et al. 2005) by adding more data available from various sources (MSA Reports 1995–2007; Jang 2011; Baek et al. 2014; Lee and Chang 2014). The eddy/mean kinetic energy ratio is superimposed on the mean current vectors at 800 m in Fig. 4.9b, and denoted by colored circles at mooring locations in the UB in Fig. 4.10b.

The mean features of the deep circulation are a basin-scale cyclonic circulation in each of the three deep basins (Fig. 4.8a), with strong currents along the basins' peripheries and weaker flows in the basins' interiors (Figs. 4.9a, 4.10). Strong

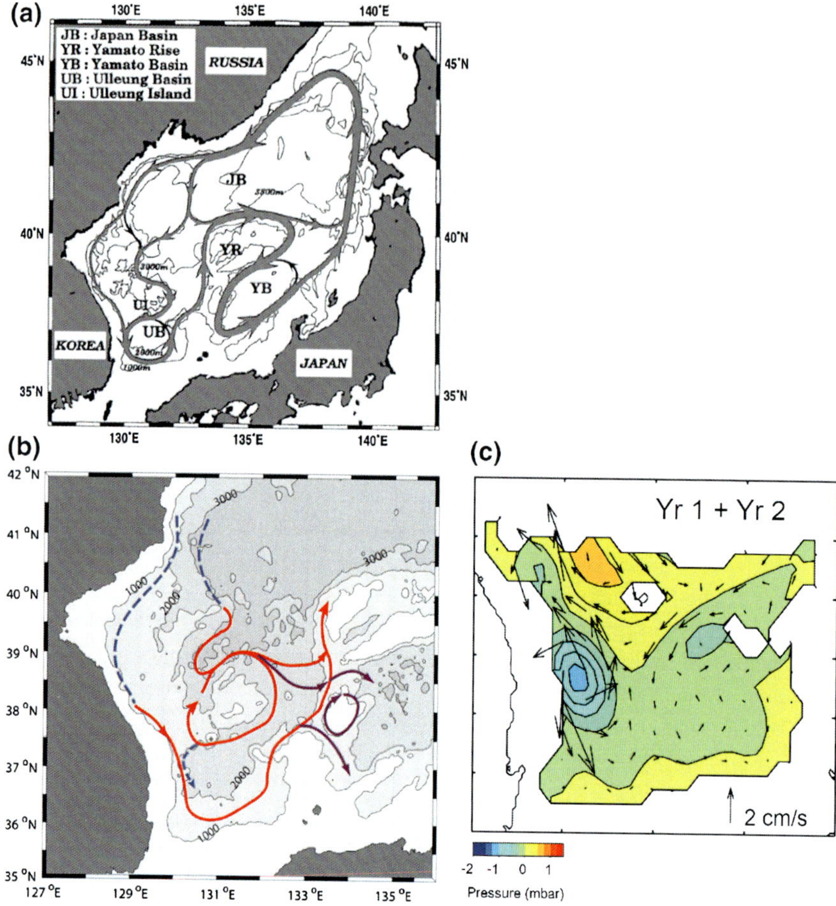

Fig. 4.8 Schematic deep circulation maps of **a** abyssal layer based on moored current measurements (from Senjyu et al. 2005), **b** 800 m layer based on trajectories of floats. *Red* and *violet lines* are drawn based on the float data, and *dotted lines* indicate the inferred continuation of subsurface currents (from Park et al. 2010), and **c** two-year average dynamic pressure field at 1000 dbar with corresponding geostrophic current vectors determined by bottom pressure and current measurements (from Teague et al. 2005)

cyclonic recirculation cells are found in the eastern JB and in the western YB (Choi and Yoon 2010; Park and Kim 2013). The deep circulation of the UB is complex with alternating cyclonic and anticyclonic sub-basin-scale cells (Fig. 4.8c) (Teague et al. 2005; Choi and Yoon 2010). Localized anticyclonic circulation at 800 m is found over the Korea Plateau (Park et al. 2004), seamounts over the Yamato Rise (Choi and Yoon 2010), and a seamount between Oki Bank and Yamato Rise (Choi and Yoon 2010) (Fig. 4.8b). Also observed is the flow at 800 m through a gap between the Oki Bank and the Yamato Rise, with mean volume transport of 1.83 Sv below 200 m according to numerical model results (violet solid arrows in

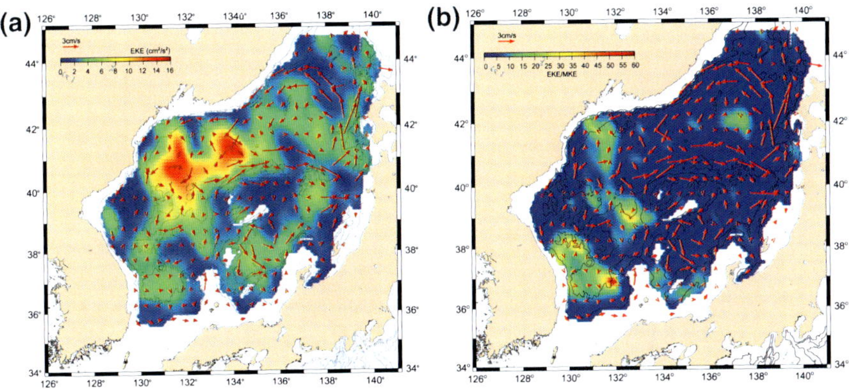

Fig. 4.9 $0.5° \times 0.5°$ bin-averaged currents at 800 m computed from float data compiled between 1999 and 2010. Superimposed (*color*) are **a** eddy kinetic energy and **b** ratio between eddy kinetic energy and mean kinetic energy, based on the same data (Redrawn from data provided by Dr. Jong-Jin Park)

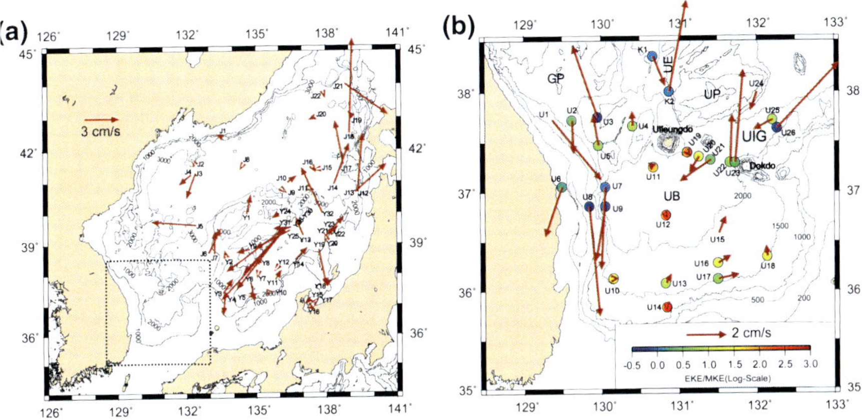

Fig. 4.10 **a** Mean abyssal currents based on long-term moored current measurements conducted in the Japan and Yamato basins between 1986 and 2007. **b** Expanded view of mean abyssal currents in the Ulleung Basin denoted by a *box* in (**a**). Ratio between eddy kinetic energy (EKE) and mean kinetic energy (MKE) in log scale is also shown as *colored dots* in (**b**). *GP, UB, UE, UIG, UP* denote Gangwon Plateau, Ulleung Basin, Ulleung Escarpment, Ulleung Interplain Gap, and Ulleung Plateau, respectively. For the detailed explanation of topographic features, see Sects. 1.1 and Chap. 16

Fig. 4.8b) (Park et al. 2010). It is argued that this throughflow in the gap is partly fed by the anticyclonic circulation over the Korea Plateau and an outflow from the UB (Park et al. 2010).

Space-time vector-averaged current in the entire East Sea based on float data at 800 m is about 2.5 cm s^{-1}, but mean current speeds in the strong recirculation cells reach 4–6 cm s^{-1} in the eastern JB and 7–8 cm s^{-1} in the western YB (Choi and Yoon 2010). The magnitude of deep currents at 800 m shows interbasin differences, strongest in the JB and weakest in the UB (Choi and Yoon 2010; Park and Kim 2013) with maximum volume transports below 800 m (assuming no velocity shear below 800 m) in the JB, YB, and UB of about 10, 2.0, and 0.4 Sv, respectively (Choi and Yoon 2010).

Mean current speeds at deeper depths from mooring data range from 0.1 to 8.2 cm s^{-1} (Fig. 4.10), and strong currents with speeds higher than 4.0 cm s^{-1} were observed on the northwestern slope region of the YB (from Y26 to Y29) and the western slope region of the UB (U7, U8) which constitute the western boundary currents of the cyclonic circulation. The strong current on the eastern boundary of the JB (J18) is part of a deep cyclonic recirculation cell, and strong currents following the southeastern slope of the Ulleung Interplain Gap (UIG) (U23, U26) are due to the Dokdo Abyssal Current (DAC) (Chang et al. 2009). The Korea Plateau splits into the Ulleung Plateau to the east and the Gangwon Plateau to the west with the deep (>2000 m) and narrow (~10 km) Usan Escarpment in between (see Fig. 1.2 for topographic features in the UB). Moored current measurements at this passage (K2) show a relatively strong mean northward flow of about 4.5 cm s^{-1} (Baek et al. 2014), indicating that part of the deep water entering the UIG from the western JB bifurcates into two branches, one branch flowing back to the JB through the Usan Escarpment and the other branch entering the UB. A weak southward abyssal flow observed at K1, north of mooring K2, is thought to be due to the cyclonic penetration of abyssal waters near the northern side of the Usan Escarpment, according to Hogan and Hurlburt (2000). Over the Gangwon Plateau, northward abyssal currents are observed which correspond to the western rim of the anticyclonic deep circulation over the entire Korea Plateau (Park et al. 2004; Teague et al. 2005). At U1 and U2, a short distance to the west from the northward currents over the Gangwon Plateau, abyssal currents in a depth range of 800–1200 m show an opposite direction, south or southeastward. The southward deep currents at U1 and U2 are thought to originate from the area farther north, merging with the deep cyclonic circulation in the UB to form the strong western boundary current observed at U7 and U8. A mean southward current is also observed at around 450 m at U6 (Jang 2011).

Analysis of float data indicates that the deep currents generally follow potential vorticity contours (f/H contours, where f is the local Coriolis parameter and H is the depth.) in the JB (Choi and Yoon 2010). On the other hand, the topography-following characteristics of deep currents are less clear in the UB (Choi and Yoon 2010) and floats at 800 m tend to move toward shallower depths (Park and Kim 2013).

Evidence of cyclonic penetration of deep waters from the JB into the UB through the UIG was provided by both float data and moored current measurements (Park et al. 2004; Teague et al. 2005; Choi and Yoon 2010). Deployment of an array of five mooring across the UIG revealed that the abyssal flow through the UIG shows a two-way circulation, weak, broad southwestward flows in the western UIG and strong, narrow northward flows of the DAC in the eastern UIG (Chang et al. 2009). The mean speed of the DAC at 2000 m is about 7.9 cm s^{-1} with a maximum recorded current speed of about 33.9 cm s^{-1}. Kim et al. (2013)

showed that the DAC extends from the bottom up to 300 m depth, while the upper currents, above 200 m, flow to the south or southwest opposite to the DAC. While the inflow transport towards the UB through the UIG below 1800 m is about 0.15 Sv, the net volume transport is negligibly small with high temporal variability (Chang et al. 2009). It suggests an insignificant upwelling of deep water across the 1800 m depth level. On the other hand, other transport estimates from 250 m to the bottom or from 800 m to the bottom, based on numerical model results and float data, yielded a net inflow transport of about 0.3–0.4 Sv (Teague et al. 2005; Choi and Yoon 2010). The inflow transports below 250 m based on the models are estimated to be 0.68 Sv (Teague et al. 2005) and 1.0 Sv (Park et al. 2010). The formation rate of the East Sea Deep Water below 1500 m determined by box models constrained by chemical tracers (Postlethwaite et al. 2005; Jenkins 2008) is about 4–5 times smaller than the directly observed transport below 1800 m in the UIG, suggesting a different generation mechanism of deep currents besides that of deep water-formation (see Sect. 4.5.4) and implying that the water mass formation rate may not represent the strength of the East Sea THC.

4.4.2.2 Temporal Variability

While the mean speeds of deep currents range 0–8 cm s^{-1}, temporal variation of these currents is large and instantaneous current speeds reach as high as 30–50 cm s^{-1} below 2000 m (Takematsu et al. 1999a; Chang et al. 2009). EKE provides insights into the variability of currents, although it is often difficult to identify the source of variability. EKE from full-depth moored current measurements in the UB from about 40 m depth down to 1500 m or deeper ranges from 1.5 to 440 cm^2 s^{-2}, and it sharply decreases at around 300 m depth near the permanent pycnocline (Kim et al. 2009a, 2013; Jang 2011). EKEs of deep currents below the pycnocline computed from mooring data at discrete locations range from 0.8 to 19.7 cm^2 s^{-2} with the ratio between EKE and mean kinetic energy (MKE) ranging from 0.5 to 497 in the UB (Chang et al. 2002, 2009; Teague et al. 2005; Kim et al. 2013). EKE of deep currents in the JB shows a similar range (Takematsu et al. 1999a). In the UB, low EKE/MKE ratios occur in regions of strong mean currents such as the DAC (U25, U23), on the western continental slope (from U7 to U9), over the Korea Plateau where the strong anticyclonic circulation exists (U3), and in the Ulleung Escarpment (K2) (Fig. 4.10b).

While the surface EKE based on surface drifters is low in the JB, except in a narrow band along the Primorye coast, EKE computed from the float data within 0.5° bins at 800 m exhibits the highest EKE (~15 cm^2 s^{-2}) in the western JB (Fig. 4.9a) where a strong dipole structure of wind-stress curl in winter is located (Park and Kim 2013). Apart from the western JB, EKEs are generally high (~5 cm^2 s^{-2}) in the interior of the UB and YB, and the eastern JB where mean currents are weak. The EKE/MKE ratio from the float data shows a high ratio in the UB and along the boundary of the dipole wind-stress curl in winter (Fig. 4.9b).

Moored current measurements in the East Sea show a distinct seasonal variation of deep currents in the JB (Takematsu et al. 1999a) with maximum velocities (~4.4 cm s^{-1}) in March and April and a minimum in November (~2.3 cm s^{-1}) (Choi and Yoon 2010). The estimation of monthly mean velocities at 800 m using the float data from the JB also shows a clear seasonal signal with a maximum (~3.5 cm s^{-1}) in March and a minimum in November (~2.3 cm s^{-1}) (Choi and Yoon 2010). On the other hand, mooring data indicate that the seasonal variability of deep currents is not obvious in the UB and YB, showing double current speed maxima, one in January and the other in July/August (Chang et al. 2002; Choi and Yoon 2010). The monthly mean velocities from float data at 800 m in the UB and YB also show reduced annual amplitude with poorly defined seasonal variation in the UB and YB (Choi and Yoon 2010).

The integral time scale of the measured deep currents ranges from 3 to 30 days, indicating the dominant time scales of deep current fluctuations. Intraseasonal variability of deep currents has been documented in the UB (Chang et al. 2002, 2009; Kim et al. 2013) with a period range of 10–60 days, and also in the YB with a shorter timescale of 2–10 days (Senjyu et al. 2005). Kim et al. (2013) showed that the variability of the DAC in the eastern UIG on timescales of 10–20 days is consistent with the propagation of topographic Rossby waves associated with warm events in the upper layer resulting from eddying processes and/or meandering of the TWC.

4.5 Dynamical Aspects

4.5.1 How Is the Tsushima Warm Current Driven?

The question of how the TWC is driven has long been an interesting theoretical problem. There are some studies showing that volume transport of the TWC is related to sea level difference between the interior and exterior of the East Sea (Toba et al. 1982; Ohshima 1994; Tsujino et al. 2008). On larger spatial and temporal scales, the TWC can be thought of as linked essentially to the wind-driven ocean circulation in the North Pacific, outside the East Sea. This idea was first explored by Minato and Kimura (1980). They considered a barotropic, linear dissipative, wind-driven ocean connected on the west to a shallow marginal sea through narrow inflow and outflow channels, and successfully showed that the transport through the marginal sea is driven by the large-scale wind field. However, the resulting volume transport was greatly overestimated because of many simplifications. The most serious of these may have been the neglect of the large barrier effect arising in the shallow inflow and outflow channels. In a similar way, Nof (1993) considered the same problem as one of geostrophic adjustment of buoyant Kuroshio water injected through a narrow gap into the marginal sea filled with deep, motionless denser water. Later, he also considered various other models applicable to the TWC (Nof 2000): a one-and-a half layer numerical model (Bleck and Smith 1990), Godfrey's island rule (Godfrey 1989), and Nof's (1993) β-controlled formula. All these models, except for the island rule, are non-linear and neglect the barrier effect. They all give a transport of about 10 Sv, much larger than the measured value of about 2.5 Sv (Teague et al. 2002; Takikawa et al. 2005; Fukudome et al. 2010). These studies indicate that non-linear effects are not important in controlling the volume transport through the marginal sea. Instead, the overestimation of volume transport may be induced by the neglect of the barrier effect arising in shallow channels, as it is in the study of Minato and Kimura (1980). Indeed, Seung (2003) has obtained a transport of about 2.5 Sv by applying the island rule, taking a bottom friction into account with a reasonable bottom friction coefficient. Hence, he parameterizes the barrier effect simply as bottom friction. According to Yang et al. (2013), the barrier effect includes the form drag associated with bottom pressure torque, generation of cross-isobathic flow by horizontal friction, and northward shifting of the forcing region affecting the through flow. All these facts indicate that the TWC is essentially driven by the oceanic wind field, and its volume transport is largely controlled by the barrier effect arising in the inflow and outflow channels. In fact, it can be understood that the downstream sea level drops occurring in the inflow and outflow channels are consequences of the barrier effect. Zheng et al. (2010), however, suggest a different view that the TWC becomes stronger and flows closer to the Japanese coast as the Kuroshio becomes stronger, based on a numerical experiment.

The volume transport of the TWC undergoes a seasonal variation with maximum in summer-fall and minimum in winter (Teague et al. 2002; Takikawa et al. 2005; Fukudome et al. 2010). An interesting feature about it is that the seasonal variation of outflow volume transport appears to be larger in the Soya Strait than in the Tsugaru Strait (Nishida et al. 2003; Fukamachi et al. 2008). This problem is addressed by Seung et al. (2012) with a simple linear barotropic model. As in the steady problem, they use the island rule with bottom friction taken into account. In the steady problem, only one outflow channel with only one island is considered. The island rule with multiple islands and multiple frictional channels gives somewhat different results from the case with a single island and two channels. They show that the volume transport through a channel is determined not only by the circulation around the adjacent two islands created by the oceanic wind field off those islands, which is the case if there is no bottom friction, but also by those around the neighboring islands further away. Note that the latter does not occur if only one island (two channels) is considered, even in the presence of bottom friction. Hence, the volume transport through each channel (strait) of the East Sea is determined by a linear combination of two components of circulation: one is the circulation around Honshu/Kyushu created by the oceanic wind field east of Honshu/Kyushu, and the other is the circulation around Hokkaido created by the oceanic wind field east of Hokkaido. For example, the transport through the Tsugaru Strait (Q_{23}) is the sum of two components $Q_{23,12}$ and $Q_{23,34}$ associated, respectively, with the wind fields off Honshu/Kyushu and Hokkaido (Fig. 4.11). In the same manner, the transport through the Soya Strait (Q_{45}) is the sum of the

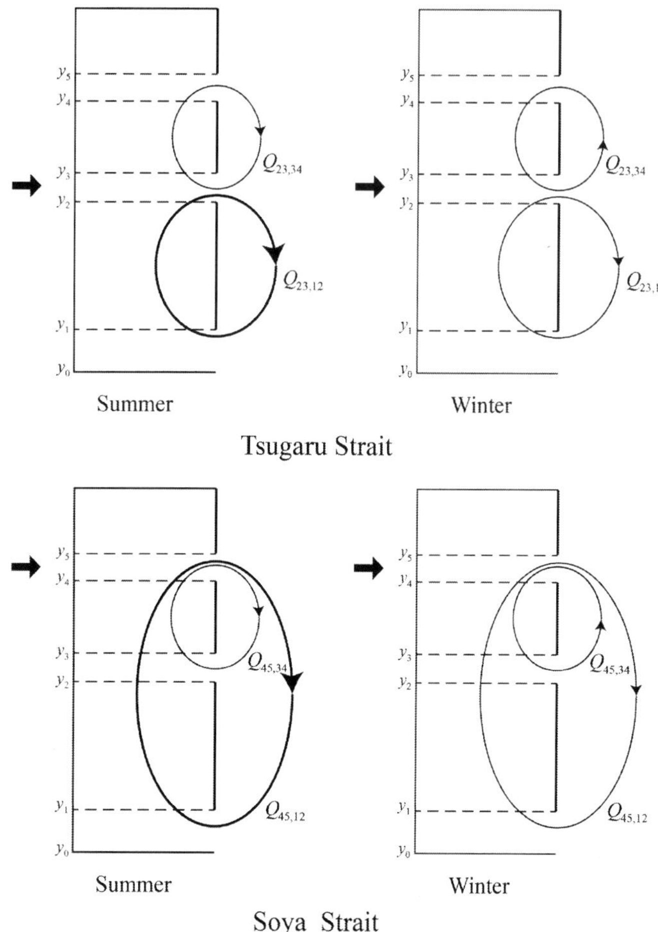

Fig. 4.11 Formation of volume transports through the Tsugaru and Soya Straits (*thick horizontal arrows*) in summer and winter by the circulations (*arrowed circles*) induced by wind fields off Honshu/Kyushu ($Q_{23,12}$ for Tsugaru Strait and $Q_{45,12}$ for Soya Strait) and Hokkaido ($Q_{23,34}$ for Tsugaru Strait and $Q_{45,34}$ for Soya Strait). The *arrowed circles* $Q_{23,12}$ and $Q_{45,12}$ are thicker in summer than in winter because they are larger in summer than in winter (from Seung et al. 2012)

two components $Q_{45,12}$ and $Q_{45,34}$ (Fig. 4.11). According to the model, the different behavior in seasonal variation of volume transports between the Tsugaru and Soya Straits is due essentially to the fact that the wind-stress curl off Hokkaido changes its sign from negative in summer to positive in winter, the latter fact being associated with the large seasonal fluctuation of zero-stress-curl latitude east of Hokkaido.

4.5.2 How Are the Branches of the Tsushima Warm Current Formed?

As described in Sect. 4.1, the TWC has traditionally been believed to split into three branches after entering the East Sea: the Nearshore Branch, the Offshore Branch, and the EKWC. In a series of numerical experiments, Yoon (1982a, b) has found that the EKWC is formed by western intensification and the Nearshore Branch is constrained by shallow bottom depth to flow along the Japanese coast. Later, Kawabe (1982) has shown that the Offshore Branch is induced by the coastally-trapped waves associated with the seasonal variation of TWC volume transport. It is already noted in Sect. 4.1 that the Nearshore and Offshore Branches are not usually identifiable separately and instead there seems to be only one broad current especially notable in summer, referred to as the Eastern Branch (Fig. 4.1). The hypothesis that the EKWC is a western boundary current is problematic in explaining formation of the Eastern Branch and seasonal variability of the EKWC which is strongest in summer and weakest in winter (cf. Lee and Niiler 2010a). Suppose that the EKWC is a western boundary current; then, seasonal variability of the EKWC requires that Rossby waves cross the East Sea basin very quickly, in a time shorter than a few months, which is not possible. Also, the Eastern Branch cannot exist if any pressure disturbance on the eastern boundary propagates westward by Rossby waves and hence disappears. The hypothesis of topographic trapping is possible, but it will be limited to a very narrow Japanese coastal region where the isobaths are discontinuous at some locations. The questions that follow are about how the EKWC and the Eastern Branch are formed and how their seasonal variabilities arise.

Recently, Spall (2002) has proposed an idea that the Eastern Branch is induced by the buoyancy difference between TWW and the other water in the East Sea, although he does not consider the seasonal variability. The buoyancy of the TWW is carried along the Japanese coast relatively quickly by Kelvin waves. It is trapped along the coast by the effect of local thermal damping, and would otherwise be propagated westward by Rossby waves. Hence the Eastern Branch can be present in both deep offshore and shallow coastal areas. This idea may be extended further in explaining the seasonal variations of the EKWC and the Eastern Branch if we take the local buoyancy forcing into consideration as an additional factor, as shown by Seung and Kim (2011). In the latter model, the East Sea basin is subject to seasonally-varying local buoyancy forcing, with larger buoyancy in summer than in winter, while the western and eastern boundaries are respectively affected by less buoyant water originating from the north and more buoyant water originating from the south. Hence, this model indicates that the seasonal variation of the EKWC and the Eastern Branch is induced by both local and remote buoyancy forcing.

It can be argued that the EKWC is driven by the inertia of the TWC incident upon the East Sea basin with seasonally-varying strength. In order to demonstrate this possibility, Seung (2005a) has considered an incident inertial jet initially in contact with the bottom and injecting into a deep basin by crossing a depth discontinuity. The grounded portion changes its flow direction being trapped over the shallow bottom while the ungrounded portion continues to flow by virtue of the inertial effect, hence resulting in the branching. Arruda et al. (2004) also share the same view in explaining the formation of the UWE. Eddies are usually expected to migrate due to the β-effect and to be advected by mean currents. Hence, the quasi-stationary characteristics of the UWE had long been an enigma. According to Arruda et al. (2004), the UWE owes its presence to the force balance in the area surrounding it. The stationary UWE exerts a southward force, referred to as β-induced force, because of latitudinal difference of the Coriolis force arising from the flow within the UWE. This β-induced force is balanced by the northward inertial force exerted by the TWC injecting into the East Sea basin. Hence this study presumes that the EKWC results from the inertia of the injecting TWC.

There are some other studies suggesting that the branching is locally triggered. Cho and Kim (2000) have proposed an analytical model suggesting that branching is caused by the creation of negative vorticity induced by intrusion of the Korea Strait Bottom Cold Water (KSBCW) along the western side of the Korea Strait in summer (see Chap. 3). Ou (2001) has shown that two current axes, one along the coast and the other along the outcrop of the buoyant layer, are generated when a single-axis current enters a deep basin after passing through a shallow channel subject to bottom friction. Both observations (Teague et al. 2002) and fine-resolution numerical models (e.g. Kim 2007) show that the branching begins just south of Tsushima Island, indicating that the branching mechanism is more or less similar to that of a jet split by an obstacle. An interesting question that follows is how far such locally-triggered branching is maintained under the effect of Earth's rotation. In conclusion, the dynamics associated with formation of the TWC branch currents are not yet definitely clear.

4.5.3 What Is the Role of Local Forcing?

The East Sea is strongly affected by local forcing, both of wind and buoyancy (Hirose et al. 1996; Yoon and Kawamura 2002). Strong northwesterly winds prevail in winter. At some locations, such as off Vladivostok, winds are orographically funneled becoming stronger than elsewhere (see Chap. 2). Such strong outbreaks of cold and dry continental air cause intense surface cooling that can lead to deep water formation (Seung and Yoon 1995a), as well as small-scale intense wind-stress curl. In the warm-water region, the behavior of the surface currents carrying TWW depends mostly on the volume transport through the inflow opening, whereas in the cold-water region, the current forms a closed cyclonic circulation, as observed by Isobe and Isoda (1997), with its strength depending on the magnitude of local forcing, as numerically demonstrated by Hogan and Hulbert (2000). Indeed, a simple linear analytical model proposed by Seung (1992) seems to be revealing. In the presence of inflow and outflow alone, western intensification predominates and most of the basin looks like the warm-water region. In the presence of local forcing alone, the basin-scale cyclonic gyre predominates and the whole basin looks like the cold-water region. This is because the local wind stress curl is generally positive and local surface cooling induces a vortex stretching of the upper layer. In the presence of both effects, the inflow-outflow and the local forcing, there appears a front separating the warm-water and cold-water regions in such a way that the two effects compete with each other. The relative importance of one effect compared to the other determines the separation latitude and the position of the front. In this simple model, however, the Eastern Branch does not appear because both bottom topography and buoyancy difference between TWW and the basin water are ignored. Additionally, it is questionable whether the local wind forcing in the cold-water region can induce any circulation intruding beneath the surface currents through the mechanism of ventilation proposed by Luyten et al. (1983). This problem has been addressed by Seung (1997) with the conclusion that the wind-forced circulation indeed pushes the lateral boundary of the surface currents southward, in agreement with the results described above, but hardly intrudes beneath the surface currents. Nevertheless, this study does not consider bottom topography, which may be an important factor in determining the circulation underlying the surface currents.

4.5.4 *How Is the Deep Layer Circulation Driven?*

As described in Sect. 4.3, the deep circulation flows cyclonically along the isobaths, often shows oscillatory and eddy-like features, is stronger in winter than in summer in the JB, and occasionally becomes stronger after an event of deep water formation. Holloway et al. (1995) have attempted to explain the formation of deep currents in terms of the eddy-topography interaction, the so-called Neptune effect. However, the results depend strongly on the parameterization coefficient of the interaction. On the other hand, Hogan and Hulbert (2000) have suspected that baroclinic instability is a candidate mechanism for deep current generation, although it was not explicitly shown to be so in their numerical experiments. In these explanations, the energy of deep currents is assumed to come ultimately from disturbances generated in the upper layer, and any effect of buoyancy forcing, such as deep water formation, is ignored. Recent numerical experiments (e.g. Kim 2007) with horizontal resolution comparable to, but vertical resolution more refined than, that of Hogan and Hulbert (2000) give nearly the same pattern of deep currents as that observed. However, the numerically obtained deep currents appear much weaker than those observed. It seems that numerical models are not yet complete enough to reproduce either the process of downward momentum transfer or realistic deep water formation. Based on the fact that most geostrophic contours are closed on themselves in the deeper part of the East Sea and that even weak forcing can generate strong abyssal currents over the closed geostrophic contours (Rhines and Young 1982), Seung (2005b, 2012) has discussed the possible roles of wind and deep water formation in generating deep currents. In winter, vertical stratification becomes very weak in some areas by intense surface cooling. If these areas, though not large, are over the closed geostrophic contours, the surface wind may be effective in generating deep currents (Seung 2005b). Deep water formation is equivalent to a volume source and can give rise to a source-driven circulation in the deep layer. If this happens in an area of sloping bottom, the pressure disturbances associated with the volume source propagate rapidly as topographical waves along the isobaths, resulting in deep currents (Seung 2012). However, it is known that the intensity of deep water formation has been decreasing in recent decades (e.g. Gamo et al. 2001; Kim et al. 2004a), implying that the resulting convection is not deep enough, while the deep currents are still present. Yoshikawa (2012) has addressed this problem with an idealized numerical model. He considers a three-layer ocean with sloping bottom, and supposes that winter convection reaches only to the intermediate layer. In the bottom layers, there still occurs a cyclonic circulation along the isobaths, created by the interaction between eddies generated during the process of water mass conversion and bottom topography. The strength of the circulation is found to depend on the volume of water mass conversion. This study indicates that deep water formation is the most probable candidate for the generation of deep circulation, although the detailed mechanism of it is not yet completely clear. Recently, Park et al. (2013) suggest that geothermal heating, being about twice that of a typical abyssal plain, is yet another possible candidate driving force for the deep circulation.

4.6 Numerical Modeling Studies of Circulation

4.6.1 Numerical Simulations

Early versions of realistic numerical models were configured on the Arakawa B-grid which gives better performance when the horizontal resolution is coarse. Seung and Kim (1993) applied Cox's (1984) model to simulate the circulation in the East Sea. Its horizontal grid size was 0.2° in both latitude and longitude and the number of vertical grid cells was 23, which was a high resolution model for that time. Surface temperature and salinity were restored by observed values from the Japan Oceanographic Data Center. The model was forced by surface wind stress from Na et al. (1992) and by constant inflow and outflow. At the open boundary, salinity and temperature were nudged by observed values from the National Fishery Research and Development Institute of Korea. Typical features of the circulation in the East Sea, such as the Nearshore Branch, EKWC, NKCC and so on, were reproduced in this model.

Seung and Kim (1997) applied the Miami Isopycnic Coordinate Ocean Model (MICOM; Bleck et al. 1992) to estimate the time scale of renewal of the upper intermediate water in the East Sea. Though their model has low horizontal resolution of 0.5° and only 4 vertical layers, it is noteworthy in applying an isopycnic coordinate model for the first time with realistic meteorological forcing and open boundary conditions. They investigated the formation of intermediate and deep waters and could estimate that the renewal timescales for upper intermediate water and for the whole water mass are about 10 and 81.4 years, respectively. From their numerical study, the renewal timescale for the whole water mass is comparable to that of Kim and Kim (1996) based on their moving box model and observations.

Kim and Seung (1999) also applied the MICOM but increased the horizontal resolution to 0.2° and vertical layers to 10 to investigate the formation and movement of East Sea Intermediate Water (ESIW). They resolved realistic surface currents in the East Sea such as the TWC and the separation of the EKWC. In particular, they reproduced the formation of ESIW by winter convection off Vladivostok and its southward extension along the Korean coast into the UB.

With the Geophysical Fluid Dynamics Laboratory Modular Ocean Model Version 2 (GFDL-MOM2) z-coordinate model, Kim and Yoon (1999) set up a relatively high resolution numerical model with horizontal resolution of 1/6° and 19 vertical levels forced by seasonally varying inflow and outflow, monthly mean meteorological wind stresses, Haney-type heat flux, and restored freshwater flux. They resolved well the surface and intermediate circulations in the East Sea in terms of seasonal variation as well as the mean features of the Nearshore Branch, EKWC, and NKCC. It is noteworthy that Kim and Seung (1999) and Kim and Yoon (1999) could reproduce the general circulation of the East Sea comparable to the observed one. Those models had higher resolution than the previous numerical models, and reproduced the branching of the TWC and suppressed separation of the EKWC to the south of 38°N, in agreement with observations. In particular, both models resolved intermediate waters in the East Sea such as the NKCC and ESIW in terms of their formation and extension.

Yoon and Kawamura (2002) applied the same model as that of Kim and Yoon (1999) except using the European Centre for Medium-range Weather Forecasts (ECMWF) wind forcing which resolves a more realistic dipole structure for the wind stress curl produced by the orographic effect off Vladivostok. Their model resolved the ESIW, UJSPW, and the High Salinity Intermediate Water although its salinity was lower and its temperature higher than the observed values. In particular, the model results suggested that UJSPW is formed off Vladivostok by deep convection in winter and is advected by the current near the base of the convection to intrude below the ESIW layer.

A model intercomparison experiment was conducted by Lee et al. (2003a) between RIAM Ocean Model (RIAMOM; Lee and Yoon 1994) and Geophysical Fluid Dynamics Laboratory Modular Ocean Model (GFDL MOM1, version 1.1). Their work was the first intercomparison study of East Sea circulation models, although only two models may not be enough to assess the models' performances. Basically, the two models have similar numeric schemes such as assuming hydrostatic and Boussinesq approximations, applying the Arakawa B-grid system, and so on, but RIAMOM has a free surface while MOM assumes a rigid-lid without a free surface. Both models have the same horizontal grid size of 1/6° and the same 19 uneven vertical levels and were forced by the monthly mean inflows and outflows, Haney-type heat flux, and monthly mean wind stresses from the re-analyzed ECMWF data. Although both models successfully reproduced the general circulation features of the East Sea, such as the Nearshore Branch, EKWC, and NKCC, they showed a tendency to overestimate (underestimate) temperature (salinity) below 50 m compared to the observed values of the Generalized Digital Environmental Model (GDEM) climatology. Nevertheless, it was shown that RIAMOM has better performance in producing a realistic East Sea flow field.

Coarse resolution models have had trouble reproducing the general circulation of the East Sea, particularly, the overshooting of the EKWC. To resolve this defect of the models, Holloway et al. (1995) introduced "topostress" to parameterize the eddy-topographic interaction in the East Sea which was poorly resolved in the coarse resolution models. They used the Modular Ocean Model (MOM, Pacanowski et al. 1991) which is very similar to that used by Seung and Kim (1993). Its horizontal resolution was 1/5° and the number of vertical levels was 25. By exploring sensitivity to an eddy parameter, the general circulation in the East Sea was improved, and in particular the overshoot of the EKWC was reduced or eliminated, bringing results close to the observations in the East Sea. It is interesting to compare their results with Hogan and Hurlburt's (2000) exploration of sensitivity to horizontal resolution, which will be introduced next.

With realistic configurations of the East Sea, a few experiments have been conducted on model sensitivities to horizontal resolution and mixing parameterization. Hogan and Hurlburt (2000) investigated numerical performance in reproducing the East Sea surface circulation, in relation to horizontal resolution and bottom topography by applying the Naval Research Laboratory's Layered Ocean Model (NLOM). In particular, it is noteworthy that Hogan and Hurlburt (2000) applied, for the first time, an Arakawa C-grid model which gives better performance in high horizontal resolution models (Haidvogel and Beckmann 1999). Hogan and Hurlburt (2000) demonstrated the sensitivity of the circulation in the East Sea to various horizontal grid sizes from 1/8° to 1/64°. In addition, the impacts of the bottom topography and surface wind forcing were discussed. From various simulations, they recommended that the high resolution model with at least 1/32° horizontal grid size is required to resolve the realistic separation of the EKWC from the Korean coast and to generate sufficient baroclinic instability to produce eddy-driven cyclonic deep mean flows. Their results also suggest that the separation latitude of the EKWC depends on the spatial structure of the wind stress curl and bottom topography as well as the model's horizontal grid size. In particular, the problems of EKWC overshooting and too-weak deep circulation, which are commonly noted in previous numerical studies in the East Sea, were discussed in this study. The sensitivity to the surface wind stress forcing was presented by Hogan and Hurlburt (2005) in more detail. Hogan and Hurlburt (2006) applied Hybrid Coordinate Ocean Model (HYCOM) with a 1/25° horizontal grid and 12 layers to reproduce the intra-thermocline eddies reported by Gordon et al. (2002) in the East Sea and suggested three mechanisms for the formation of those eddies.

Lee et al. (2011) conducted a sensitivity experiment to study tidal effects on intermediate waters in the East Sea. They applied the RIAMOM, improving the

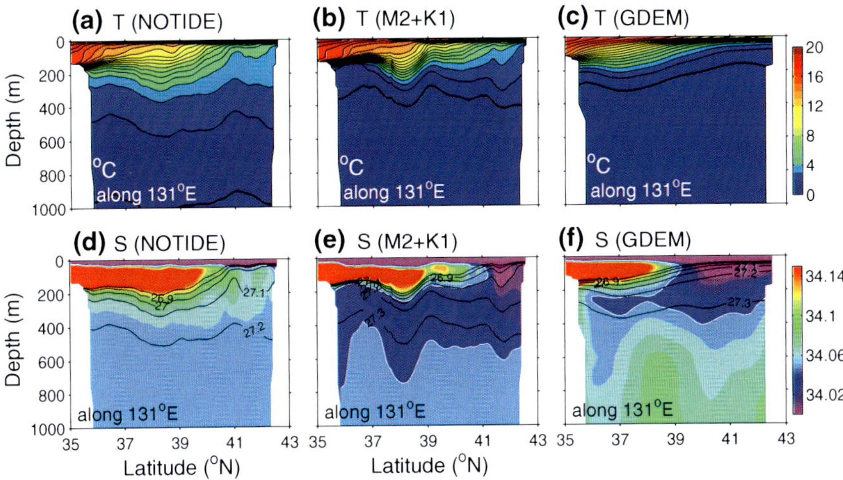

Fig. 4.12 The meridional section of potential temperature along 131°E for **a** without tide, **b** with M_2 and K_1 tides, and **c** the GDEM climatology. Contour interval is 1.0 °C, and thick contours indicate 1.0 °C. **d**–**f** corresponding salinities at the same section. Contour interval is 0.01, and *white contours* indicate S = 34.05 psu. The *solid contours* indicate isopycnal layers from $\sigma_\theta = 26.8$ kg m^{-3} to $\sigma_\theta = 27.3$ kg m^{-3} with an interval of 0.1 kg m^{-3} (from Lee et al. 2011)

horizontal grid resolution to 1/12° and increasing the number of vertical levels to 50, compared to 1/6° and 19 levels used by Lee et al. (2003a). Their study showed that direct inclusion of tide-induced processes significantly improves the representation of intermediate water mass properties and circulation in the East Sea. Figure 4.12 shows that the case with tides reproduces well the main ESIW features on the meridional section along 131°E. In addition, the domain-averaged temperature and salinity biases in their model were reduced considerably and the wind-driven southwestward flow along the coastline in the upper layer was enhanced, because of tidal mixing and internal tide processes. Their results represent the first approach to a discussion of tidal effects on the East Sea circulation.

Recently, numerical models in the East Sea have advanced to produce results comparable to observations. Luneva and Clayson (2006) and Clayson et al. (2008) simulated deep convection in the East Sea during the winters of 1999–2000 and 2000–2001 with an eddy-resolving model having high horizontal resolution. They used the University of Colorado version of the Princeton Ocean Model (POM; Kantha and Clayson 2000) with 6 km horizontal resolution and 38 depth levels (11 sigma levels in the upper 100 m and 27 z-levels at the shelf break and in the deep sea). They showed realistic surface and deep circulations, comparable with the observed surface and deep circulation reported by Lee and Niiler (2005), Takematsu et al. (1999a), Senjyu et al. (2005), and Teague et al. (2005). In particular, these studies attempted to explain the observed characteristics of the deep East Sea circulation by numerical modeling. Luneva and Clayson (2006) analyzed the effect of seasonal surface flux variability on the abyssal current and a possible mechanism for a weakening of the bottom drag coefficient due to turbulent mixing during the winter ventilation of the East Sea. Using the same model, Clayson et al. (2008) suggested that a vertical coupling between the upper and lower circulations can enhance deeper mixing through downwelling in a localized region.

Most numerical models, including those with realistic configurations, have assumed the hydrostatic approximation. To study the dynamics of the subpolar front in the East Sea, however, Yoshikawa et al. (2012) adopted a non-hydrostatic numerical model. Since the non-hydrostatic model computes vertical motion dynamically, it can represent the effect of wind stress and surface cooling on ageostrophic vertical motion and frontal subduction at the frontal region. Though the model employed a simple configuration with a rectangular domain and flat bottom, its results reproduced vertical motion by wind-generated internal gravity waves and subduction by surface cooling, which were confirmed by comparing with observational results (Lee et al. 2006)

4.6.2 Data Assimilation and Forecasting Systems

Over the last decade, interest in ocean forecasting systems has grown with increase in computing power as well as improving ocean observing systems such as satellites, Argo, CREAMS programs, and so on (see Chap. 1). Ocean data assimilation has been regarded as an essential technique in developing an ocean forecast system, since it initializes the ocean model with a sufficient amount of available data, thereby improving the forecasting performance. Ocean data assimilation also provides diagnostic tools for understanding oceanic and climatic conditions.

Initially, ocean data assimilation was applied to the East Sea by introducing nudging terms to the 3 dimensional temperature and salinity fields to reproduce realistic circulations and allow investigation of intermediate water formation and circulation processes (Seung and Yoon 1995b; Yoshikawa et al. 1999).

Hirose et al. (1999) applied the approximate Kalman Filter (Fukumori and Malanotte-Rizzoli 1995) to assimilate TOPEX/Poseidon altimeter data into a 1.5-layer, reduced gravity model. The formulation of the model is the same as that of Kim and Yoon (1996). The horizontal grid resolution is 1/6°. Though it is not a 3-dimensional model, it was a first attempt to apply a data assimilation technique. It was shown that their model was improved by data assimilation method, especially south of the subpolar front. The approximate Kalman Filter was also applied to the RIAMOM by Hirose et al. (2007) to develop an operational forecasting system for the East Sea. The model configuration is similar to Lee et al.'s (2011) eddy-resolving model but the number of vertical levels is 36. The model is forced by realistic in/outflow and daily meteorological forcing from the ECMWF. Temperature and salinity at the boundary are taken from the results of You and Yoon's (2004) 1/6° Pacific Ocean model while the velocity is diagnosed from the density field by the thermal wind relationship under the constraint of total volume transport in the Korea Strait measured by a ferryboat ADCP (Takikawa et al. 2005). Satellite measurements of SST and sea surface height (SSH) were assimilated into the model while surface salinity was restored to the climatological monthly mean. It was shown that their data assimilation system improves subsurface mesoscale variability as well as surface circulation in the East Sea. It is remarkable that averaged root-mean-square SST error between their best estimates and the radiometer data is only 1.2 °C.

Kim et al. (2009b) implemented a three-dimensional variational method (3D-Var; Weaver and Courtier 2001) to an East Sea circulation model. They adopted GFDL-MOM3 for their eddy-resolving circulation model of the East Sea. The longitudinal horizontal grid varied from 0.06° at the western boundary to 0.1° in the east of 130°E in order to resolve a realistic EKWC, but latitudinal grid spacing was fixed to 0.1°, and the number of vertical levels was 42. Surface heat flux was calculated by the bulk formula with realistic meteorological variables from ECMWF, while the surface salinity was restored to the observations from the World Ocean Atlas. For the velocity through the Korea Strait, the barotropic velocity was given by the volume transport monitored by a submarine cable (Kim et al. 2004b). Satellite measurements of SST and SSH anomaly were assimilated, together with temperature profiles, into the model. Through comparison between model results and observations, it was shown that the assimilation system enables the model to reproduce realistic mesoscale eddy variability such as the UWE and the DCE as well as the mean circulation of the East Sea (Fig. 4.13). In particular, seasonal variation of the NKCC was well reproduced: a strong southward NKCC in summer and its weakening in winter, consistent with results from hydrographical observations (Kim and Kim 1983).

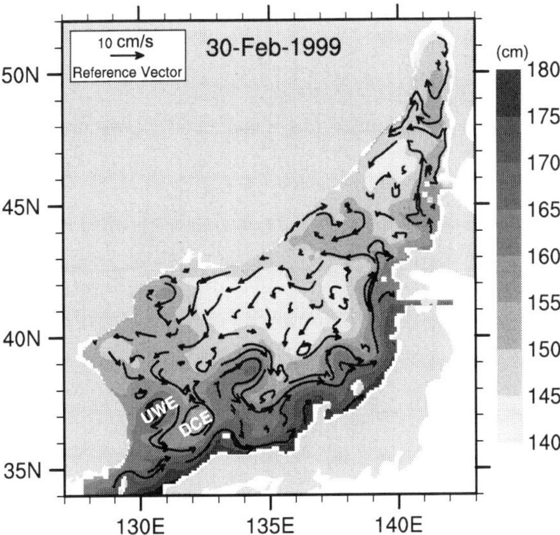

Fig. 4.13 Snapshot of sea surface height and currents on 24 February 1999 produced by a 3-dimensional variational data assimilation system. *UWE* and *DCE* respectively denote the Ulleung Warm Eddy and Dok Cold Eddy (Redrawn from Kim et al. 2009b)

Recently, numerical models of the East Sea, incorporating the data assimilation technique, have advanced to become operational ocean forecast systems. RIAMOM (Hirose et al. 2007) has already been applied to the East Sea operational ocean forecast system for the first time in the East Sea, which is available online (http://dreams-i.riam.kyushu-u.ac.jp/vwp/html/model_info.html).

4.7 Summary and Discussion

Various direct observations using satellite technology in the southern East Sea have revealed new near-surface circulation features. The EKWC and Nearshore Branch, after respectively leaving the coasts of Korea and Japan, join in the area east of the UB and form a broad mean current system named the East Sea Current (Lee and Niiler 2005). The East Sea Current, which exits the East Sea through the Tsugaru Strait, is well reproduced by numerical models (e.g. Clayson et al. 2008) and by an analytical solution in a simplified box model of the East Sea (Arruda et al. 2004). The Nearshore Branch, which is historically described as an alongshore current following the coast of Japan from the eastern channel of the Korea Strait to the Tsugaru Strait, is only observed in the summer season (May–August). The path of the flow passing through the western channel of the Korea Strait is found to be determined by the vorticity of the flow before entering the East Sea and branching of the current only occurs less than 15 % of the time. A reversal of the NKCC is also observed, and a southward alongshore NKCC is only observed during summer months (May–August). But these observations of NKCC are based on statistically insufficient drifter observations in the area due to deployment difficulties near North Korea.

It is found that eddies in the southern East Sea congregate in meridional bands determined by the sign of the mean flow vorticity, i.e. the anticyclonic and cyclonic eddies are born, grow, propagate and decay in the mean negative vorticity band and in the mean positive vorticity band, respectively. A relatively high number of anticyclonic eddies in the UB (negative vorticity zone) are found to have life spans longer than one year, being sustained by receiving kinetic energy from the meandering mean flow.

After its formation in the JB, deep water circulates cyclonically in the JB, and part of it penetrates cyclonically into the southern UB and YB through abyssal channels and gaps (Senjyu et al. 2005; Chang et al. 2009). Hence, the deep circulation of the East Sea is characterized by cyclonic circulation in each of the deep basins with strong currents of 4–8 cm s^{-1} along the perimeter of each basin and sluggish flows of 1–3 cm s^{-1} in the basins' interiors. Deep circulation in the UB is weakest and highly variable with alternating cyclonic and anticyclonic sub-basin-scale cells. Strong cyclonic recirculation cells with spatial scales of about 200 km exist in the eastern JB and the western YB. Deep currents in the JB show a distinct seasonal cycle with their respective maximum and minimum speeds in March and November. The seasonal variation of currents, however, is not obvious in the UB and YB. Instead, energetic temporal variability on timescales of 5–60 days is dominant.

The East Sea circulation is classified into two regimes: upper-layer circulation generated by bifurcation of the TWC entering the basin, and deep-layer circulation driven by local forcing and isolated in the basin. Most studies agree that the TWC is essentially driven by the oceanic wind field in the North Pacific, and its volume transport is largely controlled by the barrier effect arising at the inflow and outflow channels. The different behavior in seasonal variation of volume transports in the Tsugaru and Soya straits has been ascribed to sign changes of the wind stress curl off Hokkaido (it changes from negative in summer to positive in winter, due to the large seasonal fluctuation of the zero-stress-curl latitude east of Hokkaido), although there are not yet enough studies supporting this explanation. The formation mechanisms of the EKWC and the Eastern Branch, both being branch currents of the TWC, are diverse, as reviewed in Sect. 4.5.2. There is an indication that seasonal variation of the EKWC and Eastern Branch is due to the combined effect of local and remote buoyancy forcing, although this explanation does not find further supporting studies. There are many studies suggesting that the upper-layer circulation in the cold-water region is susceptible to local wind and buoyancy forcing, which are favorable for a basin-wide cyclonic gyre.

In the early phase of numerical studies in the East Sea, idealized simple models were introduced. Though those simple models did not simulate realistic circulations in the East Sea, they have provided valuable tools for understanding the dynamics of the East Sea circulation. With the advance of computational power, it has become possible to perform realistic numerical simulations with high resolution, real topography, meteorological forcing, and lateral boundary conditions. These realistic simulations have (1) given us pictures of what the circulation looks like in the entire East Sea, which is insufficiently revealed by the observations, (2) expanded our understanding of the extent to which numerical configurations, such as small-scale parameterization, bottom topography, and horizontal/vertical resolution, affect a model's ability to represent the East Sea circulation, (3) provided a tool to study dynamical processes in the East Sea circulation. As ocean observational data have increased abundantly with advancing observing technologies such as Argo floats, satellites, and so on, ocean data assimilation techniques have been applied to numerical models in the East Sea. The approximate Kalman Filter and 3 D-Var methods have been applied to circulation models of the East Sea, resulting in operational forecast systems for the East Sea.

Despite extensive efforts of direct current measurements by current meter moorings and tracked floats since the 1990s, as well as studies based on hydrographic properties and numerical modeling, the thermohaline circulation and associated subsurface circulation of the East Sea remains somewhat elusive.

Especially, following issues have been identified as important processes that need to be addressed through more observational efforts and analytical/numerical modeling.

- An overarching question is whether or not the East Sea MOC is slowing down, while studies using chemical tracers have reported a decline in the formation of deep waters (Gamo 1999; Jenkins 2008). This may be answered by combining numerical modeling and long-term choke-point monitoring of currents.
- The driving mechanism of the deep-layer circulation is one of the least known problems. Presently, deep-water formation seems to be a more probable candidate than surface wind. Overall, there are still a number of issues to be resolved in order to understand well the circulation dynamics of the East Sea.
- Cyclonic penetration of deep waters from the JB to the UB and YB is a major conduit for oceanic climate signals to travel southward to the UB and YB from their origin in the JB. The climate signals would appear in the JB first, so it is important to examine the variability of this circulation. It is unclear what causes the observed inter-basin difference of seasonal and intraseasonal variation of deep currents. Interannual and longer term variations have not been adequately documented.
- It is an open question why the deep circulation of the UB is weakest among three deep basins, while mesoscale eddy activities in the UB are vigorous.
- Also required is an improved understanding of the coupling of upper and deep circulations, and of the role of surface processes such as air-sea flux and mesoscale eddies on the deep circulation.

Acknowledgements DK Lee was supported by Basic Science Research Program through the National Research Foundation of Korea (NRF) funded by the Ministry of Education (NRF-2014R1A1A2054910). YB Kim was supported by KIOST project (PE99292). YH Kim was supported by KIOST project (PE99295). CW Shin was supported by KIOST project (PE98921). KI Chang was supported by Ministry of Oceans and Fisheries of Korea through EAST-I project.

References

An HS (1974) On the cold water mass around the southeast coast of Korean Peninsula. J Oceanol Soc Korea 9:10–18

Arruda WZ, Nof D, O'Brien JJ (2004) Does the Ulleung eddy owe its existence to β and nonlinearity? Deep-Sea Res Part I 51:2073–2090

Baek GN, Seo S, Lee JH et al (2014) Characteristics of the flow in the Usan Trough in the East Sea. J Korean Soc Oceanogr 19:99–108 (in Korean)

Balmaseda MA, Smith GC, Haines K et al (2007) Historical reconstruction of the Atlantic Meridional Overturning Circulation from the ECMWF operational ocean reanalysis. Geophys Res Lett 34:L23615. doi:10.1029/2007GL031645

Bleck R, Smith LT (1990) A wind-driven isopycnic coordinate model of the north and equatorial Atlantic Ocean 1. Model development and supporting experiments. J Phys Oceanogr 95:3273–3285

Bleck R, Rooth C, Hu D et al (1992) Salinity-driven thermocline transients in a wind- and thermohaline-forced isopycnic coordinate model of the North Atlantic. J Phys Oceanogr 22:1486–1505

Byun SK (1989) Sea surface cold water near the southeastern coast of Korea: wind effect. J Oceanol Soc Korea 24:121–131

Byun SK, Seung YH (1984) Description of current structure and coastal upwelling in the southwest Japan Sea-summer 1981 and spring 1982. In: Ichiye T (ed) Ocean hydrodynamics of the Japan and East China Seas. Elsevier Oceanography Series 39, Amsterdam, pp 83–93

Byun SS, Park JJ, Chang KI, Schmitt RW (2010) Observation of near-inertial wave reflections within the thermostad layer of an anticyclonic mesoscale eddy. Geophys Res Lett 37:L01606. doi:10.1029/2009GL041601

Chang KI, Hogg NG, Suk MS et al (2002) Mean flow and variability in the southwestern East Sea. Deep-Sea Res Part I 49:2261–2279

Chang KI, Kim K, Kim YB et al (2009) Deep flow and transport through the Ulleung Interplain Gap in the southwestern East/Japan Sea. Deep-Sea Res Part I 56:61–72

Cho YK, Kim K (2000) Branching mechanism of the Tsushima Current in the Korea Strait. J Phys Oceanogr 30:2788–2797

Cho CB, Nam SH, Chang KI (2014) Subtidal temperature variability in stratified shelf water off the mid-east coast of Korea. Cont Shelf Res 75:41–53

Choi YJ, Yoon JH (2010) Structure and seasonal variability of the deep mean circulation of the East Sea (Sea of Japan). J Oceanogr 66:349–361

Clayson CA, Luneva M, Cunningham P (2008) Downwelling and upwelling regimes: connections between surface fronts and abyssal circulation. Dyn Atmos Ocean 45:165–186

Cox MD (1984) A primitive equation, three-dimensional model of the ocean. Ocean Group, GFDL, Princeton, NJ, Technical Report No 1

Csanady GT (1982) On the structure of transient upwelling events. J Phys Oceanogr 12:397–419

Falkowski PG, Ziemann D, Kolber Z et al (1991) Role of eddy pumping in enhancing primary production in the ocean. Nature 353:55–58

Fukamachi Y, Tanaka I, Ohshima KI et al (2008) Volume transport of the Soya Warm Current revealed by bottom-mounted ADCP and ocean-radar measurement. J Oceanogr 46:85–392

Fukudome K, Yoon JH, Ostrovskii A et al (2010) Seasonal volume transport variation in the Tsushima Warm Current through the Tsushima Straits from 10 years of ADCP observations. J Oceanogr 66:539–551

Fukumori I, Malanotte-Rizzoli P (1995) An approximate Kalman filter for ocean data assimilation: an example with an idealized Gulf Stream model. J Geophys Res 100:6777–6793

Gamo T (1999) Global warming may have slowed down the deep conveyor belt of a marginal sea of the northwestern Pacific: Japan Sea. Geophys Res Lett 26:3137–3140

Gamo T, Momoshima N, Tolmachyov S (2001) Recent upward shift of deep convection system in the Japan Sea, as inferred from the geochemical tracers tritium, oxygen, and nutrients. Geophys Res Lett 28:4143–4146

Godfrey JS (1989) A Sverdrup model of the depth-integrated flow for the world ocean allowing for island circulations. Geophys Astrophys Fluid Dyn 45:89–112

Gordon A, Giulivi C, Lee C et al (2002) Japan/East Sea intrathermocline eddies. J Phys Oceanogr 32:1960–1974

Haidvogel D, Beckmann A (1999) Numerical ocean circulation modeling. Imperial College Press, Singapore

Han MS, Jang DH, Yang HS (1998) The ecosystem of the southern coastal waters of the East Sea, Korea. II. Promary productivity in and around cold water mass. J Korean Soc Oceanogr 33:196–204

Hirose N, Kim CH, Yoon JH (1996) Heat budget in the Japan Sea. J Oceanogr 55:217–235

Hirose N, Fukumori I, Yoon JH (1999) Assimilation of TOPEX/Poseidon altimeter data with a reduced gravity model of the Japan Sea. J Oceanogr 55:53–64

Hirose N, Kawamura H, Lee HJ et al (2007) Sequential forecasting of the surface and subsurface conditions in the Japan Sea. J Oceanogr 63:467–481

Hogan PJ, Hulbert HE (2000) Impact of upper ocean-topographical coupling and isopycnal outcropping in Japan/East Sea models with 1/8° to 1/64° resolution. J Phys Oceanogr 30:2535–2561

Hogan PJ, Hulbert HE (2005) Sensitivity of simulated circulation dynamics to the choice of surface wind forcing in the Japan/East Sea. Deep-Sea Res Part II 52:1464–1489

Hogan PJ, Hurlburt HE (2006) Why do intrathermocline eddies form in the Japan/East Sea? A modeling perspective. Oceanography 19:134–143

Holloway G, Sou T, Eby M (1995) Dynamics of circulation of the Japan Sea. J Mar Res 53:539–569

Hyun JH, Kim D, Shin CW et al (2009) Enhanced phytoplankton and bacterioplankton production coupled to coastal upwelling and an anticyclonic eddy in the Ulleung Basin, East Sea. Aquat Microb Ecol 54:45–54

Ichiye T, Takano K (1988) Mesoscale eddies in the Sea of Japan. La Mer 26:69–76

Isobe A, Isoda Y (1997) Circulation in the Japan Basin, the northern part of the Japan Sea. J Oceanogr 53:373–381

Isoda Y (1994) Warm eddy movements in the eastern Japan Sea. J Oceanogr 50:1–15

Jang JH (2011) Spatial and temporal structure of low-frequency currents off Uljin in the East Sea. Dissertation, Seoul National University

Jenkins WJ (2008) The biogeochemical consequences of changing ventilation in the Japan/East Sea. Mar Chem 108:137–147

Kang DJ, Park S, Kim YG et al (2003) A moving-boundary box model (MBBM) for oceans in change: An application to the East/Japan Sea. Geophys Res Lett 30:1299. doi:10.1029/200 2GL016486

Kantha LH, Clayson CA (2000) Numerical models of oceans and oceanic processes. International Geophysics Series, vol 66. Academic Press, New York

Katoh O, Teshima K, Kubota K et al (1996) Downstream transition of the Tsushima Current west of Kyushu in summer. J Oceanogr 52:93–108

Kawabe M (1982) Branching of the Tsushima Current in the Japan Sea. Part II: Numerical experiment. J Oceanogr Soc Japan 38:183–192

Kawamura H, Yoon JH, Ito T (2007) Formation rates of water masses in the Japan Sea. J Oceanogr 63:243–253

Kim YJ (2007) A study on the Japan/East Sea oceanic circulation using an ultra-high resolution model. Dissertation, Kyushu University

Kim CH, Kim K (1983) Characteristics and origin of the cold water mass along the east coast of Korea. J Oceanol Soc Korea 18:73–83 (in Korean)

Kim KR, Kim K (1996) What is happening in the East Sea (Japan Sea)? Recent chemical observations during CREAMS 93-96. J Korean Soc Oceanogr 31:164–172

Kim YH, Min HS (2008) Seasonal and interannual variability of the North Korean Cold Current in the East Sea reanalysis data. Ocean Polar Res 30:21–31

Kim KJ, Seung YH (1999) Formation and movement of the ESIW as modeled by MICOM. J Oceanogr 55:369–382

Kim CH, Yoon JH (1996) Modeling of the wind-driven circulation in the Japan Sea using a reduced-gravity model. J Oceanogr 52:359–373

Kim CH, Yoon JH (1999) A numerical modeling of the upper and the intermediate layer circulation in the Eat Sea. J Oceanogr 55:327–345

Kim K, Kim KR, Min DH et al (2001) Warming and structural changes in the East (Japan) Sea: a clue to future changes in global oceans? Geophys Res Lett 28:3293–3296

Kim K, Kim KR, Kim YG et al (2004a) Water masses and decadal variability in the East Sea (Sea of Japan). Prog Oceanogr 61:157–174

Kim K, Lyu SJ, Kim YG et al (2004b) Monitoring volume transport through measurement of cable voltage across the Korea Strait. J Atmos Ocean Tech 21:671–682

Kim YB, Chang KI, Kim K et al (2009a) Vertical structure of low frequency currents in the southwestern East Sea (Sea of Japan). J Oceanogr 65:259–271

Kim YH, Chang KI, Park J et al (2009b) Comparison between a reanalyzed product by the 3-dimensional variational assimilation technique and observations in the Ulleung Basin of the East/Japan Sea. J Mar Syst 78:249–264

Kim YB, Chang KI, Park JH et al (2013) Variability of the Dokdo Abyssal Current observed in the Ulleung Interplain Gap of the East/Japan Sea. Acta Oceanol Sin 32:12–23

Kuhlbrodt T, Griesel A, Montoya M et al (2007) On the driving processes of the Atlantic meridional overturning circulation. Rev Geophys 45:1–32

Kwak JH, Lee SH, Park HJ et al (2013) Monthly measured primary and new productivities in the Ulleung Basin as a biological "hot spot" in the East/Japan Sea. Biogeosciences 10:4405–4417

Lee JC (1983) Variations of sea level and sea surface temperature associated with wind-induced upwelling in the southeast coast of Korea in summer. J Oceanol Soc Korea 18:149–160

Lee JC, Chang KI (2014) Variability of the coastal current off Uljin in summer 2006. Ocean Polar Res 36:165–177 (in Korean)

Lee T, Kim IN (2003) Chemical imprints of the upwelled waters off the coast of the southern East Sea of Korea. J Korean Soc Oceanogr 38:101–110

Lee JC, Na JY (1985) Structure of upwelling off the southeast coast of Korea. J Oceanol Soc Korea 20:6–19

Lee DK, Niiler PP (2005) The energetic surface circulation patterns of the Japan/East Sea. Deep-Sea Res Part II 52:1547–1563

Lee DK, Niiler PP (2010a) Surface circulation in the southwestern Japan/East Sea as observed from drifters and sea surface height. Deep-Sea Res Part I 57:1222–1232

Lee DK, Niiler PP (2010b) Eddies in the southwestern East/Japan Sea. Deep-Sea Res Part I 57:1233–1242

Lee HC, Yoon JH (1994) On the free surface OGCM. In: Proceeding of fall meeting of Japan Oceanographical Society

Lee DK, Kwon JI, Hahn SB (1998) The wind effect on the cold water formation near Gampo-Ulgi coast. J Korean Fish Soc 31:359–371 (in Korean)

Lee HJ, Yoon JH, Kawamura H et al (2003a) Comparison of RIAMOM and MOM in modeling the East Sea/Japan Sea circulation. Ocean Polar Res 25:287–302

Lee JC, Kim DH, Kim JC (2003b) Observations of coastal upwelling at Ulsan in summer 1997. J Korean Soc Oceanogr 38:122–134

Lee CR, Park C, Moon CH (2004) Appearance of cold water and distribution of zooplankton off Ulsan-Gampo area, eastern coastal area of Korea. J Korean Soc Oceanogr 9:51–63 (in Korean)

Lee CM, Thomas LN, Yoshikawa Y (2006) Intermediate water formation at the Japan/East Sea subpolar front. Oceanography 19:110–121

Lee HJ, Park JH, Wimbush M et al (2011) Tidal effects on intermediate waters: a case study in the East/Japan Sea. J Phys Oceanogr 41:234–240

Lie HJ, Byun SK, Bang IK et al (1995) Physical structure of eddies in the southwestern East Sea. J Korean Soc Oceanogr 30:170–183

Luneva M, Clayson CA (2006) Connections between surface fluxes and deep circulations in the Sea of Japan. Geophys Res Lett 33:L24602. doi:10.1029/2006GL027350

Luyten JR, Pedlosky J, Stommel H (1983) The ventilated thermocline. J Phys Oceanogr 13:292–309

Lyu SJ, Kim K, Perkins HT (2002) Atmospheric pressure-forced subinertial variations in the transport through the Korea Strait. Geophys Res Lett 29(9):1294. doi:10.1029/2001GL014366

Martin S, Kawase M (1998) The southern flux of sea ice in the Tatarskiy Strait, Japan Sea and the generation of the Liman Current. J Mar Res 51:141–155

McCarthy G, Frajka-Williams E, Johns WE et al (2012) Observed annual variability of the Atlantic meridional overturning circulation at 26.5°N. Geophys Res Lett 39:L19609. doi:10.1029/2012GL052933

Minato S, Kimura R (1980) Volume transport of the western boundary current penetrating into a marginal sea. J Oceanogr Soc Japan 36:185–195

Mitchell DA, Teague WJ, Wimbush M et al (2005a) The Dok Cold Eddy. J Phys Oceanogr 35:273–288

Mitchell DA, Watts DR, Wimbush M et al (2005b) Upper circulation patterns in the Ulleung Basin. Deep-Sea Res Part II 52:1617–1638

Morimoto A, Yanagi T, Kaneko A (2000) Eddy field in the Japan Sea derived from satellite altimetric data. J Oceanogr 56:449–462

MSA Reports (1995–2007) Report of marine pollution surveys. Hydrographic Department, Maritime Safety Agency, Japan (in Japanese)

Na JY, Seo JW, Han SK (1992) Monthly-mean sea surface winds over the adjacent seas of the Korean Peninsula. J Oceanol Soc Korea 27:1–10

Nishida Y, Kanomata I, Tanaka I et al (2003) Seasonal and interannual variations of the volume transport through the Tsugaru Strait. Umi no Kenkyu 12:487–499 (in Japanese)

Nof D (1993) The penetration of Kuroshio water into the Sea of Japan. J Phys Oceanogr 23:797–807

Nof D (2000) Why much of the Atlantic circulation enters the Caribbean Sea and very little of the Pacific circulation enters the Sea of Japan. Prog Oceanogr 45:39–67

Ohshima KI (1994) The flow system in the Japan Sea caused by a sea level difference through shallow straits. J Geophys Res 99:9925–9940

Onitsuka G, Yanagi T, Yoon JH (2007) A numerical study on nutrient sources in the surface layer of the Japan Sea using a coupled physical-ecosystem model. J Geophys Res 112:C05042. doi:10.1029/2006JC003981

Ou HW (2001) A model of buoyant throughflow: with application to branching of the Tsushima Current. J Phys Oceanogr 31:115–126

Pacanowski RC, Dixon K, Rosati A (1991) The GFDL modular ocean model user guide. Geophysical Dynamics Laboratory, Princeton, USA, The GFDL Ocean Group, Technical Report No 2

Park YG (2007) The effects of Tsushima Warm Current on the interdecadal variability of the East/Japan Sea thermohaline circulation. Geophys Res Lett 34:L06609. doi:10.1029/200 6GL029210

Park KA, Kim KR (2010) Unprecedented coastal upwelling in the East/Japan Sea and linkage to long-term large-scale variations. Geophys Res Lett 37:L09603. doi:10.1029/2009GL042231

Park JJ, Kim K (2013) Deep currents obtained from ARGO float trajectories in the Japan/East Sea. Deep-Sea Res Part II 85:169–181

Park JH, Watts DR (2005) Near-inertial oscillations interacting with mesoscale circulation in the southwestern Japan/East Sea. Geophys Res Lett 32:L10611. doi:10.1029/2005GL022936

Park YG, Oh KH, Chang KI et al (2004) Intermediate level circulation of the southwestern part of the East/Japan Sea estimated from autonomous isobaric profiling floats. Geophys Res Lett 31:L13213. doi:10.1029/2004GL020424

Park KA, Kim K, Cornillon PC et al (2006) Relationship between satellite-observed cold water along the Primorye coast and sea ice in the East Sea (the Sea of Japan). Geophys Res Lett 33:L10602. doi:10.1029/2005GL025611

Park YG, Choi A, Kim YH et al (2010) Direct flows from the Ulleung Basin into the Yamato Basin in the East/Japan Sea. Deep-Sea Res Part I 57:731–738

Park YG, Park JH, Lee HJ et al (2013) The effects of geothermal heating on the East/Japan Sea circulation. J Geophys Res 118:1893–1905. doi:10.1002/jgrc.20161

Postlethwaite CF, Rohling EJ, Jenkins WJ et al (2005) A tracer study of ventilation in the Japan/ East Sea. Deep-Sea Res Part II 52:1684–1704

Rhines PB, Young WR (1982) A theory of the wind-driven circulation. J Mar Res 40:559–596

Senjyu T, Sudo H (1996) Interannual variation of the Upper Portion of the Japan Sea Proper Water and its probable cause. J Oceanogr 52:27–42

Senjyu T, Shin HR, Yoon JH et al (2005) Deep flow field in the Japan/East Sea as deduced from direct current measurements. Deep-Sea Res Part II 52:1726–1741

Seung YH (1974) A dynamic consideration on the temperature distribution in the east coast of Korea in August. J Oceanol Soc Korea 9(1–2):52–58 (in Korean)

Seung YH (1984) A numerical experiment of the effect of coastline geometry on the upwelling along the east coast of Korea. J Oceanol Soc Korea 19:24–30

Seung YH (1992) A simple model for separation of the East Korean Warm Current and formation of the North Korean Cold Current. J Oceanol Soc Korea 27:189–196

Seung YH (1997) Application of the ventilation theory to the East Sea. J Korean Soc Oceanogr 32:8–16

Seung YH (2003) Significance of shallow bottom friction in the dynamics of the Tsushima Current. J Oceanogr 59:113–118

Seung YH (2005a) Branching of the Tsushima Current by an abrupt increase of bottom depth. J Oceanogr 61:261–269

Seung YH (2005b) Abyssal currents driven by a local wind forcing through deep mixed layer: implication to the East Sea. Ocean Sci J 40:101–107

Seung YH (2012) Abyssal circulation driven by a periodic impulsive source in a small basin with steep bottom slope with implications to the East Sea. Ocean Polar Res 34:287–296

Seung YH, Kim K (1993) A numerical modeling of the East Sea circulation. J Oceanogr Soc Korea 28:292–304

Seung YH, Kim KJ (1997) Estimation of the residence time for renewal of the East Sea Intermediate Water using MICOM. J Korean Soc Oceanogr 32:17–27

Seung YH, Kim KJ (2011) Boundary currents in a meridional channel subject to seasonally varying buoyancy forcing: application to the Tsushima Current. J Oceanogr 67:563–575

Seung YH, Yoon JH (1995a) Some features of winter convection in the Japan Sea. J Oceanogr 51:61–73

Seung YH, Yoon JH (1995b) Robust diagnostic modeling of the Japan Sea circulation. J Oceanogr 51:421–440

Seung YH, Han SY, Lim EP (2012) Seasonal variation of volume transport through the straits of the East/Japan Sea viewed from the island rule. Ocean Polar Res 34:403–411

Shin HR, Byun SK, Kim C et al (1995) The characteristics of the structure of a warm eddy observed to the northwest of Ulleungdo in 1992. J Korean Soc Oceanogr 30:39–56 (in Korean)

Shin HR, Shin CW, Kim C et al (2005) Movement and structural variation of warm eddy WE92 for three years in the western East/Japan Sea. Deep-Sea Res Part II 52:1742–1762

Smith RL (1974) A description of current, wind and sea level variations during coastal upwelling off the Oregon coast. J Geophys Res 79:435–443

Son YT, Chang KI, Yoon ST et al (2014) A newly observed physical cause of the onset of the subsurface spring phytoplankton bloom in the southwestern East Sea/Sea of Japan. Biogeosciences 11:1–12

Spall MA (2002) Wind- and buoyancy-forced upper ocean circulation in two-strait marginal seas with application to the Japan/East Sea. J Geophys Res 107(C1). doi:10.1029/2001JC000966

Takematsu M, Nagano Z, Ostrovskii AG et al (1999a) Direct measurements of deep currents in the northern Japan Sea. J Oceanogr 55:207–216

Takematsu M, Ostrovskii AG, Nagano Z (1999b) Observations of eddies in the Japan Basin interior. J Oceanogr 55:237–246

Takikawa T, Yoon JH, Cho KD (2005) The Tsushima Warm Current through Tsushima Straits estimated from ferryboat ADCP data. J Phys Oceanogr 35:1154–1168

Talley L, Reid J, Robbins P (2003) Data-based meridional overturning streamfunctions for the global ocean. J Climate 16:3213–3224

Tanioka K (1968) On the East Korean Warm Current (Tosen Warm Current). Oceanogr Mag 20:31–38

Teague WJ, Jacobs GA, Hwang PA et al (2002) Low-frequency current observations in the Korea/Tsushima Strait. J Phys Oceanogr 32:621–1641

Teague WJ, Tracey KL, Watts DR et al (2005) Observed deep circulation in the Ulleung Basin. Deep-Sea Res Part II 52:1802–1826

Toba Y, Tomizawa K, Kurasawa Y et al (1982) Seasonal and year-to-year variability of the Tsushima-Tsugaru Warm Current system with its possible cause. La Mer 20:41–51

Tomosada A (1986) Generation and decay of Kuroshio warm-core rings. Deep-Sea Res 33:1475–1486

Tsujino H, Nakano H, Motoi T (2008) Mechanism of currents through the straits of the Japan Sea: Mean state and seasonal variation. J Oceanogr 64:141–161

Uda M (1934a) Oceanographic conditions of the Japan Sea and its adjacent waters. J Imp Fish Exp St 7:91–191 (in Japanese)

Uda M (1934b) The results of simultaneous oceanographical investigations in the Japan Sea and its adjacent waters in May and June, 1932. J Imp Fish Exp St 5:57–190 (in Japanese)

Weaver A, Courtier P (2001) Correlation modeling on the sphere using a generalizing diffusion equation. Q J Roy Meteor Soc 127:1815–1846

Yamada K, Ishizaka J, Nagata H (2005) Spatial and temporal variability of satellite estimated primary production in the Japan Sea from 1998 to 2002. J Oceanogr 61:857–869

Yang JY (1996) Current structure derived from satellite tracked drifter in the northern part of the East Sea. Dissertation, Seoul National University

Yang HS, Oh SJ, Lee HP et al (1998) Distribution of particulate organic matter in the Gampo upwelling area of the southwestern East Sea. J Korean Soc Oceanogr 33:157–167

Yang J, Lin X, Wu D (2013) Wind-driven exchanges between two basins: Some topographic and latitudinal effects. J Geophys Res 118:4585–4599. doi:10.1002/jgrc.20333

Yasuda I, Okuda K, Li J (1992) Evolution of a Kuroshio warm-core ring variability of the hydrographic structure. Deep-Sea Res 39(S1):S131–S161

Yoo S, Park J (2009) Why is the southwest the most productive region of the East Sea/Sea of Japan? J Mar Syst 78:301–315

Yoon JH (1982a) Numerical experiment on the circulation in the Japan Sea. Part I. Formation of the East Korean Warm Current. J Oceanogr Soc Japan 38:43–51

Yoon JH (1982b) Numerical experiment on the circulation in the Japan Sea. Part II. Influence of seasonal variations in atmospheric conditions on the Tsushima Current. J Oceanogr Soc Japan 38:81–94

Yoon JH, Kawamura H (2002) The formation and circulation of the Intermediate Water in the Japan Sea. J Oceanogr 58:197–211

Yoon JH, Abe K, Ogata T et al (2005) The effect of wind-stress curl on the Japan/East Sea circulation. Deep-Sea Res Part II 52:1827–1844

Yoshikawa Y (2012) An eddy-driven abyssal circulation in a bowl-shaped basin due to deep water formation. J Oceanogr 68:971–983

Yoshikawa Y, Awaji T, Akimoto K (1999) Formation and circulation processes of intermediate water in the Japan Sea. J Phys Oceanogr 29:1701–1722

Yoshikawa Y, Lee CM, Thomas LN (2012) The subpolar front of the Japan/East Sea. Part III: Competing roles of frontal dynamics and atmospheric forcing in driving ageostrophic vertical circulation and subduction. J Phys Oceangr 42:991–1011

You SH, Yoon JH (2004) Modelling of the Ryukyu Current along the Pacific side of the Ryukyu Islands. Pac Oceanogr 2:85–92

Zheng P, Wu D, Lin X et al (2010) Interannual variability of Kuroshio Current and its effect on the Nearshore Branch in Japan/East Sea. J Hydrodyn 22:305–311

Chapter 5
High-Frequency Variability: Basin-Scale Oscillations and Internal Waves/Tides

SungHyun Nam, Jae-Hun Park and Jong Jin Park

Abstract High-frequency variability captured by recent observations in the East Sea (Japan Sea) is summarized; it includes basin-scale oscillations, internal tides, near-inertial oscillations, and nonlinear internal waves. Observational and theoretical/numerical studies have been conducted on non-isostatic sea level response to atmospheric pressure forcing within the East Sea at periods of a few days to tens of days. Semidiurnal water temperature oscillations and their surface signatures are often markedly enhanced along and off the east coast of Korea due to propagation and refraction of internal tides generated at the Korea Strait shelf break; this has implications for the properties of water masses (e.g. intermediate water) and circulation (particularly off the Russian coast) in the East Sea, particularly from late summer to early winter. Research on heterogeneous and stochastic near-inertial oscillations in the open basin and coastal areas off Korea and Japan has shown a clear seasonal cycle in surface inertial motions, and interactions of near-inertial waves with mesoscale circulation, bathymetry, coastal trapped waves, and coastal/

S. Nam (✉)
Scripps Institution of Oceanography, University of California, San Diego,
CA 92093-0230, USA
e-mail: sunam@ucsd.edu; namsh6513@gmail.com; namsh@snu.ac.kr

S. Nam
School of Earth and Environmental Sciences, Seoul National University,
Seoul 151-742, Republic of Korea

J.-H. Park
Physical Oceanography Division, Korea Institute of Ocean Science and Technology,
Ansan 426-744, Republic of Korea
e-mail: jhpark@kiost.ac

J.-H. Park
Department of Ocean Sciences, Inha University, Incheon 402-751, Republic of Korea

J.J. Park
School of Earth System Sciences, Kyungpook National University, Daegu 702-701,
Republic of Korea
e-mail: jjpark@knu.ac.kr

© Springer International Publishing Switzerland 2016
K.-I. Chang et al. (eds.), *Oceanography of the East Sea (Japan Sea)*,
DOI 10.1007/978-3-319-22720-7_5

bottom boundaries. Observations of nonlinear internal waves and their packets at near-inertial and semidiurnal intervals, have suggested generation by both semidiurnal internal tides and wind-induced near-inertial waves. Investigation of high-frequency variability in the East Sea is far from closed, and the several results introduced in this chapter raise many questions rather than provided answers. A general presumption of stationary characteristics of such high-frequency variability under changing climate conditions needs to be tested; this requires long-term time series observations with high temporal sampling rates at relevant spatial resolutions.

Keywords High-frequency variability · Basin-scale oscillations · Internal tides · Near-inertial oscillations · Nonlinear internal waves · East Sea (Japan Sea)

5.1 Introduction

Over the past two decades, our understanding of high-frequency variability in the East Sea has significantly improved. This is largely due to advances in ocean observing technology, e.g. long-term or temporary surface and subsurface moorings with high sampling rates, numerous surface drifters and profiling floats, intensive measurements with pressure-sensor-equipped inverted echo sounders (PIESs), and better utilization of satellite radar altimetry and synthetic aperture radar (SAR) data (see Chap. 1). High-frequency processes, with periods of tens of days to a few days, diurnal and semidiurnal frequencies, and frequencies close to the local inertial and buoyancy frequencies, include (but are not limited to) basin-scale sea level or pressure oscillations, surface and internal tides, near-inertial (NI) oscillations, and nonlinear internal waves (NLIWs). In this chapter, we review past (but mostly recent) findings on some (though not all) of such high-frequency processes in the East Sea.

5.2 Basin-Scale Oscillations

Being connected to the Pacific Ocean through three narrow and relatively shallow (<150 m) straits (the Korea, Tsugaru, and Soya Straits, Fig. 5.1), the East Sea as a semi-enclosed deep (mean depth 1,700 m) basin has inherent basin-scale sea level oscillations that are primarily driven by high-frequency (periods generally less than 15 days) atmospheric pressure forcing. Water exchange between the East Sea and the North Pacific is hindered by the narrow straits, and sea level response within the basin to high-frequency atmospheric pressure forcing departs from the well-known inverted barometer (IB) effect, i.e. 1 hPa decrease in atmospheric pressure raises sea level by 1 cm with no phase lag. Such non-isostatic sea level responses within the basin drive subsurface pressure difference between the basin and the North Pacific, and are imprinted in the variability of volume transport (VT) across the straits, as in the Mediterranean Sea (Garrett 1983; Garrett and

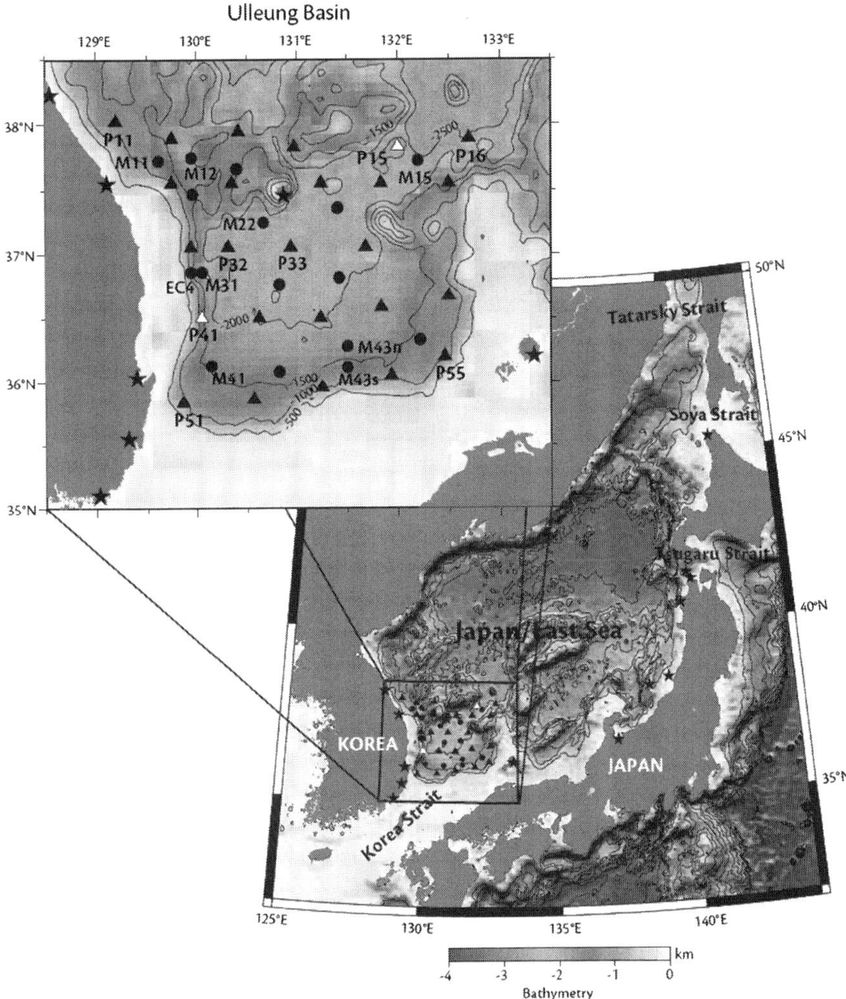

Fig. 5.1 The PIES (*black triangles*) and current meter (*black circles*) array in the East Sea. Coastal tide-gauge stations (*stars*) were used to estimate mean sea level. *White* (*open*) *triangles* indicate PIES sites where little or no data were obtained. Bathymetry data are from two-minute-resolution global sea floor topography (from Smith and Sandwell 1997) and are contoured at 500-m intervals (from Park et al. 2006)

Majaess 1984; Candela et al. 1989; Candela 1991; Le Traon and Gauzelin 1997). From records of submarine cable voltage difference between Busan, Korea and Hamada, Japan, Lyu et al. (2002) showed strong 3–5-day band transport variations through the Korea Strait. They calculated a Helmholtz-like resonance at a period of 3.12 days using a simple analytic model, and interpreted an abrupt drop of coherence between sea level and atmospheric pressure at periods of 3–5-days as the non-isostatic sea level response to atmospheric pressure forcing. Most VT variations through the straits, as well as sea level averaged over the basin or mean sea level (MSL) of the East Sea, could be reproduced using the model.

Observations of atmospheric and subsurface pressures, sea levels, and VT support non-isostatic sea level response to atmospheric pressure, as the analytic model suggests. Due to the resonance, atmospheric pressure variability dominantly drives changes in bottom pressure (BP) within the East Sea, flows (and VT) into or out of the East Sea and MSL, at periods of 2–10 days. Strong 2–10-day band fluctuations were observed in (1) BP in the southwestern East Sea ranging from 0.1 to 0.3 dbar (Nam et al. 2004a; Park and Watts 2005a; Park et al. 2006), (2) VT through the straits ranging from 1 to 3 Sv (Lyu and Kim 2005), and (3) MSL from both tide-gauge (TG) sea levels around the coast and satellite altimeter-derived (TOPEX/Poseidon; TP) sea surface height within the basin ranging from 10 to 30 cm (Lyu et al. 2002; Nam et al. 2004a; Lyu and Kim 2005).

It is particularly important to understand basin-scale oscillations within a semi-enclosed basin like the East Sea at such short periods, less than the 20–70 day resolvable periods from satellite altimeters, since they can produce substantial aliasing error in the altimetry data because of trackiness errors between crossing and neighboring tracks (Nam et al. 2004a; Park et al. 2006; Xu 2006; Xu et al. 2008). In this section, we review the results of observational studies on basin-scale high-frequency oscillations, and an analytic model with its application.

5.2.1 Observations

Time series of BP measured at 23 PIES sites within the East Sea (triangles in Fig. 5.1) are markedly correlated (correlation coefficients > 0.95) and nearly identical during the two-year experiment (June 1999–June 2001) (Park and Watts 2005a; Park et al. 2006). Although the BP measurements cover only the southwestern East Sea they could be used as a proxy for MSL all over the East Sea at periods longer than 2 days, due to the rapid adjustment of sea level disturbances (Nam et al. 2004a; Park and Watts 2005a; Park et al. 2006). The MSL estimate formed by averaging sea levels observed at coastal TG stations (21 sites including, but not limited to, the sites shown in Fig. 5.1; stars) shows high (~0.8) coherences and near-zero phases with the BP time series (Nam et al. 2004a) in spite of potential coastal effects on TG sea levels (Lyu and Kim 2005). Barotropic oscillations with periods longer than 2 days are uniform and in phase throughout the basin and account for the observed common mode among all the PIES pressure records (Park et al. 2006).

Both the BP common mode in the southwestern East Sea and the TG MSL exhibit dominant fluctuations at periods in the 2–15-day band which are highly correlated with atmospheric pressure (Nam et al. 2004a; Park and Watts 2005a; Park et al. 2006), in contrast to the BP record in the open North Pacific at a nearby site (35.6°N, 148.4°E, during 2004–2005) where virtually no BP response was

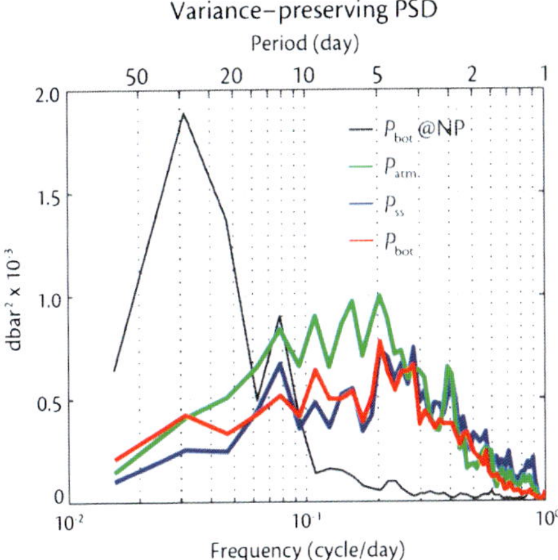

Fig. 5.2 Variance-preserving power spectra of mean bottom pressure of five sites (P_{bot}, *red line*) and mean subsurface pressure (P_{ss}, *blue line*) within the East Sea exhibit high energy in the 5–10 day period band. In contrast, little energy occurs in that same period band for a one-year bottom pressure record collected in the North Pacific near 35.6°N, 148.4°E (P_{bot}@NP, *black line*). Also shown is the power spectrum for atmospheric pressure averaged over the Ulleung Basin (P_{atm}, *green line*). The 95 percent confidence factors are (0.69, 1.60) for P_{bot}, P_{ss}, and P_{atm}, and (0.60, 2.00) for P_{bot}@NP (from Park et al. 2006)

observed similar to atmospheric pressure forcing at periods shorter than 10 days—the IB effect (Fig. 5.2, Park et al. 2006). Multiple coherence analyses of VT, TG MSL, and BP with atmospheric pressure and wind stress revealed that atmospheric pressure is the most significant forcing for VT, BP, and MSL in the 2–15 day band (Park and Watts 2005a; Lyu and Kim 2005). This dominant role of atmospheric pressure forcing in driving VT, BP, and MSL high-frequency variability has been examined and applied as a correction to satellite altimetry data.

5.2.2 Analytic Model and Applications

The barotropic response of the East Sea to atmospheric pressure forcing was examined by Lyu et al. (2002), Park and Watts (2005a), and Lyu and Kim (2005) using a model where the East Sea is considered as a deep flat-bottomed basin connected to the open ocean through only three straits. Since the East Sea is deep (~1700 m) and its horizontal length scale is small enough (~1000 km) for barotropic waves to travel around the basin within several hours or less, subsurface pressure (P) and the MSL (η_0) are considered spatially uniform and its local change ($\partial\eta_0/\partial t$) can be expressed as a sum of three VTs across the straits (Q_i where $i = K$ for Korea Strait, $i = T$ for Tsugaru Strait, and $i = S$ for Soya Strait) divided by the horizontal area (A_0), i.e., $A_0(\partial\eta_0/\partial t) = \sum Q_i$. Three momentum equations in the along-strait directions, are $\partial Q_i/\partial t = -(A_i/\rho)\,(\partial P/\partial x) - \lambda_i\,Q_i$ where A_i, ρ, and $\lambda_i Q_i$ are respectively cross-sectional area of each strait, constant water density, and linearly parameterized bottom friction as a geostrophic control (Toulany and Garrett 1984; Wright 1987). Atmospheric pressure forcing (P_a) is exerted in the form of a hydrostatic balance: $P = P_a + \rho g \eta_0$ where g is the gravity acceleration.

This simple linear system with five equations (the four plus mass conservation equation) and five unknowns can be solved for MSL, VT, and BP as frequency-dependent functions of P_a by taking all of them to behave as $\exp(i\omega t)$ where ω is frequency. The gains (in cm hPa^{-1} for η_0 and Sv hPa^{-1} for Q_i) and phases for η_0 and Q_i relative to P_a well explain the observed behavior (Lyu and Kim 2005; Park and Watts 2005a) including important departures from the IB effect in the 2 to 10-day band. Inverse Fourier transformed η_0, Q_i, and P time series show close agreement with the observations, e.g. the correlation coefficient for Q_K is 0.75 (Lyu and Kim 2005). Lyu et al. (2002) and Park and Watts (2005a) obtained results on a Helmholtz-like resonance frequency of 0.32 cpd (period of 3.12 days) where gains and phases of η_0 to P_a depart substantially from the IB response (gain 1 cm hPa^{-1} and phase of 180°).

The impacts of such high-frequency basin-scale oscillations on satellite altimetry data are rather clear. The model removes atmospheric-pressure-driven fluctuations better than the standard IB method, leaving residuals about 10 % smaller (Table 5.1). Since the two different corrections for atmospheric pressure effects on the altimetry data have maximum difference of up to 10 cm (variance 7.7 cm^2, Table 5.1), the impact of the correction choice is substantial (Nam et al. 2004a). In further examining the non-isostatic sea level response to atmospheric pressure within the East Sea, Inazu et al. (2006) found a zonally asymmetric structure, and using a model with realistic topography they successfully simulated sea level oscillations along the Japanese coast. They attributed the zonal asymmetry in the non-isostatic sea level response to the typical wavelength of the synoptic-scale atmospheric pressure system (several thousand kilometers) slightly larger than the spatial scale of the East Sea.

Table 5.1 Variance of East Sea mean sea level (MSL) with/without various corrections (from Nam et al. 2004a)

Sea levels with/without various corrections	Variance (cm^2)			
	TG MSL	BP MSL	T/P MSL	T/P track
Observation	19.0	21.2	19.4	139.7
Analytic model	13.2			
IB model	14.2			
Analytic model minus IB model	7.7			
IB model-corrected MSL	11.1	11.0	14.6	133.8
Analytic model-corrected MSL	9.2	9.6	12.4	123.5

5.3 Surface and Internal Tides

Surface tides within the East Sea are known to be weak (<10 cm, compared to ~1 m in the neighboring marginal seas or mid-ocean, Morimoto et al. 2000) except in the shallow boundary straits, yet strong internal tides are often observed within the basin as well as surrounding coastal (shallow) areas. Such internal tides are often stochastic and rather unpredictable, while surface tides are still regular and much more predictable, as in other seas (Nash et al. 2012). Recent studies have revealed how internal tides in the East Sea propagate, interacting with mesoscale circulation and bathymetric features, and having an impact on water masses and circulation. In this section, we review recent studies on the generation and propagation/refraction of internal tides in the southwestern East Sea, and their effects on intermediate water and circulation in the northern East Sea.

5.3.1 Observations

It is well known that tidal currents (M_2 is dominant among all tidal constituents) in the shallow Korea Strait are ten times larger than those inside the basin where the water depth increases sharply (Fig. 5.3). Substantial semidiurnal internal tides with 20 m vertical displacement are generated where strong M_2 tidal currents interact with the shelf break between the Korea Strait and the deep basin, and propagate northward or northeastward into the basin (Park and Watts 2006; Park et al. 2006). Park and Watts (2006) suggested that the internal tides are generated at the western Korea Strait shelf break (35.5–35.7°N, 130–131°E) where the M_2 barotropic tidal currents normal to the shelf slope are strong, and where the bottom topography slope matches the internal wave characteristics.

Acoustic travel-times (τ) measured by the 23 PIESs (triangles in Fig. 5.1) were used to estimate isotherm depth displacements. Using the isotherm time series, along with surface tide signals provided by simultaneously-measured bottom pressure, Park and Watts (2006) successfully observed the propagation of first-mode internal tides in the basin. Laterally restricted beam-like patterns of high internal tidal energy propagating from the shelf break were steered by the mesoscale circulation in the vicinity of the shelf break. Lunar monthly maps of isotherm displacement in the semidiurnal band (period from 11.50 to 12.92 h) or root-mean-squared semidiurnal internal tide amplitude were pertinent to mesoscale patterns resolved by maps of mean 5 °C depth (Park and Watts 2006; Park et al. 2006). For example, a northward propagating beam of strong semidiurnal internal tides was refracted significantly eastward when the Ulleung Warm Eddy (UWE) crossed its path (Case-I, most common), but was refracted westward when it encountered the Dok Cold Eddy (Mitchell et al. 2005) near the Korean coast (Case-II) (Fig. 5.4). When the Dok Cold Eddy was located in the vicinity of the shelf break in the Korea Strait, the beam weakened (Case-III, Fig. 5.4) as the thermocline rose and the

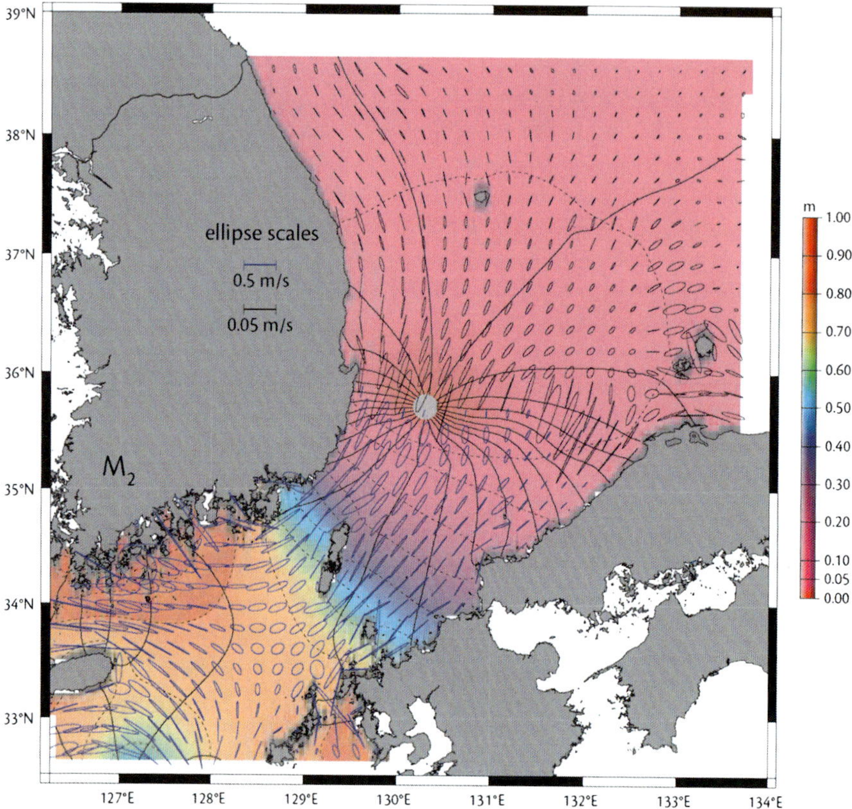

Fig. 5.3 Model-predicted tidal-current ellipses for the M_2 tide are superimposed on a model-predicted co-tidal (*solid lines*) and co-range (*dashed lines* together with colored map) chart in the Korea Strait and Ulleung Basin. Note the factor of 10 difference in scales for the tidal ellipses north (*black*) and south (*blue*) of the Korea Strait shelf break. The co-tidal lines are in increments of 15° (counter-clockwise rotation around the amphidromic point). The co-range lines and color bar are in increments of 0.1 m. The 0.05-m contour is also drawn (from Park et al. 2006)

internal wave characteristic slope (function of wave frequency, inertial frequency, and buoyancy frequency) no longer matched the shelf-break slope.

Clear evidence for Case-II (westward refracting semidiurnal internal tides) was also provided from three SAR images taken on September 2, 1996 off the east coast of Korea (see Fig. 1 of Nam and Park 2008). The curvatures of NLIW packets (Group B in Nam and Park (2008)) detected in the southwestern East Sea from the SAR images suggested they were generated in the vicinity of the Korea Strait shelf break and propagated northward, refracting westward toward the east coast of Korea since the large amplitude NLIWs are slower in the colder water. Nam and Park (2008) examined the structure of water temperature observed off the coast and near the shelf break in August and October 1996 when cold (<5 °C) waters occupied the coastal region at 100 m depth. The structure explained why

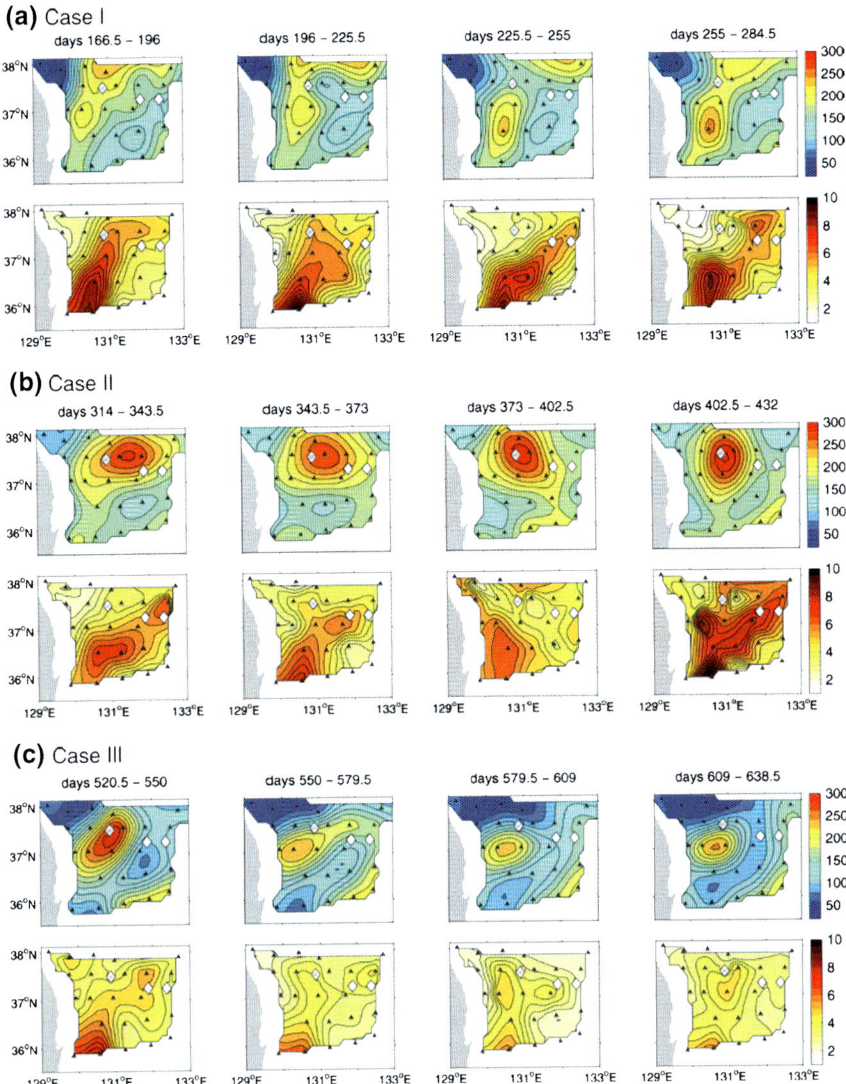

Fig. 5.4 Lunar monthly maps of mean 5 °C depth (*upper panel*) and rms semidiurnal internal tide amplitude (*lower panel*) during days **a** 166.5–284.5, **b** 314–432, and **c** 520.5–638.5. Contour intervals are 20 and 0.5 m, respectively. Dates are days since 0000 UTC 1 Jan 1999. Solid triangles indicate PIES sites (from Park and Watts 2006)

the internal tides were generated 1.0–3.5 days prior to the time the images were taken and why they showed westward refraction instead of the more usual eastward refraction. It also provided a good explanation for intermittent modulations of semidiurnal temperature oscillations which had been observed off the east coast of Korea (Lie et al. 1992).

In contrast to semidiurnal internal tides, diurnal internal tides cannot freely propagate within the East Sea since it is located north of the diurnal critical latitude (~30°N) where inertial frequencies are higher than the diurnal frequencies. Park and Watts (2006) noted that diurnal internal tidal energy was mostly trapped along the slope and shelf off the Japanese coast and not propagated northward.

5.3.2 Models and Applications

A simple geometric optics model was useful for simulating the propagation of the semidiurnal internal tide from the Korea Strait shelf break into the basin. The first-mode internal tide phase speed (C_1) can be calculated as a function of stratification, and the wave front propagation can be simulated, along with the current field (U) to include the additional effect of horizontal current shear, by calculating the distance a wave front advances (Δd) in a time step (Δt), $\Delta d = \Delta t (C_1 + |U| \cos \varphi)$ where φ is the angle between U and the wave propagation direction.

Park and Watts (2006) illustrated the essential roles of both mesoscale stratification and horizontal current shear on internal tide energy propagation, which is consistent with the two-year-long PIES observations. The beam of semidiurnal internal tides refracted eastward (westward) when a warm (cold) eddy intersected their paths north of the Korea Strait shelf break. Westward refraction toward the east coast of Korea of NLIW packets in September 1996 could also be simulated by the same model (both with and without changes in bottom topography) under the condition of cold water intrusion into the western channel of the Korea Strait as observed from hydrographic measurements (Nam and Park 2008). The ray path for semidiurnal internal tides simulated by the model refracted primarily westward regardless of bottom topography as C_1 decreased westward due to the presence of the cold water; this accounts for the observed SAR images, in contrast to the two-year mean path derived from the PIES observations (eastward refraction due to the prevailing UWE) (Fig. 5.5). Once the ray paths refract westward they tend to refract severely toward the coast as the water depth becomes shallower in the west, thus further decreasing C_1 (Fig. 5.5).

Extending the results of Park and Watts (2006) and Nam and Park (2008), the semidiurnal internal tides would ultimately reach the Russian coast in the northern East Sea (Case-I or two-year mean of the PIES observations, Fig. 5.5) unless they are significantly refracted westward toward the east coast of Korea (Case-II) or are suppressed on the shelf break near the generation area (Case-III). A recent modeling study conducted by Lee et al. (2011) reveals that an inclusion of the M_2 internal tides and resultant enhancement of turbulent mixing near the Russian coast modifies water masses in the northern East Sea (toward denser and fresher water) and affects the entire water mass in the intermediate layer of the East Sea so that the overall (domain-averaged) biases in temperature and salinity are improved considerably (see Fig. 4.12 in Chap. 4). In addition to tidal mixing effects on the intermediate water of the East Sea, they also found that coastal trapped diurnal

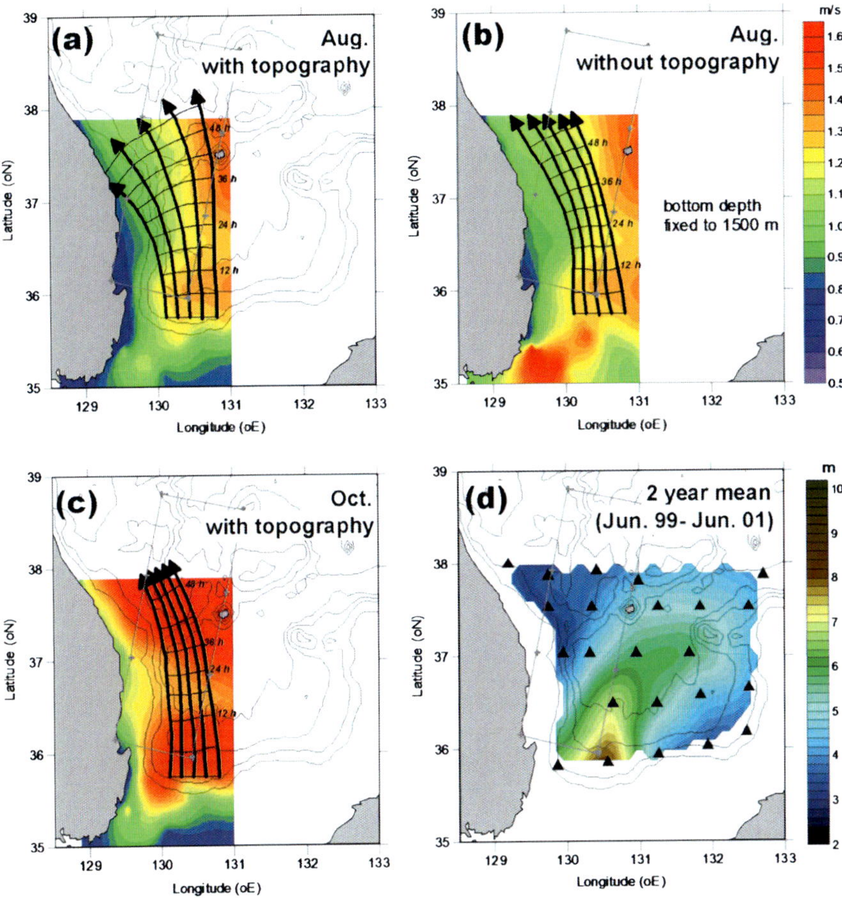

Fig. 5.5 Model simulated ray paths of semidiurnal internal tide (*thick arrows*) with **a**, **c** real and **b** constant (1500 m) bottom topography superimposed on color-contoured (scale bar in the upright) first-mode baroclinic phase speeds (C_1) in **a**, **b** August and **c** October. Wave fronts (*thin black lines*) are drawn every 6 h. **d** Two-year mean (from June 1999 to June 2001) vertical displacement of thermocline estimated from the PIES array data. PIES locations are shown as triangles. Bottom topography contours are superimposed in (**a**), (**c**), and (**d**) (from Nam and Park 2008)

(K_1) internal tides (north of the diurnal critical latitude) enhances wind-driven southward flow along the Russian coast (i.e. the Liman Current) in the upper layer via an internal-wave-induced setup. Most recently, Jeon et al. (2014) demonstrate seasonal variation of semidiurnal internal tides in the East Sea using simulation results from a data-assimilated model that includes tidal forcing of 16 major components along open boundaries. Their simulation results indicate that the semidiurnal internal tides would be able to reach the Russian coast from late summer to early winter when energetic internal tides are generated due to enhanced stratification near the generation region and propagate far from the generation region due to longer-wavelength induced less energy dissipation.

5.4 Near-Inertial Oscillations

The Coriolis effects create circular motion on a rotating Earth (in the absence of other forces), so called inertial motion, and the energy travels from the sea surface to the ocean bottom and equatorward of the source region with a small frequency shift to NI band (Garrett 2001). Inertia-gravity waves in this band, or simply NI waves, are frequently generated at the base of the mixed layer, and inertial motions are often amplified after passage of an atmospheric disturbance (e.g. D'Asaro 1985). They can propagate toward regions where their frequencies remain greater than the local inertial frequency, e.g. lower latitudes or regions of negative relative vorticity, where their energy can be transferred to the deep ocean (Kunze 1985; D'Asaro 1995; Lee and Niiler 1998; Garrett 2001). NI waves propagating through anticyclonic eddies are quickly dissipated at the bottom of the eddies (Kunze 1995, 1998), penetrate through the eddies without energy loss (Park and Watts 2005b), or are reflected back to the sea surface (Byun et al. 2010).

Although NI oscillations (manifested by particle movement inside an NI wave) are easily recognized as clockwise (anti-clockwise) rotating near-circular horizontal currents in the northern (southern) hemisphere, with frequencies slightly higher than the local inertial frequency, they still have not been well observed with sufficient resolutions in space and time, and their complex interactions with fronts, bathymetry, mesoscale circulations, and coastal boundaries have remained largely unexplored (Nam 2006). Nevertheless, recent progress in new observational techniques has enabled us to understand key aspects of inertial motions, NI oscillations/waves, and their complex interactions with circulation and topography in open and coastal areas of the East Sea, and we describe that progress in this section.

5.4.1 Observations

Lagrangian observations with surface drifters or floats have been very useful in characterizing surface inertial motions in the East Sea (Park et al. 2004) and the global ocean (Park et al. 2005, 2009). Climatology of inertial currents obtained from satellite tracked surface drifter trajectories in the East Sea shows a clear seasonal cycle having maximum amplitude in fall and somewhat patchy spatial patterns (Fig. 5.6). Strong inertial motions in fall are primarily attributed to the fact that stronger winds are starting to blow but the surface mixed layer has not yet deepened as in other (open) oceans (Park et al. 2005). Although energy transfer from wind to oceanic inertial motion reaches its maximum in winter, the inertial energy is spread over the deep mixed layer, resulting in relatively weak surface inertial currents. Such stronger fall (than in spring) inertial motions are consistent with observations in the Korea Strait (Jacobs et al. 2001). In winter, the area of strong surface inertial motions becomes somewhat limited to the northwest region

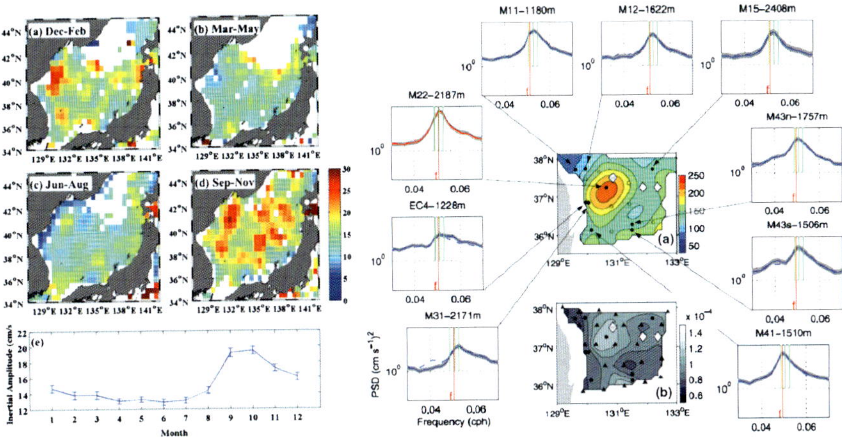

Fig. 5.6 *Left* climatological amplitudes of near-inertial motions obtained from satellite tracked surface drifter trajectories in 1990–2009. **a–d** Spatial distribution of near-inertial amplitudes averaged in 0.5° × 0.5° bins in **a** December–February, **b** March–May, **c** June–August, and **d** September–November. **e** Monthly-averaged near-inertial amplitudes over the East Sea. *Right* power spectra for zonal (*solid line*) and meridional (*dashed line*) components at nine bottom current-meter sites, where the observed current records are long enough to have more than 20 segments of 512 data points (21.33 days) overlapped by 50 %. Vertical red and green lines indicate the local Coriolis frequency *f* and three spectral estimates closest to *f* and the spectral peak. **a** 2-year mean map of 5 °C depth (Z5) (contour interval 20 m). **b** 2-year root-mean-squared amplitude map of τ_i, that is the band-pass filtered, PIES acoustic travel-time τ in the near-inertial frequency bands (contour interval 10^{-5} s). *Solid circles* and *triangles* on the maps indicate bottom current-meter moorings and PIESs, respectively (from Park and Watts 2005b)

(Fig. 5.6a) due to frequent changes in strength and direction of northwesterly monsoon winds under the orographic influence of the mountainous landmass near the coast (Nam et al. 2005b).

Deep NI oscillations are also heterogeneous and intermittent. Surface NI oscillations as a major source of deep NI oscillation are themselves very patchy and their energy can reach the depths through various routes, making their spatio-temporal distributions in the deep layer even more complicated. Long-term moored observations (Takematsu et al. 1999) showed strong deep NI oscillations mostly in the central part of the northern East Sea in the vicinity of the subpolar front, with a sharp peak at the inertial period (~18 h). Shcherbina et al. (2003) examined the full-depth current structure using lowered shipboard Acoustic Doppler Current Profilers (ADCPs), and they observed highly energetic downward propagating NI waves in localized patches along the southern edge of the subpolar front. Basin-averaged net-upward energy flux found at depths between 500 and 2500 m was discussed, raising the possibility of topographic generation of NI waves (Shcherbina et al. 2003). Mori et al. (2005), analyzing data from moored current measurements, further described NI oscillations in the deep waters of the East Sea, showing the following results.

1. More energetic NI oscillations in the area south of the subpolar front, and the Tsushima Warm Current region,
2. A seasonal/annual cycle with wintertime intensification, and
3. Different temporal variations in the Yamato Basin where the NI oscillations are less energetic in spring and more energetic in summer.

Eulerian observations using PIES or moored current measurements in the southwestern East Sea (Fig. 5.1) are well suited to address interactions of NI waves with mesoscale circulation (Park and Watts 2005b; Park et al. 2006; Byun et al. 2010). Using the acoustic travel-times measured by the 23 PIESs and deep velocities measured by a current-meter array in the southwestern East Sea (Fig. 5.1), Park and Watts (2005b) demonstrated wintertime high amplitude NI oscillations which vary with observed interannual changes in mesoscale circulation. All spectra but one showed a blue shift, indicating that the NI energy originates from the north. The exception, having the highest NI energy, was located in the center of anticyclonic circulation (UWE); it showed a spectral peak at the local inertial frequency without the blue shift (Fig. 5.6). Long (16.5 month) moored current observations in the southwestern East Sea revealed two episodes of upward (energy) propagating NI waves when the mooring was located near the center of the anticyclonic circulation (UWE) as found by Byun et al. (2010). The upward energy propagation could be explained by the reflection of the NI waves within the thermostad of the eddy where the effect of vertical shear of subinertial currents is larger than the buoyancy effect in controlling the propagation of NI waves (Byun et al. 2010).

Recent observations near the east coast of Korea and the Japanese coast of the East Sea provided better ideas on such coastal NI oscillations. NI waves off the east coast of Korea have been observed propagating both upward and downward and both shoreward and seaward (Lie 1988; Kim et al. 2001, 2005a, b, c; Nam 2006; Nam et al. 2007). Using long time-series data obtained between 1999 and 2004 (Nam et al. 2005a), Nam (2006) identified three different mechanisms primarily controlling NI current variability off the east coast of Korea: wind-induced surface inertial motion (53 %, 23 of 40 events), interaction with subinertial shear (30 %, 13 of 40 events), and forward or super-critical bottom reflection (17 %, 7 of 40 events). Igeta et al. (2009) and Igeta et al. (2011) observed NI oscillations and NI waves propagating along a peninsula and a small island located on the Japanese side of the East Sea. The data suggested scattering and trapping of NI waves by the coast (Igeta et al. 2009) and a strong coastal current derived from the NI waves combined with coastal-trapped waves (Igeta et al. 2011). Furthermore Igeta et al. (2011) observed more complicated vertical structures and time variations of the strong current in fall than in winter, and attributed this to low-mode coastal-trapped waves strengthened by NI oscillations.

5.4.2 Generation and Propagation of Near-Inertial Waves

It is known that NI waves are typically generated at the base of the mixed layer after passage of any kind of atmospheric disturbances, e.g. storms, atmospheric fronts, or tropical cyclones. An atmospheric disturbance imprints an inertial motion with a particular length scale determined by the translation speed and length scale of the disturbance (D'Asaro 1985). Once inertial motions are generated, spatially separated inertial currents differ in their oscillatory phase because of spatial variation of the local inertial frequency and relative vorticity. Such out-of-phase inertial currents produce convergence or divergence of surface currents, generating vertical undulations of isotherm surfaces with near-inertial frequency, so-called NI waves. The propagation speed of an NI wave is primarily determined by the initial forcing scale, local inertial frequency, buoyancy frequency, and mixed layer depth (Park et al. 2009).

Although NI waves are best observed after the passage of a typhoon or local wind event, significant NI oscillations are occasionally observed without a clear local wind event, mostly in cases of upward propagating NI waves resulting from bottom reflection. Lie (1988) first observed NI waves of upward propagation (downward phase propagation) off the east coast of Korea, and suggested forward or super-critical bottom reflection of remotely-generated NI waves; this has been further investigated with comprehensive observations in the same area by Kim et al. (2005c) and Nam (2006). The possibility of detecting sea surface signatures of onshore propagating NI waves from space (SAR images) has been raised by Kim et al. (2005b) and further discussed by Kim et al. (2005a). They are concerned with distinguishing these NI-wave signatures from possible footprints of atmospheric lee waves which are more commonly seen in coastal SAR images; the interpretations are still under active debate since many undulations seen on the images are interpreted as sea surface signatures of katabatic-wind-induced coastal lee waves (Gan et al. 2008; Zheng 2005).

The significance of NI waves in the East Sea was first presented by Kim et al. (2001) where a few NLIW packets were observed off the east coast of Korea at NI (rather than semidiurnal or diurnal) period intervals. Such NI-period NLIW packets, not reported in other seas, were further examined by analyzing SAR-image data taken approximately 19 h after the passage of a typhoon eye across the coast, along with simultaneously collected current and water-property data (Nam et al. 2004b, 2007).

5.5 Nonlinear Internal Waves

Internal waves are ubiquitous wherever strong stratification and tides (or winds) occur near bathymetric features. They often show (highly) nonlinear characteristics; they are then called NLIWs (or solitary internal waves), and KdV-type theories have played the primary role in elucidating the essential features of the observations (Helfrich and Melville 2006). They are sometimes prominent features seen in optical and radar satellite imagery of coastal waters, propagating over several hundred kilometers and transporting mass, momentum, and energy (Jackson 2004). They can lead to turbulent mixing, thus fertilizing the local region, significantly modifying the biology therein.

Since they were first observed in May 1999 off the east coast of Korea using vertical arrays of thermistors and a bottom-mounted ADCP (Kim et al. 2001), NLIWs in the East Sea have been investigated using both in situ and satellite remote sensing observations. Though overall features are still largely unexplored, the observations, mostly off the east coast of Korea, have revealed some aspects of the generation and propagation/refraction of NLIWs.

5.5.1 Observations

Packets of NLIWs were reported for the first time in the area by Kim et al. (2001), who observed three such packets propagating shoreward with a speed of ~0.5 m s^{-1}, based on moored observations off the east coast of Korea in May 1999 (A-C in Fig. 5.7). Each packet contained a few NLIWs as a group, each having the shape of a solitary wave of depression (downward thermocline displacement), with period of around 10 min. Simultaneously collected ADCP data showed consistent horizontal and vertical flows, e.g. oppositely-flowing horizontal currents with an undulating high vertical shear zone at thermocline depths and abrupt downward and upward vertical flows. Isotherm displacement of individual NLIWs could be explained by the combined (or extended) KdV model (Kim et al. 2001).

Such NLIWs, together with their packets, have been confirmed from a few satellite (mostly SAR) images combined with simultaneous time series observations from moorings (Jackson 2004; Nam et al. 2007; Nam and Park 2008). Nam et al. (2007) reported both shoreward and seaward propagating NLIW packets from a SAR image taken approximately 19 h after typhoon eye passage across the coast in September 2003. The SAR observations were consistent with temporal structures of water temperature observed by a mooring within the area at thermocline depths (Nam et al. 2004b, 2007). Nam and Park (2008) observed two pairs of NLIW packets from SAR images taken in September 1996; one on the western side of the southwestern East Sea and the other on the northern side. Using a geometric optics model, they accounted for the surface manifestation of these NLIW packets by westward and eastward refractions of semidiurnal internal tides generated at the Korea Strait (see Sect. 5.3.2 and Fig. 5.5). Characteristic scales of NLIWs observed in the area are shown in Table 5.2 (Jackson 2004).

5.5.2 Generation and Propagation/Refraction of Nonlinear Internal Waves

The NLIW packets observed at two thermistor moorings and one bottom-mounted ADCP off the east coast of Korea had NI-period (~19 h rather than semidiurnal or diurnal tidal) intervals in contrast to those found in most other seas. For

Fig. 5.7 **a** A time series of temperature from a thermistor chain moored at 100 m depth off the east coast of Korea. Three cycles of steep leading edges at an interval of about 19 h can be seen, mostly visible at the 25 and 35 m depth levels. **b** A 2-h time series of the temperature from 1:00 am to 3:00 am in Korea Standard Time, shows the solitary wave packet C. **c** A time series of isotherm depths from the temperature data for the same period as (**b**). The 10 °C isotherm (*white line*) represents the thermocline (from Kim et al. 2001)

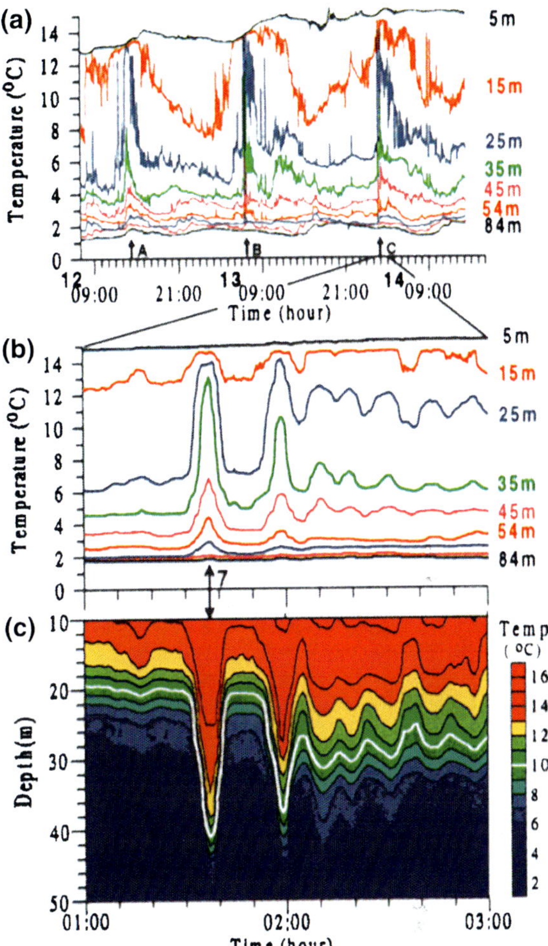

Table 5.2 Characteristic scales for NLIWs (solitons) along the east coast of Korea (from Jackson 2004)

Characteristic	Scale
Amplitude factor	−20 to −26 m
Long wave speed	0.5 ms^{-1}
Maximum wavelength	1.5 km
Wave period	3–10 min
Surface width	180 m
Packet length	2.7 km
Along crest length	>30 km
Packet separation	40–45 km

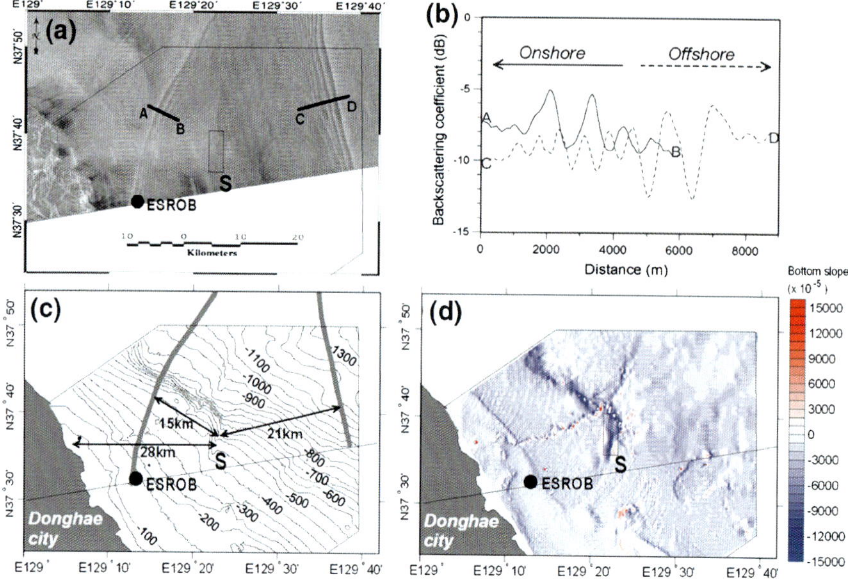

Fig. 5.8 **a** ENVISAT SAR image taken on 22:10 (13:10 UTC) September 13, 2003, **b** backscattering coefficients across the A–B and C–D sections, **c** high resolution (less than 50 m in space) bottom topography, and **d** eastward bottom slope in the coastal region off Donghae city. Surface signatures of highly nonlinear internal wave packets are marked with thick gray curves in (**c**). The region of sharply varying topography 28 km off the coast is noted with the symbol 'S' in (**a**, **c** and **d**). ESROB denotes location of the East Sea Real-time Ocean Buoy (from Nam et al. 2007)

this reason, wind-induced NI waves instead of internal tides are suggested as the source for generating NLIWs in the East Sea (Kim et al. 2001). Strong NI oscillations observed at a surface mooring, before a SAR image was taken (showing surface manifestation of two NLIW packets, Fig. 5.8) and after a typhoon eye passed by (Nam et al. 2004b), also support the claim that these are wind-induced NLIWs of NI wave origin. Nam et al. (2007) pointed out sharp bathymetric features located 28 km off the coast (Fig. 5.8) as a location of NLIW generation based on a KdV model calculation along with observations from the moored and SAR sensors. Moreover such suggestions of NLIW generation by wind-induced NI waves off the east coast of Korea are reasonable when considering the weak (~5 cm) barotropic tides in the East Sea.

However, internal temperature oscillations at tidal frequencies have also been observed off the Korean east coast; it is believed they have propagated from a remote area as the local barotropic tidal current is too weak to generate them (e.g. Lie et al. 1992). As shown in Sect. 5.3, NLIW packets with semidiurnal period intervals can reach the shallow regime when semidiurnal internal tides propagate northwestward from the source region (the shelf break in the Korea Strait) refracting toward the east coast of Korea (Fig. 5.5, Case-II in Park and Watts 2006). Cold water in the coastal area can provide a wave guide to refract the NLIW packets toward the coast (Nam and Park 2008) since they are generated in the shallow Korea Strait where tidal currents are strong (Park and Watts 2006; Park et al. 2006). Thus, it is suggested that NLIWs off the Korean east coast have two origins—both wind-induced NI waves and semidiurnal internal tidal waves can reach this coast though their interactions with mesoscale circulations. The NLIWs generated by semidiurnal internal tides generally propagate longer distances (from the shelf break in the Korea Strait) than those generated by NI waves (generation on the local shelf off the coast).

5.6 Conclusion and Remaining Issues

In this chapter, we have introduced high-frequency variability recently observed in the East Sea, including basin-scale oscillations, internal tides, NI oscillations, and NLIWs. At periods of 2–10 days, due to Helmholtz-like resonance, sea level response to atmospheric pressure forcing is non-isostatic and departs from the standard IB response, driving flows (VT) in and out of the basin via the straits and producing strong basin-scale MSL and BP fluctuations. Semidiurnal temperature oscillations can often be significantly enhanced as internal tidal beams are guided by mesoscale circulation as well as bathymetry along the ray path, refracting either eastward or westward or being suppressed after generation at the shelf break in the western channel of the Korea Strait. In spite of a clear seasonal cycle in surface inertial motion (peak in fall), NI oscillations in the interior of the East Sea are generally heterogeneous and stochastic being shaped by the subpolar front, major currents, and mesoscale circulation, as well as the wind forcing. They are further complicated by NI waves interacting with coastal and bottom boundaries, coastal-trapped waves, etc. Coexistence of both wind-induced NI waves and semidiurnal internal tides enables us to see NLIW packets at NI as well as semidiurnal intervals off the east coast of Korea.

However, investigation of high-frequency variability in the East Sea is far from complete. A few results introduced in this chapter have raised more questions than they answered. Even though the analytic model introduced in Sect. 5.2.2 addresses relatively well basin-scale MSL (and BP and VT) oscillations in the East Sea as primarily forced by atmospheric pressure forcing, a more realistic simulation is still required to address effects of other forcings such as regional-scale winds, wind-stress curl forcing, and non-uniform oscillations due to local wind, interaction with meso- and submeso-scale eddies, and coastal processes. In spite of some progress in understanding internal tides in the East Sea, further observational and theoretical works are particularly needed to address the potential seasonal cycle of tidal effects on circulation in the East Sea, and vice versa. Process studies would be helpful for better understanding of interactions between propagating internal tides and bathymetric features, fronts, meso- and submeso-scale eddies, winds, etc. Processes underlying highly variable (both in space and time) NI oscillations and NLIWs in the East Sea still need further studies. Many hypotheses on behaviors of NI waves and NLIWs under different conditions may be tested with the use of higher-resolution measurements, incorporation of more satellite data into comprehensive, simultaneous observations, and also process modeling studies.

The general presumption of stationary characteristics of internal tides/waves in the East Sea may need to be tested under changing climate conditions. For example, the following questions are of interest. Is there significant long-term change within the basin in frequency, intensity, and duration of events of enhanced semi-diurnal/diurnal or NI waves and associated generation of NLIWs? How do near-surface warming and long-term changes in upper-ocean stratification affect the generation, propagation, and dissipation of internal tides/waves in the East Sea? Does the predictability of internal tides/waves in the basin vary with long-term climate change? To answer such questions, long-term time series observations with particularly high temporal sampling rates are obviously essential.

Acknowledgements Authors would like to thank people who have dedicated their time and efforts to address high-frequency variability in the East Sea. The authors thank Profs. M. Wimbush and K.I. Chang (editors) for improving the quality of this manuscript and two reviewers for providing their ideas to improve the quality of the original manuscript. J.-H. Park is supported by KIOST in-house grants PE99293 and PE99297. S.H. Nam is partly supported by Research Resettlement Fund for the new faculty of Seoul National University and partly by Ministry of Oceans and Fisheries, Korea through "East Asian Seas Time-series (EAST-I)".

References

Byun SS, Park JJ, Chang KI et al (2010) Observation of near-inertial wave reflections within the thermostad layer of an anticyclonic mesoscale eddy. Geophys Res Lett 37:L01606. doi:10.10 29/2009GL041601

Candela J (1991) The Gibraltar Strait and its role in the dynamics of the Mediterranean Sea. Dyn Atmos Ocean 15:267–299

Candela J, Winant CD, Bryden H (1989) Meteorologically forced subinertial flows through the Strait of Gibraltar. J Geophys Res 94:12667–12679

D'Asaro EA (1985) The energy flux from the wind to near-inertial motions in the surface mixed layer. J Phys Oceanogr 15:1043–1059

D'Asaro EA (1995) Upper-ocean inertial currents forced by a strong storm. Part III: Interaction of inertial currents and mesoscale eddies. J Phys Oceanogr 25:2953–2958

Gan X, Huang WG, Li XF et al (2008) Coastally trapped atmospheric gravity waves on SAR, AVHRR and MODIS images. Int J Remote Sens 29:1621–1634

Garrett C (1983) Variable sea level and strait flows in the Mediterranean: a theoretical study of the response to meteorological forcing. Oceanol Acta 6:79–87

Garrett C (2001) What is the "near-inertial" band and why is it different from the rest of the internal wave spectrum? J Phys Oceanogr 31:962–971

Garrett C, Majaess F (1984) Non-isostatic response of sea level to atmospheric pressure in the eastern Mediterranean. J Phys Oceanogr 14:656–665

Helfrich KR, Melville WK (2006) Long nonlinear internal waves. Ann Rev Fluid Mech 38:395–425

Igeta Y, Kumaki Y, Kitade Y et al (2009) Scattering of near-inertial internal waves along the Japanese coast of the Japan Sea. J Geophys Res 114:C10002. doi:10.1029/2009JC005305

Igeta Y, Watanabe T, Yamada H et al (2011) Coastal currents caused by superposition of coastal-trapped waves and near-inertial oscillations observed near the Noto Peninsula, Japan. Cont Shelf Res 31:1739–1749

Inazu D, Naoki H, Kizu S et al (2006) Zonally asymmetric response of the Japan Sea to synoptic pressure forcing. J Oceanogr 62:909–916

Jackson CR (2004) An atlas of internal solitary-like waves and their properties, 2nd edn. Global Ocean Associates, Alexandria

Jacobs GA, Book JW, Perkins HT et al (2001) Inertial oscillations in the Korea Strait. J Geophys Res 106:26943–26957

Jeon C, Park JH, Varlamov SM et al (2014) Seasonal variation of semidiurnal internal tides in the East/Japan Sea. J Geophys Res 119:2843–2859. doi:10.1002/2014JC009864

Kim HR, Ahn S, Kim K (2001) Observations of highly nonlinear internal solitons generated by near-inertial internal waves off the east coast of Korea. Geophys Res Lett 28:3191–3194. doi:10.1029/2001GL013130

Kim HR, Nam SH, Kim DJ et al (2005a) Reply to comment by Q. Zheng on "Can near-inertial internal waves in the East Sea be observed by synthetic aperture radar?". Geophys Res Lett 32:L20607. doi:10.1029/2005GL024351

Kim DJ, Nam SH, Kim HR et al (2005b) Can near-inertial internal waves in the East Sea be observed by synthetic aperture radar? Geophys Res Lett 32:L02606. doi:10.1029/2004GL021532

Kim HJ, Park YG, Kim K (2005c) Generation mechanism of the near-inertial internal waves observed off the east coast of Korea. Cont Shelf Res 25:1712–1719

Kunze E (1985) Near-inertial wave propagation in geostrophic shear. J Phys Oceanogr 15:544–565

Kunze E (1995) The energy balance in a warm-core ring's near-inertial critical layer. J Phys Oceanogr 25:942–957

Kunze E (1998) A model for vortex-trapped internal waves. J Phys Oceanogr 28:2104–2115

Le Traon PY, Gauzelin P (1997) Response of the Mediterranean mean sea level to atmospheric pressure forcing. J Geophys Res 102:973–984

Lee DK, Niiler PP (1998) The inertial chimney: the near-inertial energy drainage from the ocean surface to the deep layer. J Geophys Res 103:7579–7591

Lee HJ, Park JH, Wimbush M et al (2011) Tidal effects on intermediate waters: a case study in the East/Japan Sea. J Phys Oceanogr 41:234–240

Lie HJ (1988) Near-inertial current oscillations off the mid-east coast of Korea. Prog Oceanogr 21:241–253

Lie HJ, Shin CW, Seung YH (1992) Internal tidal oscillations of temperature off Jukbyun on the east coast of Korea. J Oceanogr Soc Korea 27:228–236

Lyu SJ, Kim K (2005) Subinertial to interannual transport variations in the Korea Strait and their possible mechanisms. J Geophys Res 110:C12016. doi:10.1029/2004JC002651

Lyu SJ, Kim K, Perkins HT (2002) Atmospheric pressure-forced subinertial variations in the transport through the Korea Strait. Geophys Res Lett 29(9):1294. doi:10.1029/2001GL014366

Mitchell DA, Teague WJ, Wimbush M et al (2005) The Dok Cold Eddy. J Phys Oceanogr 35:273–288

Mori K, Matsuno T, Senjyu T (2005) Seasonal/spatial variations of the near-inertial oscillations in the deep water of the Japan Sea. J Oceanogr 61:761–773

Morimoto A, Yanagi T, Kaneko A (2000) Tidal correction of altimetric data in the Japan Sea. J Oceanogr 56:31–41

Nam SH (2006) Near-inertial current variability off the east coast of Korea. Dissertation, Seoul National University

Nam SH, Park JH (2008) Semidiurnal internal tides off the east coast of Korea inferred from synthetic aperture radar images. Geophys Res Lett 35:L05602. doi:10.1029/2007GL032536

Nam SH, Lyu SJ, Kim YH et al (2004a) Correction of TOPEX/Poseidon altimeter data for non-isostatic sea level response to atmospheric pressure in the Japan/East Sea. Geophys Res Lett 31:L02304. doi:10.1029/2003GL018487

Nam SH, Yun JY, Kim K (2004b) Observations on the coastal ocean response to typhoon MAEMI at the East Sea Real-time Ocean Buoy. J Korean Soc Oceangr 9:111–118

Nam SH, Kim G, Kim KR et al (2005a) Application of real-time monitoring buoy systems for physical and biogeochemical parameters in the coastal ocean around the Korean peninsula. Mar Tech Soc J 39:54–64

Nam SH, Kim YH, Park K-A et al (2005b) Spatio-temporal variability in sea surface wind stress near and off the east coast of Korea. Acta Oceanol Sin 24:107–114

Nam SH, Kim DJ, Kim HR et al (2007) Typhoon-induced, highly nonlinear internal solitary waves off the east coast of Korea. Geophys Res Lett 34:L01607. doi:10.1029/2006GL028187

Nash JD, Shroyer EL, Kelly SM et al (2012) Are any coastal internal tides predictable? Oceanography 25:80–95

Park JJ, Kim K, Crawford WR (2004) Inertial currents estimated from surface trajectories of ARGO floats. Geophys Res Lett 31:L13307. doi:10.1029/2004GL020191

Park JJ, Kim K, King BA (2005) Global statistics of inertial motions. Geophys Res Lett 32:L14612. doi:10.1029/2005GL023258

Park JJ, Kim K, Schmitt RW (2009) Global distribution of the decay timescale of mixed layer inertial motions observed by satellite-tracked drifters. J Geophys Res 114:C11010. doi:10.10 29/2008JC005216

Park JH, Watts DR (2005a) Response of the southwestern Japan/East Sea to atmospheric pressure. Deep-Sea Res Pt II 52:1671–1683

Park JH, Watts DR (2005b) Near-inertial oscillations interacting with mesoscale circulation in the southwestern Japan/East Sea. Geophys Res Lett 32:L10611. doi:10.1029/2005GL022936

Park JH, Watts DR (2006) Internal tides in the southwestern Japan/East Sea. J Phys Oceanogr 36:22–34

Park JH, Watts DR, Wimbush M et al (2006) Rapid variability in the Japan/East Sea: basin oscillations, internal tides, and near-inertial oscillations. Oceanogr 19:76–85

Shcherbina AY, Talley LD, Firing E et al (2003) Near-surface frontal zone trapping and deep upward propagation of internal wave energy in the Japan/East Sea. J Phys Oceanogr 33:900–912

Smith WHF, Sandwell DT (1997) Global seafloor topography from satellite altimetry and ship depth soundings. Science 277:1956–1962

Takematsu M, Nagano Z, Ostrovskii A et al (1999) Direct measurements of deep currents in the northern Japan Sea. J Oceanogr 55:207–216

Toulany B, Garrett C (1984) Geostrophic control of fluctuating barotropic flow through the straits. J Phys Oceanogr 14:649–655

Wright DG (1987) Comments on "Geostrophic control of fluctuating barotropic flow through the straits". J Phys Oceanogr 17:2375–2377

Xu Y (2006) Analyses of sea surface height, bottom pressure and acoustic travel time in the Japan/East Sea. Dissertation, University of Rhode Island

Xu Y, Watts DR, Park JH (2008) De-aliasing of large-scale high-frequency barotropic signals from satellite altimetry in the Japan/East Sea. J Atmos Oceanic Technol 25:1703–1709

Zheng Q (2005) Comment on "Can near-inertial internal waves in the East Sea be observed by synthetic aperture radar?" Geophys Res Lett 32:L20606. doi:10.1029/2005GL023770 (by D. J. Kim et al.)

Chapter 6
Dissolved Oxygen and Nutrients

TaeKeun Rho, Tongsup Lee and Soonmo An

Abstract Changing global temperatures may alter oceanic concentrations of both dissolved oxygen (DO), which reflects water quality, and nutrients, which are an essential factor for the growth and distribution of phytoplankton. The East Sea (Japan Sea) is often considered a "miniature ocean" and an ideal natural laboratory to estimate the impact of global climate change on the oceanic biogeochemical cycles, because of the East Sea's physical dynamics. Thus, the monitoring of nutrients and DO concentrations in the East Sea may help predict the response of biogeochemical cycles in the global ocean to changing climate. We describe and discuss the history of studies on DO in the East Sea, the change in vertical structure of DO profiles, the trend of DO inventory in the bottom layer, and the controversy regarding future projections of DO in the bottom water mass. We also summarize the history of nutrient measurements in the region, the general structure of vertical nutrient profiles, the horizontal nutrient distribution pattern on isobaric surfaces, the seasonal nutrient flux through the Korea Strait, and the recent debate on nutrient dynamics in the East Sea. The results of the studies we examine suggest that the processes controlling the structures and inventories of DO and nutrients in the East Sea are very sensitive to global climate changes.

Keywords Nutrients · Dissolved oxygen · N:P ratio · Deoxygenation · East Sea (Japan Sea)

T. Rho · T. Lee (✉) · S. An
Department of Oceanography, Pusan National University, Busan 609-735, Republic of Korea
e-mail: tlee@pusan.ac.kr

T. Rho
e-mail: tkrho@kiost.ac.kr

S. An
e-mail: sman@pnu.edu

T. Rho
Ocean Observation and Information Section, Korea Institute of Ocean Science and Technology, Ansan 426-744, Republic of Korea

© Springer International Publishing Switzerland 2016
K.-I. Chang et al. (eds.), *Oceanography of the East Sea (Japan Sea)*,
DOI 10.1007/978-3-319-22720-7_6

6.1 Introduction

The East Sea is a semi-enclosed marginal sea of the western North Pacific Ocean and has three relatively deep basins: the Ulleung Basin in the southwest, the Yamato Basin in the southeast, and the Japan Basin in the north (Fig. 1.1). The basins are separated from each other by the Yamato Rise located in the central region of the East Sea. The continental shelves are narrow except in the northern region. The East Sea is connected to the North Pacific Ocean through shallow straits: the Korea Strait in the southwest (sill depth ca. 130 m) and the Tsugaru Strait in the northeast (sill depth ca. 130 m). It is also connected to the Okhotsk Sea by even shallower straits in the north: the Soya Strait (sill depth ca. 55 m) and the Tatarsky Strait (sill depth ca. 50 m) (Fig. 1.1).

Annual mean water transport into the East Sea through the Korea Strait shows large interannual fluctuation with a mean value of 2.5 Sv. The East Sea has oceanic features such as a meridional overturning circulation, subpolar fronts, and mesoscale eddies. Thus it is frequently considered a "miniature ocean" and an ideal natural laboratory to study the response of oceanic processes to global climate changes (Kim et al. 2001).

Dissolved oxygen (DO) is one of the most commonly measured oceanographic parameters. The major sources of DO to the ocean are flux from the atmosphere through the air-sea gas exchange and in situ production by phytoplankton within the euphotic layer. The in situ respiration and remineralization of sinking organic matter (microbial respiration) are the two main processes that consume DO within the water column. Deep convection of surface water (meridional overturning) brings oxygen-rich surface water down to different layers of the ocean, depending on the density of the sinking water mass.

Recently, DO inventories in the interior of the ocean have declined significantly due to global warming and increased stratification of the upper-ocean (Shaffer et al. 2009; Keeling et al. 2010). Increased surface temperatures reduce the solubility of oxygen in the surface mixed layer and increased stratification also suppresses the transport of oxygen-rich surface water into the interior of ocean. The expansion of the intermediate oxygen minimum layer (also called oxygen minimum zone, OMZ) was conspicuous in the eastern tropical Pacific Ocean, the tropical Indian Ocean, and the tropical Atlantic Ocean during the past 50 years (Stramma et al. 2008). Models predict further decline of oxygen inventory (about 1–7 %) over the next century. However, the global ocean (due to its vastness), takes a relatively long time to respond to changing climatic conditions and the change of DO concentration is relatively small (0.09–0.34 μmol kg^{-1}), which is smaller than the precision of conventional oxygen measurement (Shaffer et al. 2009; Keeling et al. 2010). Therefore, careful monitoring of DO concentration within the interior of the ocean is required to detect the response of the ocean's DO inventory to global climate change (Joos et al. 2003).

Dissolved nutrients in seawater are an essential factor for the growth and distribution of phytoplankton, which form the base of marine ecosystems and help determine community structure through primary production (Glibert 1982). In the sunlit surface mixed layer, phytoplankton rapidly convert dissolved nutrients into particulate organic matter (POM), which is recycled by microorganisms and transferred to higher trophic levels or exported to the deep ocean as sinking particles (Eppley and Peterson 1979). Among the major elemental constituents of marine life, nitrogen (N) and phosphorus (P) are present in seawater in low concentrations and limit the growth of phytoplankton (Glibert 1982). Ammonium, nitrite, and nitrate are major forms of dissolved inorganic nitrogen, and phosphate is a major form of dissolved inorganic phosphorus. Therefore, the processes that regulate the transformations of N and P between dissolved inorganic nutrients and organic matter have been a major topic of ocean biogeochemistry. Low-nutrient surface waters, while sinking, show systematically lower nutrient concentrations than the surrounding water mass (Kim et al. 1991). This suggests that the vertical nutrient profile can serve as a water mass tracer of sinking surface waters. Recently, there have been vigorous debates on whether there is a significant change in the relative abundance of dissolved inorganic nitrogen (DIN) over dissolved inorganic phosphorus (DIP) throughout the water column, and, if so, what mechanisms are responsible for this change (Kim et al. 2011; Kim and Kim 2013; Kim et al. 2013).

In this chapter, we describe the history of DO measurements in the East Sea; the changes in DO inventory in the East Sea during the last several decades; and the relationship between DO and global climate change. We also describe, with reference to the East Sea: the history of nutrient studies; the spatial and temporal distribution patterns of nutrients in this region; factors controlling nutrient concentrations and profiles; and related major research topics.

6.2 Dissolved Oxygen

6.2.1 History of DO Measurements

The first extensive investigation of DO concentration in the East Sea started during the 1930s (Uda 1934). This investigation revealed very high and uniform concentrations of around 250 μM (5.6 ml L^{-1}) for waters below a few hundred meters over entire basins and suggested the presence of very fast ventilation in the East Sea (Uda 1934). DO was measured by Russian scientists in the northern part of the East Sea in the 1950s and in 1995 during the Vityaz and Lavrentyev expeditions, respectively. DO measurements in the eastern Japan Basin and Yamato Basin in 1969, 1977, 1979, and 1984 showed changes in vertical DO profiles and bottom DO concentrations (Gamo et al. 1986; Chen et al. 1999). These changes in DO structure were probably caused by changes in bottom water formation in the northern Japan Basin (Gamo et al. 1986). DO concentration in the Ulleung Basin has been monitored continuously since 1961 by the National Fisheries Research and Development Institute (NFRDI) of Korea and others have also collected DO time series data (Kim et al. 1991; Shim et al. 1992), but most data were collected

at shallow depths (<600 m depth). High quality DO data from a spectrophometri-
cally modified Winkler method (Pai et al. 1993) and a DO sensor (Sea Bird model
SBE13) were collected over the entire East Sea during the Circulation Research of
the East Asian Marginal Seas (CREAMS) program as part of Korea-Russia-Japan
international cooperative studies between 1993 and 1996, while most earlier data
were measured by the Winkler method (Uda 1934; Gamo 2011). The CREAMS
study reported an intrusion of surface water with high DO concentration between
600 m and 1200 m (Kim et al. 2001) and confirmed the earlier observations by
Gamo et al. (1986). During CREAMS II (1999–2000), basin-wide distributions of
DO concentrations were examined as part of Korea-USA-Russia-Japan international cooperative studies. From 2001 to 2012, DO concentrations were measured
continuously in the Ulleung Basin and the western Japan Basin for the East Asian
Seas Time Series-I (EAST-I) program as part of Korea-Russia cooperative studies
(Kim et al. 2002; Kang et al. 2010a).

6.2.2 Vertical Structure of DO Profile

In general, from the 1930s to the late 1990s, DO concentrations in the East
Sea were high (>300 μmol kg^{-1}) within the surface layer (<200 m), and
then decreased rapidly with increasing depth to the minimum value (ca.
220 μmol kg^{-1}) at about 1300 m (Fig. 6.1). DO concentrations increased
slightly between 1300 and 2100 m, and then remained almost constant (ca.
230 μmol kg^{-1}) to the bottom layer (Fig. 6.1).

The shape of the DO profile during the late 1970s reflects the expected vertical DO trend, with DO concentration controlled by the balance between air-sea
exchange and/or biological production in the surface layer and microbial consumption at depth (Fig. 6.1). The slight increase in DO concentration from the

Fig. 6.1 Vertical profiles
of dissolved oxygen (DO)
concentration observed from
the 1930s to the late 1990s
in the East Sea (from Kang
et al. 2004)

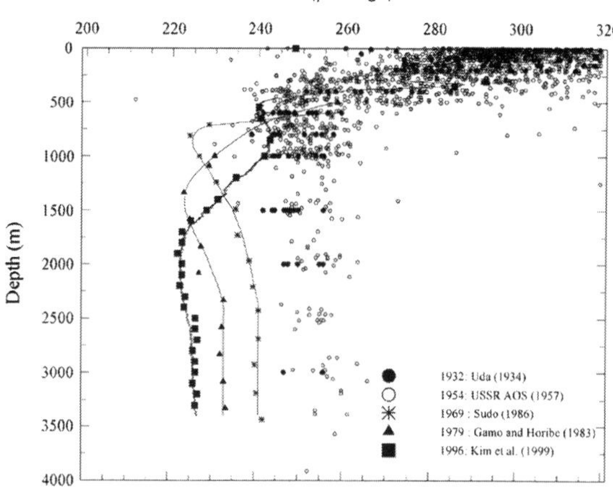

oxygen minimum zone toward the bottom layer indicates the presence of an oxygen supply mechanism to the deep layer, such as the formation of deep and bottom water in the East Sea (Kim et al. 2002; Kang et al. 2003).

During the late 1990s, the general shape of the vertical DO profile remained unchanged from that of the late 1970s (Fig. 6.1). However, the oxygen minimum layer was observed at a deeper depth than in the late 1970s, and the DO concentration in the bottom layer decreased slightly. The most noticeable change in the vertical DO profile was the increase in DO concentration between 600 and 1200 m (ca. 237 μmol kg^{-1} at a depth of 1000 m). Since there was no biological production at this depth, other mechanisms (such as an intrusion of oxygen rich surface water) must be responsible for the increase in DO concentration.

Along the Russian coast during winter, deep convection and/or brine rejection generate dense water masses, which sink down to the intermediate, central, and bottom layers and replenish oxygen in the interior of the East Sea (Kang et al. 2003; Postlethwaite et al. 2005). Thus the change in the shape of vertical DO concentration profiles is closely related to changes in the pathway of the sinking oxygen-rich surface water. During the winter of 2001, the massive injection of oxygen-rich surface water into the bottom layer resulted in a sudden increase of DO concentration in the bottom layer (Kim et al. 2002). The changes in vertical structure of DO concentration were synchronized over different parts of the East Sea (Ulleung Basin, Yamato Basin, and Japan Basin), suggesting that this process may have influenced all basins within a short time, although it begins along the Russian coast.

6.2.3 *Trend of DO Inventory in the Bottom Layer*

A significant temporal change in DO inventory has been reported for the bottom layer in the East Sea (Gamo et al. 1986; Chen et al. 1999; Kim et al. 2001). Gamo et al. (1986) reported a sharp decrease of bottom DO inventory as well as a decreasing trend in the thickness of the bottom adiabatic layer in the eastern Japan Basin and the Yamato Basin during the 1969–1984 period. DO concentrations appear to have gradually decreased in the bottom adiabatic layer from the early 1930s to the late 1990s, suggesting a gradual warming of bottom water in the East Sea (Kim et al. 2001). Long-term monitoring of DO concentrations within the interior of the East Sea revealed a decreasing linear trend with temperature increases in the upper 1000 m layer (0.1–0.5 °C) and below 2000 m (0.01 °C) during the 1950–2007 period (Chen et al. 1999; Minami et al. 1999; Kim et al. 2001; Gamo 2011). However, there were also interdecadal oscillations of DO concentrations in the intermediate water mass (1000–1500 m) superimposed on the linear trend in the East Sea (Watanabe et al. 2003; Cui and Senjyu 2010).

The decrease of DO inventory in the bottom adiabatic layer has been attributed to the shift in the mode of deep water formation, from a bottom water formation mode in the late 1970s to an intermediate water formation mode in the late 1990s

(Kim et al. 2001; Gamo 2011). Other studies proposed that the periodic interdecadal changes in DO concentrations may be closely related to global scale fluctuations of climate, such as the North Pacific Index (NPI), the monsoon index (MOI), and the Arctic Oscillation Index (AOI) (Yasuda and Hanawa 1999; Gong et al. 2001; Watanabe et al. 2003; Cui and Senjyu 2010).

6.2.4 Projecting DO Inventory in Relation to Climate Change

The expansion of the OMZ in the major oceans and the predicted continuous reduction of oxygen concentration in the next century have evoked great concern in the scientific community. There have been attempts to project the fate of DO concentration in the deep layer of the East Sea (Chen et al. 1999; Kang et al. 2003). Chen et al. (1999) projected that the deep water of the East Sea could become anoxic by the year 2200 based on the oxygen consumption rate, estimated from the mass balance method, the 1-D advection-diffusion model, and direct comparison of the 1950 and 1995 data, assuming continued stagnation of the deep water.

On the other hand, Kang et al. (2003) developed a Moving Boundary Box Model (MBBM), which describes the change of ventilation flux from the cold surface to East Sea Central Water (ESCW), East Sea Deep Water (ESDW), and East Sea Bottom Water (ESBW), respectively, and allows for structural change in relative size of each water reservoir (Fig. 6.2). The MBBM is based on several

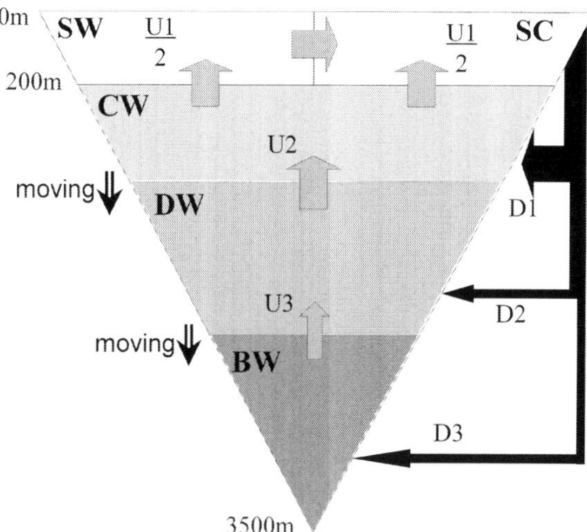

Fig. 6.2 A schematic illustration of the Moving Boundary Box Model for the East Sea. *SW* is Surface Warm Water, *SC* is Surface Cold Water, *CW* is East Sea Central Water, *DW* is East Sea Deep Water, *BW* is East Sea Bottom Water. D1, D2, D3, U1, U2, and U3 represent fluxes between boxes. Boundaries of CW/DW and DW/BW have been changed with time (from Kang et al. 2003)

assumptions: (1) deep water originates only from cold surface water (SC); (2) total rate of deep water formation (sum of D1, D2, and D3; see Fig. 6.2) is the same as upwelling to the surface (U1), which remains constant with time although the individual rates of D1, D2, and D3 change; (3) Surface Warm Water (SW) and SC are equally divided in area and U1 occurs equally in SW and SC; (4) the ESBW formation rate (D3) decreases linearly from the year 1952 onward and is compensated for by an increase of the CW formation rate, but D2 remains constant.

Kang et al. (2004) estimated convective oxygen flux from the surface into deep layers by multiplying the DO concentration in the surface layer and the biological oxygen consumption (also called oxygen utilization rate, OUR) in each water mass by comparison of the model results with the measured DO concentration (Fig. 6.3). They also assumed that the OUR remained constant with time, while the convective fluxes changed with time as the deep water formation mode changed. The MBBM projected that the CW would expand and occupy about 90 % of the deep water by the year 2020 because of changes in the deep water formation mode, but consequently the deep water will not become anoxic (Kang et al. 2004).

In contrast to models that assumed stagnation of deep water since the 1950s, several studies have reported episodic bottom water formation during the severe winter of 2000–2001 in the East Sea (Kim et al. 2002; Hahm and Kim 2008). This observation undermines one of the basic assumptions of the models, and also the models' future projections of DO concentrations in the deep water. Clearly, continued monitoring of DO concentrations is required to better understand mechanisms controlling DO concentration in the East Sea.

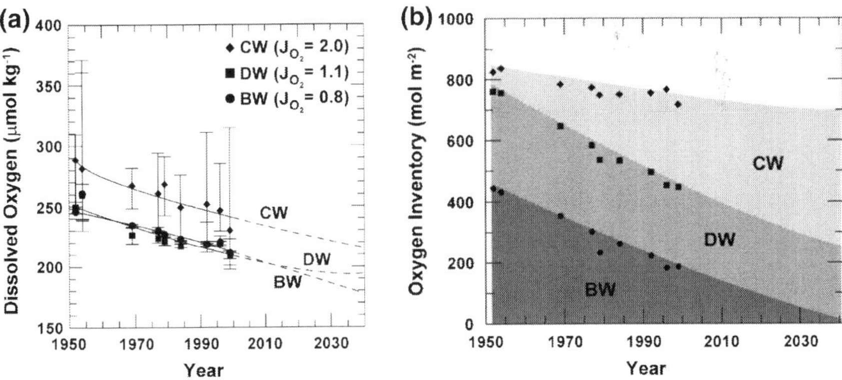

Fig. 6.3 a Temporal variation of dissolved oxygen concentrations in deep water masses and best-fit lines for the observed dissolved oxygen of each deep-water mass obtained from the MBBM. Error bars and symbols represent the maximum/minimum value and the median value of measured historical data, respectively. Dashed lines represent model results showing the secular change of dissolved oxygen in the deep-water masses of the East Sea. b Temporal variation in oxygen inventories for each water mass in the East Sea, simulated by the MBBM (from Kang et al. 2004)

6.3 Nutrients

6.3.1 History of Nutrient Studies

Since 1994, NFRDI of Korea has measured the concentration of certain nutrients (phosphate, nitrate, nitrite, and silicate) in the surface water (<100 m depth) of the East Sea. Nutrient concentration is periodically monitored at 69 stations along 8 transects perpendicular to the coastline in the Ulleung Basin (data are available at http://kodc.nfrdi.re.kr/page?id=kr_index). The Japan Meteorological Agency also periodically monitors nutrient concentrations in the Japan Basin and Yamato Basin of the East Sea (data are available at www.data.go.jp). Since 1991, concentrations of phosphate, nitrate, nitrite, ammonium, and silicate along the near-costal region have also been regularly monitored (in February, May, August, and November) by the National Marine Environmental Monitoring Array (data are available at http://www.meis.go.kr). Individual studies have also reported nutrient concentrations in various parts of the Ulleung Basin from the 1980s to the 1990s (Shim and Park 1986; Chung et al. 1989; Shim et al. 1992; Kim et al. 1991; Shim and Park 1996; Moon et al. 1996; Cho et al. 1997; Yang et al. 1997; Han et al. 1998; Moon et al. 1998). However, all the nutrient measurements mentioned were restricted to the upper 1000 m of the Ulleung Basin, which spatially limits our understanding of nutrient dynamics to one region within the East Sea.

Full vertical profiles of nutrients were measured basin-wide over the entire East Sea through the international research program, CREAMS, during the 1993–1996 period. During the CREAMS II expeditions (1999–2000), nutrient profiles of high spatial horizontal and vertical resolutions were obtained in summer and winter seasons over most of the East Sea (data are available at http://sam.ucsd.edu/onr_data/hydrography.html). These data provided a breakthrough in our understanding of nutrient dynamics in the East Sea (Talley et al. 2004; Kim and Lee 2004; Kim et al. 2010a; b). From 2000 to 2007, series of cooperative research surveys between Korea and Russia were conducted along the 132°E line in the Ulleung Basin and the western Japan Basin. Since 2007, these surveys have been succeeded by the EAST-I program. Nutrient data produced from EAST-I were used to understand the changes in N:P ratio and its influence on the composition of phytoplankton in the East Sea (Kim et al. 2010c; Kim and Kim 2013).

6.3.2 Spatial and Temporal Distribution of Nutrients

Generally, the conversion of dissolved nutrients into POM depletes nutrients in the surface euphotic layer and then nutrients increase below the surface mixed layer (SML) due to the degradation of sinking POM.

During summer in the Ulleung, Yamato, western Japan and eastern Japan basins, vertical nutrient distributions show that nitrate and phosphate are almost entirely depleted in the surface layer, followed by increases to about 25 μmol kg^{-1} for nitrate and about 2 μmol kg^{-1} for phosphate down to about 500 m depth (Fig. 6.4). Below 500 m, concentrations for both nutrients remain relatively constant (Fig. 6.4). Silicate is also depleted almost completely in the SML; it then gradually increases to >80 μmol kg^{-1} at about 2000 m depth, and then remains constant down to the bottom (3500 m depth). Among basins, there were no large differences in vertical nutrient distributions at depths greater than 750 m (Fig. 6.4). However, there was a slight decrease in both nitrate and phosphate concentrations between 300 and 500 m in the eastern and western Japan Basins when compared to the Ulleung and Yamato Basins (Fig. 6.4).

During winter, nutrient concentrations in deep waters were similar to summer concentrations. However, nutrient concentration in surface waters showed large variations, with ranges 5.8–14.8 μmol kg^{-1} for nitrate, 0.45–1.16 μmol kg^{-1} for phosphate, and 8.5–24.7 μmol kg^{-1} for silicate (Fig. 6.4). In the western Japan Basin, surface nutrient concentrations are higher than in the Ulleung and

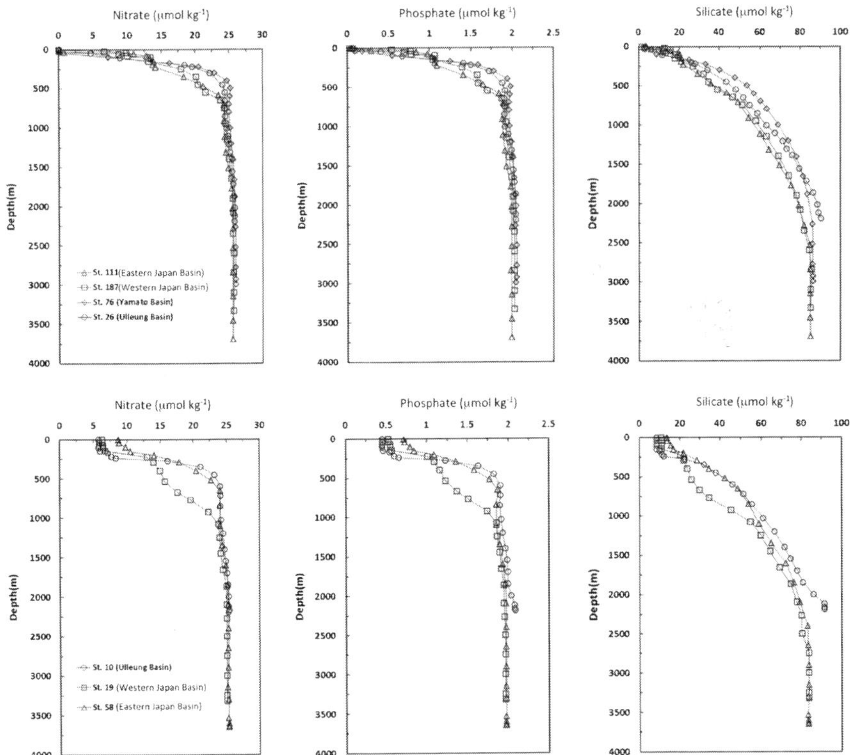

Fig. 6.4 Vertical profiles of nitrate, phosphate, and silicate in the Ulleung, Yamato, western Japan, and eastern Japan Basins during summer 1999 (*upper*) and winter 2000 (*bottom*)

eastern Japan basins. The difference is probably related to deeper vertical mixing in the western Japan Basin during winter, which decreases nutrient concentrations between about 500 m and 700 m. Nutrient concentrations in deeper layers (>200 m) of the East Sea are higher than in the North Atlantic, and lower than in the North Pacific Ocean (Kim and Kim 2013).

During summer 1999, there are strong horizontal gradients in nutrient concentrations along isobaric surfaces at pressure levels less than 200 dbar (Fig. 6.5). Along the 50 dbar surface, nutrient concentrations in the western and eastern Japan basins are higher than in the Ullueng and Yamato basins. However, nutrient concentrations are elevated along the coastal region of the Korean Peninsula, probably because of frequent coastal upwelling in this region during summer (Lee and Kim 2003; Yoo and Park 2009; Rho et al. 2010).

Horizontal gradients along isobaric surfaces diminish as water column depth increases (Fig. 6.5). Nutrients concentrations along the 1500 dbar surface show no apparent difference over the entire East Sea.

Episodic events such as coastal upwelling, the passage of typhoons, and the presence of mesoscale eddies can modify the vertical structure of nutrient profiles in the surface layer (Son et al. 2006; Hyun et al. 2009). Hyun et al. (2009) reported that high nutrient inventories were maintained outside an anticyclonic eddy, whereas low and homogeneous water masses extended from the surface to a depth of 200 m in the middle of a warm eddy in the Ulleung Basin.

6.3.3 Factors Controlling Nutrient Concentrations in the Surface Layer

Recently, it has been suggested that the most of the primary production in the southern part of the East Sea could be supported by the nutrients transported through the Korea Strait (Lee and Rho 2013). The basin-wide distribution of the subsurface chlorophyll maximum was closely related to the High Salinity Tsushima Warm Water (HSTWW) passing through the Korea Strait during the stratified season (Rho et al. 2012). Nutrient flux through the Korea Strait is determined both by the amount of water passing through the Korea Strait and by the nutrient concentration of the water mass. Transport through the Korea Strait has been estimated by various methods (Na et al. 2009). Here, we will describe the nutrient concentration in the water passing through the Korea Strait and estimate its contribution to the nutrient inventory of the surface ocean.

Major factors controlling nutrient concentrations in the East Sea's surface waters are (1) the input of nutrients through the Korea Strait in the southwestern region; (2) the export of nutrients through the Tsugaru and Soya straits in the northern region; (3) atmospheric deposition; (4) processes within the East Sea, such as coastal upwelling (Yanagi 2002; Lee and Kim 2003; Onitsuka and Yanagi 2005; Jenkins 2008; Jang et al. 2013).

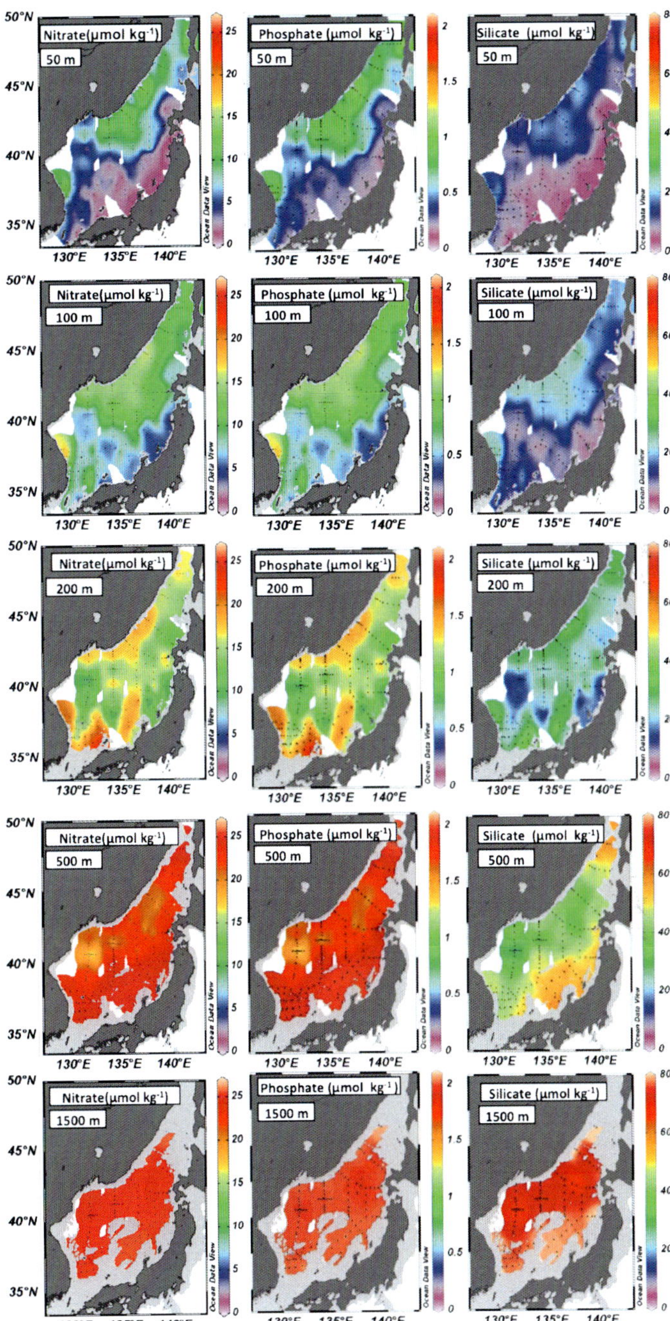

Fig. 6.5 Horizontal distributions of nitrate, phosphate, and silicate in the East Sea at the 50, 100, 200, 500, and 1500 dbar isobaric surfaces

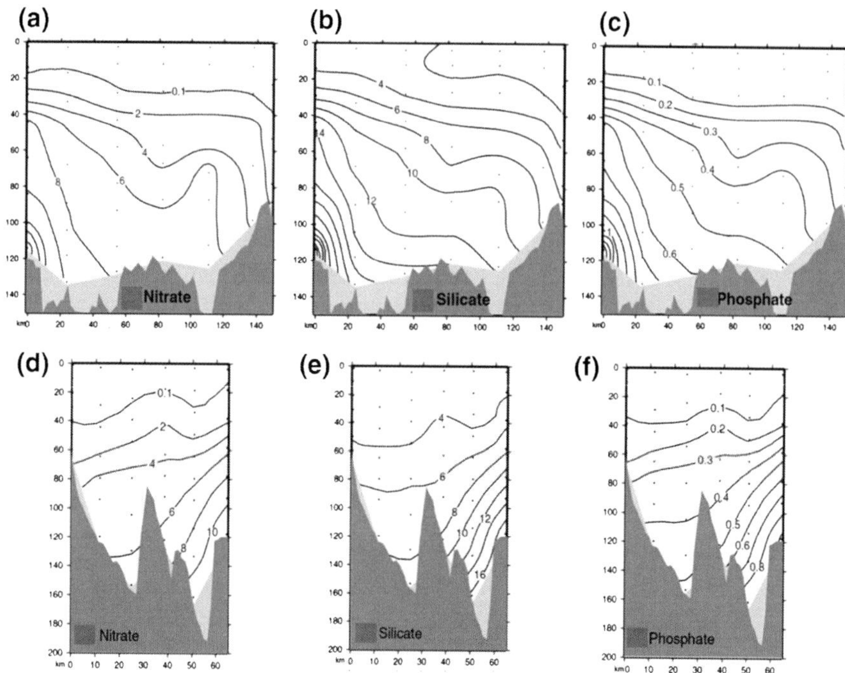

Fig. 6.6 Vertical cross sections of nitrate, silicate, and phosphate in the Korea Strait (**a–c**) and the Tsugaru Strait (**d–f**) during summer 1999 (from Talley et al. 2004)

Vertical cross sections of nutrient concentrations in the Korea and Tsugaru straits show that nutrients are almost depleted in the SML (<30 m) during the stratified season, whereas they are enriched below the SML (Fig. 6.6). Long-term time series measurements of nutrient concentrations have been conducted at station M (34° 46′ 30″N, 129° 07′ 59″E) in the Korea Strait since 2006. Vertical nutrient concentrations from March 2006 to January 2008 revealed strong seasonality throughout the water column (Fig. 6.7). Combined nitrate and nitrite concentrations ranged approximately from 0.01 to 13.8 μM (4.0 ± 3.6 μM). Phosphate concentrations were approximately between 0.01 and 1.19 μM (with a mean of 0.25 ± 0.27 μM). Silicate concentrations ranged approximately from 0.93 to 29.1 μM (with a mean of 10.9 ± 6.4 μM).

Vertical cross sections of nutrient concentrations showed that nutrients were vertically homogenous, with relatively low nutrient concentrations during the winter, when waters were well mixed (blue boxes in Fig. 6.7). During the winter, mean nutrient concentrations were around 8 μM for silicate, 4 μM for nitrate + nitrite, and 0.3 μM for phosphate. The concentrations of nutrients contained in the water mass passing station M at about 100 m depth were approximately 800 mmol m^{-2} for silicate, 400 mmol m^{-2} for nitrate + nitrite, and 30 mmol m^{-2} for phosphate. Nutrient concentrations in the surface layer decreased when thermal stratification developed (Fig. 6.7).

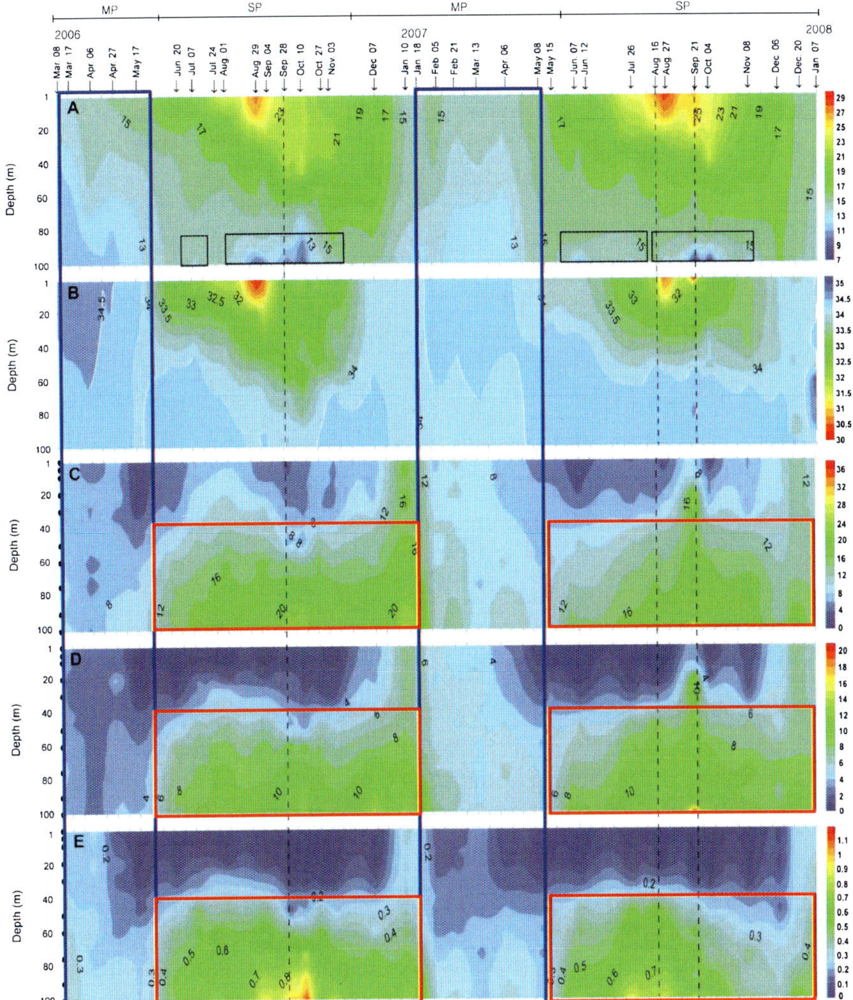

Fig. 6.7 Seasonal variation at station M (34° 46′ 30″N, 129° 07′ 59″E) in the Korea Strait, of **a** temperature (°C), **b** salinity(psu), **c** silicate (μM), **d** nitrate + nitrite (μM), and **e** phosphate (μM), from March 2006 to January 2008 (from Jang et al. 2013)

Hydrographic conditions clearly showed the seasonality of water column structure in the Korea Strait. Temperature and salinity during the winter were around 13–15 °C and 34.3–34.5, respectively (Fig. 6.7). During the stratified summer, surface waters were warmer (temperatures > 17 °C) and less saline (30–33.5), similar to Changjiang Diluted Water (CDW). In contrast, bottom waters were cooler (about 13–15 °C) and more saline (about 34–34.5), similar to HSTWW (Rho et al. 2010). Water masses below the HSTWW during the stratified season were cold (<10 °C) and saline (about 34.2 psu), similar to Korea Strait Bottom Cold Water (KSBCW) flowing out from the interior of the East Sea.

During the stratified season, nutrient concentrations showed a two-layer structure. Nutrients were almost depleted in the surface layer (<40 m depth) and mean nutrient concentrations in the surface layer were approximately 4 μM for silicate, 1 μM for nitrate + nitrite, and 0.1 μM for phosphate. High nutrient concentrations were observed in the lower layer (>40 m depth, red box in Fig. 6.7) and mean nutrient concentrations in the lower layer were about 16 μM for silicate, 8 μM for nitrate + nitrite, and 0.6 μM for phosphate. This suggests that the surface water mass passing through the Korea Strait contained approximately 160 mmol m^{-2} of silicate, 40 mmol m^{-2} of nitrate + nitrite, and 4 mmol m^{-2} of phosphate; while the lower water mass contained 960 mmol m^{-2} for silicate, 480 mmol m^{-2} nitrate + nitrite, and 36 mmol m^{-2} for phosphate. During the stratified season, total concentrations of nutrients in the water column at the Korea Strait were 1120 mmol m^{-2} for silicate, 520 mmol m^{-2} for nitrate + nitrite, and 40 mmol m^{-2} for phosphate.

During winter, the water mass originating from the Kuroshio occupied the entire water column of the Korea Strait (100 m depth), with relatively low nutrient concentrations. During summer, however, the water mass originating from the Kuroshio occupied the lower layer (>40 m depth) of the Korea Strait with higher nutrient concentrations. The upper layer (<40 m depth) of the Korea Strait was occupied by less saline CDW water with low nutrient concentrations. The total amount of nutrients contained in the water mass passing through the Korea Strait was higher during the stratified summer season than during the well-mixed winter season.

6.3.4 Major Research Topics Involving Nutrients

The East Sea is located downwind of northern China, the world's biggest anthropogenic N source over the last few decades. In the East China Sea, anthropogenic N flux into the surface ocean may be comparable to the riverine inputs of the Changjiang (Nakamura et al. 2005). A portion of the anthropogenic N emitted from China was transported to marginal seas in the northwest Pacific (Ohara et al. 2007). Additional N flux from atmospheric deposition contributes approximately 10–15 % of annual new production over the southern East Sea (Onitsuka et al. 2009; Kang et al. 2010b).

One interesting problem regarding nutrient dynamics in the East Sea is whether the N inventory in the southwestern part of this sea has changed, and if it has, what might be the source of N into this region (Kim et al. 2011, 2013; Kim and Kim 2013). Kim et al. (2011) argued that there was an increase in relative abundance of N to P in the southwestern region of the East Sea (Ulleung Basin) because of the enhanced atmospheric N input from China during the last three decades. They suggested that N flux via atmospheric processes enhanced integrated N* (N $-$ R$_{N:P}$ $\times$ P, here R$_{N:P}$ is Redfield N:P ratio) from the surface to a depth of 750 m, and that this increase may result in P-limitation instead of the traditionally observed N-limitation of phytoplankton growth during the past decades (Kim et al. 2011). In a later study, Kim et al. (2013) suggested that nutrients transported by ocean currents through the Korea Strait could be a more crucial factor for the fluctuating ratio of dissolved inorganic nitrogen to phosphorus in the region.

The N:P ratio observed in the deep water showed no apparent seasonality in the entire East Sea region while there were large seasonal difference in the surface layer, suggesting biological modification of the N:P ratio in the surface layer. During the CREAMS II and EAST-I cruises in the East Sea, observed N:P ratios were 8.7 ± 5.5 in the surface layer (<200 m), and 12.7 ± 0.14 in the deep layer (>200 m) (Fig. 6.8). Kim and Kim (2013) reported that the N:P ratios of the surface water mass (<200 m) and the deep water mass (>200 m) in the Ulleung Basin and western Japan Basin were 7.0 ± 6.2 and 13 ± 1, respectively, during 2001–2009. N:P ratios observed in the Ulleung Basin in March 2013 were 16.0 ± 3.0 in the surface layer and 12.9 ± 0.31 in the deep layer (Fig. 6.8). Therefore, the N:P ratio in the surface layer may not be appropriate for the comparison of long-term changes. There was no apparent temporal change in N:P ratio of the deep layer during last decade. Kim and Kim (2013) argued that there was no shift from N-limitation to P-limitation and predicted no increase of nitrogen within the water column caused by atmospheric nitrogen input in the next century. Time series of vertical nitrate + nitrite and phosphate concentrations show similar vertical structures and no apparent changes in nutrient concentration between July 1999 and March 2013 in the East Sea (Fig. 6.8). Nevertheless, there should be continuous monitoring of N:P ratios to detect any changes that may result from anthropogenic perturbation.

A fundamental but unanswered question is why the N:P ratio of the East Sea is much lower than the Redfield ratio. Yanagi (2002) noted that the N:P ratios of riverine sources and water masses passing through the Korea Strait were 16.4 and 16.6, respectively, although in the interior of the East Sea, the N:P ratio was low (11.3).

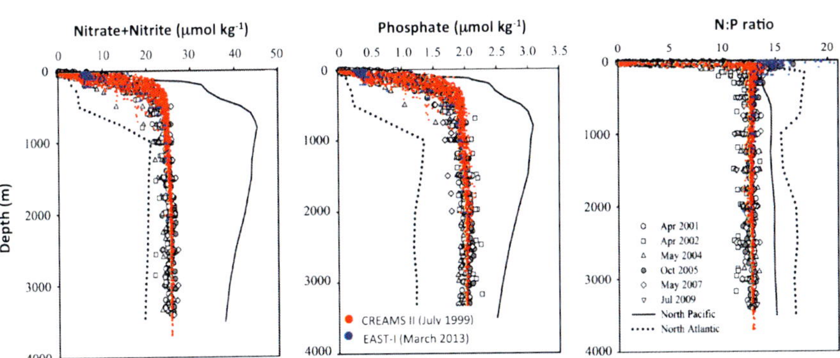

Fig. 6.8 Vertical profiles of nitrate + nitrite, phosphate concentrations, and N:P ratio in the East Sea. *Black* symbols represent data used by Kim and Kim (2013). *Red* and *blue circles* respectively indicate data collected over the entire East Sea during July 1999 and in the Ullueng Basin alone during March 2013

He argued that this may be caused by a selective nitrogen removal process (such as denitrification) within the water column. Kim et al. (2012) also suggested the possibility of denitrification in the deep layer of the East Sea based on low N:P ratios (<12.4) and the occurrence of nitrite peaks despite high DO concentrations. Denitrification rates estimated by Kim et al. (2012) were similar to those measured in sediments near Dokdo (Dok Island) in the Ulleung Basin (Jeong et al. 2009).

In addition to denitrification, anaerobic ammonium oxidation (anammox) may be an important fixed nitrogen sink in marine sediments (Rysgaard et al. 2004; Dalsgaard et al. 2005; Engström et al. 2005; 2009; Nicholls and Trimmer 2009) and anoxic water columns (Dalsgaard et al. 2003; Kuypers et al. 2003; 2005). Anammox and denitrification measured in sediments collected from stations in the Korea Strait and the Dokdo shelf were about 0.35 nmole N ml^{-1} wet sediment h^{-1} for anammox and 1.1 nmole N ml^{-1} wet sediment h^{-1} for denitrification in the Korea Strait, and about 0.002–0.11 nmole N ml^{-1} wet sediment h^{-1} for anammox and about 0.01–0.61 nmole N ml^{-1} wet sediment h^{-1} for denitrification on the Dokdo shelf. The contribution of anammox to the total N production increased with water column depth and showed a maximum value of 47 % at about 2200 m. This implies that anammox in sediments may play a large role in the removal of fixed nitrogen in the East Sea. It is also possible that the nitrogen loss (denitrification and/or anammox) in surface sediments may be responsible for the nitrogen deficit in the overlying water column and for the low N:P ratio in the East Sea.

However, other studies argue against the occurrence of in situ denitrification in the interior of the East Sea because of the high DO concentrations in deep waters (Lee et al. 2009; Kim and Kim 2013). Kim and Kim (2013) proposed that the low N:P ratio in the deep layer of the East Sea may result from physical processes, such as rapid ventilation of the surface water mass with a low N:P ratio (<10) in the deep ocean. The low N:P ratio (<8) in surface waters during spring and summer could be related to the biological utilization of N and the presence of excess P in the surface water column (Kim et al. 2010c). Therefore, the relative contributions of various processes should be evaluated in future studies, to provide a better explanation for the low N:P ratios observed in the overlying water column.

6.4 Summary and Future Challenges

As changing global climate causes ocean warming and increases ocean stratification, DO inventories may decline in the ocean interior. According to model predictions, the DO inventory will decrease by 1–7 % over the next century. This decline will likely increase the area and volume of the OMZ; thus it may significantly impact macrofauna and also cause profound changes in biogeochemical cycling. The DO inventory of the ocean interior is controlled by two processes: (1) ventilation that supplies atmospheric gases to the ocean interior, including air-sea gas exchange, mixing between the surface mixed layer and the immediate subsurface layer, and circulation in the ocean interior; (2) utilization within the water column, including biological consumption and microbial consumption in the degradation of sinking organic matter from the surface euphotic zone.

Changes in nutrient concentration in the ocean interior are tightly related to changes in DO. Recent reports on changes in the nitrogen content of the East Sea have kindled our interest in the sources and fates of nutrients in the East Sea. The relative importance of different nitrogen fluxes to the East Sea (atmospheric input and transport by advection through the Korea Strait) and resultant changes in nutrient dynamics are under vigorous debate. The mechanisms resulting in the low N:P ratios observed in the interior of the East Sea are also uncertain.

Because of the close relationship between nutrient concentrations and DO levels, understanding these processes is key to accurately projecting future DO inventories. It is also important to monitor DO concentrations continuously with increased precision, so that changes on the order of a few μmol kg^{-1} decade^{-1} can be detected. To settle the ongoing debate on the nutrient dynamics in the East Sea, nutrient concentrations should be continuously monitored in the Korea Strait and interior of the East Sea, and atmospheric fallout must also be measured with high precision and accuracy. This may also provide a clue as to whether changes in DO concentration result from increased biological consumption within the water column, or from the decrease in deep water formation, or both.

References

Chen CTA, Bychkov AS, Wang SL et al (1999) An anoxic Sea of Japan by the year 2200? Mar Chem 67:249–265

Cho HJ, Moon CH, Yang HS et al (1997) Regeneration processes of nutrients in the polar front area of the East Sea III. Distribution patterns of water masses and nutrients in the middle-northern East Sea of Korea in October, 1995. J Korean Fish Soc 30:442–450

Chung CS, Shim JH, Park YC et al (1989) Primary productivity and nitrogenous nutrient dynamics in the East Sea of Korea. J Oceanol Soc Korea 24:52–61

Cui Y, Senjyu T (2010) Interdecadal oscillations in the Japan Sea Proper Water related to the arctic oscillation. J Oceanogr 66:337–348

Dalsgaard T, Canfield DE, Petersen J et al (2003) N$_2$ production by the anammox reaction in the anoxic water column of Golfo Dulce, Costa Rica. Nature 422:606–608

Dalsgaard T, Thamdrup B, Canfield DE (2005) Anaerobic ammonium oxidation (anammox) in the marine environment. Res Microbiol 156:457–464

Engström P, Dalsgaard T, Hulth S et al (2005) Anaerobic ammonium oxidation by nitrite (anammox): implications for N$_2$ production in coastal marine sediments. Geochim Cosmochim Acta 69:2057–2065

Engström P, Penton CR, Devol AH (2009) Anaerobic ammonium oxidation in deep-sea sediments off the Washington margin. Limnol Oceanogr 54:1643–1652

Eppley RW, Peterson BJ (1979) Particulate organic matter flux and planktonic new production in the deep ocean. Nature 282:677–680

Gamo T (2011) Dissolved oxygen in the bottom water of the Sea of Japan as a sensitive alarm for global climate change. Trend Anal Chem 30:1308–1319

Gamo T, Nozaki Y, Sasaki H et al (1986) Spatial and temporal variations of water characteristics in the Japan Sea bottom layer. J Mar Res 44:781–793

Glibert PM (1982) Regional studies of daily, seasonal, and size fraction variability in ammonium remineralization. Mar Biol 70:209–222

Gong DY, Wang SW, Zhu JH (2001) East Asian winter monsoon and Arctic Oscillation. Geophys Res Lett 28:2073–2076

Hahm D, Kim KR (2008) Observation of bottom water renewal and export production in the Japan Basin, East Sea using tritium and helium isotopes. Ocean Sci J 43:39–48

Han MS, Jang DH, Yang HS (1998) The ecosystem of the southern coastal water of the East Sea, Korea II. Primary productivity in and around cold water mass. J Oceanol Soc Korea 33:196–204

Hyun JH, Kim D, Shin CW et al (2009) Enhanced phytoplankton and bacterioplankton production coupled to coastal upwelling and an anticyclonic eddy in the Ulleung Basin, East Sea. Aquat Microb Ecol 54:45–54

Jang PG, Shin HH, Baek SH et al (2013) Nutrient distribution and effects on phytoplankton assemblages in the western Korea/Tsushima Strait. New Zealand J Mar Freshwater Res 47:21–37

Jenkins WJ (2008) The biogeochemical consequences of changing ventilation in the Japan/East Sea. Mar Chem 108:137–147

Jeong JH, Kim DS, Lee TH et al (2009) High remineralization and denitrification activity in the shelf sediments of Dok Island, East Sea. J Korean Soc Oceanogr 14:80–89

Joos F, Plattner GK, Stocker TF et al (2003) Trends in marine dissolved oxygen: Implications for ocean circulation changes and carbon budget. EOS Trans Am Geophys Union 84:197

Kang DJ, Park Y, Kim YG et al (2003) A moving-boundary box model (MBBM) for oceans in change: an application to the East/Japan Sea. Geophys Res Lett 30:1229. doi:10.1029/200 2GL016484

Kang DJ, Kim JY, Lee T et al (2004) Will the East/Japan Sea become an anoxic sea in the next century? Mar Chem 91:77–84

Kang DJ, Kim YB, Kim KR (2010a) Dissolved oxygen at the bottom boundary layer of the Ulleung Basin, East Sea. Ocean Polar Res 32:439–448

Kang J, Cho BC, Lee CB (2010b) Atmospheric transport of water-soluble ions (NO^{3-}, NH^{4+}, and nss-SO_4^{2-}) to the southern East Sea (Sea of Japan). Sci Total Environ 408:2369–2377

Keeling RF, Körtzinger A, Gruber N (2010) Ocean deoxygenation in a warming world. Annu Rev Mar Sci 2:199–229

Kim TH, Kim G (2013) Changes in seawater N:P ratios in the northwestern Pacific ocean in response to increasing atmospheric N deposition: results from the East (Japan) Sea. Limnol Oceanogr 58:1907–1914

Kim IN, Lee T (2004) Summer hydrographic features of the East Sea analyzed by the optimum multiparameter method. Ocean Polar Res 26:581–594

Kim KR, Rhee TS, Kim K et al (1991) Chemical characteristics of the East Sea intermediate water in the Ulleung Basin. J Oceanol Soc Korea 26:278–290

Kim K, Kim KR, Min DH et al (2001) Warming and structural changes in the East (Japan) Sea: a clue to future changes in the global ocean? Geophys Res Lett 28:3293–3296

Kim KR, Kim G, Kim K et al (2002) A sudden bottom-water formation during the severe winter 2000-2001: the case of the East/Japan Sea. Geophys Res Lett 29:1234. doi:10.1029/200 1GL014498

Kim BG, Lee T, Kim IN (2010a) Phosphate vs. silicate discontinuity layer developed at mid-depth in the East Sea. Ocean Polar Res 32:331–336

Kim IN, Min DH, Kim DH et al (2010b) Investigation of the physicochemical features and mixing of East/Japan Sea intermediate water: an isopycnic analysis approach. J Mar Res 68:799–818

Kim TH, Lee YW, Kim G (2010c) Hydrographically mediated patterns of photosynthetic pigments in the East/Japan Sea: low N:P ratios and cyanobacterial dominance. J Mar Syst 82:72–79

Kim TW, Lee K, Najjar RG et al (2011) Increasing N abundance in the northwestern Pacific ocean due to atmospheric nitrogen deposition. Science 334:505–508

Kim IN, Min DH, Lee T (2012) Deep nitrate deficit observed in the highly oxygenated East/Japan Sea and its possible cause. Terr Atmos Ocean Sci 23:671–683

Kim SK, Chang KI, Kim B et al (2013) Contribution of ocean current to the increase in N abundance in the northwestern Pacific marginal seas. Geophys Res Lett 40:143–148

Kuypers MMM, Sliekers OA, Lavik G et al (2003) Anaerobic ammonium oxidation by anammox bacteria in the Black Sea. Nature 422:608–611

Kuypers MMM, Lavik G, Woebken D et al (2005) Massive nitrogen loss from the Benguela upwelling system through anaerobic ammonium oxidation. Proc Natl Acad Sci 102:6478–6483

Lee T, Kim IN (2003) Chemical imprints of the upwelled waters off the coast of the southern East Sea of Korea. J Oceanol Soc Korea 38:101–110

Lee T, Rho T (2013) Contribution of nutrient flux through the Korea Strait to a primary production in the warm region of the East Sea. J Korean Soc Oceanogr 18:65–69

Lee JY, Kang DJ, Kim IN et al (2009) Spatial and temporal variability in the pelagic ecosystem of the East Sea (Sea of Japan): a review. J Mar Syst 78:288–300

Minami H, Kano Y, Ogawa K (1999) Long-term variations of potential temperature and dissolved oxygen of the Japan Sea Proper Water. J Oceanogr 55:197–205

Moon CH, Yang HS, Lee KW (1996) Regeneration processes of nutrients in the polar front area of the East Sea I. Relationships between water mass and nutrient distribution pattern in autumn. J Korean Fish Soc 29:503–526

Moon CH, Yang SR, Yang HS et al (1998) Regeneration processes of nutrients in the polar front area of the East Sea IV. Chlorophyll *a* distribution, new production and the vertical diffusion of nitrate. J Korean Fish Soc 31:259–266

Na H, Isoda Y, Kim K et al (2009) Recent observations in the straits of the East/Japan Sea: a review of hydrography, currents and volume transports. J Mar Syst 78:200–205

Nakamura T, Matsumoto K, Uematsu M (2005) Chemical characteristics of aerosols transported from Asia to the East China Sea: an evaluation of anthropogenic combined nitrogen deposition in autumn. Atmos Environ 39:1749–1758

Nicholls JC, Trimmer M (2009) Widespread occurrence of the anammox reaction in estuarine sediments. Aquat Microb Ecol 55:105–113

Ohara T, Akimoto H, Kurokawa J et al (2007) An Asian emission inventory of anthropogenic emission sources for the period 1980–2020. Atmos Chem Phys 7:4419–4444

Onitsuka G, Yanagi T (2005) Differences in ecosystem dynamics between the northern and southern parts of the Japan Sea: analyses with two ecosystem models. J Oceanogr 61:415–433

Onitsuka G, Uno I, Yanagi T et al (2009) Modeling the effects of atmospheric nitrogen input on biological production in the Japan Sea. J Oceanogr 65:433–438

Pai S, Gong G, Liu K (1993) Determination of dissolved oxygen in seawater by direct spectrophotometry of total iodine. Mar Chem 41:343–351

Postlethwaite CF, Rohling EJ, Jenkins WJ et al (2005) A tracer study of ventilation in the Japan/East Sea. Deep-Sea Res Pt II 52:197–205

Rho T, Kim YB, Park JI et al (2010) Plankton community response to physico-chemical forcing in the Ulleung Basin, East Sea during summer 2008. Ocean Polar Res 32:269–289

Rho T, Lee T, Kim G et al (2012) Prevailing subsurface chlorophyll maximum (SCM) layer in the East Sea and its relation to the physico-chemical properties of water masses. Ocean Polar Res 34:413–430

Rysgaard S, Glud RN, Risgaard-Petersen N et al (2004) Denitrification and anammox activity in arctic marine sediments. Limnol Oceanogr 49:1493–1502

Shaffer G, Olsen SM, Pedersen JOP (2009) Long-term ocean oxygen depletion in response to carbon dioxide emissions from fossil fuels. Nature geosci 2:105–109. doi:10.1038/NGEO420

Shim JH, Park YC (1986) Primary productivity measurement using carbon-14 and nitrogenous nutrient dynamics in the southeastern Sea of Korea. J Oceanol Soc Korea 21:13–24

Shim JH, Park YC (1996) Primary production system in the southern waters of the East Sea, Korea III. Vertical distribution of the phytoplankton in relation to chlorophyll maximum layer. J Oceanol Soc Korea 31:196–206

Shim JH, Yeo HG, Park JG (1992) Primary production system in the southern waters of the East Sea, Korea I. Biomass and productivity. J Oceanol Soc Korea 27:91–100

Son SH, Platt T, Bouman H et al (2006) Satellite observation of chlorophyll and nutrients increase induced by Typhoon Megi in the Japan/East Sea. Geophys Res Lett 33:L05607. doi:10.1029/2005GL025065

Stramma L, Johnson GC, Sprintall J et al (2008) Expanding oxygen minimum zones in the tropical oceans. Science 320:655–658

Talley LD, Tishchenko P, Luchin V et al (2004) Atlas of Japan (East) Sea hydrographic properties in summer, 1999. Prog Oceanogr 61:277–348

Uda M (1934) The results of simultaneous oceanographic investigations in the Japan Sea and its adjacent waters in May and June, 1932. J Imp Fish Exp St 5:57–190 (in Japanese)

Watanabe YW, Wakita M, Maeda N et al (2003) Synchronous bidecadal periodic changes of oxygen, phosphate and temperature between the Japan Sea deep water and the North Pacific intermediate water. Geophys Res Lett 30:2273. doi:10.1029/2003GL018338

Yanagi T (2002) Water, salt, phosphorus and nitrogen budgets of the Japan Sea. J Oceanogr 58:797–804

Yang HS, Moon CH, Oh SJ et al (1997) Regeneration processes of nutrients in the polar front area of the East Sea II. Distribution of particulate organic carbon and nitrogen in winter, 1995. J Korean Fish Soc 30:442–450

Yasuda T, Hanawa K (1999) Composite analysis of North Pacific subtropical mode water properties with respect to the strength of the wintertime East Asian monsoon. J Oceanogr 55:531–541

Yoo SJ, Park J (2009) Why is the southwest the most productive region of the East Sea/Sea of Japan? J Mar Syst 78:301–315

Chapter 7
Natural and Anthropogenic Carbon Cycling

Il-Nam Kim, Kitack Lee and Jeomshik Hwang

Abstract The East Sea (Japan Sea) is experiencing changes in water temperature, oxygen content, and deep water circulation pattern. These changes, in turn, have affected and will continue to affect carbon cycling in the East Sea, directly and indirectly. As the physical dynamics of the East Sea resembles that of the open ocean, studying inorganic and organic carbon cycling in the East Sea may improve our understanding of global carbon cycling, and consequently enable us to predict more accurately the response of the carbon cycle to global climate change. In our review of inorganic carbon cycling, we focus on the uptake of anthropogenic carbon by the East Sea and the shoaling of the saturation depths of the aragonite and calcite caused by acidification. The saturation depths for aragonite and calcite in the East Sea have shoaled by about 80–220 and 500–700 m, respectively, compared to pre-industrial values. Anthropogenic CO_2 in the bottom water of the Japan Basin ranged from 15 to 20 μmol kg^{-1}. The largest water-column inventory of anthropogenic CO_2, about 80 mol C m^{-2}, was found in the Japan Basin. The uptake rate of anthropogenic CO_2 decreased from 0.6 $\pm$ 0.4 mol C m^{-2} year^{-1} during the 1992–1999 period to 0.3 $\pm$ 0.2 mol C m^{-2} year^{-1} during the 1999–2007 period, potentially reflecting the slowing of deep water ventilation in the East Sea. In our review of organic carbon cycling, we focus on sinking particulate

I.-N. Kim
Department of Marine Science, Incheon National University, Incheon 406-772,
Republic of Korea
e-mail: ilnamkim@inu.ac.kr

K. Lee
School of Environmental Science and Engineering,
Pohang University of Science and Technology, Pohang 790-784, Republic of Korea
e-mail: ktl@postech.ac.kr

J. Hwang (✉)
School of Earth and Environmental Sciences, Seoul National University, Seoul 151-742,
Republic of Korea
e-mail: jeomshik@snu.ac.kr

© Springer International Publishing Switzerland 2016
K.-I. Chang et al. (eds.), *Oceanography of the East Sea (Japan Sea)*,
DOI 10.1007/978-3-319-22720-7_7

organic carbon (POC) and dissolved organic carbon (DOC). Sinking POC flux in the East Sea generally reflected conspicuous events in biological production at the surface. Sinking POC flux in the East Sea was higher than in the nearby Kuroshio region. DOC concentration in the Ulleung Basin of the East Sea was higher than in the open oceans. The high content of the lithogenic component and low radio-carbon content in sinking particles in deep waters suggests that the contribution of resuspended sediment, and potentially other allochthonous sources of organic carbon, is considerable in the East Sea.

Keywords Dissolved inorganic carbon · Anthropogenic CO_2 · Carbon uptake · Organic carbon · Biological pump · East Sea (Japan Sea)

7.1 Inorganic Carbon Cycling

7.1.1 Introduction

The use of fossil fuels for energy has caused a rapid increase in the concentrations of many greenhouse gases (such as CO_2, CH_4, and N_2O) in the atmosphere since the industrial revolution (IPCC 2007). During the anthropocene (1800 to present), the rate of CO_2 increase in the atmosphere is approximately equal to the rate of increase in the use of fossil fuels (Millero 2007). However, the amount of CO_2 remaining in the atmosphere is only about half of the total CO_2 produced from anthropogenic activities. The rest has been absorbed by the ocean and the land carbon reservoirs (IPCC 2007). As the ocean contains about 50 times more carbon than the atmosphere (Sabine and Tanhua 2010), the capacity of the ocean to take up CO_2 is especially critical for accurate predictions of future climate changes.

Although the East Sea is a semi-enclosed marginal sea of the western North Pacific Ocean at mid-latitudes, it is frequently referred to as a "Miniature Ocean" (Kim et al. 2001), because its physical dynamics, which include deep-water formation, eddies, subpolar front, and gyre circulation, resemble those of the open oceans (Talley et al. 2006). Of these processes, deep-water formation is particularly significant in terms of atmospheric CO_2 levels, because deep-water formation quickly transports anthropogenic CO_2 from the surface to the interior (Park et al. 2006). This drawing down of CO_2 is frequently referred to as the "Solubility Pump".

Annual primary production in the East Sea is high and comparable to that of the upwelling regions (Kwak et al. 2013a). The rate of biological carbon removal (also called the "Biological Pump") is likely to be high in the East Sea, and the East Sea may be an efficient reservoir for the uptake of anthropogenic CO_2. These features, in addition to the sea's ocean-like dynamics, make it an optimal location to study the response of the carbon cycle to climate change (Kim et al. 2001; Lee et al. 2011).

We begin this section with brief explanations of ocean carbonate chemistry and anthropogenic CO_2 estimation methods, to help readers who are not familiar with these subjects. Then we describe the spatial features of inorganic carbon chemistry in the East Sea and the dynamics of anthropogenic CO_2 in this region. We end by discussing the important problem of ocean acidification, which has generated much recent interest.

7.1.2 Ocean Carbonate Chemistry

Ocean carbonate chemistry controls the pH (acidity/basicity) of seawater. Consequently, carbonate chemistry determines the buffering capacity of seawater and thus the seawater's capacity to regulate atmospheric CO_2 levels (Emerson and Hedges 2008).

When gaseous CO_2 (CO_2 (g)) dissolves in seawater, it exists as four distinct inorganic carbon species: aqueous CO_2 (CO_2 (aq)), carbonic acid (H_2CO_3), bicarbonate (HCO_3^-), and carbonate (CO_3^{2-}). Since it is difficult to distinguish CO_2 (aq) from H_2CO_3 analytically, they are usually combined and their concentration is expressed as $H_2CO_3^*$ ($=CO_2$ (aq) $+ H_2CO_3$).

Chemical reactions among the inorganic carbon species in seawater may be summarized in three steps as follows:

$$CO_2(g) + H_2O \leftrightarrow H_2CO_3^* \quad K_0 = \frac{\left[H_2CO_3^*\right]}{\left[CO_2(g)\right]} \tag{7.1}$$

$$H_2CO_3^* \leftrightarrow H^+ + HCO_3^- \quad K_1 = \frac{\left[H^+\right]\left[HCO_3^-\right]}{\left[H_2CO_3^*\right]} \tag{7.2}$$

$$HCO_3^- \leftrightarrow H^+ + CO_3^{2-} \quad K_2 = \frac{\left[H^+\right]\left[CO_3^{2-}\right]}{\left[HCO_3^-\right]}, \tag{7.3}$$

where K represents the equilibrium constant, and the unit of inorganic carbon species in the brackets is given as total concentration. [CO_2 (g)] can be, and often is, replaced by pCO_2, which is the partial pressure of CO_2.

In order to determine the CO_2 system, we need to know at least two variables. Since pCO_2 and pH ($= -\log_{10}[H^+]$) are easily influenced by biological activities, dissolved inorganic carbon (DIC) and total alkalinity (TA) are widely used to characterize the oceanic CO_2 system (Sarmiento and Gruber 2006). DIC and TA are defined as

$$DIC = \left[H_2CO_3^*\right] + \left[HCO_3^-\right] + \left[CO_3^{2-}\right] \approx \left[HCO_3^-\right] + \left[CO_3^{2-}\right] \tag{7.4}$$

$$TA = \left[HCO_3^-\right] + 2\left[CO_3^{2-}\right] + \left[OH^-\right] - \left[H^+\right] + \left[B(OH)_4^-\right] \approx \left[HCO_3^-\right] + 2\left[CO_3^{2-}\right]. \tag{7.5}$$

The concentrations of bicarbonate and carbonate can be estimated by combining the two above Eqs. (7.4 and 7.5) as follows:

$$\left[\text{HCO}_3^-\right] \approx 2 \times DIC - TA \tag{7.6}$$

$$\left[\text{CO}_3^{2-}\right] \approx TA - DIC. \tag{7.7}$$

Since CO_2 species are interconnected through chemical reactions, if we know two variables among pCO_2, pH, DIC, and TA, we can estimate the other two species using this set of Eqs. (7.1 to 7.7).

Note that we used simplified definitions of DIC and TA. For more detail, readers are referred to Sarmiento and Gruber (2006). Also note that programs such as CO_2SYS are available for calculation of the undetermined variables (http://cdiac.ornl.gov).

DIC operates as a natural buffer and maintains seawater pH within relatively narrow limits. However, as pCO_2 in the ocean's surface increases (as increasing amounts of CO_2 are absorbed by seawater), this buffering capacity will decrease, the ocean's acidity will increase, and the ocean's ability to act as a sink for atmospheric anthropogenic CO_2 will decrease.

7.1.3 Methods for Estimation of the Oceanic Anthropogenic CO_2 Content

We briefly describe below the basic concepts of methods to distinguish the anthropogenic CO_2 from the DIC. More details are provided in Sabine and Tanhua (2010).

The first estimation of anthropogenic CO_2 was carried out by Brewer (1978) and Chen and Millero (1979) using directly measured DIC and TA concentrations. This approach is based on the assumption that the ocean has been at steady state in terms of circulation and biological pump since preindustrial times. The basic concept is given as

$$C_{ant} = DIC_{mea} - \Delta DIC_{bio} - DIC_{eq}, \tag{7.8}$$

where DIC_{mea} is the measured DIC concentration for a seawater sample, ΔDIC_{bio} represents the contributions of remineralization of organic matter and dissolution of $CaCO_3$, and DIC_{eq} is the preformed preindustrial DIC concentration in equilibrium with the preindustrial atmospheric CO_2.

However, this approach had large uncertainties and was not widely accepted (Shiller 1981). Later, Gruber et al. (1996) proposed an improved approach for estimating anthropogenic CO_2 (called the ΔC^* technique). This approach reflects the fact that surface waters are in disequilibrium with the atmosphere (ΔDIC_{diseq}) as follows:

$$C_{ant} = DIC_{mea} - \Delta DIC_{bio} - DIC_{eq} - \Delta DIC_{diseq} = \Delta C^* - \Delta DIC_{diseq}. \tag{7.9}$$

A different back-calculation approach known as the 'TrOCA' method (TrOCA $=$ $O_2 + 1.2 \cdot DIC - 0.6 \cdot TA$) was introduced by Touratier and Goyet (2004). This approach estimates anthropogenic CO_2 from the difference between the measured TrOCA conservative tracer and the preformed TrOCA tracer. However, a study comparing observed data with modeling results (Yool et al. 2010) showed that the TrOCA method had substantial biases.

Various recent methods have been introduced to estimate the anthropogenic CO_2. One new method, called the transit time distribution (TTD), is a time-dependent tracer-based approach (Hall et al. 2004; Waugh et al. 2004, 2006; Tanhua et al. 2008). The TTD method is based on the fact that the surface concentration of anthropogenic CO_2 propagates into the ocean interior, and assumes that the TTD in the ocean's interior position is determined by the inverse Gaussian function (Waugh et al. 2003). Another approach, known as the extended multiple linear regression (eMLR), compares the observed carbon inventory changes between two cruises separated in time (Tanhua et al. 2007). The eMLR method takes advantage of the exponential increase of anthropogenic CO_2 and does not require many assumptions, unlike the back-calculation methods. Each method has its advantages and disadvantages, and there is no consensus on which method works best.

Note that anthropogenic CO_2 data presented in Sect. 7.1.5 were estimated using the ΔC^* and eMLR methods (Park et al. 2006, 2008).

7.1.4 Distribution of CO$_2$ Variables (TA, DIC, and pH)

The highest concentration of TA in the surface layer (>2270 μmol kg^{-1}) is observed in the southern East Sea (south of 40° N). This region is covered by the Tsushima Warm Water (TWW), which enters the East Sea through the Korea Strait (Fig. 7.1a, c) and is characterized by a salinity maximum (Kim et al. 2004).

Since the concentration of TA in surface waters is significantly correlated with salinity (Lee et al. 2000), the distributions of TA and salinity in the surface layer of the East Sea are quite similar (Fig. 7.1a, c). Biological activities such as photosynthesis and remineralization greatly influence pH and DIC. In surface waters, pH is high as a result of photosynthesis (Fig. 7.1b), whereas DIC is high in the intermediate/deep waters as a result of remineralization (Fig. 7.1d). Overall, the concentrations of TA and DIC increase gradually with increasing depth; conversely, pH decreases with increasing depth (Fig. 7.1b–d).

The increase in TA with depth (about 15 μmol kg^{-1}) is mostly associated with the dissolution of $CaCO_3$ in the water column and/or in sediments. However, the increase in DIC with depth (about 90 μmol kg^{-1}) is not only because of the dissolution of $CaCO_3$ but also because of remineralization (Fig. 7.1c–d). The increase in TA (about 15 μmol kg^{-1}) caused by $CaCO_3$ dissolution is equivalent to an approximately 7.5 μmol kg^{-1} increase in DIC (see Eqs. 7.4 and 7.5). Therefore, the rain ratio, which is the relative contribution of the effects of remineralization (about 82.5 μmol kg^{-1}) and $CaCO_3$ dissolution (about 7.5 μmol kg^{-1}) to

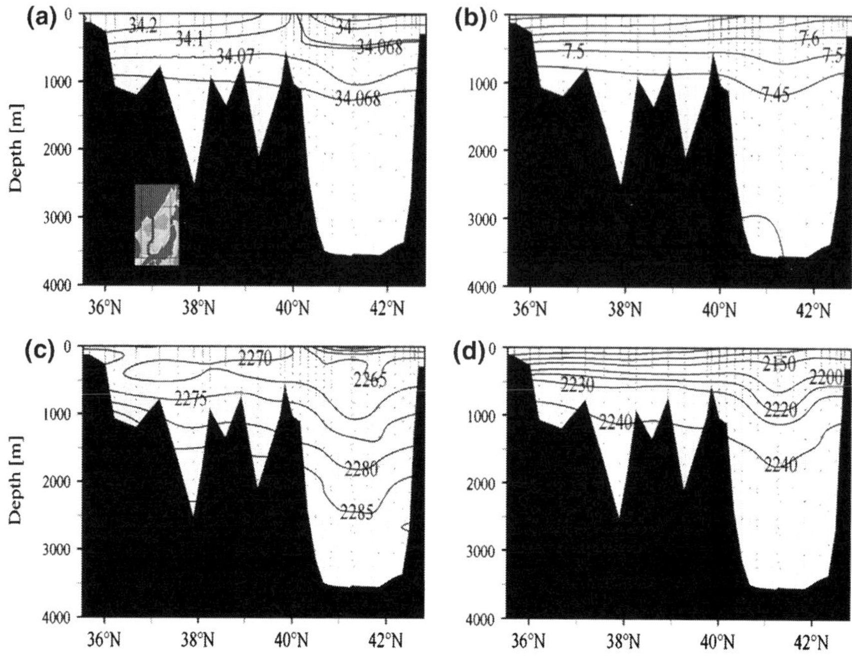

Fig. 7.1 Meridional sections of **a** salinity, **b** pH, **c** total alkalinity (TA), **d** total dissolved inorganic carbon (DIC) along 134° E in the East Sea (from Park et al. 2006). *Dots* indicate locations of measured data. The *inset* in **a** shows the path of the cruise track

observed DIC increase, is approximately 11 (=82.5/7.5) in the East Sea. This estimate for the East Sea is within the range for the global rain ratio (4–14) at the bottom of the seasonal mixed layer (Lee 2001).

7.1.5 Dynamics of Anthropogenic CO₂

The East Sea is connected to the western North Pacific Ocean via three shallow straits with depths less than 150 m (the Korea, Tsugaru, and Soya straits). The formation of deep-water in the western Japan Basin is a key feature of the East Sea's circulation (Kim et al. 2002; Talley et al. 2003). Another important feature is the sea's weak vertical stability, relative to the open ocean and the other marginal seas (Kim et al. 2001). Both the deep-water formation system and the weak vertical stability are closely associated with the East Sea's uptake of anthropogenic CO_2 (Lee et al. 2011), and these two features combine to facilitate active and effective transfer of surface waters laden with anthropogenic CO_2 to the interior of the basin.

The highest concentrations of anthropogenic CO_2 (about 50–55 μmol kg^{-1}, using the eMLR method) were observed in the upper layer of the southern East

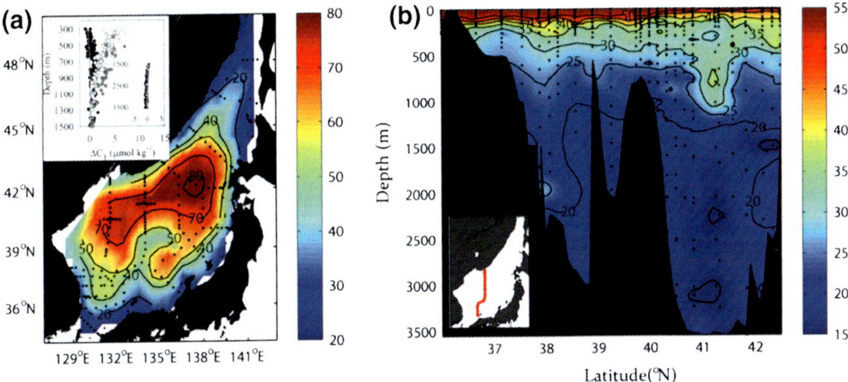

Fig. 7.2 **a** Water column anthropogenic CO_2 (C_{ant}) inventory per unit area (mol C m^{-2}) in the East/Japan Sea, as of 1999 (using the ΔC^* method) and **b** C_{ant} (μmol kg^{-1}) in the section indicated in the *inset* (using the eMLR method). The *inset* shows the amounts of C_{ant} accumulated during the periods 1992–1999 (*open circles*) and 1999–2007 (*solid circles*) as a function of depth (using the eMLR method). The *gray circles* denote data within the anticyclonic eddies (Reproduced from Lee et al. 2011 with permission from The Royal Society of Chemistry)

Sea (Fig. 7.2b). In the southern East Sea, where vertical stratification is strongly developed between surface waters and subsurface waters, anthropogenic CO_2 is densely accumulated in the surface layer. However, the vertical gradient in anthropogenic CO_2 in the surface layer of the northern East Sea is relatively small because of active vertical mixing. In the Japan Basin, anthropogenic CO_2 reached the bottom (about 3500 m) and levels of 15–20 μmol kg^{-1} were observed (Park et al. 2006). This feature is similar to that of the North Atlantic Ocean where anthropogenic CO_2 penetrated to depths of around 3000 m or greater (Lee et al. 2003). However, anthropogenic CO_2 reaches only about 1500 m depth at similar latitudes in the North Pacific Ocean (Sabine et al. 2002). The largest water-column inventory of anthropogenic CO_2 was found in the Japan Basin (Fig. 7.2b) with a mean value of 80 mol C m^{-2} (Park et al. 2006). This value is 2–3 times higher than the observed value of approximately 20–30 mol C m^{-2} in the adjacent North Pacific Ocean (Sabine et al. 2002). This large difference between the two basins is attributed primarily to deep-water formation.

Recent changes in dissolved O_2 and temperature in the East Sea's intermediate and deep waters indicate that the mode of deep-water formation has shifted from bottom to intermediate water formation, likely because of global warming and climate change (Kim et al. 1996; Gamo 1999; Kim et al. 2001; Kang et al. 2003; Chae et al. 2005). However, sudden bottom-water formation occurred in the Japan Basin in winter 2000–2001 (Kim et al. 2002; Talley et al. 2003). These findings imply that the ventilation system of the East Sea is sensitive to climate change (Kim et al. 2010).

The mean uptake rate of anthropogenic CO_2 in the East Sea for the period of 1992–1999, estimated from observational data and the eMLR method, was

0.6 ± 0.4 mol C m^{-2} year^{-1}. Surprisingly, however, from 1999 to 2007 the uptake rate fell (Fig. 7.2a) to 0.3 ± 0.2 mol C m^{-2} year^{-1} (Park et al. 2008), implying that anthropogenic CO_2 was not accumulating efficiently in the water column (>300 m) during this period. The rapid and substantial reduction in the uptake rate of anthropogenic CO_2 in the East Sea for the 1999–2007 period may have been caused by climate change and consequent slowing of water-column ventilation in the East Sea (Gamo 1999; Kim et al. 2001).

A key piece of evidence suggesting that water-column ventilation in the East Sea has slowed, is the progressive decrease in the water-column O_2 inventory in the Japan Basin, the location of deep-water formation (Kim et al. 2002; Talley et al. 2003), assuming that O_2 consumption by remineralization remained constant. The rate of decrease in the water-column O_2 inventory of the Upper Portion of the Japan Sea Proper Water (UPJSPW) (300–800 m) was two times higher for the period 1999–2007 than for the period 1992–1999 (Park et al. 2008). This layer covers about 80 % of the total anthropogenic CO_2 accumulated in the East Sea since 1992. The discernible reduction in the water-column O_2 inventory may be a response to the two-fold increase in the rate of heat content increase in this layer for the period of 1999–2007 relative to 1992–1999 (Park et al. 2008). The changes suggest that ventilation of the UPJSPW is weakening and consequently reducing the transport of anthropogenic CO_2 from the surface to this layer.

In summary, the following consecutive processes are likely to provide evidence of the East Sea's changes: warming and climate change $\rightarrow$ increasing heat content $\rightarrow$ weakening of the water-column ventilation $\rightarrow$ decreasing water-column O_2 inventory and reducing the transport of anthropogenic CO_2 from the surface to the interior. However, more observational studies are required to elucidate the governing mechanism(s) in the future.

7.1.6 Acidification of Seawater and Saturation State of Aragonite and Calcite

When CO_2 enters the seawater, it reacts with the carbonate ion (CO_3^{2-}) to produce the bicarbonate ion (HCO_3^-):

$$H_2CO_3^* + CO_3^{2-} \leftrightarrow 2HCO_3^-. \tag{7.8}$$

Thus, the invasion of anthropogenic CO_2 into the seawater results in the reduction of CO_3^{2-} concentration, and consequently decreases the saturation state of both calcite and aragonite (Millero 2007). Seawater pH has decreased by about 0.1 units during the anthropocene (IPCC 2007), and is expected to decrease by as much as about 0.4 units by the end of the 21st century (Caldeira and Wichett 2003). Such a decrease in pH will cause a reduction of about 60 % in the CO_3^{2-} concentration and, in turn, decrease the saturation states of aragonite and calcite in seawater (Feely et al. 2004; Doney et al. 2009).

The saturation states of aragonite (Ω_{ara}) and calcite (Ω_{cal}) can be calculated as follows:

$$\Omega_{ara} = \frac{\left[CO_3^{2-}\right]\left[Ca^{2+}\right]}{K_{sp}^{ara}} \tag{7.9}$$

$$\Omega_{cal} = \frac{\left[CO_3^{2-}\right]\left[Ca^{2+}\right]}{K_{sp}^{cal}}, \tag{7.10}$$

where $[CO_3^{2-}]$ and $[Ca^{2+}]$ are the observed carbonate and calcium concentrations, and K_{sp}^{ara} and K_{sp}^{cal} are solubility products of aragonite and calcite, respectively.

The saturation depths ($\Omega = 1$) of aragonite and calcite in the East Sea are approximately 400 and 1000 m, respectively (Fig. 7.3). The aragonite saturation depth in the East Sea appears similar to that of the mid-latitude North Pacific

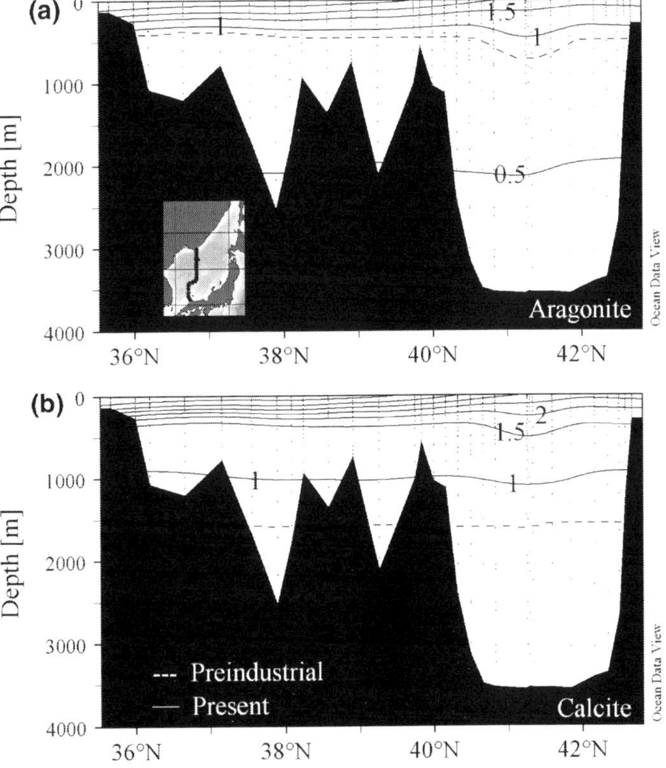

Fig. 7.3 Meridional distributions of the degree of seawater saturation with respect to **a** aragonite and **b** calcite at the present time (*solid lines*), nominally along 134° E in the East Sea (from Park et al. 2006). The dotted lines denote the aragonite and calcite saturation horizons ($\Omega = 1$) during the preindustrial era. The *inset* in **a** shows the path of the cruise track

Ocean, whereas the calcite saturation depth in the East Sea appears to be about 300–600 m deeper than in the mid-latitude North Pacific Ocean (Feely et al. 2002). During the anthropocene, the saturation depths of aragonite and calcite in the East Sea have shoaled by approximately 80–220 and 500–700 m, respectively, compared to preindustrial depths of about 500 m for aragonite and about 1500 m for calcite (Fig. 7.3). These upward shifts in saturation depths are roughly five times greater than upward shifts observed in the mid-latitude North Pacific Ocean (Feely et al. 2002, 2004). Because anthropogenic input of CO_2 is increasing, these dramatic changes in aragonite and calcite saturation depths in the East Sea will likely continue. Additionally, Kim et al. (2014) reported that surface waters in the Ulleung Basin of the East Sea have acidified by about 0.04 pH unit decade^{-1}, which is approximately twice as fast as the global mean for acidification (0.02 pH unit decade^{-1}). These data reiterate the need to regularly survey the CO_2 variables (DIC, TA, pH, and pCO$_2$) and collect the hydrographic data necessary for assessing the impacts of human perturbations on the East Sea ecosystem and its carbon cycle.

7.2 Organic Carbon Cycling

7.2.1 Introduction

Organic carbon cycling in the ocean includes biological production in the surface water, burial in the sediment, and intermediate processes that POC and DOC undergo from production to removal. Primary production in surface waters and processes in the sediment are discussed in detail elsewhere (see Chaps. 9 and 10). Here, the cycling of POC is discussed, with a focus on the sinking flux of POC and its potential link to primary production in surface waters and remineralization in the water column. Some interesting recent discoveries on DOC cycling are also briefly described.

7.2.2 Primary Production

Basin-wide primary production has been estimated based on satellite observation (Yamada et al. 2004, 2005). Yamada et al. (2005) provide a comprehensive data set of basin-wide primary production based on satellite data from 1998 to 2002. The southwestern part of the East Sea is reportedly the most productive (annual average primary production = 222 gC m^{-2} year^{-1}) followed by southeastern part (191 gC m^{-2} year^{-1}). There have been several attempts to explain the processes behind the high primary production in the Ulleung Basin, including strong upwelling along the east coast of the Korean Peninsula (Hyun et al. 2009; Yoo and Park 2009). The middle of the Japan Basin showed the lowest annual primary production, 161 gC m^{-2} year^{-1}. It should be noted that interannual variability was high: for example, the maximum difference in primary production in the south-eastern part was more than 25 gC m^{-2} during a 4 months study period, which is >40 % of the corresponding primary production in the least productive year (Yamada et al. 2005). Because of the high interannual variability, spatial variations in sinking POC flux obtained at different times must be interpreted with caution. Also note that recent studies suggest that subsurface production accounts for a considerable portion of primary production (Rho et al. 2012; Kwak et al. 2013b), and hence satellite-based results may provide a limited understanding of primary production.

7.2.3 *Particulate Organic Carbon Flux*

Sinking POC flux has been determined or estimated by several methods. In 1984, five cylindrical sediment traps that directly collected sinking particles were deployed on a single mooring for two weeks near the Japanese coast in the Japan Basin (Masuzawa et al. 1989). Since then, time-series sediment traps have been deployed intermittently at several locations (Hong et al. 1997a, b; Suk 2002; Otosaka et al. 2004). Sinking POC flux and export production have also been estimated by indirect approaches: using the flux of radionuclides, stable isotopes of helium, oxygen consumption rate and the tritium-based ages of water masses (Hahm and Kim 2001; Kim et al. 2003; Hahm and Kim 2008; Kim et al. 2011).

Limited data on sinking POC flux at depths of about 1000 m suggests that sinking POC flux tracks conspicuous events in biological production such as spring and fall plankton blooms (Fig. 7.4). Otosaka et al. (2004, 2008) reported time-series (1–2 year duration) sediment trap results from the Japan Basin and the

Fig. 7.4 Particulate organic carbon (POC) flux at ~1000 m depth in the major basins of the East Sea (see text for references)

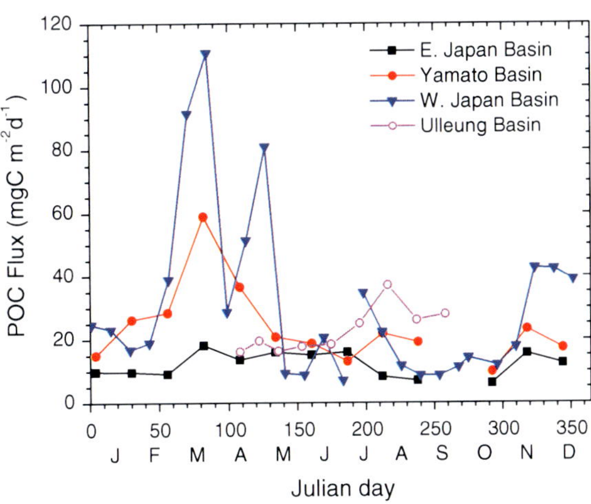

Yamato Basin from 2000 to 2002. In the western Japan Basin and the Yamato Basin, high sinking POC flux was observed from March to April. In contrast, in the eastern Japan Basin the spring bloom signal was not as distinct. The fall bloom signal was not as conspicuous as the spring bloom signal, except in the western Japan Basin where a large peak in POC flux was observed in November and December.

The annual average POC flux was 31 mgC m^{-2} day^{-1} in the western Japan Basin, 12 mgC m^{-2} day^{-1} in the eastern Japan Basin, and 24 mgC m^{-2} day^{-1} in the Yamato Basin (Fig. 7.4 and Table 7.1). Sinking POC flux data are scarcer for the Ulleung Basin and mainly exist in technical reports such as, for example, Suk (2002). According to Hong (1997; Hong et al. 2008a), the sinking POC flux ranged between 16 and 37 mgC m^{-2} day^{-1} (with a mean value of 23 mgC m^{-2} day^{-1}) at a depth of 1000 m, at a station (located at 37° 00'N and 131° 00'E) in the Ulleung Basin from April to September 1996. Preliminary results from a nearby station (37° 19'N, 131° 29'E), also in the Ulleung Basin, show that sinking POC flux from March 2011 to February 2012 ranged between 3 and 53 with an annual mean of 20 mgC m^{-2} day^{-1} (unpublished data). Sinking POC fluxes in the East Sea are significantly higher than sinking POC fluxes in the nearby Kuroshio region, where values (normalized to the 2000 m depth) range between 1.8 and 7.7 mgC m^{-2} day^{-1} (Honjo et al. 2008 and references therein).

A question of interest is whether the observed sinking POC flux reflects the spatial variation in primary production. It is hard to explore the spatial variability of sinking POC flux because of a lack of synchronous sinking POC flux data. However, when limited data at about 1000 m were analyzed, sinking POC flux in the western Japan Basin appeared higher than in the eastern Japan Basin or other major basins (Fig. 7.4 and Table 7.1). This result was inconsistent with satellite-based productivity data that suggest the Ulleung Basin is the most productive region in the East Sea (Yamada et al. 2005). Further study is needed to determine whether this discrepancy was caused by temporal variability, or spatial variability in biogeochemical processes such as degradation efficiency during the vertical transport of sinking POC.

Annual mean POC export flux from the upper 200 m layer in the Ulleung Basin, estimated by the ^{234}Th/^{238}U disequilibrium method, was 161 mgC m^{-2} day^{-1} (Kim et al. 2011). The ratio of export to primary production (ThE ratio) was high (34 % annually), with an especially high value (48 %) in summer.

Export production was also estimated from the oxygen consumption rate, based on the tritium age of the water mass and apparent oxygen utilization in the western Japan Basin (Hahm and Kim 2008). An estimated 99 gC m^{-2} year^{-1} (271 mgC m^{-2} day^{-1}) was regenerated (adopting the Redfield ratio) between 200 and 1600 m of the water column. This carbon regeneration value may be considered a minimum value for export production from the overlying water column, assuming that oxygen is used only for the degradation of sinking POC in this depth horizon (Hahm and Kim 2008).

Table 7.1 Sampling location and average particulate organic carbon (POC) flux data

Location	Lat (N)	Long (E)	Bottom depth (m)	Trap depth (m)	Sampling period (mm/dd/yy)	Average POC flux mgC m^{-2}day^{-1} (gC m^{-2}year^{-1})	Reference
W. Japan Basin	69° 40′	132° 24′	3300	2800	7/19/94–7/18/95	8.0 ± 9.5 (2.9)	Hong et al. (1997a)
W. Japan Basin (St. MS)	41° 14′	132° 21′	3424	927	7/11/01–7/9/2002	30.7 ± 26.9 (11.2)	Otosaka et al. (2008)
				2746	8/6/00–7/9/02	17.3 ± 23.0 (6.3)	
E. Japan Basin (St. JN)	42° 28′	138° 30′	3642	1057	10/6/00–9/8/01	12.1 ± 3.9 (4.4)	Otosaka et al. (2008)
				3043	10/6/00–9/8/01	5.6 ± 3.2 (2.0)	
Yamato Basin (St. JS)	38° 01′	135° 02′	2900	1175	10/6/00–9/8/01	23.8 ± 12.7 (8.7)	Otosaka et al. (2008)
				2100	10/6/00–9/8/01	17.9 ± 12.4 (6.5)	
Ulleung Basin	37° 00′	131° 00′	2100	1000	4/11/96–9/25/96	23.4 ± 6.6	Hong et al. (2008a)
				2000	4/11/96–12/18/96	19.0 ± 11.6	Hong et al. (1997b), Hong (1997)

Another estimate of export production was obtained by examining ^{3}He flux into the mixed layer from below (Hahm and Kim 2001). Using the relationship between nitrate and ^{3}He concentrations, ^{3}He flux was converted to nitrate flux, and this value, in turn, was converted to new production. According to this method, export production was 64 gC m^{-2} year^{-1} (175 mgC m^{-2} day^{-1}) in the Ulleung Basin (and the corresponding f-ratio was $64/222 = 0.3$).

Export production was also estimated by a box model using the oxygen and phosphorous budget (Kim et al. 2003). The estimated export production was about 0.016 PgC year^{-1} (16 gC m^{-2} year^{-1} or 44 mgC m^{-2} day^{-1}) for the entire East Sea.

Allochthonous POC may account for a considerable fraction of sinking POC in continental margin settings (Hwang et al. 2010). Allocthonous POC sources include riverine input, eolian input, and resuspension of sediment from the seafloor (and subsequent lateral sediment transport). Riverine input of POC is minor compared to autochthonous production in the East Sea (Hong et al. 1997b). The seasonal input of Asian dust (also known as yellow dust or Kosa) may be an important source of iron and ballast minerals (Jo et al. 2007). POC supplied from the resuspension of surface sediment on the shelf and the slope may also be significant. An early study suggested that suspended POC concentrations on the eastern side of the East Sea (20–40 µgC L^{-1} in the upper layer and 10–20 µgC L^{-1} in the deeper layer) were comparable to suspended POC concentrations in other East Asian seas (Ichikawa 1982). However, the possibility for sediment resuspension was demonstrated by light transmission profiles: a dramatic decrease in light transmission was observed in the bottom 100–200 m thick layers in the slope region of the Yamato Basin (Fujita et al. 2010). These bottom nepheloid layers (where light transmission is low) were associated with increased concentrations of total iron. Decreased light transmission was also observed in intermediate nepheloid layers in the Yamato Basin. Thus, it appears that the East Sea may be considerably influenced by resuspended sediment, perhaps because this sea has relatively strong deep currents (Takematsu et al. 1999).

In a study using U-Th radionuclides, Hong et al. (2008b) observed an approximately 40 % surplus of ^{230}Th to the deeper trap in the western Japan Basin, suggesting the contribution of particles from lateral advection to this site. Otosaka et al. (2008) observed that the Δ^{14}C values of sinking POC at three stations in the Japan and Yamato Basins were significantly lower than expected, if the only source of POC was freshly produced organic matter in the surface water. Based on ^{14}C isotope mass balance, Otosaka et al. (2008) estimated that allochthonous POC contributed up to 53 ± 15 % to sinking POC, with much higher contributions to deeper traps in the Japan and Yamato Basins. Deeper traps also had higher aluminum fluxes compared to shallower traps, suggesting that the allochthonous POC originated from sediment resuspension (Otosaka et al. 2008). The correlation between Δ^{14}C values of sinking POC and Al concentration of sinking particles was weaker in the East Sea than at other locations (Hwang et al. 2010), suggesting either that the ^{14}C-age of resuspended sediment is not homogeneous, or that there are other sources of allochthonous POC associated with lithogenic particles such as Kosa, Asian loess, and particles from the Island Arc (Otosaka and Noriki 2005). Lithogenic matter collected in the sediment traps deployed at 1000 m in the eastern part of the East Sea accounted for about 44 % of sinking particles (Otosaka and Noriki 2005). In the Ulleung Basin, the Al concentration of sinking particles increased slightly (4.2 to 4.9 %) from 1000 to 2000 m (Hong 1997; Hong et al. 2008a), indicating that lithogenic material may account for an even higher proportion (50–60 %) of sinking particles in the southwestern part of the East Sea.

7.2.4 Particulate Organic Carbon Budget in the Water Column

The rate of organic carbon decomposition in the East Sea has mostly been estimated indirectly, from the consumption rate of dissolved O_2. Oxygen consumption rates estimated by various methods agreed well with one another, and were in the range 1.1–1.4 μmol kg^{-1} year^{-1} below 2000 m (Chen et al. 1996; Nakayama et al. 2007). Using the Redfield model for organic matter production/decomposition, the organic carbon decomposition rate would be 10–13 mgC m^{-3} year^{-1}, which corresponds to 10–13 gC m^{-2} year^{-1} between 2000 and 3000 m. Considering the small decrease in DOC concentration in the deep layer, dissolved O_2 can be assumed to be consumed mostly by heterotrophic decomposition of POC. The estimated regeneration rate of organic carbon appears somewhat high compared to the observed sinking POC flux at and below 2000 m (Table 7.2). The addition of POC from resuspended sediment may, in part, account for the discrepancy. However, this discrepancy shows that our understanding of the organic carbon budget is far from complete and underlines our need for more data.

Table 7.2 Production rates, fluxes, and burial rates of particulate organic carbon (POC) in three major basins of the East Sea. The values are given in gC m^{-2} year^{-1}

	Western Japan Basin	Yamato Basin	Ulleung Basin
Primary production	161[a]	191[a]	222[a]
Export flux	99[b]		64[c]
Sinking POC flux at ~1000 m depth	11.2[d]	8.7[d]	7.4[e]
Sinking POC flux at deeper depths	6.3 (2746 m)[d] 2.9 (2800 m)[f]	6.5 (2100 m)[d]	9.1 (2300 m)[e]

Values are based on following references
[a]Yamada et al. (2004)
[b]Hahm and Kim (2008)
[c]Hahm and Kim (2001)
[d]Otosaka et al. (2008)
[e]unpublished data
[f]Hong et al. (1997a)

7.2.5 Dissolved Organic Carbon Cycling

Compared to POC data, very limited data are available on DOC cycling in the East Sea, because studies on DOC have only just begun. A few recent studies suggest that DOC cycling in the East Sea may be different from DOC cycling in the major oceans (Kim and Kim 2010, 2013). This difference may stem from the much shorter time scale for deep water circulation in the East Sea, compared to deep water circulation in major oceans (Kim and Kim 2010, 2013). DOC concentrations measured by a high temperature combustion oxidation method in the upper 100 m along a north-south transect in the western part of the East Sea ranged between 60 and 83 μM; these values are comparable to DOC concentrations in the major oceans (Kim and Kim 2010, 2013). In contrast, DOC concentration in the deep Ulleung Basin between 200 and 1500 m ranged between 54 and 64 μM; this is higher than DOC concentrations observed in the Pacific and Atlantic Oceans (Kim and Kim 2010, 2013). The reason for this high DOC concentration in the deep water of the Ulleung Basin is not well understood. Kim and Kim (2010) suggest that a large portion of the semi-labile DOC is present in the deep water because of its distinctly short turnover time. Biological factors such as low efficiency for bacterial degradation of DOC and/or refractory nature of some portion of DOC of terrestrial origin may contribute to high concentrations of DOC in the deep water in the Ulleung Basin. If, in addition to the Ulleung Basin, high DOC concentrations are confirmed in the other major basins of the East Sea, the East Sea's capacity for carbon storage will be of great interest (Kim and Kim 2010). Further studies (using approaches such as molecular characterization of DOC and determination of ^{14}C-age of DOC, for example) are needed to better understand DOC cycling in the East Sea.

7.3 Summary and Remaining Issues

Only a few studies have examined the inorganic carbon system in the East Sea. There is a special dearth of basin-wide examinations in this region. Sinking POC flux has been determined intermittently at a few locations. Studies of DOC have just begun and data for deep waters are limited to the Ulleung Basin. Interesting findings are summarized below:

- The highest concentration of anthropogenic CO_2 was observed in the upper layer of the southern part of the East Sea. In contrast, in the northern part, active vertical mixing has transported anthropogenic CO_2 into the sea's interior.
- The largest water-column inventory of the anthropogenic CO_2, about 80 mol C m^{-2}, was found in the Japan Basin. This is 2–3 times greater than in the adjacent north Pacific. Levels of anthropogenic CO_2 in the bottom water in the Japan Basin range from 15 to 20 μmol kg^{-1}.

- The uptake rate of the anthropogenic CO_2 decreased from 0.6 ± 0.4 mol C m^{-2} year^{-1} during the 1992–1999 period to 0.3 ± 0.2 mol C m^{-2} year^{-1} during the 1999–2007 period. This reduction may reflect the slowing of deep water ventilation in the East Sea.
- The saturation depths for aragonite and calcite in the East Sea have shoaled by about 80–220 m and 500–700 m, respectively, compared to pre-industrial values.
- Sinking POC flux (determined by direct particle collection at approximately 1000 m depth) tracked conspicuous events in biological production at the surface, such as spring and fall blooms.
- Sinking POC flux in the East Sea was significantly higher than in the nearby Kuroshio region.
- Both the high lithogenic content and the low radiocarbon content of sinking particles suggest that contributions from allochthonous sources of organic carbon (such as resuspended sediment) are considerable in deep waters.
- DOC concentration in the deep water of the Ulleung Basin was considerably higher than in the deep water of the open ocean. DOC concentrations are yet to be determined for other Basins in the East Sea.

The current understanding of carbon cycling in the East Sea is limited. Several interesting questions remain unanswered. Suggested topics for further research are listed below:

- The uptake rate of anthropogenic CO_2 appears to have decreased in time (from 1992 to 2007) and this rate needs to be monitored at regular time intervals to understand how the system is responding to changes in physical conditions at the air-sea boundary. A thorough examination of the CO_2 variables, in association with hydrographic data and other biogeochemical tracers (such as hydrofluorocarbons), must continue.
- As seawater acidification progresses, the shoaling of saturation horizons for calcite and especially for aragonite, should be monitored. In addition, the potential physiological responses of phytoplankton (such as pteropods) should be studied.
- Sinking POC flux should be measured synchronously in the major basins, accompanied by satellite observations, to obtain basin-wide pictures of the biological pump system in the East Sea.
- Sinking POC flux data must be collected in the southwestern part of the East Sea, both to understand the mechanisms responsible for the high primary production in this region, and to understand how this high production is transferred to the basin interior and to the seafloor.
- A high concentration of DOC was observed in the Ulleung Basin, suggesting that other major basins in the East Sea may also have high DOC concentrations. Basin-wide examinations are needed to confirm DOC levels in all the basins, because high DOC concentrations have important implications in terms of carbon storage and the impact of deep water circulation on DOC cycling.

- Potential causes for the observed high DOC concentration include: the rapid turnover of deep water, inefficient bacterial degradation, and/or the refractory nature of DOC. Future studies must seek to understand why DOC concentration in this region is significantly higher than in the open ocean, by using approaches such as molecular characterization of DOC and determination of its ^{14}C-age.
- Carbon flux (in particulate and dissolved forms) through the Korea Strait must be determined to better understand the carbon budget in the East Sea.

References

Brewer PG (1978) Direct observation of the oceanic CO_2 increase. Geophys Res Lett 5:997–1000

Chae YK, Seung YH, Kang SK (2005) Mode change of deep water formation deduced from slow variation of thermal structure: one-dimensional model study. Ocean Polar Res 24:115–123

Chen GT, Millero FM (1979) Gradual increase of oceanic CO_2. Nature 277:205–206

Chen CTA, Dong GC, Wang SL et al (1996) Redfield ratios and regeneration rates of particulate matter in the Sea of Japan as a model of closed system. Geophys Res Lett 23:1785–1788

Caldeira K, Wickett ME (2003) Anthropogenic carbon and ocean pH. Nature 425:365

Doney SC, Fabry VJ, Feely RA et al (2009) Ocean acidification: the other CO_2 problem. Annu Rev Mar Sci 1:169–192

Emerson SR, Hedges JI (2008) Chemical oceanography and the marine carbon cycle. Cambridge University Press, New York

Feely RA, Sabine CL, Lee K et al (2002) In situ calcium carbonate dissolution in the Pacific Ocean. Global Biogeochem Cycles 16:1144. doi:10.1029/2002GB001866

Feely RA, Sabine CL, Lee K et al (2004) Impact of anthropogenic CO_2 on the $CaCO_3$ system in the oceans. Science 305:362–366

Fujita S, Kuma K, Ishikawa S et al (2010) Iron distribution in the water column of the Japan Basin and Yamato Basin (Japan Sea). J Geophys Res 115:C12001. doi:10.1029/2010JC006123

Gamo T (1999) Global warming may have slowed down the deep conveyor belt of a marginal sea of the northwestern Pacific: Japan Sea. Geophys Res Lett 26:3137–3140

Gruber N, Sarmiento JL, Stocker TF (1996) An improved method for detecting anthropogenic CO_2 in the oceans. Global Biogeochem Cycles 10:809–837

Hahm D, Kim KR (2001) An estimation of the new production in the southern East Sea using helium isotopes. J Korean Soc Oceanogr 36:19–26

Hahm D, Kim KR (2008) Observation of bottom water renewal and export production in the Japan Basin, East Sea using tritium and helium isotopes. Ocean Sci J 43:39–48

Hall TM, Waugh DW, Haine TWN et al (2004) Estimates of anthropogenic carbon in the Indian Ocean with allowance for mixing and time-varying air-sea CO_2 disequilibrium. Global Biogeochem Cycles 18:GB1031. doi:10.1029/2003GB002120

Hong GH (1997) Development of monitoring technology for the wastes disposal sea areas. Korea Ocean Research and Development Institute, Technical Report BSPN 96340-00-1012-4 (in Korean)

Hong GH, Choe S-M, Suk M-S et al (1997a) Annual biogenic particle fluxes to the interior of the East/Japan Sea, a large marginal sea of the Northwest Pacific. In: Tsunogai S (ed) Biogeochemical processes in the North Pacific. Japan Marine Science Foundation, pp 300–321

Hong GH, Kim SH, Chung CS et al (1997b) ^{210}Pb-derived sediment accumulation rates in the southwestern East Sea (Sea of Japan). Geo-Mar Lett 17:126–132

Hong GH, Kim YI, Baskaran M et al (2008a) Distribution of ^{210}Po and export of organic carbon from the euphotic zone in the southwestern East Sea (Sea of Japan). J Oceanogr 64:277–292

Hong GH, Baskaran M, Lee HK et al (2008b) Sinking fluxes of particulate U-Th radionuclides in the East Sea (Sea of Japan). J Oceanogr 64:267–276

Honjo S, Manganini SJ, Krishfield RA et al (2008) Particulate organic carbon fluxes to the ocean interior and factors controlling the biological pump: a synthesis of global sediment trap programs since 1983. Prog Oceanogr 76:217–285

Hwang J, Druffel ERM, Eglinton TI (2010) Widespread influence of resuspended sediments on oceanic particulate organic carbon: insights from radiocarbon and aluminum contents in sinking particles. Global Biogeochem Cycles 24:GB4016. doi:10.1029/2010GB003802

Hyun JH, Kim D, Shin CW et al (2009) Enhanced phytoplankton and bacterioplankton production coupled to coastal upwelling and an anticyclonic eddy in the Ulleung Basin, East Sea. Aquat Microb Ecol 54:45–54

Ichikawa T (1982) Particulate organic carbon and nitrogen in the adjacent seas of the Pacific Ocean. Mar Biol 68:49–60

IPCC (2007) Climate change: the physical science basis. Contribution of working group i to the fourth assessment report of the intergovernmental panel on climate change. Cambridge University Press, Cambridge

Jo CO, Lee JY, Park KA et al (2007) Asian dust initated early spring bloom in the northern East/Japan Sea. Geophys Res Lett 34:L05602. doi:10.1029/2006GL027395

Kang DJ, Lee KE, Kim KR (2003) Recent developments in chemical oceanography of the East (Japan) Sea with an emphasis on CREAMS findings: a review. Geosci J 7:179–197

Kim TH, Kim G (2010) Distribution of dissolved organic carbon (DOC) in the southwestern East Sea in summer. Ocean Polar Res 32:291–297

Kim TH, Kim G (2013) Factors controlling the C:N: P stoichiometry of dissolved organic matter in the N-limited, cyanobacteria-dominated East/Japan Sea. J Mar Syst 115–116:1–9

Kim K, Kim KR, Kim YG et al (1996) New findings from CREAMS observations: water masses and eddies in the East Sea. J Korean Soc Oceanogr 31:155–163

Kim K, Kim KR, Min DH et al (2001) Warming and structural changes in the East (Japan) Sea: a clue to future changes in global oceans? Geophys Res Lett 28:3293–3296

Kim KR, Kim G, Kim K et al (2002) A sudden bottom-water formation during the severe winter 2000-2001: the case of the East/Japan Sea. Geophys Res Lett 29(8):1234. doi:10.1029/2001GL014498

Kim JY, Kang DJ, Kim E et al (2003) Biological pump in the East Sea estimated by a box model. J Korean Soc Oceanog 8:295–306 (in Korean)

Kim K, Kim KR, Kim YG et al (2004) Water mass and decadal variability in the East Sea (Sea of Japan). Prog Oceanogr 61:157–174

Kim IN, Min DH, Kim DH et al (2010) Investigation of the physicochemical features and mixing of East/Japan sea intermediate water: an isopycnic analysis approach. J Mar Res 68:799–818

Kim D, Choi MS, Oh HY et al (2011) Seasonal export fluxes of particulate organic carbon from $^{234}Th/^{238}U$ disequilibrium measurements in the Ulleung basin (Tsushima basin) of the East Sea (Sea of Japan). J Oceanogr 67:577–588

Kim JY, Kang DJ, Lee T et al (2014) Long-term trend of CO_2 and ocean acidification in the surface water of the Ulleung Basin, the East/Japan Sea inferred from the underway observational data. Biogeosciences 11:2443–2454

Kwak JH, Lee SH, Park HJ et al (2013a) Monthly measured primary and new productivities in the Ulleung Basin as a biological "hot spot" in the East/Japan Sea. Biogeosciences 10:4405–4417

Kwak JH, Hwang J, Choy EJ et al (2013b) High primary productivity and f-ratio in summer in the Ulleung basin of the East/Japan Sea. Deep-Sea Res Pt I 79:74–85

Lee K (2001) Global net community production estimated from annual cycle of surface water total dissolved inorganic carbon. Limnol Oceanogr 46:1287–1297

Lee K, Millero FJ, Byrne RH et al (2000) The recommended dissociation constants of carbonic acid for use in seawater. Geophys Res Lett 27:229–232

Lee K, Choi SD, Park GH et al (2003) An updated anthropogenic CO_2 inventory in the Atlantic Ocean. Global Biogeochem Cycles 17(4):1116. doi:10.1029/2003GB002067

Lee K, Sabine CL, Tanhua T et al (2011) Roles of marginal seas in absorbing and storing fossil fuel CO_2. Energy Environ Sci 4:1133–1146

Masuzawa T, Noriki S, Kurosaki T et al (1989) Compositional change of settling particles with water depth in the Japan Sea. Mar Chem 27:61–78

Millero FJ (2007) The marine inorganic carbon cycle. Chem Rev 107:308–341

Nakayama N, Obata H, Gamo T (2007) Consumption of dissolved oxygen in the deep Japan Sea, giving a precise isotopic fractionation factor. Geophys Res Lett 34:L20604. doi:10.1029/200 7GL029917

Otosaka S, Noriki S (2005) Relationship between composition of settling particles and organic carbon flux in the western north Pacific and the Japan Sea. J Oceanogr 61:25–40

Otosaka S, Togawa O, Baba M et al (2004) Lithogenic flux in the Japan Sea measured with sediment traps. Mar Chem 91:143–163

Otosaka S, Tanaka T, Todawa O et al (2008) Deep sea circulation of particulate organic carbon in the Japan Sea. J Oceanogr 64:911–923

Park GH, Lee K, Tishchenko P et al (2006) Large accumulation of anthropogenic CO_2 in the East (Japan) Sea and its significant impact on carbonate chemistry. Global Biogeochem Cycles 20:4013. doi:10.1029/2005GB002676

Park GH, Lee K, Tishchenko P (2008) Sudden, considerable reduction in recent uptake of anthropogenic CO_2 by the East/Japan Sea. Geophys Res Lett 35:L23611. doi:10.1029/200 8GL036118

Rho T, Lee T, Kim G et al (2012) Prevailing subsurface chlorophyll maximum (SCM) layer in the East Sea and its relation to the physico-chemical properties of water masses. Ocean Polar Res 34:413–430

Sabine CL, Tanhua T (2010) Estimation of anthropogenic CO_2 inventories in the ocean. Annu Rev Mar Sci 2:175–198

Sabine CL, Feely RA, Key RM et al (2002) Distribution of anthropogenic CO_2 in the Pacific Ocean. Global Biogeochem Cycles 16:1083. doi:10.1029/2001GB001639

Sarmiento JL, Gruber N (2006) Ocean biogeochemical dynamics. Princeton University Press, Princeton

Shiller AM (1981) Calculating the oceanic CO_2 increase: a need for caution. J Geophys Res 86:11083–11088

Suk M (2002) Marine ecosystem responses to climate variability in the East Sea. Korea Ocean Research and Development Institute, Technical Report BSPE 817-00-1396-1 (in Korean)

Takematsu M, Nagano Z, Ostrovskii AG et al (1999) Direct measurements of deep currents in the Northern Japan Sea. J Oceanogr 55:207–216

Talley LD, Lobanov V, Ponomarev V et al (2003) Deep convection and brine rejection in the Japan Sea. Geophys Res Lett 30(4):1159. doi:10.1029/2002GL016451

Talley LD, Min DH, Lobanov VB et al (2006) Japan/East Sea water masses and their relation to the sea's circulation. Oceanography 19:32–49

Tanhua T, Körtzinger A, Friis K et al (2007) An estimate of anthropogenic CO_2 inventory from decadal changes in oceanic carbon content. Proc Natl Acad Sci 104:3037–3042

Tanhua T, Waugh DW, Wallace DWR (2008) Use of SF_6 to estimate anthropogenic CO_2 in the upper ocean. J Geophys Res 113:C04037. doi:10.1029/2007JC004416

Touratier F, Goyet C (2004) Definition, properties, and Atlantic Ocean distribution of the new tracer TrOCA. J Mar Syst 46:169–179

Waugh DW, Hall TM, Haine TWN (2003) Relationship among tracer ages. J Geophys Res 108:3138. doi:10.1029/2002JC001325

Waugh DW, Haine TWN, Hall TM (2004) Transport times and anthropogenic carbon in the subpolar North Atlantic Ocean. Deep-Sea Res Pt I 51:1475–1491

Waugh DW, Hall TM, McNeil BI et al (2006) Anthropogenic CO_2 in the oceans estimated using transit time distributions. Tellus B 58:376–389

Yamada K, Ishizaka J, Yoo S et al (2004) Seasonal and interannual variability of sea surface chlorophyll-a concentration in the Japan/East Sea (JES). Prog Oceanogr 61:193–211

Yamada K, Ishizaka J, Nagata H (2005) Spatial and temporal variability of satellite primary production in the Japan Sea from 1998 to 2002. J Oceanogr 61:857–869

Yoo S, Park J (2009) Why is the southwest the most productive region of the East Sea/Sea of Japan? J Mar Syst 78:301–315

Yool A, Oschlies A, Nurser AJG et al (2010) A model-based assessment of the TrOCA approach for estimating anthropogenic carbon in the ocean. Biogeosciences 7:723–751

Chapter 8
Uranium Series Radionuclides

Tae-Hoon Kim, Jeonghyun Kim and Guebuem Kim

Abstract The activities of ^{228}Th, ^{230}Th, ^{234}Th, ^{226}Ra, ^{228}Ra, ^{210}Po, and ^{210}Pb have been reported in the East Sea (Japan Sea) over the past few decades. The activities of thorium isotopes over the entire depth of the East Sea ranged from 0.4 to 8.3 dpm 100 L^{-1} for ^{228}Th, 0.06 to 0.37 dpm 1000 L^{-1} for ^{230}Th, and 0.4 to 2.1 dpm L^{-1} for ^{234}Th. Thorium isotopes in the East Sea were used for determining the collection efficiency of sediment traps (using ^{230}Th) and the annual vertical flux of particulate organic carbon (using ^{234}Th). The activities of radium isotopes over the entire depth of the East Sea ranged from 0.3 to 11 dpm 100 L^{-1} for ^{228}Ra and 6.4 to 16 dpm 100 L^{-1} for ^{226}Ra. Radium isotopes in the East Sea were used for determining the vertical eddy diffusion coefficient (using ^{228}Ra) and the residence time of deep water (using ^{226}Ra). The activities of ^{210}Pb over the entire depth ranged from 4.4 to 27 dpm 100 L^{-1}, and the activities for ^{210}Po over the entire depth ranged from 1.5 to 11 dpm 100 L^{-1}. The ^{210}Pb and ^{210}Po in the East Sea were used for estimating the residence time of trace elements in the dissolved, colloid, and particulate phases. In summary, U-series radioisotopes have provided useful information on various physical and biogeochemical processes in the East Sea.

Keywords U-series radionuclides · Thorium · Radium · Polonium · Lead · East Sea (Japan Sea)

T.-H. Kim (✉)
Department of Earth and Marine Sciences, Jeju National University,
Jeju 690-756, Republic of Korea
e-mail: thkim@jejunu.ac.kr

J. Kim · G. Kim
School of Earth and Environmental Sciences, Seoul National University,
Seoul 151-742, Republic of Korea
e-mail: e927913@snu.ac.kr

G. Kim
e-mail: gkim@snu.ac.kr

© Springer International Publishing Switzerland 2016
K.-I. Chang et al. (eds.), *Oceanography of the East Sea (Japan Sea)*,
DOI 10.1007/978-3-319-22720-7_8

8.1 Introduction

The use of U/Th decay-series radionuclides as tracers has greatly increased our understanding of various oceanographic phenomena, such as water circulation (Broecker 1963), particle scavenging (Coale and Bruland 1985), marine export production (Cochran et al. 1995; Buesseler 1998), and the geochronology of marine sediments (Goldberg and Koide 1962; Koczy 1963).

U is conservative in seawater. So, in seawater, the activities of ^{238}U and ^{234}U are calculated using their relationships with salinity (Chen et al. 1986; Owens et al. 2011). Thorium isotopes (^{228}Th, $t_{1/2} = 1.9$ year; ^{230}Th, $t_{1/2} = 7.5 \times 10^4$ year; ^{234}Th, $t_{1/2} = 24.1$ day) are very reactive to particles, and are therefore used to determine ocean scavenging of particles (Li et al. 1979; Coale and Bruland 1985; Cochran et al. 1995) and sedimentation rates (Goldberg and Koide 1962; Koczy 1963). The short-lived ^{234}Th has been used to determine the export of organic matter in the euphotic zone (Murray et al. 1989) and the residence time of colloids in the surface ocean (Baskaran et al. 1992; Moran and Buesseler 1992; Guo et al. 1997). The long half-life of ^{230}Th lends itself to open-ocean scale studies (Moran et al. 2001), and the moderate half-life of ^{228}Th has led to its successful use in basin-scale-studies (Trimble et al. 2004).

Radium isotopes (^{223}Ra, $t_{1/2} = 11.4$ day; ^{224}Ra, $t_{1/2} = 3.6$ day; ^{226}Ra, $t_{1/2} = 1.6 \times 10^3$ year; ^{228}Ra, $t_{1/2} = 5.8$ year) are very soluble in seawater and are used to determine the transport and mixing of seawater (Cochran 1992). The two radium isotopes with longer half-lives (^{228}Ra and ^{226}Ra) are used mainly for basin- and regional-scale studies, and the short-lived pair (^{223}Ra and ^{224}Ra) are used for coastal-ocean scale studies. Radium isotopes are also used extensively to determine the magnitude of submarine groundwater discharge (Kim et al. 2005; Moore et al. 2008). Radon (^{222}Rn, $t_{1/2} = 3.8$ day) is a noble gas. Radon is used to determine air-sea gas exchange rates (Roether and Kromer 1978) and also to gauge submarine groundwater discharge in the coastal ocean (Kim and Hwang 2002; Kim et al. 2011b).

^{210}Po ($t_{1/2} = 138$ day) and ^{210}Pb ($t_{1/2} = 22.3$ year), both of which are particle-reactive, are used to study particle scavenging in the ocean (Cherry et al. 1975; Cochran et al. 1983; Fowler and Knauer 1986; Nozaki et al. 1997; Masqué et al. 2002). In addition, ^{210}Pb is used to gauge the input of dust to the ocean (Settle et al. 1982; Turekian et al. 1989), while ^{210}Po is used to estimate new production of the ocean because of its strong affinity for biota (Stewart and Fisher 2003).

In this chapter, we review studies on ^{228}Th, ^{230}Th, ^{234}Th, ^{226}Ra, ^{228}Ra, ^{210}Po, and ^{210}Pb in the East Sea. We do not include U isotopes and ^{222}Rn since directly measured data are not yet available for those isotopes in this region.

8.2 Thorium Isotopes

The activities of ^{228}Th ranged from 0.6 to 8.3 dpm 100 L^{-1} in the surface layer (0–200 m) and from 0.4 to 2.6 dpm 100 L^{-1} in the deep layer (below 200 m) (Fig. 8.1a). The maximum activity of ^{228}Th showed in the subsurface layer (100–600 m) (Nozaki and Yamada 1987). The ^{228}Th deficiency relative to ^{228}Ra in the upper layer can be attributed to rapid scavenging of Th, and ^{228}Th-^{228}Ra was nearly in equilibrium below a depth of 500 m (Nozaki and Yamada 1987). Moon et al. (2000) suggested that the total activities of total ^{228}Th over entire depths in the East Sea were higher than in the Pacific, due to higher activities of ^{228}Ra in East Sea water. Vertical fluxes of ^{228}Th increased from the Korea Strait toward the center of the East Sea, as ^{228}Ra originates from the coastal and continental shelf sediments in the Yellow Sea, East China Sea, and the southern sea of Korea (Moon et al. 2000).

The activities of ^{230}Th ranged from 0.06 to 0.34 dpm 1000 L^{-1} in the surface layer and from 0.14 to 0.37 dpm 1000 L^{-1} in the deep layer (Fig. 8.1b). The relatively constant vertical distribution of ^{230}Th in the bottom layer of the East Sea may be because of rapid mixing at the sediment-water interface (Nozaki and Yamada 1987). Hong et al. (2008a) reported the collection efficiency of the sediment trap at about 3000 m in the Japan Basin by comparing particulate ^{230}Th fluxes with in situ production rate from decay of ^{234}U, after subtracting for ^{230}Th fluxes derived from detrital and atmospheric deposition. The calculated trap efficiency was 142 ± 14 % or higher, due to significantly higher fluxes of authigenic ^{230}Th (from lateral inputs). The higher activities of ^{230}Th relative to ^{228}Th above 500 m in the East Sea, compared to those below 500 m, may be due to additional inputs of Th, with lower ^{228}Th/^{230}Th ratios, from river, coastal sediment diagenesis, and/or eolian dust transport (Nozaki and Yamada 1987).

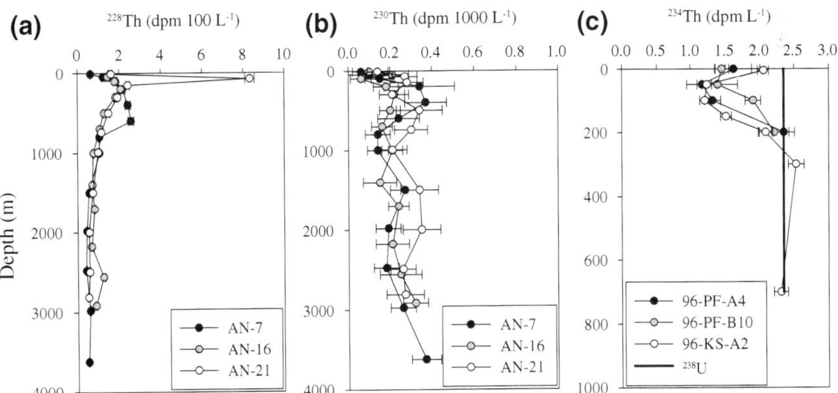

Fig. 8.1 Vertical distributions of activities of **a** ^{228}Th, **b** ^{230}Th, and **c** ^{234}Th in the East Sea (Redrawn from Nozaki and Yamada 1987; Moon et al. 2000). Station AN-7 is located at 42° 49′N, 138° 06′E; AN-16 is located at 38° 17.5′N, 135° 29′E; AN-21 is located at 38° 40.6′N, 132° 47.8′E. 96-PF-A4 is located at 38° 00′N, 133° 40′E; 96-PF-B10 is located at 40° 20′N, 135° 00′E; 96-KS-A2 is located at 35° 47′N, 130° 10′E. The *solid line* in **c** represents ^{238}U activities calculated using the relationship with salinity

The activities of ^{234}Th ranged from 0.4 to 2.1 dpm L^{-1} in the 0–800 m layer (Fig. 8.1c). Using ^{234}Th-^{238}U disequilibria, Moon et al. (2000) showed that 36.3 % of ^{234}Th was removed by scavenging processes in the surface layer. Kim et al. (2011a) measured export fluxes of particulate organic carbon (POC) by multiplying the POC/^{234}Th ratio of the sinking particles by the flux through 100 m depth. The annual vertical fluxes of POC in the East Sea were about 161 mg C m^{-2} day^{-1} with little seasonal variation, except for autumn, when vertical POC flux dropped to 49.3 mg C m^{-2} day^{-1}.

8.3 Radium Isotopes

The activities of ^{228}Ra ranged from 4.7 to 11 dpm 100 L^{-1} in the surface layer (0–200 m) and from 0.3 to 2.4 dpm 100 L^{-1} in the deep layer (200–3300 m) (Fig. 8.2a). Okubo (1980) reported that the activities of ^{228}Ra in the East Sea were ten times higher than in the Pacific Ocean, because the East Sea receives large inputs of bottom sediments from the Yellow Sea, East China Sea, and the southern sea of Korea. The vertical eddy diffusion coefficient, estimated from the vertical distributions of ^{228}Ra, was about 2.0 cm^2 s^{-1}.

The activities of ^{226}Ra ranged from 6.4 to 10 dpm 100 L^{-1} in the surface layer and from 13 to 16 dpm 100 L^{-1} in the deep layer (Fig. 8.2b). The activities of ^{226}Ra in the surface layer were comparable to activities observed in the Korea Strait (7 dpm 100 L^{-1}, Harada and Tsunogai 1986) and the western North Pacific

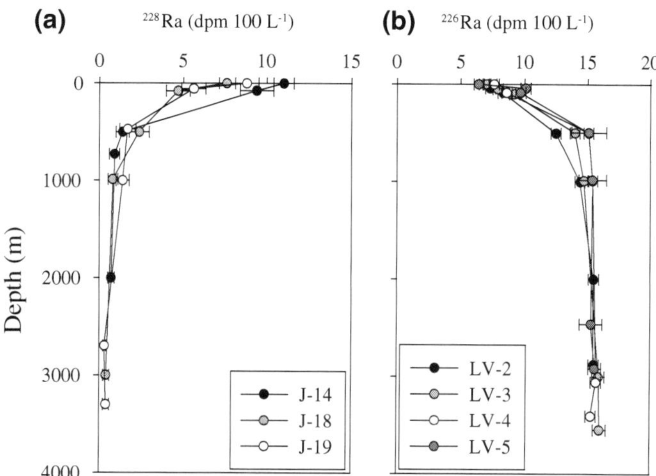

Fig. 8.2 Vertical distributions of **a** ^{228}Ra and **b** ^{226}Ra in the East Sea (Redrawn from Okubo 1980; Harada and Tsunogai 1986). Stations J-14 and LV-2 are located at 37° 44′N, 135° 12′E; J-18 and LV-3 are located at 41° 20′N, 137° 20′E; J-19 and LV-4 are located at 43° 04′N, 138° 33′E; LV-5 is located at 38° 35′N, 134° 45′E

(8 dpm 100 L^{-1}, Tsunogai and Harada 1980). Based on a 3-box model coupled with simultaneous equations for ^{226}Ra and ^{14}C, Harada and Tsunogai (1986) estimated that the residence times of the deep water and of the entire water in the East Sea were about 300–400 years and 700–1000 years, respectively.

8.4 Lead and Polonium Isotopes

The activities of the truly dissolved (<10 kDa; hereafter referred to as dissolved), colloidal (10 kDa–0.45 μm), and particulate (>0.45 μm) phases of ^{210}Pb ranged from 2.0 to 10 dpm 100 L^{-1}, 3.3 to 13 dpm 100 L^{-1}, and 0.8 to 3.7 dpm 100 L^{-1}, respectively, in the surface layer (Fig. 8.3). The dissolved, colloidal, and particulate phases constituted 32–38 % (with a mean of 35 %), 39–53 % (with a mean of 48 %), and 11–26 % (with a mean of 17 %), respectively, of the total ^{210}Pb (Kim and Kim 2012). Nozaki (1974) reported that the activities of total ^{210}Pb in the deep layer ranged from 4.4 to 7.0 dpm 100 L^{-1} (with a mean of 5.7 dpm 100 L^{-1}), with nearly constant values below 500 m. The maximum activity of the total dissolved ^{210}Pb (22 dpm 100 L^{-1}) in surface waters was observed in the middle of the Japan Basin. This activity is comparable to that in the North Pacific (20 dpm 100 L^{-1}, Nozaki et al. 1976) and the Kuroshio Current (27 dpm 100 L^{-1}, Nozaki et al. 1990). Similar to other parts of the open ocean, the activity of total ^{210}Pb decreased with increasing depth in the East Sea (Bacon et al. 1988; Nozaki et al. 1990; Kim 2001; Chung and Wu 2005). Since the activities of ^{226}Ra were within 10 dpm 100 L^{-1} in the surface layer of the East Sea (as mentioned in the previous section), these excess activities of ^{210}Pb relative to ^{226}Ra in the surface layer are likely due to inputs from the atmosphere (Nozaki 1974). Using a simple box-model and assuming a steady state condition, Nozaki (1974) estimated that the residence time of ^{210}Pb in the East Sea is approximately 15 years.

The activities of dissolved, colloidal, and particulate ^{210}Po ranged from 0.4 to 2.7 dpm 100 L^{-1}, 0.9 to 4.0 dpm 100 L^{-1}, and 1.5 to 4.6 dpm 100 L^{-1}, respectively, in the surface layer (0–200 m) (Fig. 8.3). Of the total ^{210}Po, the dissolved, colloidal, and particulate phases constituted 17–20 % (with a mean of 19 %), 29–42 % (with a mean of 36 %), and 38–51 % (with a mean of 45 %), respectively (Kim and Kim 2012). Hong et al. (2008b) reported that the total ^{210}Po activity in the deep layer (200–2000 m) ranged from 1.5 to 6.6 dpm 100 L^{-1}. The lowest activities of total ^{210}Po were observed in the subsurface layer (25–50 m) where the chlorophyll a maximum was also observed, suggesting effective biological uptake of ^{210}Po. The highest activities of total ^{210}Po were observed below the chlorophyll a maximum, probably because ^{210}Po was released when phytoplankton decomposed (Hong et al. 2008b). In the dissolved and colloidal phases at all stations, ^{210}Po activities were lower than ^{210}Pb activities. Particulate ^{210}Po showed almost uniform activity over the entire depth. Similar to the open ocean, particulate ^{210}Po activity was slightly higher than particulate ^{210}Pb activity (Bacon et al. 1976; Chung and Craig 1983; Chung 1987; Chung and Finkel 1988).

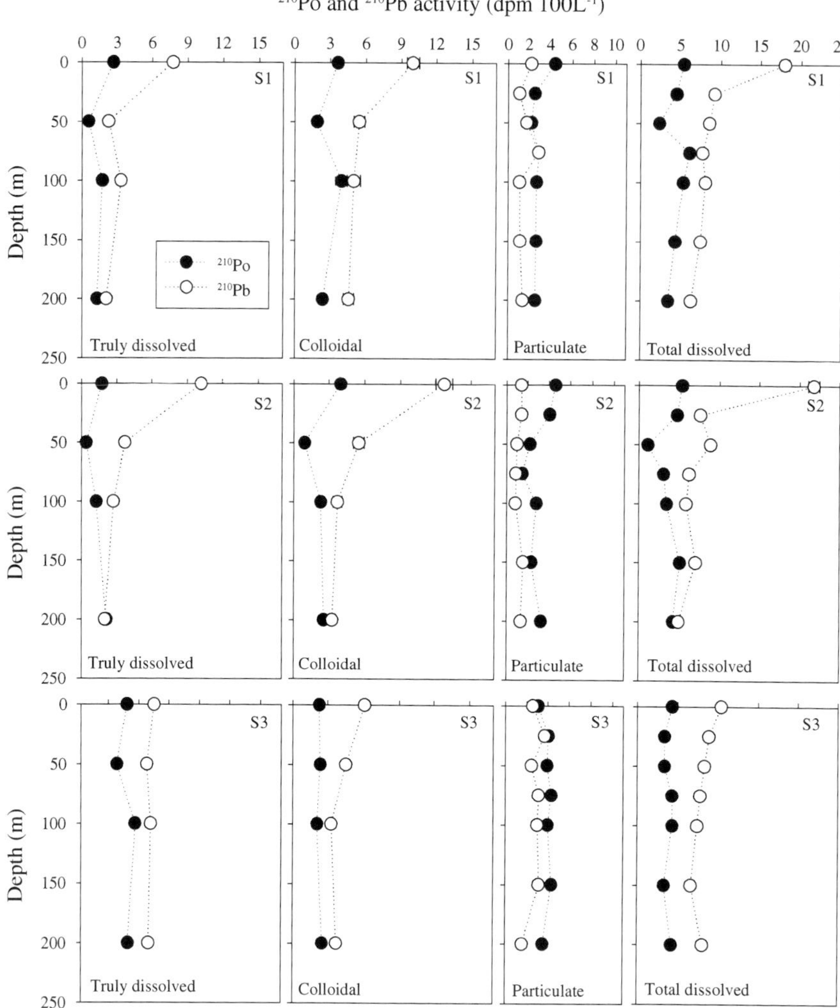

Fig. 8.3 Vertical distributions of ^{210}Pb (*open circles*) and ^{210}Po (*closed circles*) in the truly dissolved (<10 kDa), colloidal (10 kDa–0.45 μm), particulate (>0.45 μm), and total dissolved (<0.45 μm) phases in the East Sea (Kim and Kim 2012). Station S1 is located at 42° 10′N, 136° 20′E; S2 is located at 40° 60′N, 132° 20′E; S3 is located at 37° 02′N, 130 ° 56′E

Yang et al. (1995) observed the activities of ^{210}Po for the upper 100 m layer in the eddy region. In the periphery of the warm eddy, above the main thermocline, the activities of ^{210}Po were deficient relative to its parent, ^{210}Pb. However, the station in the center of the warm eddy showed no excess of ^{210}Po below 50 m. Yang et al. (1995) attributed this to the sinking of the warm surface water into the subsurface layer in this eddy region. The residence time of ^{210}Po in the surface mixed layer in the warm eddy was 0.4 years, much shorter than at stations on the periphery of the warm eddy (Yang et al. 1995). The residence time of ^{210}Po in the upper 1000 m of the water column was about 4.0–5.5 years (Hong et al. 2008b).

Kim and Kim (2012) were the first to report the activities of truly dissolved, colloidal, and particulate phases of ^{210}Po in the ocean using samples from the East Sea. Using a net residence time model, which accounts for biological uptake and remineralization, the residence times of ^{210}Po in the upper 100 m layer were estimated at 92 $\pm$ 41, 63 $\pm$ 14, and 166 $\pm$ 45 days for the truly dissolved, colloidal, and particulate phases, respectively (Kim and Kim 2012). The residence time of colloidal ^{210}Po was several-fold longer than typical residence times (<10 days) for high-molecular-weight dissolved organic carbon and colloidal ^{234}Th in the surface water. On the basis of this comparison, Kim and Kim (2012) suggested that ^{210}Po turns over several times through the colloidal phase, perhaps together with other bio-reactive elements, before settling down from the upper ocean.

References

Bacon M, Spencer D, Brewer P (1976) ^{210}Pb/^{226}Ra and ^{210}Po/^{210}Pb disequilibria in seawater and suspended particulate matter. Earth Planet Sci Lett 32(2):277–296

Bacon MP, Belastock RA, Tecotzky M, Turekian KK, Spencer DW (1988) Lead-210 and polonium-210 in ocean water profiles of the continental shelf and slope south of New England. Cont Shelf Res 8(5):841–853

Baskaran M, Santschi P, Benoit G, Honeyman B (1992) Scavenging of thorium isotopes by colloids in seawater of the Gulf of Mexico. Geochim Cosmochim Acta 56(9):3375–3388

Broecker W (1963) Radioisotopes and large-scale oceanic mixing. In Hill MN (ed) The sea: ideas and observations on progress in the study of the seas 2:88–108

Buesseler KO (1998) The decoupling of production and particulate export in the surface ocean. Global Biogeochem Cycles 12(2):297–310

Chen JH, Lawrence Edwards R, Wasserburg GJ (1986) ^{238}U, ^{234}U and ^{232}Th in seawater. Earth Planet Sci Lett 80(3–4):241–251

Cherry R, Fowler S, Beasley T, Heyraud M (1975) Polonium-210: its vertical oceanic transport by zooplankton metabolic activity. Mar Chem 3(2):105–110

Chung Y (1987) ^{210}Pb in the western Indian Ocean: distribution, disequilibrium, and partitioning between dissolved and particulate phases. Earth Planet Sci Lett 85(1):28–40

Chung Y, Craig H (1983) ^{210}Pb in the Pacific: the GEOSECS measurements of particulate and dissolved concentrations. Earth Planet Sci Lett 65(2):406–432

Chung Y, Finkel R (1988) ^{210}Po in the western Indian Ocean: distributions, disequilibria and partitioning between the dissolved and particulate phases. Earth Planet Sci Lett 88(3):232–240

Chung Y, Wu T (2005) Large ^{210}Po deficiency in the northern South China Sea. Cont Shelf Res 25(10):1209–1224

Coale KH, Bruland KW (1985) ^{234}Th:^{238}U disequilibria within the California current. Limnol Oceanogr 30(1):22–33

Cochran J (1992) The oceanic chemistry of the uranium- and thorium-series nuclides. In Ivanovich M, Harmon RS (eds) Uranium-Series disequilibrium: applications to earth, marine, and environmental sciences, pp 334–395

Cochran JK, Bacon MP, Krishnaswami S, Turekian KK (1983) ^{210}Po and ^{210}Pb distributions in the central and eastern Indian Ocean. Earth Planet Sci Lett 65(2):433–452

Cochran JK, Barnes C, Achman D, Hirschberg DJ (1995) Thorium-234/uranium-238 disequilibrium as an indicator of scavenging rates and particulate organic carbon fluxes in the Northeast Water Polynya. Greenland. J Geophys Res 100(C3):4399–4410

Fowler SW, Knauer GA (1986) Role of large particles in the transport of elements and organic compounds through the oceanic water column. Prog Oceanogr 16(3):147–194

Goldberg ED, Koide M (1962) Geochronological studies of deep sea sediments by the ionium/thorium method. Geochim Cosmochim Acta 26(3):417–450

Guo L, Santschi PH, Baskaran M (1997) Interactions of thorium isotopes with colloidal organic matter in oceanic environments. Colloids Surf A 120(1):255–271

Harada K, Tsunogai S (1986) ^{226}Ra in the Japan Sea and the residence time of the Japan Sea water. Earth Planet Sci Lett 77(2):236–244

Hong GH, Baskaran M, Lee HK, Kim SH (2008a) Sinking fluxes of particulate U-Th radionuclides in the East Sea (Sea of Japan). J Oceanogr 64(2):267–276

Hong GH, Kim YI, Baskaran M, Kim SH, Chung CS (2008b) Distribution of ^{210}Po and export of organic carbon from the euphotic zone in the southwestern East Sea (Sea of Japan). J Oceanogr 64(2):277–292

Kim D, Choi MS, Oh HY, Song YH, Noh JH, Kim KH (2011a) Seasonal export fluxes of particulate organic carbon from ^{234}Th/^{238}U disequilibrium measurements in the Ulleung Basin (Tsushima Basin) of the East Sea (Sea of Japan). J Oceanogr 67(5):577–588

Kim G (2001) Large deficiency of polonium in the oligotrophic ocean's interior. Earth Planet Sci Lett 192(1):15–21

Kim G, Hwang DW (2002) Tidal pumping of groundwater into the coastal ocean revealed from submarine ^{222}Rn and CH_4 monitoring. Geophys Res Lett 29(14):23-1–23-4

Kim G, Kim JS, Hwang DW (2011b) Submarine groundwater discharge from oceanic islands standing in oligotrophic oceans: implications for global biological production and organic carbon fluxes. Limnol Oceanogr 56(2):673–682

Kim G, Ryu JW, Yang HS, Yun ST (2005) Submarine groundwater discharge (SGD) into the Yellow Sea revealed by ^{228}Ra and ^{226}Ra isotopes: implications for global silicate fluxes. Earth Planet Sci Lett 237(1):156–166

Kim TH, Kim G (2012) Important role of colloids in the cycling of ^{210}Pb and ^{210}Po in the ocean: results from the East/Japan Sea. Geochim Cosmochim Acta 95:134–142

Koczy F (1963) Age determination in sediments by natural radioactivity. In Hill MN (ed) The sea: ideas and observations on progress in the study of the seas 3:816–831

Li YH, Feely HW, Santschi PH (1979) ^{228}Th-^{228}Ra radioactive disequilibrium in the New York Bight and its implications for coastal pollution. Earth Planet Sci Lett 42(1):13–26

Masqué P, Sanchez-Cabeza J, Bruach J, Palacios E, Canals M (2002) Balance and residence times of ^{210}Pb and ^{210}Po in surface waters of the northwestern Mediterranean Sea. Cont Shelf Res 22(15):2127–2146

Moon DS, Kim KH, Noh I (2000) The vertical fluxes of particles and radionuclides in the East Sea. J Korean Soc Oceanogr 35(1):16–33

Moore WS, Sarmiento JL, Key RM (2008) Submarine groundwater discharge revealed by ^{228}Ra distribution in the upper Atlantic Ocean. Nat Geosci 1(5):309–311

Moran SB, Buesseler KO (1992) Short residence time of colloids in the upper ocean estimated from ^{238}U-^{234}Th disequilibria. Nature 359(6392):221–223

Moran SB, Shen CC, Weinstein SE, Hettinger LH, Hoff JH, Edmonds HN, Edwards RL (2001) Constraints on deep water age and particle flux in the Equatorial and South Atlantic Ocean based on seawater ^{231}Pa and ^{230}Th data. Geophys Res Lett 28(18):3437–3440

Murray JW, Downs JN, Strom S, Wei CL, Jannasch HW (1989) Nutrient assimilation, export production and ^{234}Th scavenging in the eastern equatorial Pacific. Deep-Sea Res 36(10):1471–1489

Nozaki Y (1974) Radium-226, Lead-210 and polonium-210 in sea water and their use in oceanography, Ph.D. thesis, Hokkaido Univ., p 148

Nozaki Y, Ikuta N, Yashima M (1990) Unusually large ^{210}Po deficiencies relative to ^{210}Pb in the Kuroshio Current of the East China and Philippine Seas. J Geophys Res 95(C4):5321–5329

Nozaki Y, Thomson J, Turekian KK (1976) The distribution of ^{210}Pb and ^{210}Po in the surface waters of the Pacific Ocean. Earth Planet Sci Lett 32(2):304–312

Nozaki Y, Yamada M (1987) Thorium and protactinium isotope distribution in waters of the Japan Sea. Deep-Sea Res 34(8):1417–1430

Nozaki Y, Zhang J, Takeda A (1997) ^{210}Pb and ^{210}Po in the equatorial Pacific and the Bering Sea: the effects of biological productivity and boundary scavenging. Deep-Sea Res II 44(9):2203–2220

Okubo T (1980) Radium-228 in the Japan Sea. J Ocean Soc Japan 36(5):263–268

Owens SA, Buesseler KO, Sims KWW (2011) Re-evaluating the ^{238}U-salinity relationship in seawater: implications for the ^{238}U–^{234}Th disequilibrium method. Mar Chem 127(1):31–39

Roether W, Kromer B (1978) Field determination of air-sea gas exchange by continuous of radon-222. Pure appl Geophys 116(2–3):476–485

Settle D, Patterson C, Turekian K, Cochran J (1982) Lead precipitation fluxes at tropical oceanic sites determined from ^{210}Pb measurements. J Geophys Res Oceans (1978–2012) 87(C2):1239–1245

Stewart GM, Fisher NS (2003) Experimental studies on the accumulation of polonium-210 by marine phytoplankton. Limnol Oceanogr 48(3):1193–1201

Trimble SM, Baskaran M, Porcelli D (2004) Scavenging of thorium isotopes in the Canada Basin of the Arctic Ocean. Earth Planet Sci Lett 222:915–932

Tsunogai S, Harada K (1980) ^{226}Ra and ^{210}Pb in the western North Pacific. In: Goldberg E, Horibe Y, Saruhashi K (eds) Isotope marine chemistry, pp 165–191

Turekian KK, Graustein WC, Cochran JK (1989) Lead-210 in the SEAREX program: an aerosol tracer across the Pacific. In: Riley JP, Chester R, Duce RA (eds) Chemical oceanography, vol 10, pp 51–81

Yang H, Kim S, Lee J (1995) Effect of eddy on the cycle of ^{210}Po and ^{234}Th in the central region of Korean East Sea. J Korean Soc Oceanogr 30(4):279–287

Chapter 9
Distribution of Chemical Elements in Sediments

Hyun Ju Cha

Abstract Geochemical aspects of sediments in the East Sea (Japan Sea), including spatial distributions of chemical elements and processes governing these distributions, are reviewed. The chemical composition of sediments in the East Sea shows distinct geographical and geomorphological variation and is greatly affected by water temperature and circulation pattern. The existence of a subpolar front at around 38°N–40°N separates the East Sea into northern and southern parts. This separation is reflected by differentiation in the sediments, which are silica-rich, with low levels of organic carbon in the north and carbonate-rich with high levels of organic carbon in the south. The southern part of the East Sea, especially the southwestern East Sea, is characterized by high productivity and high sedimentation rate. Consequently, post-depositional diagenetic processes greatly influence the chemical composition of sediments in the south and southwest. As a result of such redox processes, sediments in the southern part of the East Sea are highly enriched in manganese oxides. Other important factors determining the distributions of chemical elements in sediment are chemical and biological processes in the water column and the sediment's proximity to continents. Generally, the chemical composition of coastal sediments, including upper slope sediments, in the East Sea is more influenced by the supply, transport, and preservation of detrital sediments than by diagenetic processes in the sediments. In contrast, the chemical composition of basin sediments is more influenced by post-depositional diagenetic and authigenic processes than by the supply, transport and preservation of detrital sediments.

H.J. Cha (✉)
Research Institute of Marine Sciences, Chungnam National University,
Daejeon 305-764, Republic of Korea
e-mail: hcha80@chol.com

© Springer International Publishing Switzerland 2016
K.-I. Chang et al. (eds.), *Oceanography of the East Sea (Japan Sea)*,
DOI 10.1007/978-3-319-22720-7_9

Keywords Biogechemical process · Metal · Organic matter · Sediment · East Sea (Japan Sea)

9.1 Introduction

Marine sediments are composed of minerals and organic particles supplied through fluvial, atmospheric, and hydrothermal pathways from continents, nearby islands, and overlying waters (Chester 2000). Sediments display considerable geographic variation, because the lithogenic and/or volcanogenic, biogenic and authigenic source particles may change greatly during lateral transport, sinking and sedimentation, due to physico-chemical conditions and hydrodynamic interactions during their formation, as well as post-depositional processes (Fütterer 2006). Thus, the chemical composition of sediment reflects and is controlled by the type of sedimentary environment, which, in turn, depends on its proximity to continents, particle-water interactions, water depth, biological activity, sedimentation rate, the physico-chemical properties of bottom water and benthic biogeochemical dynamics (Calvert and Pedersen 1993; Fütterer 2006; Sarmiento and Gruber 2006; Cha et al. 2007; Arndt et al. 2013). Consequently, the distribution of chemical elements in sediment sheds light on the sedimentary sources and the processes affecting sedimentary chemical composition.

The East Sea is a semi-enclosed marginal sea surrounded by the East Asian continent and the Japanese Islands. It is characterized by a narrow continental shelf and three deep basins (the Japan, Yamato, and Ulleung basins), which are separated by the Korea Plateau, the Oki Bank, the Yamato Rise and the Kita-Yamato Bank (Chough et al. 2000; Lee et al. 2011). The floors of the Japan and Yamato basins are generally smooth and flat, except for a few seamounts and hills. The South Korea Plateau is characterized by ridges, seamounts, and troughs (Chough et al. 2000). The East Sea consists of two oceanographic regions called the cold and warm regions, divided by the subpolar front (SPF) (Chough et al. 2000; Lee et al. 2009). The SPF runs east-west at around 38°N–40°N and divides the upper layer (upper 200 m) of the East Sea into northern and southern waters. Northern waters are characterized by subpolar temperatures and overlie cold siliceous sediments; southern waters are characterized by subtropical temperatures and overlie warm carbonaceous sediments (Chough et al. 2000; Yanagi 2002; Lee et al. 2009; Chang et al. 2004). Ocean-like thermohaline circulation starts from the northwestern part of the East Sea in winter (Kim et al. 1991, 2002; Seung 1992; Senjyu et al. 2002; Talley et al. 2003). This thermohaline circulation greatly affects the chemical properties of both water (Kim and Kim 1996; Tishchenko et al. 2007) and sediments (Otosaka et al. 2004; Cha et al. 2005, 2007).

9.2 Organic and Inorganic Carbon and Silica

Spatially, the distribution of chemical elements in recent sediments of the East Sea appears to be controlled by organic production, climatic zonation, hydrologic conditions (such as currents and circulation patterns), and the bottom morphology of the basin (Solov'yev 1962; Niino et al. 1969; Yin et al. 1989; Takematsu et al. 1993; Cha et al. 2007; Ohta et al. 2013).

The sea floor of the East Sea is covered with fine-grained sediments that are mainly composed of silt and clay. Sediment grain size changes with depth; sediments are generally coarser in shallow areas and finer in deep areas, with some exceptions (Solov'yev 1962; Niino et al. 1969; Khim et al. 1997; Cha et al. 2007). Glacial-marine sediments composed of gravels and pebbles scattered by floating ice are common in the northwest and in areas close to the Tatarsky Strait (Solov'yev 1962). Much coarser sediments, composed of sand and gravel, are distributed on the banks, continental shelves and islands (Niino et al. 1969). The floors of deep basins are covered with fine-grained sediments (Ichikura and Ujiie 1976; Cha et al. 2007). In particular, at water depths greater than 2000 m, the floor of the Ulleung Basin is mostly covered with clays (Solov'yev 1962; Cha et al. 2007). Sandy sediments with lots of relict carbonate shells and rock fragments are found in the Korea Strait, the gateway through which the Warm Tsushima Current flows into the East Sea (Choi and Park 1993; Cha et al. 2007). However, sediments stretching from the northeast of the Korea Strait to the Japanese coasts contain 20–80 % mud at water depths of 130–170 m. This mud belt extends to the southeastern slope of the Ulleung Basin (Yin et al. 1989; Ohta et al. 2013). In contrast a fine-grained sediment belt extends in coastal areas along the Korean-side from Pusan to Pohang (Cha et al. 2007).

Brownish mud occurs extensively at depths greater than 1000–2000 m in the East Sea. Sediment colors vary by water depth. Sediments of deep basins (deeper than 2500 m) are yellowish brown to olive brown, whereas continental slope, shelf and bank sediments are greenish yellow to yellowish green. Nearly all ocean floors are covered with brown or red sediments at about 3000–4000 m in the East Sea, in contrast to the Pacific Ocean, where such sediments are usually found below 5000 m (Niino et al. 1969; Yin et al. 1989). Red or brown sediments may occur at shallower depths in the East Sea because fast ventilation supplies highly oxygenated water to the bottom (Talley et al. 2003).

9.2.1 Distribution of Organic Carbon

Sedimentary organic carbon levels in the East Sea show great spatial variation around the SPF, at 38°N–40°N (Fig. 9.1a). The northern part of this region is characterized by low levels of sedimentary organic carbon, whereas the southern part

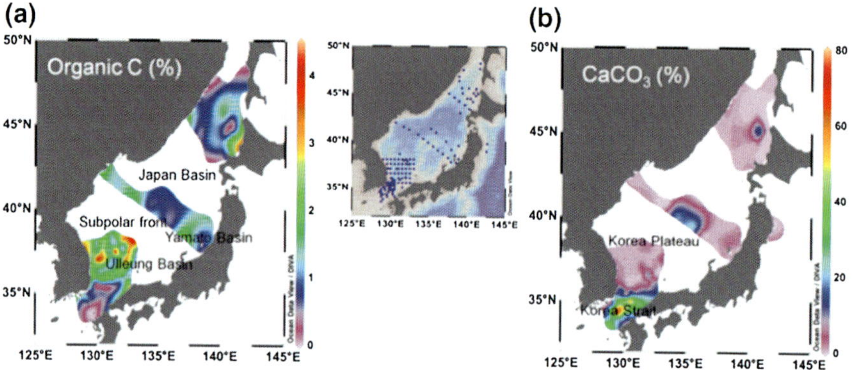

Fig. 9.1 Distribution of **a** organic and **b** inorganic carbon in the East Sea. Data taken from Cha et al. (2007) for the southwestern part (including the Korea Strait) and from Niino et al. (1969) for the central, eastern and northern East Sea

has high sedimentary organic carbon levels. High levels of sedimentary organic carbon are also found in the northern part of the sea, in shallow coastal bays and in fine-grained deposits along the continental slope. In the northern part of the sea, the fine-grained sediments of closed bays near Hokkaido contain the highest concentration of organic carbon (about 2 %). Most continental slope sediments have organic carbon concentrations of 1–2 %. The level of organic carbon decreases to less than 1 % in basin areas. Sediments near islands and continental shelves also have less than 1 % organic carbon (Solov'yev 1962; Niino et al. 1969). Because sedimentation rates are low in basin areas and because organic matter arrives at the sea floor after decomposing during its long transport through the water column, organic matter is more susceptible to oxidative degradation in basin areas. Moreover, oxidative decomposition is enhanced by the highly oxygenated bottom water of the East Sea (Kim et al. 1991; Gamo 2011).

In the southern part of the sea, the spatial distribution of organic carbon in sediments differs from that seen in the northern part. High concentrations of organic carbon (>2.5 %) are found over the Korea Plateau and continental slope and the basin plain of the Ulleung Basin (Fig. 9.1a), where the highest value reported is greater than 4 % (Cha et al. 2007; Choi et al. 2009; Cha and Choi 2013). Surprisingly, although these sediments are from the deep sea, their organic carbon concentrations are higher than those of the open ocean and even comparable to concentrations in upwelling areas (Sarmiento and Gruber 2006). In contrast to the very high organic carbon levels in the Korea Plateau and Ulleung Basin sediments, levels are very low on the shelves around the Korea Strait, where relict sands and gravels are present.

The southwestern part of the East Sea is a fairly productive region, despite little riverine input of nutrients (Hong et al. 1997; Yamada et al. 2005; Lee et al. 2008, 2009, 2010; Yoo and Park 2009; Kim et al. 2011). Using satellite data, Yoo and Park (2009) proposed that biological productivity is enhanced mainly by wind-driven upwelling along the Korean coast. They also suggested that advection of upwelled water to deep basins leads to the high productivity in the Ulleung Basin. Lee et al. (2008) suggested the high levels of organic carbon in the Ulleung Basin were caused by high export production and low dilution by inputs of terrestrial materials and calcium carbonate. Choi et al. (2009) attributed the high level of organic carbon in the basin sediments to both the low dilution by calcium carbonate and the high organic carbon sinking flux in the Ulleung Basin. The high organic carbon sinking flux may result from a combination of enhanced primary production because of upwelling in the Kampo-Ulgi area, and transport by the East Korean Warm Current (Hyun et al. 2009).

In southern East Sea sediments, organic carbon accumulation rates range from 2.0 to 60 gC m^{-2} year^{-1} and burial fluxes range from 1.0 to 11.5 gC m^{-2} year^{-1} (Lee et al. 2008, 2010; Cha and Choi 2013). This burial flux of organic carbon (1.0–11.5 gC m^{-2} year^{-1}) in southern sediments is very high (about 1 % of total primary production) compared to other areas with similar water depth (Lee et al. 2010 and references therein), although most of the organic carbon accumulated in the sea floor is returned to the water column via regeneration. Organic carbon accumulation rates vary greatly with environmental setting and rates are much higher in slope sediments than in basin sediments (Cha et al. 2007). Thus, the basins of the southwestern East Sea appear to have characteristics of hemipelagic rather than deep sea sediments (Cha et al. 2007). So, organic matter deposited on the sea floor in the southwestern East Sea may experience less degradation, and certain processes may prevent organic matter from decomposing. Sedimentation rate is generally positively related to organic carbon content, probably because high sedimentation rates rapidly remove and protect organic matter from oxic respiration or because high sedimentation rates protect organic matter by increasing its adsorption on mineral surfaces (Rullkötter 2006 and references therein).

9.2.2 Distributions of Calcium Carbonate and Silica

Distributions of calcium carbonate and silica around the SPF appear to depend on water temperature. Calcium carbonates occur mainly in the southern East Sea where the Tsushima Warm Current warms surface waters (Fig. 9.1b). Furthermore, calcium carbonates accumulate chiefly in sandy sediments of the southern East Sea and the Yamato Rise. Concentrations of calcium carbonate show great spatial variation (from less than 1 % to greater than 70 %) with highest values in the Korea Strait. Sediments on relatively flat basin floors have the lowest calcium carbonate concentrations (<1 %) due to the shallow carbonate compensation depth (CCD) in the East Sea (Chen et al. 1995; Lee et al. 2001). Based on total alkalinity and pH data, Lee et al. (2001) suggested that the crossover from supersaturation to oversaturation for calcium carbonate occurs at depths of about 200-400 m for calcite and about 100–300 m for aragonite. Chen et al. (1995) reported CCD of 300 m for aragonite and 1300 m for calcite, using total alkalinity and CO$_2$

data. Ichikura and Ujiie (1976) estimated CCD between 1500 and 2000 m, from the occurrence of planktonic foraminifera and/or red clay. In the deep basins and lower continental slope sediments of the East Sea, however, high levels of calcium carbonates (greater than 10 %) have been reported (Lee et al. 2001; Hyun et al. 2007). Lee et al. (2001) suggested that the supply of terrestrial materials from adjacent continents may keep calcium carbonate from dissolving in deep areas, despite the shallow CCD in the East Sea. The high levels of calcium carbonates (>30 %) in sandy sediments from the Korea Strait are likely due to relict carbonate shells (Niino et al. 1969; Choi and Park 1993; Cha et al. 2007).

The sediments of the East Sea are rich in amorphous silica. Silica concentrations range from less than 3 % to greater than 20 %. Silica accumulates in the northern East Sea, especially in the areas with low surface water temperatures. Particularly high levels of silica (>20 %) are found in the northern part of the Tatarsky Strait, where fine clays are deposited (Solov'yev 1962). Diatoms are the most common phytoplankton group in the East Sea and very common in East Sea sediments (Chough et al. 2000; Lee et al. 2009). Ichikura and Ujiie (1976) reported that diatomaceous yellowish brown or brown clays are widely distributed over the deep sea floors of the Japan, Yamato and Ulleung basins; the Yamato Rise; and the continental slope connected to the Yamato Rise. In the southern East Sea, the percentage of opaline (biogenic) silica varies from ca. 5 % to 12 %, with highest values in the Ulleung Basin and the Korea Plateau (Lee et al. 2003; Hyun et al. 2007; Gorbarenko and Southon 2000).

9.3 Metals

Sediment chemical composition is influenced not only by sediment source, particle-water interactions during transport from source areas to depositional sites, water column processes on the way to sea floor, and post-depositional reactions after sedimentation (Calvert and Pedersen 1993), but also by sedimentation rate, sediment size, minerals present, depth of the redox boundary, organic and inorganic carbon content, etc. (Calvert 1976; Bodur and Ergin 1994; Morford and Emerson 1999; Cho et al. 1999; Chester 2000). In the East Sea, chemical composition of sediments varies with geography and geomorphology. For the settling particles in the East Sea, Masuzawa et al. (1989) classified chemical elements into four groups based on the relationship between their concentrations and the change in metal/Al ratios with water depth: refractory (Al, Sc, La, Th, Hf, V, Ta, K, Rb, and Cs); biogenic (Ba, Ca, and Sr); scavenged (Mn, Ce, Fe, and Co); and biogenic-scavenged (As, Sb, Se, Ag, Zn, and Br). Cha et al. (2007) also categorized elements into four groups based on the relationship between elemental concentration and the change in element/Al ratios with water depth. As classified by Cha et al. (2007), Group 1 elements were lithophilic and abundant in detrital minerals, similar to the refractory elements classified by Masuzawa et al. (1989).

The concentrations of Group 1 elements decreased with water depth but element/ Al ratios remained unchanged (Sc, Cr). Group 2 elements according to Cha et al. (2007) are the same as the biogenic elements as classified by Masuzawa et al. (1989). Concentrations of Group 2 elements, as well as element/Al ratios, decrease with increasing water depths (Ca, Sr, $CaCO_3$). For Group 3 elements, concentrations and element/Al ratios increase steadily with water depth (organic carbon, V, Cu, Pb). Hence, these elements correspond to the scavenged or authigenic type. Similar to Group 3 elements, concentrations and element/Al ratios both increase greatly with water depth for Group 4 elements (Fe, Mn, P, Co, and Ni). Given that Fe and Mn have redox-related cycles within sediments and accumulate greatly in (oxy)hydroxide forms at the redox boundary and/or near the sediment surface, Group 4 elements are considered diagenetic (Froelich et al. 1979; Slomp et al. 1996; Marin and Giresse 2001; Cha et al. 2005; Choi et al. 2009).

9.3.1 Aluminum

Aluminum is a main component of clay minerals and occurs at high concentrations in fine-grained sediments and deep sea clays. Concentrations of aluminum (Al) in surface sediments of the East Sea range from 4.8 to 9.8 % (Table 9.1). High Al concentrations are found in the Japan Basin (8.9–9.8 %), followed by the Yamato Basin (8.6 %) and the Ulleung Basin (4.8 %; Ulleung Basin[b] in Table 9.1). The high concentration of Al in the Japan Basin may be due to the input of Asian dust in winter and spring (Otosaka et al. 2004; Takata et al. 2008). The Ulleung Basin is characterized by high clay and high organic carbon content and has much lower Al concentrations than the Japan and Yamato basins, probably because the Ulleung Basin receives a low supply of detrital materials and a high supply of biogenic materials (Cha et al. 2007). Ulleung Basin sediments also show very high concentrations of Mn, P, Co and Ni, as well as high Fe/Al ratios.

9.3.2 Iron

Iron concentrations vary from 3.2 to 5.4 % in East Sea sediments (Table 9.1). The distribution of Fe in surface sediments of the East Sea reflects the bathymetry of the sea floor. Similar to Al concentrations, among the three deep basins, highest concentrations of Fe occur in the sediments of the Japan Basin, while the sediments of the Ulleung Basin have relatively low Fe concentrations compared to northern areas. Within the Japan Basin, highest levels of Fe occur in the central part, followed by the western part, with lowest Fe levels occurring in the eastern Japan basin. Sediments of the eastern Japan Basin, the Yamato Basin and the Ulleung Basin show low Fe concentrations (Table 9.1).

Table 9.1 Elemental composition of surface sediments

	Japan Basin[a]				Yamato Basin[a]	Ulleung Basin[a]	Ulleung Basin[b]	Deep sea clay[c]
	Western	Central	Eastern 1 (SE)	Eastern 2 (E)				
Al (%)	8.88	9.75	9.48	8.55	8.56	9.45	4.8	9.5
Ca (%)	0.78	1.04	0.83	1.79	0.83	0.61	0.4	1.0
Mn (ppm)	5143	8175	4744	2982	5121	8250	11,000	6000
Fe (%)	5.24	5.42	5.19	4.11	4.47	4.85	3.2	6.0
Ni (ppm)	463	596	469	259	436	474	54	200
Co (ppm)	20.3	56.4	30.0	22.0	29.3	26.5		55
Cu (ppm)	47.5	121.0	62.1	47.6	63.0	65.8		200
Zn (ppm)	130	207	142	94.3	199	222	120	120
Y (ppm)	47.5	30.7	34.5	40.2	23.3	23.7		
Pb (ppm)	48.2	53.7	72.3	46.9	47.0	48.8	43.0	200
Fe/Al	0.59	0.56	0.55	0.48	0.52	0.51	0.67	0.63
Chemical composition of settling particles near bottom								
Al (%)	4.19			3.55	3.61			
Mn (ppm)	2787			2312	2796			

[a]Data from Otosaka et al. (2004)
[b]Average values in the Ulleung Basin surface sediments (Cha et al. 2007)
[c]Chester (2000)

In a review of sedimentation processes in the East Sea, Solov'yev (1962) presented the spatial distribution of Fe in the East Sea sediments, excluding the Ulleung Basin and southern areas. High concentrations of Fe were found in the central part of the sea and overlapped roughly with the occurrence of clay deposits (Solov'yev 1962). In the southern East Sea, Fe distribution in sediments is dependent on water depth and higher Fe concentrations occur in deep areas around the Ulleung Basin and Oki Trough (Yin et al. 1989; Cha et al. 2007). High concentrations of sedimentary Fe are also found in coastal and shelf areas in the Tatarsky Strait (close to the Russian coast made of basic magmatic rocks). Southern areas with high Fe concentrations include submarine outcrops and volcanic banks (Solov'yev 1962). Continental shelves on the Japanese side, northeast of the Korea Strait, have fine sands with low mud content that are subjected to strong bottom currents from the Tsushima Warm Current. The fine sandy shelf sediments are also enriched in Fe (as well as P, Ti, Nb, REEs, Ta, Th, and U). This enrichment of Fe in shelf sediments is attributed mainly to the accumulation of P- and Ti-bearing heavy minerals (such as apatite, monazite, rutile, and sphene) as a result of the strong bottom currents that also contain high levels of Fe (Ohta et al. 2013).

Iron is the fourth most abundant element in continental crust and it is transported to marine sediments through riverine, atmospheric, hydrothermal and glacial pathways (Haese 2006). The surface water of the East Sea is characterized by low nutrient and high Fe concentrations. The vertically integrated total dissolvable Fe inventories in the upper 200 m (5–200 m) are very high (284–359 μmol Fe m^{-2}) in the northeastern and mid-eastern regions of the East Sea compared to the western and eastern North Pacific Ocean, probably because of high atmospheric Fe input to the nutrient-depleted surface water of the East Sea near the Asian continent (Takata et al. 2008 and references therein). Using sediment traps in the western Japan, Yamato and Ulleung basins, Otosaka et al. (2004) collected settling particles at several depths and calculated that the mass flux was the highest in the western Japan Basin, both at depths of 1 km and near the bottom. The mass flux data and data on the chemical composition of settling particles, suggest there is significant input of atmospheric Fe in the western Japan Basin. The relatively enriched Fe in the extracted fraction of the brownish mud probably has a hydrogenous origin, given that the surface water of the East Sea is very high in total dissolvable Fe concentration compared to surface waters in the western and eastern North Pacific Ocean, likely due to high atmospheric Fe input to East Sea surface waters near the Asian continent (Yin et al. 1989; Takata et al. 2008).

9.3.3 *Manganese and Other Trace Elements*

The transition elements (Mn, Cu, Ni, Co, Zn, V and Cr) in the surface sediments of the East Sea show distinct geographical distributions (Table 9.1). Mn, Ni, Cu and Zn are greatly enriched in the southern East Sea, particularly in the Ulleung Basin and Oki Trough sediments, as well as the central Japan Basin. Sediment concentrations of Pb are higher in the eastern Japan Basin than in other parts of the East Sea (Yin et al. 1989; Otosaka et al. 2004; Cha et al. 2007).

Mn concentrations vary widely (ranging from 0.02 % to greater than 2 %) with a regional distribution pattern (Yin et al. 1989; Otosaka et al. 2004; Cha et al. 2007; Choi et al. 2009). Mn concentrations generally increase as water depth increases, with concentrations spiking to their highest levels below a depth of about 1000 m in the deep sea sediments of the Ulleung Basin and the Oki Trough, where the sea floor is covered with brownish or brownish yellow mud (Yin et al. 1989; Cha et al. 2007).

9.3.4 Redox Cycling of Manganese and Iron

In sediments, Mn distribution seems to be controlled by the redox cycle. In enriched basin sediments, Mn exists in the form of oxides and/or oxyhydroxides of suboxic diagenesis origin (Tsunogai et al. 1979; Yin et al. 1989; Cha et al. 2005; Choi et al. 2009). Mn oxides, like Fe oxides, are involved in early diagenetic processes (as an intermediate oxidant). Thus, Mn oxides experience redox cycling in sediments: burial in the particulate phase, reductive dissolution at the reducing zone, upward migration in the dissolved phase and then reprecipitation as oxides at the surface oxidizing zone. Reprecipitation may also enrich concentrations of P and transition elements (such as Co and Ni) in marine surface sediments at or near the redox boundary (Froelich et al. 1982; Sherwood et al. 1987; Berner et al. 1993; Lucotte et al. 1994; Slomp et al. 1996) as seen in basin sediments (Tsunogai et al. 1979; Dymond et al. 1984; Yin et al. 1989; Cha et al. 2005).

The amount of Fe and/or Mn oxides reprecipitated in the surface oxidized zone depends on the type of sedimentary environment, with more reprecipitation in basins and less in slope sediments. For example, in the Ulleung Basin, surface sediments are extremely rich in Mn oxides (over 2 %) but slope sediments are not enriched. Cha et al. (2007) suggest that enrichment of Mn in the surface layer of basin sediments may result from a combination of the high organic carbon supply and the high sedimentation rate in the Ulleung Basin. Even in deep basin areas, high sedimentation rates prevent dissolved oxygen from penetrating deep into sediment and thereby oxidizing organic matter, and consequently organic matter reaching the bottom may be well preserved in sediment. Dissolved Mn released by the reductive dissolution of Mn oxides during the degradation of organic matter in oxygen deficient sediment may be transported upward to the sediment surface, where it reprecipitates as Mn oxides (Cha et al. 2005, 2007). Alternatively, Choi et al. (2009) suggest that Mn enrichment results from a combination of high organic carbon content in surface sediments and optimum sedimentation rates in the basin area. According to them, Mn oxides may be used as oxidants during the degradation of organic matter in sediments containing high organic carbon; dissolved Mn released by the reductive dissolution of Mn oxides during organic matter degradation may then be transported upward if sedimentation rates are optimal; finally, this dissolved Mn may reprecipitate in surficial sediments. Choi et al. (2009) suggest another explanation, similar to what occurs in Danish coastal sediments (Canfield et al. 1993): the dissolved Mn released from organic matter decomposition in continental slope areas is transported to basin areas, where it is reprecipitated because there is no Mn enrichment in the continental slope sediments (where Mn concentrations are usually 0.006–0.03 % at the sediment surface). These explanations for the Mn enrichment in the basin sediments invoke the redox cycling of Mn in sediments overlain by oxygen-rich bottom water.

Manganese reduction occurs in all three basins (the Japan, Yamato and Ulleung basins) of the East Sea (Solov'yev 1962; Masuzawa and Kitano 1983; Yin et al. 1989; Otosaka et al. 2004; Cha et al. 2005, 2007; Choi et al. 2009; Ohta et al.

2013). Redox boundaries are present 80–120 cm below the sediment/water interface for the central Japan Basin, but they are present just below the sediment/water interface in the northeastern Japan Basin, the Ulleung Basin and the Yamato Basin (Masuzawa and Kitano 1983; Yin et al. 1989; Cha et al. 2005). The shallow redox boundaries may be the reason why Mn is most enriched in the southern East Sea. Sulfate reduction may affect concentrations of Fe and Mn in surface sediments and Masuzawa and Kitano (1983) classified sulfate reduction into three types:

(1) Areas where Mn oxide reduction is appreciable and sulfate reduction is not appreciable, as in the northern Japan Basin. In such cases, there is no mechanism to keep dissolved Fe from migrating upward. So, similar to Mn oxides, Fe oxides will enrich surface sediments.

(2) Areas where sulfate reduction takes place but hydrogen sulfide is not observed, such as the central Japan Basin. Here again, surface sediments will be enriched in Fe oxides.

(3) Areas where sulfate reduction is pronounced and hydrogen sulfide is observed, as in the Yamato and Ulleung basins. If iron sulfides are formed, a large portion of the Fe released by reductive dissolution of Fe oxides will precipitate as iron sulfide phases, so that only small amounts of dissolved Fe can diffuse upward through the interstitial water. In the East Sea, sulfate is actively reduced by microbes for organic matter degradation both in continental slope and basin sediments; sulfate reduction is especially active in the Ulleung Basin (Lee et al. 2008; Hyun et al. 2010; You et al. 2010). Therefore, the enrichment of Fe and other trace elements (Co, Ni, Cu, Zn) in deep basin areas (Japan, Yamato and Ulleung basins) may be of authigenic origin.

9.3.5 Sources and Geochemical Features of Sediments

Using La/Yb and Mn/Al ratios, Otosaka et al. (2004) suggested three sources of lithogenic material for the Japan and Yamato basins of the East Sea: atmospheric input of "fresh" Asian dust; lateral transport of "old" Asian dust from the East China Sea by the Tsushima Warm Current; and lateral transport from the Island arc of the Sakhalin and the Japanese Islands. Atmospheric input from the Asian continent in winter and spring contributed 84 % of the annual lithogenic fluxes at a depth of 1 km in the western Japan Basin; the Tsushima Warm Current transports Asian dust from the East China Sea to the southern margin of the East Sea (Yamato Basin); and lateral transport from the Japanese Islands accounts for 87 % of the annual lithogenic flux in the eastern Japan Basin (Otosaka et al. 2004).

Cha et al. (2007) studied the geochemical features of fine-grained sediments on the upper slope and continental shelf off the Korean coast and proposed that they were a mixture of Korean and Chinese stream sediments. Spatial distributions of elements from the eastern side of the Korea Strait (including the Japanese coast) to the southeastern continental slope of Ulleung Basin show the same general trend, elemental concentrations increasing with increasing mud content. The dominant flow system, the Tsushima Warm Current, plays an important role in the distribution of clay minerals over the Korea Strait region, whereas continental sources are more important for fine-grained sediments along the Korean coasts and in slope areas (Khim et al. 1997). Along the Korean coast from Pusan to Pohang and in the upper part of the continental slope south of the Ulleung Basin, fine-grained sediments are characterized by high sedimentation rates and high concentrations of Al, Fe, K, Sc, and Cr that are abundant in detrital minerals (Cho et al. 1999). The concentrations of these elements were comparable to typical nearshore mud deposits (Cho et al. 1999; Chester, 2000). Using $(Mg + Fe + K + Ti)/Al$ and $(V + Cr + Co + Ni + Zn)/Al$ ratios and isotopic compositions, Cha et al. (2007) reported that these fine-grained sediments are mixtures of sediments from the Changjiang and Nakdong Rivers.

9.4 Summary

The chemical composition of sediments in the East Sea shows distinct geographical and geomorphological variation. Factors that determine the chemical composition of sediment are: proximity to continents, water depth, reactions between particles and water, biological activity in the overlying water, sedimentation rates, biogeochemical processes in sediments and physico-chemical conditions of the overlying water (including large scale circulation and ventilation processes). Post-depositional diagenetic processes in sediments greatly influence whether a material is preserved in and buried with sediment or returned to the ocean as benthic flux (in a changed chemical form).

The chemical composition of coastal sediments (including upper slope sediments) in the East Sea was more affected by the supply, transport, and preservation of detrital sediments than by diagenetic processes in the sediments. In contrast, the chemical composition of basin sediments was more influenced by post-depositional diagenetic and authigenic processes than by the settling of detrital sediments.

Finally, it is interesting to note that although the Ulleung Basin is very deep and is overlain by highly oxygenated bottom water, high levels of organic matter are preserved in Ulleung Basin sediments, where sulfate, Fe and Mn oxides are used for organic matter degradation.

Acknowledgments This work was supported by a National Research Foundation of Korea Grant funded by the Korean Government [NRF-2009-353-C00077]. This work was also supported by the "Study on Environmental Impact Analysis for Gas Hydrate Production Test" project of the Korea Institute of Geoscience and Mineral Resources.

References

Arndt S, Jörgensen BB, LaRowe DE et al (2013) Quantifying the degradation of organic matter in marine sediments: a review and synthesis. Earth-Sci Rev 123:53–86

Berner RA, Ruttenberg KC, Ingall ED et al (1993) The nature of phosphorus burial in modern marine sediments. In: Wollast R, Mackenzie FT, Chou L (eds) Interactions of C, N, P and S biogeochemical cycles and global change. Springer, Melreux, pp 365–378

Bodur MN, Ergin M (1994) Geochemical characteristics of the recent sediments from the Sea of Marmara. Chem Geol 115:73–101

Calvert SE (1976) Mineralogy and geochemistry of nearshore sediments. In: Riley JP, Chester R (eds) Chemical oceanography 6, 2nd edn. Academic Press, London

Calvert SE, Pedersen TF (1993) Geochemistry of recent oxic and anoxic marine sediments: implications for the geological record. Mar Geol 113:67–88

Canfield DE, Thamdrup B, Hansen JW (1993) The anaerobic degradation of organic matter in danish coastal sediments; iron reduction, manganese reduction, and sulfate reduction. Geochim Cosmochim Acta 57:3867–3883

Cha HJ, Choi MS (2013) Origin and burial of organic carbon depending upon the environmental setting in the southwestern East/Japan Sea. In: Goldschmidt, Florence, Italy

Cha HJ, Lee CB, Kim BS et al (2005) Early diagenetic redistribution and burial of phosphorus in the sediments of the southwestern East Sea (Japan Sea). Mar Geol 216:127–143

Cha HJ, Choi MS, Lee CB et al (2007) Geochemistry of surface sediments in the southwestern East/Japan Sea. J Asian Earth Sci 29:685–697

Chang KI, Teague WJ, Lyu SJ et al (2004) Circulation and currents in the southwestern East/Japan Sea: overview and review. Prog Oceanogr 61:105–156

Chen CTA, Wang SL, Bychkov AS (1995) Carbonate chemistry of the Sea of Japan. J Geophys Res 100:10737–10745

Chester R (2000) Marine Geochemistry, 2nd edn. Blackwell, Oxford

Cho YG, Lee CB, Choi MS (1999) Geochemistry of surface sediments off the southern and western coasts of Korea. Mar Geol 159:111–129

Choi JY, Park YA (1993) Distributions and textural characters of the bottom sediments on the continental shelves. Korea J Oceanol Soc Korea 28(4):259–271

Choi YJ, Kim DS, Lee TH et al (2009) Estimate of manganese and iron oxide reduction rates in slope and basin sediments of Ulleung Basin, East Sea. J Korean Soc Oceanogr 14(3):127–133 (in Korean)

Chough SK, Lee HJ, Yoon SH (2000) Marine geology of Korean Seas, 2nd edn. Elsevier, Amsterdam

Dymond J, Lyle M, Finney B et al (1984) Ferromanganese nodules from MANOP Sites H, S, and R—Control of mineralogical and chemical composition by multiple accretionary processes. Geochim Cosmochim Acta 48:931–949

Froelich PN, Klinkhammer GP, Bender ML et al (1979) Early oxidation of organic matter in pelagic sediments of the eastern equatorial Atlantic: Suboxic diagenesis. Geochim Cosmochim Acta 43:1075–1090

Froelich PN, Bender ML, Luedtke NA et al (1982) The marine phosphorus cycle. Am J Sci 282:474–511

Fütterer DK (2006) The solid phase of marine sediments. In: Schulz HD, Zabel M (eds) Marine geochemistry. Springer, Berlin, pp 1–25

Gamo T (2011) Dissolved oxygen in the bottom water of the Sea of Japan as a sensitive alarm for global climate change. Trend Anal Chem 30(8):1308–1319

Gorbarenko SA, Southon JR (2000) Detailed Japan Sea paleoceanography during the last 25 kyr: constraints from AMS dating and $\delta^{18}O$ of planktonic foraminifera. Palaeogeogr Palaeoclimatol Palaeoecol 156:177–193

Haese RR (2006) The biogeochemistry of iron. In: Schulz HD, Zabel M (eds) Marine geochemistry. Springer, Berlin, pp 241–270

Hong GH, Kim SH, Chung CS et al (1997) ^{210}Pb-derived sediment accumulation rates in the southwestern East/Japan Sea (Sea of Japan). Geo-Mar Lett 17:126–132

Hyun SM, Bahk JJ, Suk BC et al (2007) Alternative modes of quaternary pelagic biosiliceous and carbonate sedimentation: a perspective from the East Sea (Japan Sea). Palaeogeogr Palaeoclimatol Palaeoecol 247:88–99

Hyun JH, Kim DS, Shin CW et al (2009) Enhanced phytoplankton and bacterioplankton production coupled to coastal upwelling and an anticyclonic eddy in the Ulleung basin, East Sea. Aquat Microb Ecol 54:45–54

Hyun JH, Mok JS, You OR et al (2010) Variations and controls of sulfate reduction in the continental slope and rise of the Ulleung Basin off the Southeast Korean upwelling system in the East Sea. Geomicrobiol J 27(2):212–222

Ichikura M, Ujiie H (1976) Lithology and planktonic foraminifera of the Sea of Japan piston cores. Bull Nat Sci Mus C 2(4):151–182

Khim BK, Shin DH, Han SJ (1997) Organic carbon, calcium carbonate, and clay mineral distributions in the Korea Strait region, the southern part of the East Sea. J Korean Soc Oceanogr 32(3):128–137

Kim KR, Kim K (1996) What is happening in the East Sea (Japan Sea)? recent chemical observations during CREAMS 93–96. J Korean Soc Oceanogr 31(4):164–172

Kim K, Kim KR, Chung JY et al (1991) Characteristics of physical properties in the Ulleung Basin. J Oceanol Soc Korea 26(1):83–100

Kim KR, Kim G, Kim K et al (2002) A sudden bottom-water formation during the severe winter 2000-2001: the case of the East/Japan Sea. Geophys Res Lett 29(8):1234. doi:10.1029/200 1GL014498

Kim DS, Yang EJ, Kim KH et al (2011) Impact of an anticyclonic eddy on the summer nutrient and chlorophyll a distributions in the Ulleung Basin, East Sea (Japan Sea). J Mar Syst 69(1):23–29

Lee KE, Kang DJ, Kim KR (2001) Degree of saturation of $CaCO_3$ in the East Sea. Sea J Korean Soc Oceanogr 6(4):234–241

Lee KE, Bahk JJ, Narita H (2003) Temporal variations in productivity and planktonic ecological structure in the East Sea (Japan Sea) since the last glaciation. Geo-Mar Lett 23:125–129

Lee T, Hyun JH, Mok JS et al (2008) Organic carbon accumulation and sulfate reduction rates in slope and basin sediments of the Ulleung basin, East Sea/Japan Sea. Geo-Mar Lett 28:153–159

Lee JY, Kang DJ, Kim IN et al (2009) Spatial and temporal variability in the pelagic ecosystem of the East Sea (Sea of Japan): a review. J Mar Syst 78:288–300

Lee T, Kim DS, Khim B et al (2010) Organic carbon cycling in Ulleung Basin sediments. East Sea Ocean Polar Res 32(2):145–156

Lee GH, Yoon YH, Nam BH et al (2011) Structural evolution of the southwestern margin of the Ulleung Basin, East Sea (Japan Sea) and tectonic implications. Tectonophysics 502:293–307

Lucotte M, Mucci A, Hillaie-Marcel C et al (1994) Early diagenetic processes in deep Labrador Sea sediments: reactive and nonreactive iron and phosphorus. Can J Earth Sci 31(1):14–27

Marin B, Giresse P (2001) Particulate manganese and iron in recent sediments of the Gulf of Lions continental margin (north-western Mediterranean Sea): deposition and diagenetic process. Mar Geol 172:147–165

Masuzawa T, Kitano Y (1983) Interstitial water chemistry in deep-sea sediments from the Japan Sea. J Oceanogr Soc Japan 39:171–184

Masuzawa T, Noriki S, Kurosaki T et al (1989) Compositional change of settling particles with water depth in the Japan Sea. Mar Chem 27:61–78

Morford JL, Emerson S (1999) The geochemistry of redox sensitive trace metals in sediments. Geochim Cosmochim Acta 63:1735–1750

Niino H, Emery KO, Kim CM (1969) Organic carbon in sediments of Japan Sea. J Sed Pet 39(4):1390–1398

Ohta A, Imai N, Terashima S et al (2013) Regional spatial distribution of multiple elements in the surface sediments of the eastern Tsushima Strait (Southwestern Sea of Japan). Appl Geochem 37:43–56

Otosaka S, Togawa O, Baba M et al (2004) Lithogenic flux in the Japan Sea measured with sediment traps. Mar Chem 91:143–163

Rullkötter J (2006) Organic matter: the driving force for early diagenesis. In: Schulz HD, Zabel M (eds) Marine geochemistry. Springer, Berlin, pp 125–168

Sarmiento JL, Gruber N (2006) Ocean biogeochemical dynamics. Princeton University Press, Princeton

Senjyu T, Aramaki T, Otosaka S et al (2002) Renewal of the bottom water after the winter 2000–2001 may spin-up the thermohaline circulation in the Japan Sea. Geophys Res Lett 29(7):1149. doi:10.1029/2001GL014093

Seung YH (1992) A simple model for separation of East Korean Warm Current and formation of North Korean Cold Current. J Oceanol Soc Korea 27(3):189–196

Sherwood BA, Sager SL, Holland HD (1987) Phosphorus in formainiferal sediments from North Atlantic Ridge cores and in pure limestones. Geochim Cosmochim Acta 51:1861–1866

Slomp CP, Van der Gaast SJ, Van Raaphorst W (1996) Phosphorus binding by poorly crystalline iron oxides in North Sea sediments. Mar Chem 52:55–73

Solov'yev AV (1962) Typical features of the sedimentation process in the Sea of Japan. Int Geol Rev 4(1):17–23

Takata H, Kuma K, Isoda Y et al (2008) Iron in the Japan Sea and its implications for the physical processes in deep water. Geophys Res Lett 35:L02666. doi:10.1029/2007GL031794

Takematsu N, Sato Y, Kato Y et al (1993) Factors regulating the distribution of elements in marine sediments predicted by a simulation model. J Oceanogr 49:425–441

Talley LD, Lobanov V, Ponomarev V et al (2003) Deep convection and brine rejection in the Japan Sea. Geophys Res Lett 30(4):1159. doi:10.1029/2002GL016451

Tishchenko PY, Talley LD, Lobanov VB et al (2007) The influence of geochemical processes in the near-bottom layer on the hydrochemical characteristics of the waters of the Sea of Japan. Oceanology 47:350–359

Tsunogai S, Yonemaru I, Kusakabe M (1979) Post depositional migration of Cu, Zn, Ni Co, Pb and Ba in deep sea sediments. Geochem J 13:239–252

Yamada K, Ishizaka J, Nagata H (2005) Spatial and temporal variability of satellite primary production in the Japan Sea from 1998 to 2002. J Oceanogr 61:857–869

Yanagi T (2002) Water, salt, phosphorus and nitrogen budgets of the Japan Sea. J Oceanogr 58:797–804

Yin JH, Kajiwara Y, Fujii T (1989) Distribution of transition elements in surface sediments of the southwestern margin of Japan Sea. Geochem J 23:161–180

Yoo S, Park JS (2009) Why is the southwest the most productive region of the East Sea/Sea of Japan? J Mar Syst 78:301–315

You OR, Mok JS, Kim SH et al (2010) Comparison of sulfate reduction rates associated with geochemical characteristics at the continental slope and basin sediments in the Ulleung Basin. East Sea Ocean Polar Res 32(3):299–307

Chapter 10
Phytoplankton and Primary Production

Joong Ki Choi, Jae Hoon Noh, Tatiana Orlova, Mi-Ok Park, Sang Heon Lee, Young-Je Park, Seunghyun Son, Inna Stonik and Dong Han Choi

Abstract The northern part of the East Sea (Japan Sea) is dominated by the cold waters of the Liman Current, whereas in the southern part, the dominant current is the Tsushima Warm Current. Together with these major current systems, subpolar fronts and mesoscale eddies that form in the region exert a strong influence on the phytoplankton ecosystem in the East Sea. These conspicuous physical forcings are manifested as often clearly discernible features in temporal and spatial distributions, composition, abundance, biomass and production of phytoplankton. In short, diatoms and dinoflagellates are found to be the most diverse phytoplankton groups and

J.K. Choi (✉)
Department of Oceanography, Inha University, Incheon 22212, Republic of Korea
e-mail: jkchoi@inha.ac.kr

J.H. Noh · D.H. Choi
Biological Oceanography and Marine Biology Division, Korea Institute of Ocean Science and Technology, Ansan 15627, Republic of Korea
e-mail: jhnoh@kiost.ac.kr

D.H. Choi
e-mail: dhchoi@kiost.ac.kr

T. Orlova
Zhirmunsky Institute of Marine Biology, Far East Division, Russian Academy of Sciences, Vladivostok 690059, Russia
e-mail: torlova06@mail.ru

M.-O. Park
Department of Oceanography, Pukyong National University, Busan 48513, Republic of Korea
e-mail: mopark@pknu.ac.kr

S.H. Lee
Department of Oceanography, Pusan National University, Busan 46241, Republic of Korea

© Springer International Publishing Switzerland 2016
K.-I. Chang et al. (eds.), *Oceanography of the East Sea (Japan Sea)*,
DOI 10.1007/978-3-319-22720-7_10

are occasionally responsible for blooms in coastal waters. Diatoms are also abundant in frontal areas and in the rings of warm core eddies. Picophytoplankton groups are also found to be important phytoplankton in the East Sea, especially in warm seasons, and photosynthetic picoeukaryotes and *Synechococcus* show distinct seasonal and vertical distribution patterns. Recent field measurements indicated that the spatial distribution of primary productivity in the Ulleung Basin (UB) of the East Sea ranged from 172 to 358 g C m^{-2} year^{-1}. This range of primary productivity is relatively higher than in other regions in the East Sea. The East Sea is a body of dynamic "non-oceanic" water with its own particular oceanic characteristics. Coastal upwelling and mesoscale eddies over a wide range of horizontal scales contribute to the high primary productivity in the UB. This vibrant primary production provides the foundation for a biological "hot" spot and strong support for an energetic biological pump cycle in the East Sea. Despite much progress in expanding knowledge of phytoplankton ecology in the East Sea, more studies on diversity, productivity, niche, and physiological adaptation to dynamic environments should be conducted to better understand ecological roles of phytoplankton in changing oceans.

Keywords Phytoplankton · Primary production · Chlorophyll *a.* · Ulleung basin · East Sea (Japan Sea)

10.1 Introduction

Since phytoplankton plays an important role as a major primary producer of organic carbon in pelagic ecosystems, it is necessary to study phytoplankton ecology for a better understanding of marine ecosystems. Early studies on phytoplankton in the East Sea were initiated by Russian and Japanese scientists in the 1920s and 1930s (Ostroumoff 1924; Skvortzow 1931; Aikawa 1936; Yamada 1938). Aikawa (1936) and Yamada (1938) grouped phytoplankton of the East Sea into warm water, cold water, and coastal water species. Then, in the 1940–1960s, the distribution of phytoplankton was studied in the open waters of the East Sea (Kisselew1947; Park 1956; Ohwada and Ogawa 1966; Uhm and Yoo 1967).

Y.-J. Park
Korea Ocean Satellite Center, Korea Insitute of Ocean Science and Technology,
Ansan 15627, Republic of Korea
e-mail: youngjepark@kiost.ac.kr

S. Son
Cooperative Institute for Research in the Atmosphere, Colorado State University, Fort Collins, CO 20523, USA
e-mail: oceancolor.son@gmail.com

I. Stonik
Zhirmunsky Institute of Marine Biology, Far East Division, Russian Academy of Sciences,
Vladivostok 690041, Russia
e-mail: innast2004@mail.ru

However, limited studies on the biological processes controlling spatial and seasonal variability of phytoplankton have been reported. Community structure and spatial distribution of phytoplankton (Park et al. 1991) and the distribution of chlorophyll *a* (chl-*a*) (Kano et al. 1984; Moon et al. 1998) were studied in the subpolar frontal region of the East Sea. Recently, the impact of mesoscale eddies on the vertical distribution of phytoplankton and nutrients were investigated in the Ulleung Basin (UB) (Kang et al. 2004; Hyun et al. 2009; Kim et al. 2012). In addition, with the improvement in satellite technology, applications of satellite observation on chlorophyll have enhanced our understanding of the temporal and spatial distributions as well as mesoscale physical processes affecting those distributions in the East Sea (Yamada et al. 2005; Yoo and Park 2009).

The East Sea has also experienced dramatic changes with global warming during the last 50–60 years (Kim and Kim 1996; Gamo 1999), which may be associated with the climate regime shift observed in long-term data from the PM line of the Japan Meteorological Agency and regular observation lines of the National Fisheries Research and Development Institute of Korea in the East Sea (Chiba and Saino 2002; Rebstock and Kang 2003). In accordance with long-term climate change, Yamada et al. (2004) suggested that seasonal and interannual variability of sea surface chl-*a* in the East Sea are largely affected by variability of the global climate such as El Niño-Southern Oscillation (ENSO) events.

Several studies described the species composition and phytoplankton community in the southwestern area (Shim and Lee 1983; Lee and Shim 1990; Shim et al. 1992, 1995), Korean coastal waters (Choe 1972; Shim and Bae 1985; Kang and Choi 2001, 2002), and upwelling area (Shim et al. 2008) of the East Sea. In addition, the distribution of photosynthetic pigments was investigated in the East Sea (Kim et al. 2010). The abundance and seasonal dynamics of planktonic microalgae, as well as episodes of harmful algae blooms (HABs) and eutrophication were studied in Russian coastal waters (Shuntov 2001). In the last 20 years, special attention has been paid to problems of eutrophication of coastal waters and HAB events (Stonik and Selina 1995; Stonik and Orlova 1998, 2002; Kim et al. 1999; Orlova et al. 2002).

Since the first report of the importance and populations of picoplankton in the East Sea (Shim and Bae 1985), several studies regarding abundance, biomass and primary productivity of the picophytoplankton, and physico-chemical factors controlling the distribution of picoplankton have been carried out in the East Sea (Suk et al. 2002; Hashimoto and Shiomoto 2002; Kang et al. 2004; Shiomoto et al. 2004; Park 2006; Roh et al. 2010). Recently, *Synechococcus* lineage diversity was studied from culture isolates (Choi and Noh 2009). In addition, a study on seasonal changes of *Synechococcus* and *Prochlorococcus* showed high diversity and obvious seasonal variation in the East Sea (Choi et al. 2013).

Regarding studies of primary production in the southwestern region of the East Sea, most reports from the 1980s to the early 1990s were based on in situ field measurements (Shim and Park 1986; Shim et al. 1992). However, environmental complexity of the UB such as coastal upwelling (Hyun et al. 2009; Yoo and Park 2009) and various scales of eddies (Hyun et al. 2009; Kim et al. 2012; Lim et al. 2012) made it difficult to ascertain the spatial and temporal variability of phytoplankton through field study. Nevertheless, recent application of satellite observation of chlorophyll a and the model-based estimation of primary production have enhanced our understanding of the temporal and spatial distributions of primary production in the East Sea (Yamada et al. 2005; Yoo and Park 2009).

In this chapter, analyses on temporal-spatial changes in chl-*a* concentration are included based on the Sea-viewing Wide Field-of-view Sensor (SeaWiFS) and shipboard observation data in the East Sea. We summarized original and literature-based data on species composition, dominant species, HABs and environmental conditions of phytoplankton blooms in the coastal waters. We also presented new information on the distribution and long-term changes of phytoplankton in the offshore and open waters of the East Sea. In addition, we explored environmental factors which may influence the spatio-temporal distribution of primary production in the UB.

10.2 Chlorophyll *a*

10.2.1 Spatial and Seasonal Variability Based on SeaWiFS Observations

We describe spatial and temporal variability in chl-*a* concentration based on the SeaWiFS data. The SeaWiFS level 3 mapped chl-*a* data with the global area coverage at 9-km resolution were retrieved from http://oceancolor.gsfc.nasa.gov/ for the period 1998–2007. Satellite-derived chl-*a* values are often used as an indicator of phytoplankton biomass in the surface layer (e.g., Racault et al. 2012).

Monthly climatology chl-*a* derived from the SeaWIFS sensor ranges from 0.2 to 2.0 μg L^{-1} in the most of the East Sea area (Fig. 10.1). Over the whole East Sea, the spring maximum occurs in April with a median value of 1.06 μg L^{-1} and 1 standard deviation (1-SD) range 0.69–1.65 μg L^{-1}, and a summer minimum in August with a median of 0.27 μg L^{-1} and 1-SD range 0.17–0.45 μg L^{-1}. Fall maximum chl-*a* is observed in November with a median of 0.68 μg L^{-1} and 1-SD range 0.47–1.00 μg L^{-1}, and winter minimum in January with a median of 0.44 μg L^{-1} and 1-SD range 0.31–0.62 μg L^{-1}. The annual minimum observed in August is approximately the same as the global mean value found in the open ocean (Conkright and Gregg 2003).

Time-series of chl-*a* show a well-defined seasonal cycle as a result of the seasonal oscillation of physical forcings similar to those observed in other temperate and higher latitude ocean environment. In the East Sea, satellite chl-*a* reveals a seasonal pattern with double peaks in the spring and autumn (Vinogradov et al. 1996; Kim et al. 2000; Yamada et al. 2004) as is typically seen in mid-latitude temperate oceans.

The spring bloom in March-May is stronger than the autumn bloom in October-December. Of the two seasonal minima, the summer chl-*a* under nutrient depletion is lower than the winter minimum under light limitation (Figs. 10.2 and 10.3). As indicated in Fig. 10.1, the spring bloom takes place earlier in coastal waters, including the East Korea Bay and south off Vladivostok. In offshore waters, an early

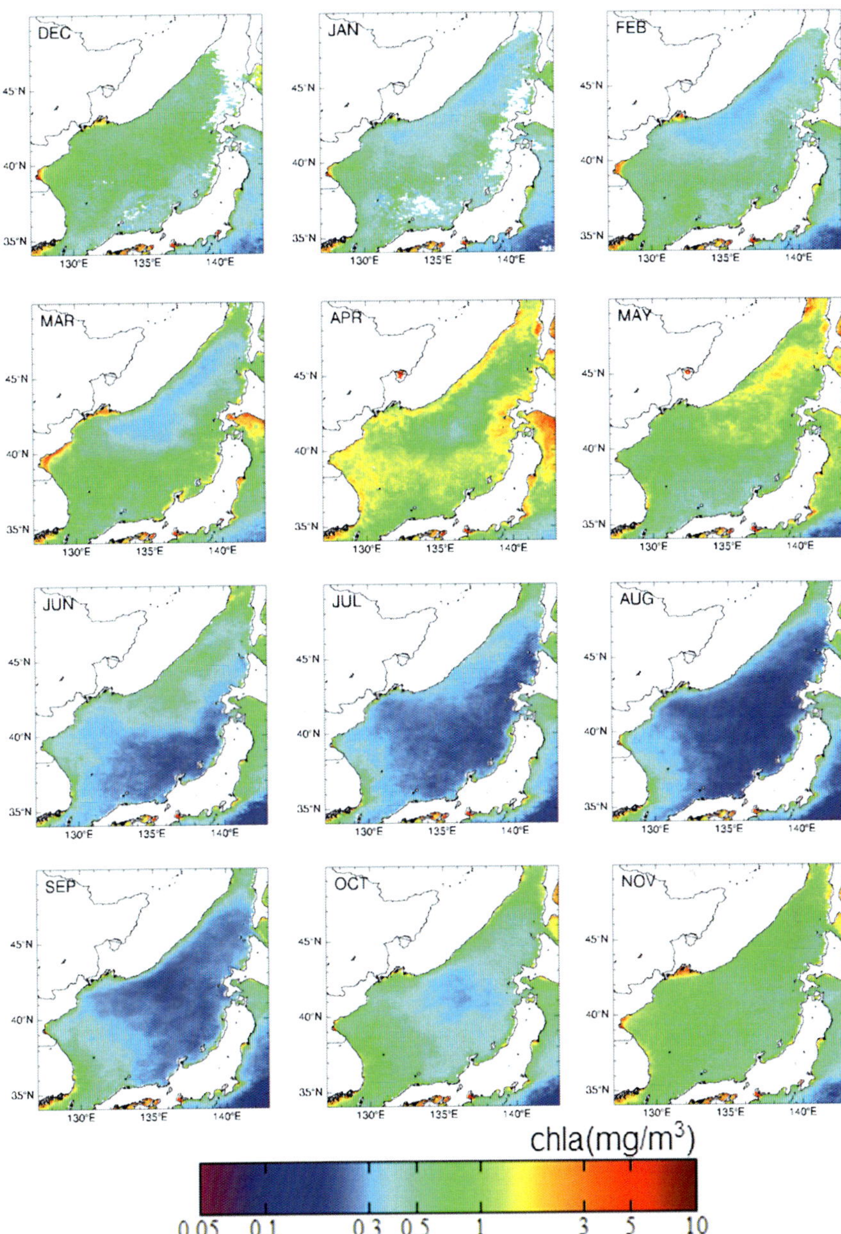

Fig. 10.1 Monthly mean chl-*a* derived from Sea WiFS data for 1998–2007

bloom onset is often observed in the subpolar front area which occurs almost simultaneously along the east-west zonal line (Fig. 10.3). This early bloom initiation could be attributed to mesoscale eddies (Lee and Niiler 2005) but it deserves further

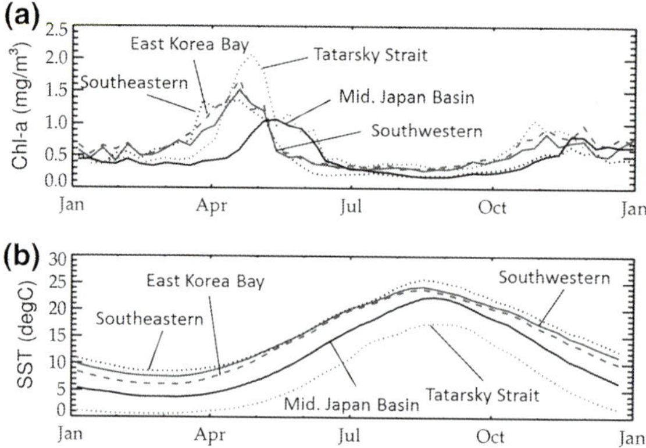

Fig. 10.2 Seasonal cycle of **a** chl-*a* and **b** sea-surface temperature for selected areas-south-western area, southeastern area, East Korea Bay, middle of the Japan Basin, and Tatarsky Strait

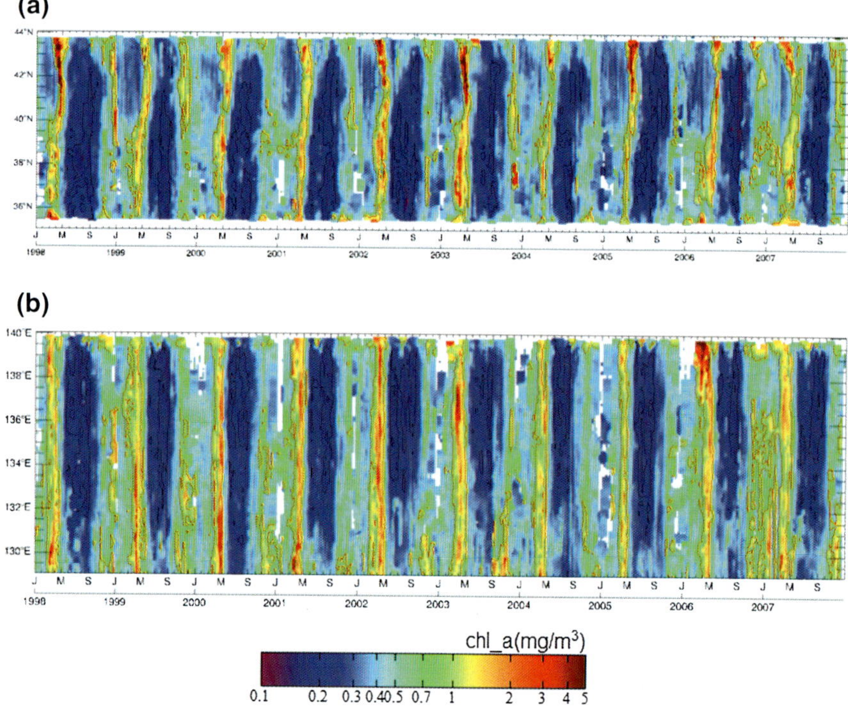

Fig. 10.3 Time evolution of chl-*a* at **a** the 135°E line and **b** the 39°N line in the East Sea

study. The spring bloom progresses south of the subpolar front (lower than 39°N) in March (Fig. 10.3). One and a half months later a bloom takes place in the central Japan Basin and further north as reported in Yamada et al. (2004). This north-south contrast in the central East Sea is also clearly visible in Figs. 10.1 and 10.2. To understand the cause of this apparent north-south difference across the subpolar front, further investigation will be required.

When the chl-*a* is at an annual minimum around August, the chl-*a* in the southwestern area is higher than that in the southeastern area, as shown in Fig. 10.2. This pattern can be attributed to the introduction of nutrient rich waters from the southern coast of Korea and the East China Sea (ECS), with frequent wind-driven upwelling along the east coast of Korea, which is visible in ocean color and sea-surface temperature imagery (Yoo and Park 2009).

Seasonal and spatial patterns of the surface chl-*a* in the East Sea showed significant interannual variability due to variability in the physical conditions. Yamada et al. (2004) reported that the spring bloom onset varies in correspondence with the wind speed. Yoo and Kim (2004) showed that the weak Tushima Warm Current (TWC) in 1997 seemed to suppress the spring bloom in the southwestern East Sea. Occasionally, an early spring bloom can be initiated in the northern East Sea after wet deposition of Asian dust (Jo et al. 2007). Recently, Park et al. (2014) reported the interannual variability of chl-*a* along the Primorye coast was significantly correlated with sea ice concentrations in the Tatarsky Strait in the previous winter.

10.2.2 Shipboard Measurements

During the shipboard offshore surveys, well-defined seasonal and temporal variations of chl-*a* concentration were observed several times (Table 10.1). For example, during a north-south transect survey of the East Sea, high chl-*a* concentrations were reported in the southern subpolar front region in April (Kang et al. 2004) whereas in May, high chlorophyll regions were observed in the Japan Basin further north of the subpolar zonal line (Kim et al. 2010). These events reflect time-lapse succession of spring blooms. And such progression in the East Sea is well observed through the remote sensing data (Fig. 10.1). Also the combination of coastal upwellings and their entrainments by the Ulleung Warm Eddy (UWE) often produce a ring-shape trajectory of high chl-*a* water mass along the periphery of the UWE (Hyun et al. 2009).

In October, the high chl-*a* water masses appear in the southern region of the East Sea. This pattern, with reference to Sect. 6.3.2, can be attributed to the introduction of nutrient-rich waters from the Korea Strait (Moon et al. 1998; Kang et al. 2004; Yoo and Park 2009), and in November, high chl-*a* is observed along the frontal zone where the East Korea Warm Current (EKWC) and the UWE converge (Kang et al. 2004).

As shown above, the East Sea is characterized by seasonal changes as well as various physical processes such as currents, mesoscale eddies, fronts, and upwelling manifesting in dynamic changes in chlorophyll concentrations of the region (Yamada et al. 2004; Yoo and Park 2009; Hyun et al. 2009; Kim et al. 2010; also refer to Sect. 6.3.2).

Table 10.1 Offshore shipboard measurements of chl-*a* concentrations in the East Sea

Sampling time		Sampling area	chl-*a* concentration		High chl-*a* conc. area (latitude)	Reference
Year	Month/Day	Lat./Long.	Surface (mg/m^3)	Σ Mixed layer (mg/m^2)		
2001	Apr/10–24th	35°–38°30′N/130°30′E		20.1–80.3	Ulleung basin (37°–38°30′N)	Kang et al. (2004)
2006	Apr/3–25th	37°N/130°–132°E	0.91–4.03		Periphery of the UWE	Hyun et al. (2009)
2004	May/3–19th	36°–42°N/130°–131°30′E	0.1–1.0		Japan basin 41°N	Kim et al. (2010)
1995	Oct./9–19th	37°–41°30′N/133°40′E	−0.71		Southern area (37°—37°30′N)	Moon et al. (1998)
2001	Oct./12–21th	35°′–38°30′N/130°30′E		14.2–72.1	Southern area (35—35°45′N)	Kang et al. (2004)
2000	Nov/3–10th	35°–38°10′N/130°30′E		9.4–26.9	Frontal area of UWE (36°–36°45′N)	Kang et al. (2004)

However, Yamada et al. (2004) suggested that seasonal and inter-annual variability of the sea surface chl-*a* in the East Sea are largely affected by variability of the global climate such as ENSO events. But less information is available for the East Sea to fully satisfy this suggestion of the changes in chl-*a* concentrations largely affected by global climate change.

Long-term chl-*a* measurements of Korean coastal waters have been carried out since 1992 in the waters near the three nuclear power plants located in the southeastern Korean Peninsula. The seasonal change for surface chl-*a* concentrations for these waters, on average, are 0.94–4.43 μg L^{-1} in the Gori area, 0.91–4.51 μg L^{-1}in the Weolsung area, and 0.71–3.98 μg L^{-1} in the Wooljin area. High concentrations of chl-*a* are measured during the spring phytoplankton blooms (Kang and Choi 2002). In the southeastern coastal waters of the East Sea, chl-*a* concentrations ranged from 0.29 μg L^{-1} in December to 3.07 μg L^{-1} in April (Shim and Bae 1985), showing a typical seasonal fluctuation pattern found in other areas of the East Sea. Consistently, long-term chl-*a* concentration (0.21–8.26 μg L^{-1}) measured during the period from 1961 to 1990 in the southwestern East Sea also showed a fluctuation within a similar range found in other studies (Park et al. 1998). Thus, chl-*a* concentration may fluctuate within roughly an order of magnitude in the East Sea.

10.2.3 Measurements of Algal Pigments

According to the analysis of photosynthetic pigments for identifying phytoplankton groups using the high-performance liquid chromatography (HPLC) analysis (Kim et al. 2010), cyanobacteria dominated (40–60 %) in the surface frontal zone (0–20 m) in spring, summer, and autumn. This contribution decreased to less than 10 % in depths greater than 50 m. Diatom pigments were higher in the warm water zone than in the cold water zone in the upper 100 m layer in spring, summer and autumn. Diatom pigments contributed about 40 and 15 % of the phytoplankton pigments in the surface layer in spring and summer, respectively, with the highest proportions (~80 %) in the eddy zone in spring. The higher diatom proportions in May 2004 were associated with the spring diatom bloom in the East Sea. Pelagophyte pigments were relatively higher in the surface of the cold water zone. Pelagophytes (picoplankton) composed 10–40 % of the total biomass in the surface layer of the frontal zone and the cold water zone of the East Sea in spring (Kim et al. 2010). Prymnesiophyte pigments were relatively higher in the surface layer of the cold water and frontal zone in spring and the contribution of prymnesiophytes to the total phytoplankton biomass was almost constant (5–15 %) in the surface of the East Sea (Kim et al. 2010). Cryptophytes which were small nanoplankton contributed 5–10 % of all phytoplankton in the entire surface waters of the East Sea. Dinoflagellate pigments were relatively higher in the warm water zone in spring, summer and autumn. However, contributions of dinoflagellates to the total phytoplankton were generally less than 5 % in the upper layer (0–100 m)

of the East Sea. Concentrations of chlorophyll *b*, a pigment of green algae including prasinophytes, were highest in the frontal zone with a higher trend in the cold
water zone than those in the warm water zone. From these results, Kim et al.
(2010) concluded that in the warm water zone, diatoms and prasinophytes dominated the phytoplankton community in spring. In the frontal zone, cyanobacteria and pelagophytes dominated the phytoplankton community. In the cold water
zone, diatoms, pelagophytes and prasinophytes dominated in spring, while pelagophytes and cyanobacteria dominated in summer and autumn.

10.3 Distribution of Micro- and Nano-Phytoplankton Abundance and Harmful Algal Blooms

10.3.1 Korean Coastal Waters

In Korean coastal waters, phytoplankton abundance was reported in the ranges of
10^3–10^4 cells L^{-1} in the 1960s and 10^3–10^6 cells L^{-1} in the 1970s. Phytoplankton
abundance sharply increased to the range of 10^4–10^7 cells L^{-1} in the 1980s, and
densities remained high, in the range of 10^5–10^7 cells L^{-1} in the 1990s and 2000s
due to eutrophication in the coastal waters (Choi et al. 2011).

In these eutrophic coastal waters, four species (*Skeletonema* spp., *Thalassiosira
decipiens*, *Chaetoceros debilis*, and *Prorocentrum* spp.) predominate all year
round. Together with these species, *Leptocylindrus danicus* and *Pseudo-nitzschia*
co-dominated in spring, *Asterionellopsis glacialis*, *Chaetoceros lorenzianus*
and *Ch. compressus* in summer, *Probascia alata*, *Nitzschia longissima* and
Distephanus speculum in autumn, and *Chaetoceros socialis*, *Cylindrotheca closte-
rium* and *Paralia sulcata* in winter (Kang and Choi 2002).

10.3.2 Russian Coastal Waters

In Russian coastal waters, the main features of the seasonal succession of phytoplankton are diatom blooms in winter, early summer and autumn, and dinoflagellate blooms in late spring and early autumn (Zuenko et al. 2006) (Fig. 10.4). These
blooms were initiated by 14 genera including diatoms (*Thalassiosira*, *Coscinodiscus*,
Pseuonitzschia, *Chaetoceros*, *Thalassionema*, *Rhizosolenia*, *Dactyliosolen*, *Ditylum*,
Dipopsalis and *Skeletonema*) and dinoflagellates (*Gymnodinium*, *Prorocentrum*
and *Protoperidinium*) and euglenophyta (*Eutreptia*). There were three different
zones of water quality in the Russian coastal waters: hypereutrophic, eutrophic
and mesotrophic zones. The highest abundance of micro-phytoplankton (17.9–
31.1×10^6 cells L^{-1}) was observed in the hypereutrophic zone, whereas the lowest
abundance (1.9×10^6 cells L^{-1}) was in the mesotrophic zone. Maximum abundance of

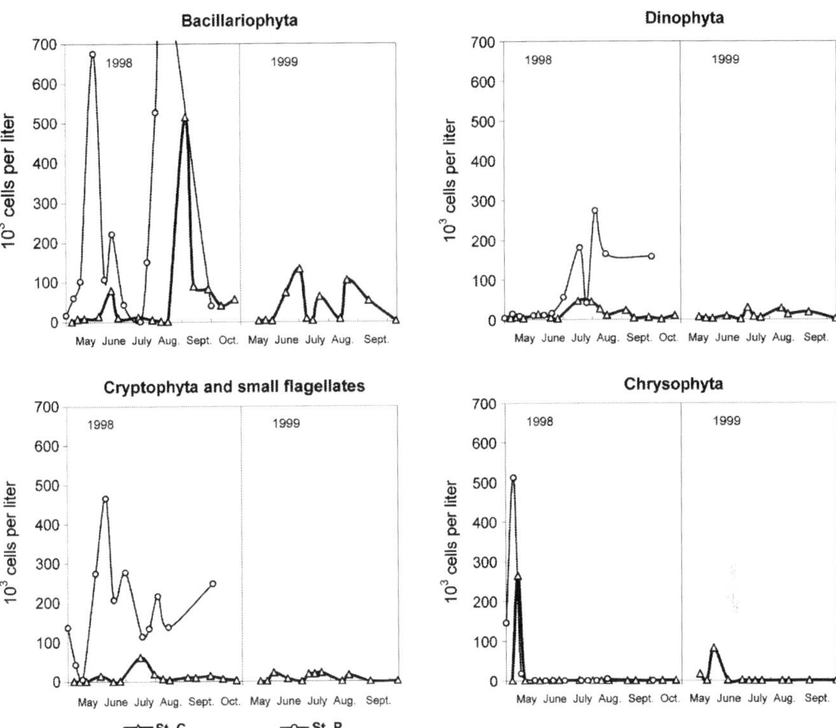

Fig. 10.4 Abundance of the main groups of phytoplankton at the sea surface in May–October for two stations near Vladivostok (from Zuenko et al. 2006)

Skeletonema coastatum (8.8 and 17.4 $\times$ 10^6 cells L^{-1}) was observed in the eutrophic area, whereas the lowest value (1.6 $\times$ 10^6 cells L^{-1}) was recorded in the mesotrophic area (Stonik and Orlova 2002).

10.3.3 Japanese Coastal Waters

In Japanese coastal waters, the abundance of phytoplankton was less than that in Korean and Russian coastal waters. According to Ohwada and Ogawa (1966), microphytoplankton is abundant in spring in the TWC near the frontal zone between the TWC and the cold waters, with population numbers of up to 10^5 cells L^{-1}, while it is less abundant in the offshore cold waters. In summer, a dense population of phytoplankton was found in the northern part of the TWC. In autumn and winter, micro-phytoplankton abundances were still higher in the TWC than in the offshore cold waters. The appearance of high abundance in the TWC area is due to the input of nutrients by the TWC, which introduces high-nutrient water from the coastal waters of Korea (Lee and Shim 1990) and the ECS

(Ohwada and Ogawa 1966). Ohwada and Ogawa (1966) reported that *Chaetoceros affinis*, *Ch. curvisetus*, and *Ch. lorenzianus* dominated in Japanese neritic waters during summer, while *Ch. debilis*, *Ch. socialis* and *Skeletnema costatum* were dominant during winter. Moreover, diatom abundances in the TWC area were one order of magnitude higher than those of the northern cold water (Kuroda 1987).

In the TWC area, warm water species including *Dactyliosolen mediterraneus*, *Rhizosolenia alata* f. *gracillima*, *R. styliformis* v. *latissima* and *Climacodium biconcavum* dominated, while *Chaetoceros atlanticus*, *Ch. concavicornis* and *Rhizosolenia hebetata* f. *semispina* were dominant in the cold waters (Ohwada and Ogawa 1966).

10.3.4 Offshore Waters

The anticyclonic Ulleung Warm Eddy (UWE) is a well-known physical phenomenon in the UB (Lie et al. 1995; Chang et al. 2004; see Sect. 4.3). Depth-integrated phytoplankton biomass and production were higher in the moderately stratified water in the periphery of the UWE than those in the vertically well mixed warm core of the eddy and in the stratified region outside of the eddy (Hyun et al. 2009). Likewise, diatom abundances in the periphery of the UWE were two or three orders of magnitude higher than those in the warm core and outside of the UWE during the spring bloom, which indicated that a phytoplankton bloom triggered by coastal upwelling was delivered into the center of the UB by the UWE (Hyun et al. 2009). In another period in summer, high phytoplankton biomass was also observed in the UWE core below the surface mixed layer as well as at the edge of the eddy (Kim et al. 2012). Therefore, the UWE could play a key ecological role in supporting substantial phytoplankton biomass in the nutrient depleted surface waters in summer and maintaining high benthic mineralization in the deep sea sediments of the UB (Kim et al. 2012).

In the subpolar frontal area, phytoplankton abundance was in the range of 10^3–10^4 cells L^{-1} in the surface layer, and higher abundance at the depth of 50 m, during summer (Park et al. 1991). The low abundance in the surface layer of the subpolar frontal zone in summer is due to nutrient depletion by the formation of a strong thermocline at depths near 50 m. In the subpolar frontal waters, cold water species such as *Chaetoceros convolatus* and *Ch. atlanticus* occurred as dominant species at 20 and 50 m depths in the northern part, while dominant species in the southern part were mostly warm water species. In the mixing zone of 50 m depth, *Ch. curvesitus*, *Ch. lorenzianus*, *Ch. convolutes*, *Ch. decipiens* and *Corethron criophilum* are dominant (Park et al. 1991).

According to the long term survey of the East Sea's PM line, spring phytoplankton abundance was higher in the 1980s than in the 1970s and 1990s, whereas summer phytoplankton was less in the 1980s than in the 1970s and 1990s, which seems to be associated with climate change (Chiba and Saino 2002).

10.3.5 Harmful Algal Blooms

As in the ECS, the frequency of HAB in the East Sea has increased since the 2000s in coastal waters, especially in the Russian coastal area (Orlova et al. 2002). The presence of harmful algae and mass blooms of some species can lead to the critical accumulation of toxins in farmed shellfish. In Russian coastal waters, 27 species belonging to *Prorocentrum, Cochlodinium, Ostreopsis, Karenia, Heterosigma, Chattonella, Alexandrium, Pseudo-nitzschia,* and *Dinophysis,* known as potential producers of various types of phytotoxins, were found (Orlova et al. 2002). Recently, highly toxic clones of diatom genus *Pseudo-nitzschia* (amnesic shellfish poisoning toxin) and dinoflagellate genera *Dinophysis* and *Prorocentrum* were commonly found in summer during periods of maximum warming and stratification of the water column (Orlova et al. 2008).

In coastal waters of Korea, blooms of *Cochlodinium polykrikoides,* dominant HABs in southern coastal waters, have been observed frequently along eastern coastal waters since 1995 (Kim et al. 1999). As eutrophication and climate change proceed, these HAB events could be an ecological disaster in coastal waters of the East Sea.

10.4 Distribution of Picophytoplankton

10.4.1 Picophytoplankton Abundance

Picophytoplankton (0.2–2 or 3 μm in size) composed of *Synechococcus, Prochlorococcus,* and pico-eukaryote groups are responsible for the majority of primary productivity in oligotrophic regions (Jardillier et al. 2010), and also contributed greatly to the biomass of total phytoplankton in the East Sea (Shim et al. 1995). Significant contributions of picophytoplankton to total chl-*a* have been reported by several studies (Shim et al. 1985; Hashimoto and Shimoto 2002) elucidating the importance of picophytoplankton as a primary producer in the East Sea. During studies on phytoplankton distribution associated with the UWE, Suk et al. (2002) revealed that picophytoplankton (<3 μm) contributed 52 % to total chl-*a* in November 2000, and 35.9 % in April 2001 (Fig. 10.5). In April, with weak stratification and an increased supply of nutrients from the subsurface layer, total chl-*a* increased up to 2.5 μg L^{-1} and there was enhanced growth of the netphytoplankton, which resulted in the decreased contribution of picophytoplankton to the total chl-*a*. In summer of 2000, the picoplankton fraction of the total chl-*a* near the shore of Japan (Shiomoto et al. 2004) was 69 % on average (16–89 %) and had a strong positive relationship with total chl-*a*. Among picophytoplankton, eukaryotic picophytoplankton contributed substantially to the regional variation in the pico-sized chl-*a* in the nearshore area. Picophytoplankton accounted for 56 % of the total chl-*a* in June 2006 near Ulleungdo (Shim et al. 2008). The high contribution

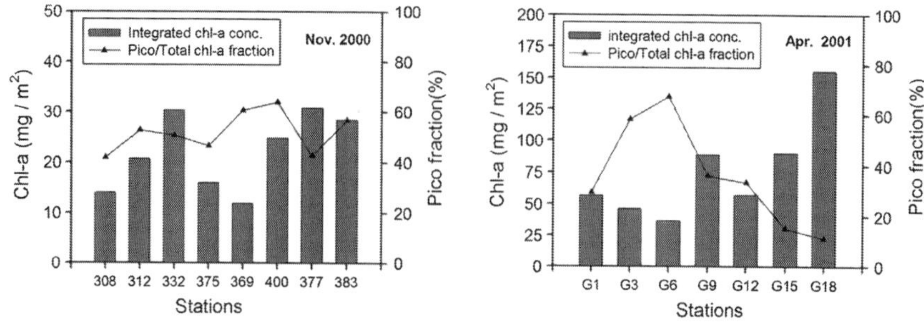

Fig. 10.5 Intergrated chl-*a* concentrations down to 100 m (bar) and contribution of picophytoplankton (<3 μm) to total chl-*a* in November 2000 and April 2001 (from Suk et al. 2002)

of picophytoplankton in June was due to the relatively low total chl-*a* (less than 0.5 μg L^{-1}) and low nutrients, a condition which persisted until early summer. In addition, the high contribution of picophytoplankton to total chl-*a* (52.0–54.4 %) in autumn was attributed to the low biomass of nano- and micro-sized phytoplankton in the surface mixed layer under the strong stratification conditions.

Cyanobacteria seem to be the most important picophytoplankton in summer. Roh et al. (2010) reported that cyanobacteria were dominant in the offshore waters of the East Sea in summer, contributing 70 % of total chl-*a*. This high biomass of cyanobacteria in the UWE and East Sea was attributed to the low nitrate and high temperature in summer (Roh et al. 2010; Kim et al. 2010).

Ecological investigation of phytoplankton along the transect across the UWE revealed that the average cell abundance for *Synechococcus* within the euphotic depths was 2.4 × 10^{12} cells m^{-2} in November 2000, while its average cell abundance decreased to 6.2 × 10^{11} cells m^{-2} in April 2001 (Suk et al. 2002). In sea waters around Ulleungdo, abundances of *Prochlorococcus*, *Synechococcus*, and photosynthetic picoeukaryotes showed a clear seasonal variation, and *Synechococcus* showed the most prominent variation. According to Shim et al. (2008), maximal abundance (average 2.51 × 10^5 cells mL^{-1}) of *Synechococcus* was observed in summer and the lowest in autumn (average 9.2 × 10^3 cells mL^{-1}) in 2006. On the contrary, *Prochlorococcus* showed the opposite pattern. The abundance of *Prochlorococcus* was highest (average 6.8 × 10^4 cells mL^{-1}) in autumn, more abundant than in summer (average 1.6 × 10^4 cells mL^{-1}). The seasonal variation of the abundance of *Synechococcus* in the East Sea was in the order of summer > autumn > spring based on the report by Shim et al. (2008), but the distribution of *Prochlorococcus* abundances seems to be related to warm water fluxes into the East Sea. Noh et al. (2006) reported that abundance of *Prochlorococcus* in the East Sea was less than that in the tropical and subtropical western Pacific, but higher than that in the continental shelf region of the ECS. *Prochlorococcus* generally did not appear at the continental shelf of the ECS and the East Sea with low salinity and low temperature (Choi et al. 2013). The average abundance of

Synechococcus and *Prochlorococcus* showed a uniform distribution in the surface mixed layer but rapidly decreased below the surface mixed layer. Conversely, picoeukaryotes showed a subsurface maximum at 20–30 m in the early summer (Suk et al. 2002). In autumn, abundance of picophytoplankton was high in the surface mixed layer (SML) and decreased rapidly below the SML, while in spring (with weak stratification) high cell abundance was found in the surface layer and then decreased slowly with little difference in cell abundance through the water column. According to a study on environmental factors regulating picophytoplankton distribution (Shiomoto et al. 2004), both *Prochlorococcus* and *Synechococcus* showed positive correlations with temperature and salinity, but picoeukaryotes showed negative correlations with temperature and salinity. It is known that (1) *Prochlorococcus* is more abundant in more oligotrophic conditions than *Synechococcus*, (2) *Prochlorococcus* uses low irradiances more effectively than *Synechococcus*, and (3) eukaryotic picophytoplanktons are more abundant in low temperature and more eutrophic conditions than the other picophytoplankton (Shiomoto et al. 2004). Thus, different responses to environmental conditions were derived from the distinct seasonal and vertical distribution patterns of each picophytoplankton group. In many surveys, the phototrophic picoeukaryotes are abundant in frontal regions of a nutrient rich environment. In nearshore areas around Japan, *Synechococcus* was abundant (10^4–10^5 cells mL^{-1}) in the surface waters of the Kuroshio and the TWC where the nutrients were very low compared to the nutrient rich cold water of the Oyashio area (10^2–10^3 cells mL^{-1}). Shiomoto et al. (2004) suggested that there are two populations of *Synechococcus* in the Kuroshio and the TWC: one is adapted to oligotrophic conditions and the other to mesotrophic conditions.

10.4.2 *Picocyanobacterial Diversity*

Among picophytoplankton, only picocyanobacterial diversity has been intensively studied in the East Sea (Choi and Noh 2009; Choi et al. 2013). The pyrosequencing method of 16S-23S internal transcribed spacer sequences was developed and applied in order to elucidate seasonal variability of picocyanobacterial diversity (Choi et al. 2013). In the study, 17 picocyanobacteria clades/ecotypes were retrieved throughout the year in the East Sea, showing a great variation of picocyanobacterial diversity in temperate water of the open East Sea (Fig. 10.6; Choi et al. 2013). Among them, clades I, II and IV were the most dominant clades, exhibiting fractions of over 50 % depending on the season. In addition, both clade 5.3-I belonging to *Synechococcus* subcluster 5.3 and *Prochlorococcus* HLII ecotype were the next dominant picocyanobacteria (over 10 %) in the East Sea. The coastal and cold-water-adapted *Synechococcus* clades (clades I and IV) were dominant during winter and spring, whereas the warm-water-adapted *Synechococcus* clade (clade II) and *Prochlorococcus* ecotypes were dominant during the summer and autumn (Choi et al. 2013; Fig. 10.7). In December, the

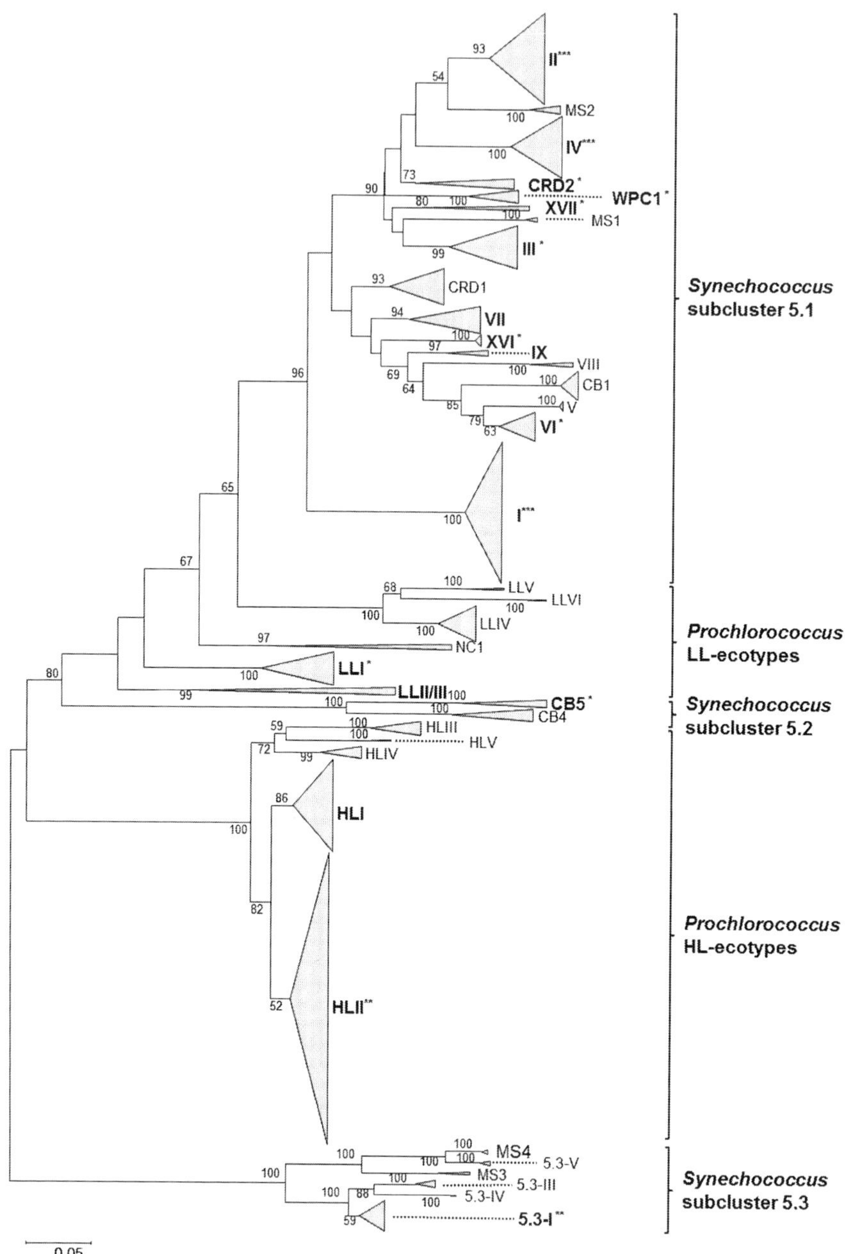

Fig. 10.6 Compressed neighbor-joining tree of 16S-23S rDNA internal transcribed spacer sequences showing the phylogenetic diversity of picocyanobacteria. Clades found in the East Sea were shown in bold. Asterisks following the clade name represent the clades appearing at fractions greater than 50 % (***), 10 % (**) and 1 % (*) in at least 1 sample (Redrawn from Choi et al. 2013)

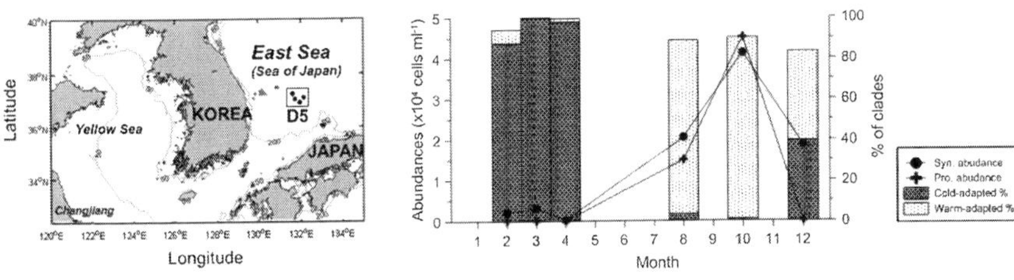

Fig. 10.7 Seasonal variation of *Synechococcus* (Syn) and *Prochlorococcus* (Pro) abundances and % of cold water- (*Synechococcus* clades I and IV) and warm water-adapted genotypes (*Synechococcus* clade II and *Prochlorococcus* ecotypes) in the D5 area of the East Sea (Redrawn from Choi et al. 2013)

abundances of both warm-water- and cold-water-adapted picocyanobacterial genotypes were similar, probably due to the decrease of warm-water-adapted picocyanobacteria as well as to growth of cold-water-adapted picocyanobacteria as temperature declines seasonally (Fig. 10.7). Thus, water temperature seems to exert a strong influence over seasonal picocyanobacterial diversity in temperate waters.

10.5 Species Composition

In recent studies, more than 700 species, varieties and forms of phytoplankton have been reported in the Russian coastal waters of the East Sea with a predominance of dinoflagellates and diatoms (Konovalova and Selina 2010; Stonik et al. 2011). In a long-term study from 1991 to 2006 in coastal waters (Orlova et al. 2009), a total of 357 species of planktonic microalgae belonging to eight divisions, Bacillariophyta (157 species), Dinophyta (143), Chlorophyta (22 species), Euglenophyta (11), Cyanophyta (8), Chrysophyta (8), Cryptophyta (5) and Raphidophyta (3) were identified. The increase of species numbers observed in the Russian coastal waters in recent studies, as compared with the data of late 1960s and early 1970s, is at least partly due to human factors: the overall number of taxonomic studies of phytoplankton with the use of modern techniques increased significantly. Also, it could be due to eutrophication and introduction of invasive species through fouling and/or from ballast waters and currents.

As shown in Table 10.2, the number of diatom species reported for different areas in the East Sea ranged from 48 to 301 and dinoflagellates species numbers ranged from 33 to 88 species. Kang and Choi (2001) reported 409 species in the Korean coastal waters of the East Sea from 1992 to 1996. These species include Bacillariophyta (301), Dinophyta (88), Chrysophyta (6), Cyanophyta (2), and Chlorophyta (2). Also, increases in phytoplankton diversity appeared in the 1980s and 1990s probably due to factors similar to those mentioned above for Russian waters.

Table 10.2 The number of phytoplankton species found in the East Sea

Area	Sampling time	Total number	Diatoms	Dinoflagellates	Reference
Korea Strait	1964. 5	48	48	n.a.	Uhm and Yoo(1967)
Southwestern	1981. 9	185	124	56	Shim and Lee (1983)
Southwestern	1981–1984	235	155	72	Lee and Shim (1990)
Subpolar front	1990. 8	95	60	33	Park et al. (1991)
Coastal waters (Uljin, Gori)	1992–1996	409	301	88	Kang and Choi (2001)

n.a. not available

Most diatom species reported in the East Sea are centric diatoms. The diversity of pennate diatoms occurring in the East Sea is limited because of the deep basin and limited tidal flats (Choi et al. 2011). Dinoflagellate species are diverse in the East Sea and even more diverse in the Russian coastal waters. This high diversity of dinoflagellates in the East Sea might be due to microtidal action and the influence of the warm current (Choi et al. 2011).

Attempts have been made to group phytoplankton species by water mass of different origin: neritic species, warm water species from the Tsushima Warm Current (TWC) and cold water species from the Liman Current (Table 10.3). Warm water species belonging to the TWC group are widely distributed all over the East Sea, and even exert a great influence on the overall phytoplankton community in the northern East Sea (Kisselew 1947). In contrast, cold water species occurred in the northern cold water areas (Ohwada and Ogawa 1966). In the subpolar frontal area,

Table 10.3 Warm water species and cold water species found in the East Sea

Group	Species name
Warm water species	**Diatoms** *Bacteriastrum comosum, Ceratulina pelagica, Chaetoceros coarctatus, Ch. furca, Ch. lorenzianus, Ch. peruvianus, Ch. messanensis, Ch. tetrastichon, Ch. anastomosans, Ch. dichaeta, Ch. eibenii, Ch. paradoxum, Ch. radicans, Climacodium biconcavum, Cl. frauenfeldianum, Ditylum sol, Eucampia cornuta, Pseudo-eunotia doliolus, Lioloma elongatum, L. delicatulum, Thalassiothrix gibberula, Planktoniella sol, Thalassiosira leptopus, Hemidiscus cuneiformis* **Dinoflagellates** *Ceratium bucephalum, C. macroceros, C. tripos, C. pentagonum, C. massilense, Cochlodinium catenatum, Cochl. polykrikoides, Ornithocercus magnificus, Phyrophacus horologium, Prorocentrum gracile* **Cyanobacteria** *Richelia intracellularis* within diatom, *Trichodesmium thiebautii*
Cold water species	**Diatoms** *Chaetoceros atlanticus, Ch. concavicornis, Rhizosolenia hebetata* f. *semispina, Thalassiosira nordenskioldii, Thalassiothrix longissima*

both warm water species and cold water species were observed (Park et al. 1991; Choi et al. 2011). In addition, the numbers of species belonging to diatoms and dinoflagellates varied seasonally in Korean coastal waters during the period from 1992 to 1996: 91–179 species in winter, 128–189 species in spring, 122–178 species in summer, and 128–185 species in autumn (Kang and Choi 2001).

10.6 The Relationship Between Phytoplankton and Environmental Factors

In Russian coastal waters, environmental conditions favorable for the growth of the main groups of phytoplankton were determined and estimated quantitatively for coastal and estuarine areas with the use of principal component analysis. The analysis revealed three periods of mean development of phytoplankton: spring, summer and autumn blooms, with successive change of species composition. The dominant contributions to principal components were made by two environmental parameters. Sea surface temperature (SST) and mixed layer depth (MLD) had the most influence on the seasonal succession of the phytoplankton community. SST-MLD combinations favorable for each case of mass development were determined (Fig. 10.8). The driving mechanisms of the seasonal succession

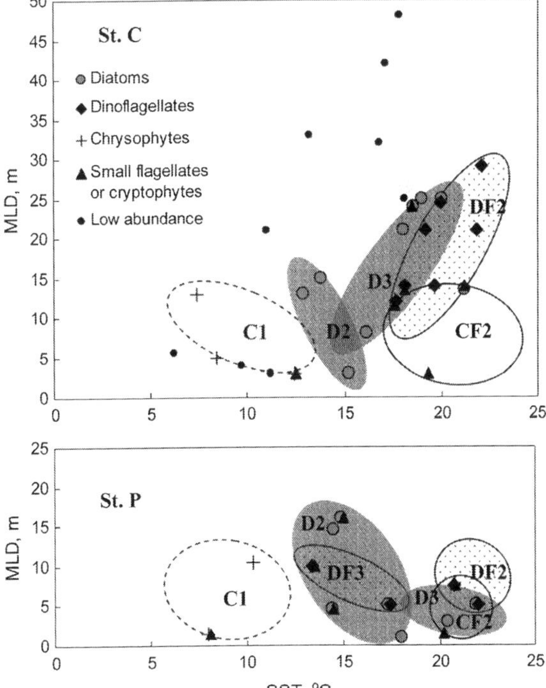

Fig. 10.8 Conditions of blooms of the main groups of phytoplankton at the sea surface (indicated by *ellipses*). *C1* spring bloom of chrysophytes, *D2* summer bloom of diatoms, *DF2* summer bloom of dinoflagellates, *CF2* summer bloom of small flagellates or cryptophytes, *D3* autumn bloom of diatoms, *DF3* autumn bloom of dinoflagellates (from Zuenko et al. 2006)

of phytoplanktonin Russian coastal waters include nutrient transport through the seasonal pycnocline by turbulent mixing, terrestrial nutrient supply by monsoon floods, nutrient supply by upwelling, and light control by the thickness of the upper mixed layer (Zuenko et al. 2006).

The East Sea has experienced dramatic changes in its environmental conditions during the last 50–60 years (Kim and Kim 1996). Chiba and Saino (2002) revealed a distinct change in diatom community structure in spring in the offshore TWC region during the 1980s, following the 1976/77 climate regime shift. They showed that interdecadal changes in the physico-chemical environment within the upper water column could alter the ecosystem dynamics of lower trophic levels in the East Sea. Nitrogen is generally regarded as a primary limiting factor to phytoplankton production in the East Sea (Shim et al. 1989; Kim et al. 2010). During October and February, nitrate concentration was generally depleted and the N:P ratio was lower than the Redfield ratio (16:1) in the surface mixed layer of the Korean coastal waters (Shim and Park 1986; Chung et al. 1989). A ratio of N:P lower than 16 was also observed in most water masses of the East Sea (Yanagi 2002; Talley et al. 2004). In a north-south transect of the East Sea, the N:P ratios in the mixed layer were less than 11 in April and less than 6 in July and October, and those deeper than 100 m were all approximately 13 (Kim et al. 2010). The very low N:P ratios in the mixed layer could be due to the active Redfieldian consumption of N and P since the initial N:P ratio in the entire East Sea was about 13 (Talley et al. 2004; Kim et al. 2010). The contribution of cyanobacteria to total phytoplankton biomass was relatively high in low N:P ratio waters. Nitrogen depletion is favorable for the growth of cyanobacteria (Kim et al. 2010). The dominance of cyanobacteria in the frontal zone could be due to the relatively low N:P ratios observed after the spring bloom (Kim et al. 2010). In contrast, more preferentially in high N:P ratio areas, diatoms dominated in the phytoplankton community of the spring bloom. The nutrient supply into the East Sea is also closely related to the amount of volume transport through the Korea Strait because the volume transport of the TWC through the Korea Strait shows seasonal and interannual variability (Lee et al. 2009). The volume transport is typically large in summer and autumn, but small in winter and spring (Teague et al. 2002).

The interannual variability of the development of the spring phytoplankton bloom in the southern East Sea was closely related to wind speed in spring: weak winds caused an earlier spring bloom while strong winds resulted in a late spring bloom, which may be affected by large scale climate variation such as ENSO events (Yamada et al. 2005; Lee et al. 2009). The strong transport of the nutrient poor TWC resulted in earlier transition of summer-like nutrient-poor conditions in the upper layer, which suppressed the development of the spring bloom (Yoo and Kim 2004). Sometimes, after the passage of Asian dust over the East Sea accompanied by precipitation, the initiation of the spring bloom occurred a month earlier than in a normal year (Jo et al. 2007).

10.7 Overview of Primary Production Studies

Field measurements of primary production began in the East Sea during the 1960s generally using ^{13}C and ^{14}C methods (Table 10.4). Sorokin (1977) reported that the average primary production values were about 720 mg C m^{-2} d^{-1} in a wide area of the East Sea traversing from the southeast to the northwest. Primary production studies in the southwestern region of the East Sea including the coastal waters of the southeastern Korean Peninsula and the UB cover mostly the 1980s to the early 1990s (Shim and Park 1986; Park et al. 1991; Shim et al. 1992). These studies suggested that daily primary production in the southwestern East Sea is relatively high compared to those in other regions of the East Sea. Values of daily primary production measured in situ in the southwestern East Sea were about 1300 mg C m^{-2} d^{-1} in autumn (October) (Shim and Park 1986), while primary production in summer (August) in the UB was estimated to be about 1400 mg C m^{-2} d^{-1} (Park et al. 1991). Based on primary production measurements from consecutive field surveys from spring to early summer (Shim et al. 1992), average primary production values in the UB as well as some stations in the Korean coastal waters were estimated to be about 2500, 2000, and 1900 mg C m^{-2} d^{-1} in April, May, and June, respectively. In the central area of the East Sea (e.g. Yamato Basin), primary production values varied from 45 to 1100 mg C m^{-2} d^{-1}, depending on the season (Nagata 1998). A clear seasonal variability of daily primary production was shown in the northeastern East Sea (the west coast of Hokkaido), varying from 254 to 2494 mg C m^{-2} d^{-1}, with an increasing trend in primary production during the phytoplankton spring bloom in April (Yoshie et al. 1999). However, it is hard to estimate primary production for the entire East Sea since these studies were temporally and spatially limited.

Satellite ocean color observations provide an important means of estimating phytoplankton production on an ocean basin and global scale with high temporal and spatial resolution (Table 10.5). There have been studies using satellite ocean color data to investigate temporal and spatial variability of phytoplankton biomass in large or entire areas of the East Sea since the 2000s (Kim et al. 2000; Yamada et al. 2004, 2005). In addition, satellite-based model studies estimated primary production on a global scale for ocean waters (Platt and Sathyendranath 1988; Longhurst et al. 1995; Antoine et al. 1996; Behrenfeld and Falkowski 1997). The standard primary production models basically rely on satellite ocean color (chlorophyll) data as well as other environmental data such as surface irradiance and SST. Model-based studies of primary production for satellite estimation in the East Sea have been achieved since the mid-2000s (Kameda and Ishizaka 2005; Yamada et al. 2005). Yamada et al. (2005) first investigated spatial and temporal variability of primary production in wide areas of the East Sea using a satellite-based primary production model for the period of 1998 to 2002. Their results provided mean annual primary production values of about 170, 161, 191, and 222 g C m^{-2} year^{-1} in the Russian coastal area, middle of the Japan Basin, southeastern East Sea, and southwestern East Sea, respectively. It has been also shown that there is a strong seasonality of primary production in the East Sea.

Table 10.4 Measurements of daily primary production in the East Sea by ^{13}C and ^{14}C methods prior to 2000

Region	Lat./Long.	Year	Month	Data points	PP (mgC m^{-2} day^{-1})	Reference
South-north cross sectional line	37°40′–43°00′N	1972	June	8	721 ± 443	Sorokin (1977)
Southwestern area	34°50′–35°40′N 129°07′–130°30′E	1985	October	5	1312 ± 668	Shim and Park (1986)
Southwestern area	37°00′–38°13′N 128°50′–130°00′E	1990	August	10	1400 ± 668	Park et al. (1991)
Southwestern area	35°03′–37°47′N 129°27′–133°04′E	1988–1990	April May July	9 6 6	2478 ± 116 2034 ± 194 1900 ± 1306	Shim et al. (1992)
Yamato Basin	39°00–39°20′N 134°20′–135°00′E	1994–1996	January April August October	1 3 2 1	44 1082 ± 252 353 ± 213 154	Nagata (1998)
West coast of Hokkaido	43°30′N 140°40′E	1997	March April July September	1 1 2 1	345 2492 359 ± 149 254	Yoshie et al. (1999)

Table 10.5 Measurements of annual primary production in the East Sea after 2000

Location	Note	Lat./Long.	Methods	PP (g C m^{-2} year^{-1})	Reference
East Sea	Russian coast	46.5°N–139°E	Satelite-based estimation	170	Yamada et al. (2005)
	Japan basin	41°N–136°E		161	
	Southeastern area	38°N–136°E		191	
	Southwestern area	38°N–131°E		222	
Ulleung basin (Southwestern area)	Middle of Ulleung basin	37°N, 129°45′–131°E	13C	270	Kwak et al. (2013b)
	Dokdo area	131°38′–132°04′ 37°08′–37°24′N	14C	172	Noh (unpublished data)
	Upwelling area	36°N–130°E	14C	358	Noh (unpublished data)

Among different regions in the East Sea, primary production in the UB was highest except in the southern upwelling area, as shown in Table 10.4. The UB has been considered a relatively highly productive region compared to adjacent regions such as the Japan Basin (Yamada et al. 2005; Hyun et al. 2009; Yoo and Park 2009; Lee et al. 2009), notwithstanding the prevalence of the nutrient-poor surface water mass of the TWC (Yoo and Kim 2004). Several potential reasons, such as coastal upwelling (Hyun et al. 2009; Yoo and Park 2009) and various scales of eddies (Hyun et al. 2009; Kim et al. 2012; Lim et al. 2012) have been proposed to explain the high productivity observed in the UB. Supported by satellite observations from 1998 to 2006, frequent wind-driven upwelling along the Korean coast in all seasons except winter was proposed to sustain the high productivity in the UB (Yoo and Park 2009). In addition, there are several recent findings for positive effects of eddies on primary productivity in the UB. Lim et al. (2012) found that upward water-flux movement at the periphery of the anticyclonic eddies may prevent phytoplankton from sinking and retain them in the euphotic zone, thus sustaining high productivity in the UB during early spring seasons before strong stratification in the upper water column develops for the seasonal spring blooms. During summer seasons, normally with nutrient-depleted conditions, eddy-induced upwelling of nutrients enhances high chl-a concentrations in the UB (Kim et al. 2012). Recently, Kwak et al. (2013b) proposed that nitrate supplies from upward flux through the pycnocline sustains the high phytoplankton productivity in the UB during summer periods. However, all these aforementioned and discussed factors could act together to sustain the high productivity of phytoplankton (Yamada et al. 2005; Hyun et al. 2009; Yoo and Park 2009; Lee et al. 2009; Kwak et al. 2013b).

Primary production data from the three different areas (Kwak et al. 2013b; Noh, unpublished data) were compared in detail to understand the spatial distribution of primary production in the UB (Fig. 10.9). Annual primary production measured in the Dokdo area was 172 g C m^{-2} year^{-1}, while substantially higher primary production of 358 g C m^{-2} year^{-1} was measured in the southern upwelling area (Table 10.5). According to Kwak et al. (2013b), annual primary production in the UB along the 37°N line was about 270 g C m^{-2} year^{-1}. The sampling points (36°N, 130°E) in the southern upwelling area along the southern Korean coast are subjected to frequent wind-driven upwelling events (Yoo and Park 2009), which bring up major inorganic nutrients through upward flux and subsequently sustain the observed high primary production in the area. Moreover, enhanced biomass and diversity of phytoplankton moving from the coastal water mass may also contribute to the high primary production. Primary production of 172 g C m^{-2} year^{-1} estimated from the Dokdo area is about half of the primary production values in the upwelling area. These spatial differences may also be attributed to physical processes such as coastal upwelling (Hyun et al. 2009; Yoo and Park 2009) and various scales of eddies in the UB (Hyun et al. 2009; Kim et al. 2012; Lim et al. 2012).

Primary production in the Dokdo area has a bimodal pattern (Fig. 10.9), showing the highest value in April with another small peak in November. The highest

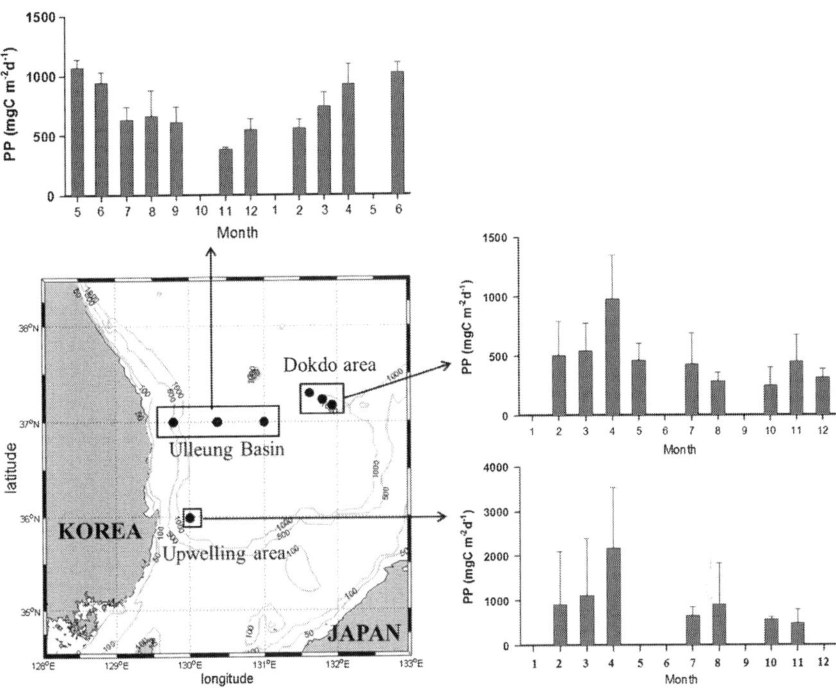

Fig. 10.9 Monthly variation of primary production in the Ulleung Basin, upwelling and Dokdo areas (Noh JH, unpublished data)

primary productivity (980 ± 367 mg C m^{-2} days^{-1}) is observed in April as the spring bloom develops. A relatively high rate (509 ± 284 mg C m^{-2} days^{-1}) was observed in winter (February) when a lower primary production rate is normally expected. It is suggested that the relatively high primary production in winter is strongly influenced by fluctuating wind and light conditions. Spatially, the highest primary production rate (2173 ± 1368 mg C m^{-2} days^{-1}) is observed in the upwelling area in the southwestern East Sea during the spring season (April), while relatively high production rates are observed in February (916 ± 185 mg C m^{-2} days^{-1}) and March (1118 ± 1285 mg C m^{-2} days^{-1}). The large standard deviations suggest that there is a strong interannual variation and that even for low temperature water conditions, primary production in the area could be increased depending on the weather conditions and distribution of the TWC. Primary production in August is relatively higher (925 ± 911 mg C m^{-2} days^{-1}) due to the upwelling effect, which is approximately three times higher than that measured in the Dokdo area at the same period. The seasonal variation pattern of primary production observed in the middle of the Ullueng Basin area is similar to that in the two other locations mentioned above. However, for this location, primary production rates in May and June are similar to or higher than those in April (Kwak et al. 2013b).

"Biological hot spots" regarded as high biological activity regions generally occur in ocean regions with discernible physical oceanographic features such as eddies, meanders, and fronts (Palacios et al. 2006). Subpolar fronts are usually found around 36°–42°N, 130°–140°E in the UB although the locations vary according to seasonal and inter-annual ocean conditions (Chiba et al. 2008). Frequent occurrence of eddies induces dynamic physico-chemical interactions which enhance biological production in the UB (Kim et al. 2012; Lim et al. 2012). Moreover, high phytoplankton productivity and biological pump efficiency are sustained by the vertical supply of major nutrients through hydrographic conditions (Kwak et al. 2013a). Considering those factors, Kwak et al. (2013a, b) proposed that the UB is a biological "hot spot" in the East Sea and must be carefully administered for resource management and conservation (Palacios et al. 2006). To date, very limited knowledge of the current baseline status and structure of the pelagic ecosystem has been obtained in the UB. Therefore, more seasonal and annual field measurements under dynamic environmental conditions should be continuously collected for better understanding of the current status and future responses of marine ecosystems in the UB and also in the East Sea as a whole due to the potential for alterations from ongoing climate and environmental change.

References

Aikawa H (1936) The planktological properties of the principal sea areas surrounding Japan. Bull Japan Sci Fish 5(1):33–41

Antoine D, André JM, Morel A (1996) Oceanic primary production: 2. Estimation at global scale from satellite (coastal zone color scanner) chlorophyll. Global Biogeochem Cycles 10(1):57–69

Behrenfeld MJ, Falkowski PG (1997) Photosynthetic rates derived from satellite-based chlorophyll concentration. Limnol Oceanogr 42(1):1–20

Chang KI, Teague WJ, Lyu SJ et al (2004) Circulation and currents in the southwestern East/Japan Sea: overview and review. Prog Oceanogr 61(2):105–156

Chiba S, Saino T (2002) Interdecadal change in the upper water column environment and spring diatom community structure in the Japan Sea: an early summer hypothesis. Mar Ecol Prog Ser 231:23–35

Chiba S, Aita MN, Tadokoro K et al (2008) From climate regime shifts to lower-trophic level phenology: synthesis of recent progress in retrospective studies of the western North Pacific. Prog Oceanogr 77(2):112–126

Choe S (1972) Studies on the seasonal variations of plankton organism and suspended particulate matter in the coastal area of Ko-Ri. J Korean Soc Oceanogr 7(2):47–57

Choi DH, Noh JH (2009) Phylogenetic diversity of *Synechococcus* strains isolated from the East China Sea and the East Sea. FEMS Microbiol Ecol 69(3):439–448

Choi DH, Noh JH, Shim JS (2013) Seasonal changes in picocyanobacterial diversity as revealed by pyrosequencing in temperate waters of the East China Sea and the East Sea. Aquat Microb Ecol 71:75–90

Choi JK, Kang YS, Noh JH, Shim JH (2011) The characteristics of phytoplankton distribution in the East Sea. In: Choi JK (ed) The Plankton Ecology of Korean Coastal Waters, Donghwa, p48–p67

Chung CS, Shim JH, Park YC et al (1989) Primary productivity and nitrogenous dynamics in the East Sea of Korea. J Korean Soc Oceanogr 24(1):52–61

Conkright ME, Gregg WW (2003) Comparison of global chlorophyll climatologies: in situ, CZCS, Blended in situ-CZCS and SeaWiFS. Int J Remote Sens 24(5):969–991

Gamo T (1999) Global warming may have slowed down the deep conveyor belt of a marginal sea of the northwestern Pacific: Japan Sea. Geophys Res Lett 26(20):3137–3140

Hashuimoto S, Shiomoto A (2002) Regional distribution of size fractionated chlorophyll a concentration at the sea surface water adjacent to Japan in May–June 2000. Bull Japan Soc Fish Oceanogr 66:148–154

Hyun JH, Kim D, Shin CW et al (2009) Enhanced phytoplankton and bacterioplankton production coupled to coastal upwelling and an anticyclonic eddy in the Ulleung Basin. East Sea. Aquat Microb Ecol 54(1):45–54

Jardillier L, Zubkov MV, Pearman J et al (2010) Significant CO_2 fixation by small prymnesiophytes in the subtropical and tropical northeast Atlantic Ocean. ISME J 4(9):1180–1192

Jo CO, Lee JY, Park K et al (2007) Asian dust initiated early spring bloom in the northern East/Japan Sea. Geophys Res Lett 34(5):L05602. doi:10.1029/2006GL027395

Kameda T, Ishizaka J (2005) Size-fractionated primary production estimated by a two-phytoplankton community model applicable to ocean color remote sensing. J Oceanogr 61(4):663–672

Kang JH, Kim WS, Chang KI et al (2004) Distribution of plankton related to the mesoscale physical structure within the surface mixed layer in the southwestern East Sea, Korea. J Plankton Res 26(12):1515–1528

Kang YS, Choi JK (2001) Ecological characteristics of phytoplankton communities in the coastal waters of Gori, Wulseong, Uljin and Youngkwang. I. Species composition and distribution (1992–1996). Algae 16(1):85–111

Kang YS, Choi JK (2002) Ecological characteristics of phytoplankton community in the coastal waters of Gori, Wulseong, Uljin and Youngkwang. II. The distribution of standing stocks and environmental factors (1992–1996). J Korean Soc Oceanogr 7(3):108–128

Kano Y, Baba N, Ebara S (1984) Chlorophyll a and primary production in the Japan Sea. Oceanography 34:31–49

Kim D, Yang EJ, Kim KH et al (2012) Impact of an anticyclonic eddy on the summer nutrient and chlorophyll a distributions in the Ulleung Basin, East Sea (Japan Sea). ICES J Mar Sci 69(1):23–29

Kim HG, Lee SG, Jeong CS et al (1999) Harmful algal blooms in Korean coastal waters. In: PICES 8th annual meeting abstracts, p 27

Kim KR, Kim K (1996) What is happening in the East Sea (Japan Sea)?: recent chemical observations during CREAMS 93-96. J Korean Soc Oceanogr 31:164–172

Kim SW, Saitoh SI, Ishizaka J et al (2000) Temporal and spatial variability of phytoplankton pigment concentrations in the Japan Sea derived from CZCS images. J Oceanogr 56(5):527–538

Kim T, Lee W, Kim G (2010) Hydrographically mediated patterns of photosynthetic pigments in the East/Japan Sea: low N: P ratios and cyanobacterial dominance. J Mar Syst 82(1):72–79

Kisselew IA (1947) Phytoplankton of the Far Eastern Seas as indicator of characteristics of their hydrological regime. TrGoin 113:189–212

Konovalova GV, MS Selina (2010) Dinophyta biota of the Russian waters of the Sea of Japan. In: Adrianov AV (ed) vol 2, Dal'nauka, Vladivostok, p 352

Kuroda K (1987) Chlorophyll distribution in the Kuroshio region, south of Japan. Kousuiken note. Sora to Umi 9:19–29

Kwak JH, Hwang J, Choy EJ et al (2013a) High primary productivity and f-ratio in summer in the Ulleung Basin of the East/Japan Sea. Deep-Sea Res I 79:74–85

Kwak JH, Lee SH, Park HJ et al (2013b) Monthly measured primary and new productivities in the Ulleung Basin as a biological "hot spot" in the East/Japan Sea. Biogeosci 10:4405–4417

Lee DK, Niiler PP (2005) The energetic surface circulation patterns of the Japan/East Sea. Deep-Sea Res II 52(11):1547–1563

Lee JY, Kang DJ, Kim IN et al (2009) Spatial and temporal variability in the pelagic ecosystem of the East Sea (Sea of Japan): a review. J Mar Syst 78:288–300

Lee WH, Shim JH (1990) Distributions of phytoplankton standing crop and the associated TS properties in the southwestern East Sea (Sea of Japan). J Korean Soc Oceanogr 25:1–7

Lie HJ, Byun SK, Bang I et al (1995) Physical structure of eddies in the southwestern East Sea. J Korean Soc Oceanogr 30(3):170–183

Lim JH, Son S, Park JW et al (2012) Enhanced biological activity by an anticyclonic warm eddy during early spring in the East Sea (Japan Sea) detected by the geostationary ocean color satellite. Ocean Sci J 47(3):377–385

Longhurst A, Sathyendranath S, Platt T et al (1995) An estimate of global primary production in the ocean from satellite radiometer data. J Plankton Res 17(6):1245–1271

Moon CH, Yang SR, Yang HS et al (1998) Regeneration processes of nutrients in the Polar front area of the East Sea. IV. Chlorophyll a distribution, new production and the vertical diffusion of nitrate. Bull Korean Fish Soc 31:259–266

Nagata H (1998) Seasonal changes and vertical distributions of chlorophyll *a* and primary productivity at the Yamato Rise, central Japan Sea. Plankton Biol Ecol 45(2):159–170

Noh JH, Yoo SJ, Kang SH (2006) The summer distribution of picophytoplankton in the western Pacific. Korean J Environ Biol 24(1):67–80

Ohwada M, Ogawa F (1966) Plankton in the Japan Sea. Ocenogr Mag 18:39–42

Orlova TY, Konovalova GV, Stonik IV et al (2002) Harmful algal blooms on the eastern coast of Russia. In: Harmful algal blooms in the PICES region of the North Pacific PICES SciRep 23, pp 47–58

Orlova TY, Stonik IV, Aizdaicher NA et al (2008) Toxicity, morphology and distribution of *Pseudo-nitzschia calliantha, P. multistriata* and *P. multiseries* (Bacillariophyta) from the northwestern Sea of Japan. Bot Mar 51(4):297–306

Orlova TY, Stonik IV, Shevchenko OG (2009) Flora of planktonic microalgae of Amursky Bay, Sea of Japan. Russian J Mar Biol 35(1):60–78

Ostroumoff A (1924) *Noctiluca miliaris* in symbiose mitgrünen. Algen Zool Anz 68:162

Palacios DM, Bograd SJ, Foley DG et al (2006) Oceanographic characteristics of biological hot spots in the North Pacific: a remote sensing perspective. Deep-Sea Res Pt II 53:250–269

Park JG, Shim JH, Lee JB (1998) Long-term variation of phytoplankton biomass and implication in the East and the South Sea, Korea. Algae 13:123–133

Park JS, Kang CK, An KH (1991) Community structure and spatial distribution of phytoplankton in the polar front region off the east coast of Korea in summer. Bull Korean Fish Soc 24(4):237–247

Park KA, Kang CK, Kim KR et al (2014) Role of sea ice on satellite-observed chlorophyll-*a* concentration variations during spring bloom in the East/Japan sea. Deep-Sea Res I 83:34–44

Park MO (2006) Composition and distribution of phytoplankton with size fraction results at southwestern East Sea. Ocean Sci J 41(4):301–313

Park TS (1956) On the seasonal change of the plankton at Korea Strait. Pusan Fish Univ Res 1:1–19

Platt T, Sathyendranath S (1988) Oceanic primary production: estimation by remote sensing at local and regional scales. Science 241(4873):1613–1620

Racault MF, Le Quéré C, Buitenhuis E et al (2012) Phytoplankton phenology in the global ocean. Ecol Indic 14(1):152–163

Rebstock GA, Kang YS (2003) A comparison of three marine ecosystems surrounding the Korean peninsula: responses to climate change. Prog Oceanogr 59(4):357–379

Roh TK, Kim YB, Park JL et al (2010) Plankton community response to physico-chemical forcing in the Ulleung Basin, East Sea during summer of 2008. Ocean Polar Res 32(3):209–289

Shim JH, Bae SJ (1985) The distribution of phytoplankton in Yeong-il Bay. J Korean Soc Oceanogr 20:49–60

Shim JH, Lee WH (1983) Plankton study in the southeastern sea of Korea (1). Phytoplankton distribution in September, 1981. J Korean Soc Oceanogr 18(2):91–103

Shim JH, Park YC (1986) Primary productivity measurement using carbon-14 and nitrogenous nutrient dynamics in the southeastern sea of Korea. J Korean Soc Oceanogr 21:13–24

Shim JH, Yang SR, Lee WH (1989) Phytohydrography and the vertical pattern of nitracline in the southern waters of the Korean East Sea in early spring. J Korean Soc Oceanogr J Kor Soc Oceanogr 24(1):15–28

Shim JH, Yeo HG, Park JG (1992) Primary production system in the southern waters of the East Sea, Korea. I. Biomass and productivity. J Korean Soc Oceanogr 27:91–100

Shim JH, Yeo HG, Park JG (1995) Primary production system in the southern waters of the East Sea, Korea. II. The structure of phytoplankton community. J Korean Soc Oceanogr 30:163–169

Shim JM, Yun SH, Hwang JD et al (2008) Seasonal variability of picoplankton around Ulneung Island. J Environ Sci 17(11):1243–1253

Shiomoto A, Tanaka H, Hashimoto S et al (2004) Regional distribution of picophytoplankton in near-shore areas around Japan in early summer. Plankton Biol Ecol 51(2):71–81

Shuntov VP (2001) Biology of the far Eastern seas of Russia. TINRO-Centre Press, Vladivostok

Skvortzow BW (1931) Pelagic diatoms of Korean Strait of the Sea of Japan. Philip J Sci 46:95–122

Sorokin YI (1977) The heterotrophic phase of plankton succession in the Japan Sea. Mar Biol 41(2):107–117

Stonik IV, Orlova TY (1998) Summer-autumn phytoplankton in Amursky Bay, Sea of Japan. Russian J Mar Biol 24(4):207–213

Stonik IV, Orlova TY (2002) Phytoplankton of the coastal waters off Vladivostok city (the north-western part of the East Sea) under eutrophic conditions. Ocean Polar Res 24(4):359–365

Stonik IV, Orlova TY, Lundholm N (2011) Diversity of *Pseudo-nitzschia* from the western North Pacific. Diatom Res 26(1):121–134

Stonik IV, Selina MS (1995) Phytoplankton as indicator of eutrophic water levels in Peter the Great Bays, Sea of Japan. Russian J Mar Biol 21(6):356–359

Suk MS, Chang KI, Kim SH et al (2002) Marine ecosystem responses to climate variability in the East Sea. Korea Ocean Research and Development Institute, Report No. BSPE 817-00-1396

Talley LD, Tishchenko P, Luchin V et al (2004) Atlas of Japan (East) Sea hydrographic properties in summer, 1999. Prog Oceanogr 61(2):277–348

Teague WJ, Jacobs GA, Perkins HT et al (2002) Low-frequency current observations in the Korea/Tsushima Strait. J Phys Oceanogr 32(6):1621–1641

Uhm GB, Yoo KI (1967) Diatoms in the Korea Strait. In: Reports from the Institute of marine biology. Seoul National University, Seoul, pp 1–6

Vinogradov ME, Shushkina EA, Vedernikov VI (1996) Characteristics of the epipelagic ecosystem in the Pacific Ocean based on satellite and expeditionary data. Primary production and their seasonal changes. Oceanography 36(2):201–249

Yamada K, Ishizaka J, Nagata H (2005) Spatial and temporal variability of satellite primary production in the Japan Sea from 1998 to 2002. J Oceanogr 61(5):857–869

Yamada K, Ishizaka J, Yoo S et al (2004) Seasonal and interannual variability of sea surface chlorophyll a concentration in the Japan/East Sea (JES). Prog Oceanogr 61:193–211

Yamada T (1938) Report of the adjacent seas of Korea. In: Annual report of hydrographical observations. Fish Exp Station 8, pp 11–90

Yanagi T (2002) Water, salt, phosphorus and nitrogen budgets of the Japan Sea. J Oceanogr 58(6):797–804

Yoo S, Kim HC (2004) Suppression and enhancement of the spring bloom in the southwestern East Sea/Japan Sea. Deep-Sea Res II 51:1093–1111

Yoo S, Park J (2009) Why is the southwest the most productive region of the East Sea/Sea of Japan. J Mar Syst 78(2):301–315

Yoshie N, Shin KH, Noriki S (1999) Seasonal variations of primary productivity and assimilation numbers in the western North Pacific. In: Special reports on the regional studies of North-East Eurasia and North Pacific. Hokkaido University, Hokkaido, pp 49–62

Zuenko Y, Selina M, Stonik I (2006) On conditions of phytoplankton blooms in the coastal waters of the north-western East/Japan Sea. Ocean Sci J 41(1):31–41

Chapter 11
Microbial Ecology and Biogeochemical Processes in the Ulleung Basin

Jung-Ho Hyun

Abstract Since heterotrophic bacteria rapidly respond to variations in physico-chemical conditions, monitoring the role of bacteria in C cycles is important in the East Sea where has been known as the rapid warming occurred during the last 2 decades. Major microbiological and biogeochemical processes associated with oceanographic conditions of the water column and geochemical properties of the sediment in the Ulleung Basin (UB) are covered in this chapter. Overall, bacterial parameters in the water column are largely affected by the occurrence of coastal upwelling and formation of the anticyclonic Ulleung Warm Eddy. Heterotrophic bacterial production is closely coupled to phytoplankton biomass, but the role of bacteria either as a trophic link within the microbial food web process or as a C sink for photosynthetically fixed organic carbon varies with the physico-chemical conditions of the water column. In the sediment, high organic content (>2.5 %, dry wt.) is attributed to the high C mineralization by sulfate reduction in the UB compared to that of other parts of the East Sea (i.e. Japan Basin and Yamato Basin). In addition, the observation that manganese-oxides are accumulated at high levels (>150 μmol cm^{-3}) in the surface sediment of the UB indicates that C oxidation by manganese reduction seems to be the major C oxidation pathway in the center of the UB, as was supported by the identification of acetate-oxidizing bacteria related to *Colwellia*, *Oceanospirillaceae* and *Arcobacter* that potentially reduce manganese. Finally, inventories of culture-dependent and culture-independent *Bacteria* and *Archaea*, mostly reported in the UB, are summarized at the final section of this chapter.

Keywords East Sea (Japan Sea) · Ulleung Basin · Bacterial production · Upwelling · Ulleung Warm Eddy · Sulfate reduction · Manganese oxides · Benthic C oxidation

J.-H. Hyun (✉)
Department of Marine Sciences and Convergent Technology,
Hanyang University, 55 Hanyangdaehak-ro, Sangnok-ku, Ansan 15588, Republic of Korea
e-mail: hyunjh@hanyang.ac.kr

© Springer International Publishing Switzerland 2016
K.-I. Chang et al. (eds.), *Oceanography of the East Sea (Japan Sea)*,
DOI 10.1007/978-3-319-22720-7_11

11.1 Introduction

Heterotrophic prokaryotes including the domains *Bacteria* and *Archaea* (hereafter "bacteria" or "bacterioplankton" as a conventional ecological term) in marine environments play a significant role in microbial food web processes and biogeochemical element cycles (Cotner and Biddanda 2002; Azam and Worden 2004). Bacteria biomass comprises approximately 50 % of the total microbial biomass (Li et al. 1992; Caron et al. 1995), and often exceeds the phytoplankton biomass in the oligotrophic open ocean (Cho and Azam 1990). Bacteria production in the open ocean accounts for 20–30 % of primary production (Cole et al. 1988). Both high bacterial biomass and production indicates that a substantial amount of photosynthetically fixed organic carbon is channeled through bacterial biomass and then transferred to upper trophic levels via the microbial loop (Azam et al. 1983; Sherr et al. 1986).

Meantime, bacteria are a potent biogeochemical agent regulating most element cycles. For example, bacteria are responsible for the mineralization of organic matter from various sources (Nagata 2000), and for uptake and regeneration of inorganic nutrients such as N and P (Kirchman 2000). Post-mortem dissolution of silicate from diatom cell walls is expedited by the bacterial secretion of protease (Bidle and Azam 1999), which ultimately enhances primary production and consequently controls biogeochemical C and Si cycles. Bacterial respiration accounts for approximately 50 % of total community respiration in the surface ocean (Robinson 2008), which implies that bacteria are a major biological sink for photosynthetically fixed organic carbon. Bacterial metabolic activities on sinking particles also affect the vertical flux of particulate organic carbon from the surface to the sea floor, which controls global carbon cycles (Simon et al. 2002). Recently, it has also been proposed that bacteria are responsible for the sequestration of organic carbon by transforming labile dissolved organic matter (LDOM) to recalcitrant dissolved organic matter (RDOM) via the microbial carbon pump (MCP) (Jiao et al. 2010; Jiao and Zheng 2011). Thus, measuring bacterial parameters and elucidating controls of each parameter are essential for interpreting the C flux through microbial processes in the water column.

Although continental margins including the continental shelf, slope and rise occupy only 20 % of the ocean's surface (Kennett 1982), this narrow domain of the ocean's fringe, especially the continental shelf and slope, is characterized by a high rate of biological production (Antoine et al. 1996), high vertical fluxes of organic matter from the water column (Romankevich 1984), and high lateral transport of organic matter from the adjacent shelves (Jahnke et al. 1990; Wollast 1991). Consequently, sediments of continental margins are a significant place for deposition and mineralization of organic matter and regeneration of inorganic carbon and nutrients (Liu et al. 2000, 2010; Reimers et al. 1992; Walsh 1991; Jahnke and Jahnke 2000). Therefore, understanding the seasonal and inter-annual fluctuation of carbon fluxes and nutrient cycles associated with physical and biogeochemical variations and evaluating the importance of carbon deposition and benthic processes have been one of the key objectives in international interdisciplinary ocean research programs such Joint Global Ocean Flux Studies (JGOFS), Land-Ocean Interaction in the Coastal Zone(LOICZ) and International Geosphere-Biosphere Programme (IGBP) toward increased understanding of biogeochemical processes on the continental margin (Liu et al. 2000).

The East Sea is a typical marginal sea surrounded by Russia, Japan and Korea in the northwest Pacific (see Chaps. 1 and 16 for a more detailed description of the geology and oceanography of the East Sea). Because of its unique thermohaline circulation similar to 'the Great Ocean Conveyor Belt' (Broecker 1991), the East Sea has often been referred to as a miniature ocean (Kim and Kim 1996; Kim et al. 1996). Long-term oceanographic observation in the East Sea has revealed that the warming of seawater from 0.08 to 0.24 °C at 1000 m depth during the period 1969–1999 (Kim et al. 2001) was comparable to the temperature increase of ca. 0.3 °C during the last 5 decades in the world ocean (Levitus et al. 2000). Recently, a rapid increase of SST (1.09 °C) in the East Sea over the last 2 decades (1982–2006) has been recorded, which is the fourth highest among the 18 large marine ecosystems in the world ocean (Belkin 2009). Thus, the East Sea has drawn scientific attention as a natural laboratory for evaluating and predicting the variations in major oceanographic processes associated with global climate change (Kim et al. 2001).

Since microbiological components rapidly respond to variations in the physico-chemical conditions of water regimes resulting from variations in global-scale biogeochemical carbon cycles and climatic changes (Karl 1999; Bidle et al. 2002; Hoppe et al. 2008), it is important to understand the response of microbial components to the variations in the water regime in the East Sea. Such studies of the response of microbial components include both qualitative (i.e. composition) and quantitative (i.e. biomass, production and respiration) parameters together with physico-chemical and biological controls. Recently, a substantial database has been established on the composition of the microbial community in the East Sea, but ecological and biogeochemical processes associated with the heterotrophic bacterial community and its major oceanographic controls have not been extensively elucidated, except for a few cases mostly associated with biological controls in the Ulleung Basin (UB) (Cho et al. 2000; Hwang and Cho 2002; Choi et al. 2005; Hyun et al. 2009). This chapter consists of three major sections. First, I discuss the distribution of bacterial biomass, production and respiration associated with oceanographic parameters such as coastal upwelling and the anticyclonic eddy in the UB. Second, by integrating oceanographic and biogeochemical parameters, I attempt to elucidate the major biogeochemical processes with special emphasis on the rates and pathways of organic carbon oxidation in the sediment of the UB. Finally, I summarize inventories of cultured and uncultured microorganisms reported in the water column and sediment and those associated with micro- and macrobenthos in the East Sea.

11.2 Microbiological Oceanography

The two most well-known oceanographic properties associated with biological processes in the UB are the frequent occurrence of coastal upwelling along the southeast Korean peninsula (Lee 1983; Lee and Na 1985) and the formation of the anticyclonic Ulleung Warm Eddy (UWE) (Lie et al. 1995; Shin et al. 2005) (see Chaps. 3 and 4). Wind-driven coastal upwelling occurs frequently in all seasons except winter (Yoo and Park 2009). The UWE in winter and spring can generally be defined as a cold core wrapped by award filament (Chang et al. 2004; Fig. 11.1).

Despite its well-known physical and oceanographic properties, it was not until a decade ago that the impact of upwelling and the UWE on biological processes in the East Sea was evaluated in earnest with the development of remote sensing. Yoo and Kim (2003) used Coastal Zone Color Scanner (CZCS) data to calculate the annual primary production of 240 g C m^{-2} year^{-1} in the East Sea. Based on the combination of satellite images and in situ measurement from 1998 to 2002, Yamada et al. (2005) revealed that annual primary production was highest in the

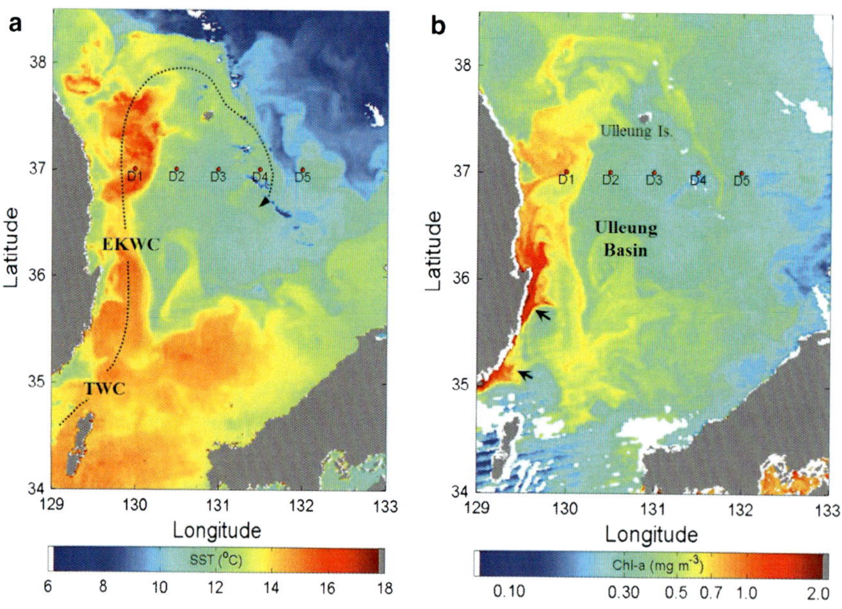

Fig. 11.1 Satellite images of sea surface temperature (SST) (panel **a**) and chlorophyll *a* (chl-*a*) (panel **b**), in the Ulleung Basin on 3 April 2006. The Ulleung Warm Eddy (UWE) is formed by the northward meandering of the East Korea Warm Current (EKWC), a branch of the Tsushima Warm Current (TWC). The paths of the TWC, EKWC, and UWE are shown with dotted-line in panel (**a**), and sites of upwelling are indicated by thick arrows in panel (**b**). High chl-*a* water in panel (**b**) corresponds to cold water in panel (**a**) along the southeast coast of Korea (Redrawn from Hyun et al. 2009)

southwestern part of the East Sea (i.e. UB). Further investigation demonstrated that wind-driven coastal upwelling along the southeast coast of the Korean peninsula was responsible for enhanced primary production and phytoplankton biomass in the UB, and that this chlorophyll a (chl-a) enhanced water moves into the center of the UB via many pathways, depending on its initial location, the direction of the East Korea Warm Current (EKWC), the location of the UWE, and the wind field (Yoo and Park 2009).

11.2.1 Microbiological Parameters Associated with Coastal Upwelling and UWE

It has been well established that bacterial abundance and production, as well as the role of bacteria in biogeochemical carbon cycles, are largely affected by water column structures that control the physical stability and nutrient conditions (Kiørboe et al. 1990; Cho et al. 1994; Hyun and Kim 2003; Hyun and Yang 2005). However, the impact of enhanced phytoplankton production and biomass associated with coastal upwelling on the heterotrophic bacterial processes has not been well understood in the UB. Hyun et al. (2009) measured the variations of bacterial parameters, together with other biological parameters such as primary production and carbon biomass of phytoplankton and heterotrophic protozoa, associated with the occurrence of upwelling and the path of the UWE (Figs. 11.1 and 11.2). First, water column structures differed greatly across sites. The coastal site (D1) and offshore sites (D4 and D5) were slightly stratified, whereas the center of the UWE (D3) was vertically well mixed. Phytoplankton carbon biomass, converted from chl-a, was higher at D1 and D4 than at the center of the UWE (D3) and outside the UWE (D5). Furthermore, the phytoplankton size distribution was similar at D1 and D4 where large phytoplankton (>5 μm) comprised respectively 66 and 67 % of total phytoplankton carbon biomass. This similarity in size distribution at the two widely separated sites (D1 and D4), located more than 100 km apart, indicated that the growth of large phytoplankton, mainly diatoms, was stimulated by high nutrient conditions in the upwelled coastal waters (D1), and then transported to the center of the basin (D4) after becoming entrained into the UWE.

In accordance with the phytoplankton biomass distribution, as revealed by the satellite images (Fig. 11.1), bacterial production was higher in the coastal upwelled water (D1) and in the offshore flow of the UWE (D4) than in the eddy center (D3) and outside the eddy (D5) (Fig. 11.2). These results indicated that heterotrophic bacterial production is closely coupled to organic substrates originated from phytoplankton (Cole et al. 1988). Unlike the spatial difference in bacterial metabolic activity, however, the difference in bacterial cell number was not as great among the sites. Since heterotrophic protozoan carbon biomass (HPCB) was higher in upwelled water (D1) and in the eddy stream (D4), a small spatial difference in bacterial biomass suggested that the heterotrophic protozoa effectively control the bacterial biomass in the eddy stream. Based on investigation

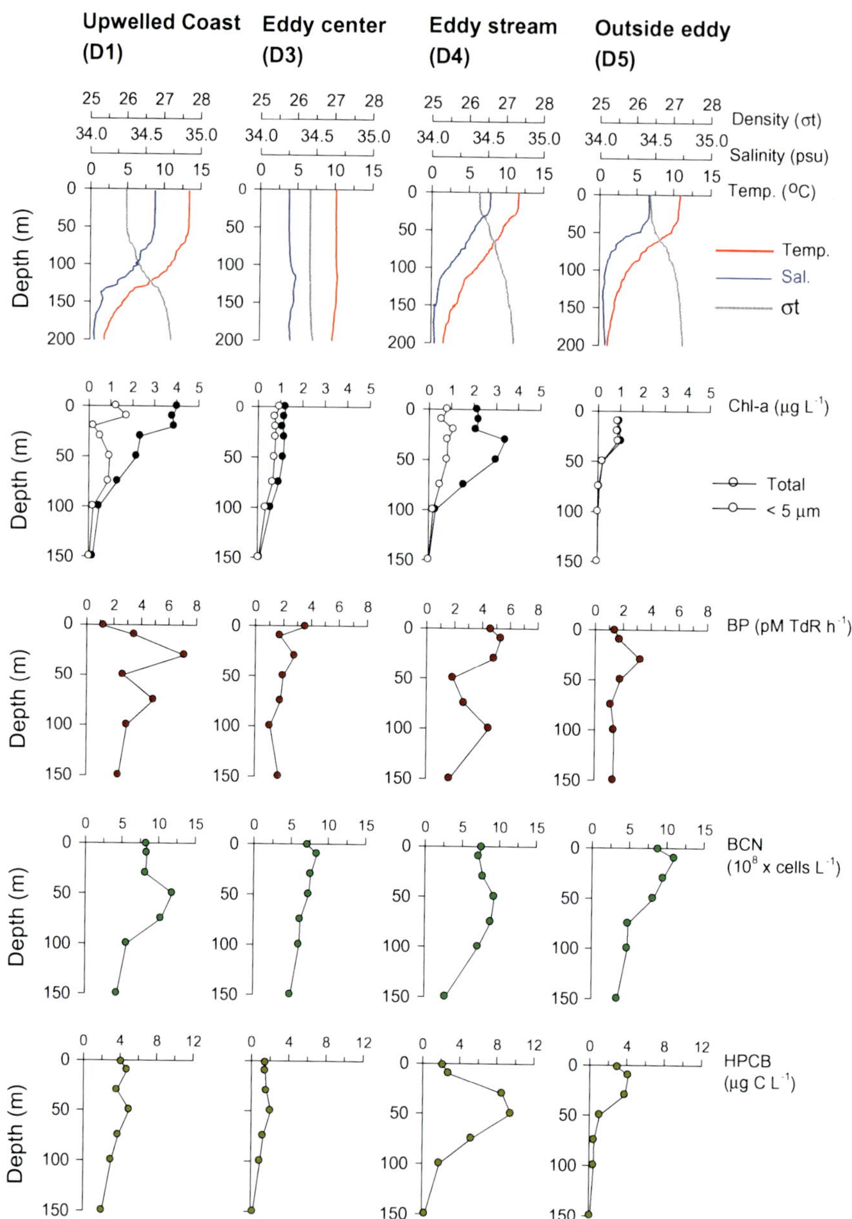

Fig. 11.2 Vertical profiles of physical and microbiological parameters from Stn. D1–D5. *Chl-a* Chlorophyll-a; *BP* bacterial production; *BCN* bacterial cell number; *HPCB* heterotrophic protozoa carbon biomass (Redrawn from Hyun et al. 2009)

of the epi-and mesopelagic zone of the UB, Cho et al. (2000) also reported that heterotrophic nanoflagellate grazing on bacteria depended significantly on bacterial abundance, but was not significantly correlated with bacterial production. The results obtained during the spring bloom in the UB (Hyun et al. 2009) are consistent with the general conclusion that biomass of heterotrophic bacteria in highly productive regions is predominantly controlled by the grazing of heterotrophic protozoa, whereas heterotrophic bacterial production is largely dependent on the availability of dissolved organic substrates that are released from phytoplankton and various sources (Ducklow 2000; Hyun and Yang 2005; Nagata 2000).

11.2.2 Role of Bacteria in Biogeochemical Carbon Cycles

Besides understanding the causes of enhanced primary production in the water column of the UB, another important question related to the microbial ecology and biochemistry of the region would be the fate of photosynthetically fixed organic carbon in the water column. This question is ultimately related to the quantitative role of heterotrophic bacteria (or the microbial loop) in biogeochemical carbon cycles within the microbial food web. If the photosynthetically fixed organic carbon is mostly respired to CO_2 within the microbial loop, then the C cycles via the microbial loop should be a carbon sink (Ducklow et al. 1986). In contrast, if the organic carbon taken up by bacteria is passed on to higher trophic levels, the bacteria act as a trophic link between dissolved organic carbon (DOC) and the metazoa via protozoan flagellates and ciliates (Azam et al. 1983; Sherr and Sherr 1988; also referred to Fig. 12.7). Elucidating the relative significance of link or sink in the fate of phytoplankton production is also crucial to understanding the role of bacteria in controlling the efficacy of the biological pump (i.e. biological control of the ocean as a carbon sink for atmospheric carbon).

The quantitative role of bacteria in utilizing organic carbon originated from primary production in the UB was evaluated during the spring bloom for two different water regimes according to primary production and phytoplankton biomass. The eddy stream (D4) exhibited relatively higher primary production and chl-*a* concentration compared to the less productive region outside the eddy (D5) (Table 11.1). Within the euphotic zone, bacterial production was slightly higher (ca. 20 %) at the productive eddy stream (9.3 mmol C m^{-2} d^{-1}) than at the less productive region outside the eddy (7.2 mmol m^{-2} d^{-1}), and thus bacterial production (BP) was responsible for 13–17 % of primary production (PP) at both sites (Table 11.1). However, prominent differences were observed for bacterial respiration (BR) and bacterial growth efficiency (BGE) at the two sites. Bacterial growth efficiency was higher at the highly productive eddy stream (D4, 21.6 %) than at the less productive region outside the eddy (D5, 8.3 %), which was largely determined by the difference in respiration at the eddy stream (33.8 mmol C m^{-2} d^{-1}) and outside the eddy (79.2 mmol C m^{-2} d^{-1}) (Table 11.1). Finally, bacterial carbon demand (BCD) accounted for 58 and 211 % of primary production at the eddy

Table 11.1 Depth integrated (down to 50 m, euphotic depth) bacterial production (BP), bacterial respiration (BR), bacterial carbon demand (BCD) and bacterial growth efficiency (BGE) and significance of bacterial parameters in processing organic carbon produced by primary production (PP) during the spring bloom in the Ulleung Basin

Site	St. ID	PP (mmol C m^{-2} d^{-1})	BP (mmol C m^{-2} d^{-1})	BR[a] (mmol O_2 m^{-2} d^{-1})	BCD[b, c] (mmol C m^{-2} d^{-1})	BGE[d] (%)	BP/PP	BCD/PP
Eddy stream	D4	73.9	9.3	38.0	43.1	21.6	0.13	0.58
Outside eddy	D5	40.8	7.2	89.0	86.4	8.3	0.18	2.11

[a]BR = 0.45 × total community respiration (Robinson 2008)

[b]BCD = BP + BR

[c]Carbon based BR was calculated from O_2 based BR using respiration quotient (RQ) of 0.89 (Robinson 2008)

[d]BGE = BP/(BCD) × 100

stream and outside the eddy, respectively, which demonstrated that a large fraction of the primary production is channeled via dissolved organic matter (DOM) and heterotrophic bacterial metabolic process.

A relatively high BCD:PP ratio exceeding 1 (i.e. heterotrophic system) as is found for the region outside the eddy (D5) is typical in less-productive oceans (Fouilland and Mostajir 2010), and implies that bacterial carbon demand is not solely dependent on the organic carbon produced by phytoplankton. In such a heterotrophic system where bacterial carbon demand exceeds organic carbon produced by phytoplankton, the subsequent question concerns how these high heterotrophic metabolic activities are sustained. Logically there should be other sources of DOM to support heterotrophic metabolism exceeding primary production. Diverse DOM sources, such as zooplankton sloppy feeding and excretion and viral lysis, have been proposed for the water column (Nagata 2000). The sources and biogeochemical cycles of DOC associated with bacterial processes are not well understood in the East Sea. Recently, however, it has been revealed that DOC in the surface water of the UB was higher (Kim and Kim 2012) than the same depth range of the Equatorial Pacific, Ross Sea polynya and Sargasso Sea (Carlson and Ducklow 1995; Hansell and Carlson 1998). In addition, Hwang and Cho (2002) demonstrated that bacterial mortality due to viral lysis in the East Sea was substantial even in oligotrophic conditions in summer, accounting for 9.8–19.2 % of bacterial production, from which a substantial amount of DOM could be released. DOM released by bacterial mortality due to viral lysis may further enable a role of bacteria as a C sink by enhancing bacterial production and respiration using organic matter that, otherwise, would be available to higher trophic levels (Fuhrman 1999). Considering that the areal extent of the eddy stream is relatively small, the UB, before the occurrence of the basin-scale spring bloom, is likely a heterotrophic system. In such a case, most DOC taken up by bacteria is respired to CO_2, and thus bacteria in most of the area of the UB may act as a C sink for photosynthetically fixed carbon.

In contrast, the relatively low BCD:PP ratio of 0.58 together with higher growth efficiency in the productive eddy stream (D4) indicates that the organic carbon produced by primary production is transferred to bacteria and then to higher trophic levels (i.e. metazoans) via protozoan grazing on bacteria before being respired by bacterial metabolism. Indeed, Yang et al. (2009) demonstrated that the copepods, *Neocalanus plumchrus* and *Calanus sinicus*, preferentially consumed ciliates and heterotrophic dinoflagellates over autotrophic protozoa, and estimated that ciliates contributed 47 % of copepod diet in highly productive conditions (phytoplankton C biomass >40 μg C). The results indicated a strong trophic link between copepods (i.e. metazoan food web) and microbial food webs including bacterial and heterotrophic protozoa in the UB (Sherr and Sherr 1988; also referred to Fig. 12.7), especially at relatively productive sites or during productive seasons. Finally, it should be emphasized that the variable role of bacteria presented here, as either coupled or uncoupled to primary production in the UB, is just one example derived from only two different conditions in spring, and thus more extensive investigation should be conducted to generalize the role of heterotrophic bacteria in biogeochemical carbon cycles at different times and locations in the UB.

Microbes are a relevant biological parameter for predicting the effect of climate change at the biological level (Rivkin and Legendre 2001; Hoppe et al. 2002, 2008). Hoppe et al. (2002) demonstrated that the ratio between bacterial production and primary production changed significantly from about 2 % in the cold and temperate climate zones to 40 % in the tropics, which indicates a large-scale temperature-dependent shift from net-autotrophic to increasingly heterotrophic conditions in the surface of the Atlantic Ocean. Based on a mesocosm experiment, Hoppe et al. (2008) repeatedly revealed that the relative significance of microbial fractions less than 3 μm (i.e. bacteria and picoplankton) in the total microbial community respiration increased from 59 to 77 % as water temperature increased from 2 to 8 °C. Their results demonstrate that the role of heterotrophic bacteria as a C sink for organic carbon increases in importance as sea water temperature increases. Increased CO_2 regeneration by bacteria (i.e. microbial respiration) will ultimately attenuate the efficacy of the biological pump, thereby reducing the ocean's capacity to mitigate increases in atmospheric CO_2. In this regard, monitoring changes in the relative role of bacteria in C cycles in response to environmental shifts resulting from large-scale climatic changes is particularly important in the East Sea where the increase of SST (1.09 °C) during the last 2 decades was one of the largest recorded for 18 large marine ecosystems in the world (Belkin 2009).

11.3 Benthic Biogeochemical Processes

Early diagenesis refers to the whole range of post-depositional processes that take place in aquatic sediments, coupled either directly or indirectly to the degradation of organic matter (Jørgensen and Kasten 2006) that is ultimately processed by microorganisms with specific biogeochemical functions (Jørgensen 2006). Since sediments are mainly heterotrophic systems, in which the photosynthetically produced organic matter is mineralized and recycled again to support the photoautotrophs, the quantity of particulate organic matter input is the most important factor controlling biogeochemical processes in sediments (Fenchel et al. 1998).

Organic particles that reach the deep-sea floor are quickly mineralized in the surface sediment (Cole et al. 1987). The mineralization of organic matter occurs mainly through microbial processes consuming an array of different electron acceptors such as O_2, NO_3^-, MnO_2, FeOOH and sulfate (Froelich et al. 1979; Jørgensen 2006). The relative significance of each carbon oxidation pathway is largely determined by the availability of organic carbon and electron acceptors. In general, due to the abundance of sulfate (ca. 28 mM) in marine environments, sulfate reduction accounts for up to 50 % of the total carbon oxidation in continental margins with high organic matter flux (Jørgensen 1982; Jørgensen and Kasten 2006) and thus plays a prominent role in benthic carbon mineralization. However, in Mn- and Fe-oxide-rich marine sediments, either manganese or iron reduction is regarded as a dominant C oxidation pathway as well (Aller 1990; Canfield et al. 1993a, b; Thamdrup and Canfield 2000; Jensen et al. 2003; Vandieken et al. 2006; Hyun et al. 2007, 2009b). Distribution and flux of geochemical constituents such as nitrogen, manganese, iron, sulfur and inorganic nutrients in the sediments are largely regulated by redox conditions of the sediment and overlying water column (Pakhomova et al. 2007; Slomp et al. 1997; Kristensen et al. 2002; Magen et al. 2011). Thus, benthic (microbial) biogeochemistry is characterized by studies: (1) to estimate the rate and pathways of organic matter mineralization including re-oxidation of inorganic elements generated by the mineralization processes of organic matter, (2) to identify the function and diversity of microbes related to the mineralization pathways, and (3) to elucidate physico-chemical and biological controls, such as organic matter input, redox conditions, bottom turbidity current, bioturbation and bioirrigation.

11.3.1 High Benthic Carbon Oxidation Rates in the Ulleung Basin

Recent geochemical investigations have shown that the sediments of the UB, despite water depths exceeding 2000 m, were characterized by high organic content (>2.5 %, dry wt.) (Lee et al. 2008a). This high organic content has rarely been found in deep-sea sediments, except for the Black Sea which receives large amounts of river discharge (Cociasu et al. 1996; Reschke et al. 2002), and the Chilean upwelling region which has high primary productivity (Shubert et al. 2000; Böning et al. 2005). At a glance, such high organic content in the UB was surprising because there are no major rivers flowing into the East Sea (Hong et al. 1997). Several observations explain the reason for high organic content in the sediment. First, phytoplankton communities in upwelled water (D1) and in the eddy stream (D4) flowing into the center of the basin (Fig. 11.2) consist of mostly large diatoms (Hyun et al. 2009a). Since the size and species composition of plankton communities largely determine the downward flux of particulate organic carbon (Karl 1999; Cotner and Biddanda 2002), large diatoms that are delivered into the center of the UB should be particularly significant in mediating vertical flux during bloom periods and in upwelling systems (Hyun et al. 2009a; Kwak et al. 2013). Actually, a C:N ratio of 6.98 measured in the center of the UB was close to the Redfield ratio (6.63), indicating that the organic matter deposited in the center of the UB is predominantly of marine origin (Lee et al. 2008a; Shubert et al. 2000). Second, the organic C accumulation rates in the UB also exhibited high values, exceeding $2 \text{ g C m}^{-2} \text{ year}^{-1}$ (Lee et al. 2008a), which has rarely been found for deep-sea sediments below 2000 m, except in regions of intense upwelling. Consequently, a C:N ratio of the sediment that is similar to that of the water column, together with the high organic carbon accumulation rates, indicated that the labile organic matter is rapidly delivered to the sediment via eddy

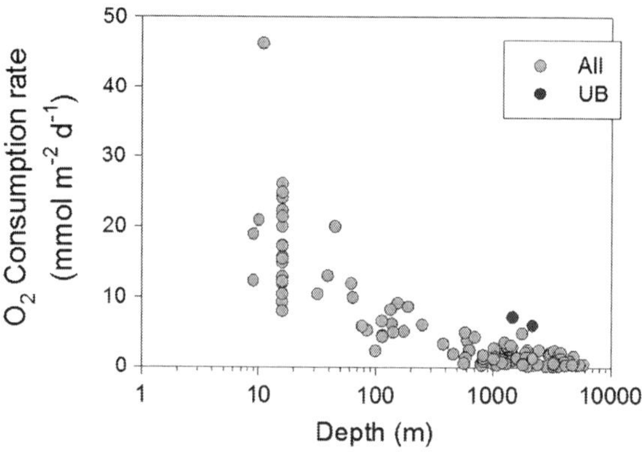

Fig. 11.3 Diffusive oxygen utilization (DOU) rates plotted as a function of the water depth (Redrawn from Glud 2008). *Dark color-filled circles* indicate the DOU measured at the slope and rise of the UB (Hyun unpublished data)

processes before being mineralized in the water column. In addition to the vertical sinking flux, the lateral transport of organic matter from the highly productive adjacent shelf should be a significant source for the high organic content along the southeastern slope of the UB (Jahnke et al. 1990; Lee et al. 2008a).

As a result of high organic carbon content in the sediment, benthic microbial metabolic activities measured by diffusive O_2 utilization (DOU) rates exhibited the highest values in the UB compared to the DOU measured at the same depth range in other continental margins (Fig. 11.3). The DOU plotted for the UB (6.0–7.1 mmol m^{-2} d^{-1}) was measured on shipboard by an oxygen microprofiler, which may overestimate, by approximately 50 %, the O_2 consumption rates compared to the DOU measured in situ at the same depth range (Glud 2008). Even taking the factor of 50 % overestimation into account, the DOU value at the UB (4.0–4.7 mmol m^{-2} d^{-1}) was among the highest compared to those reported from other continental margins. Hyun et al. (2010) also reported that carbon oxidation processed by sulfate reduction rates (SRR) showed clear seasonal and spatial variations (Fig. 11.4; Table 11.2). Higher SRRs were observed in spring when primary production was higher (Table 11.2). Spatially, the SRRs were always higher at the slope sites where upwelling occurs (UB1) and at the offshore slope (UB3) than at the deep basin sites (UB2) (Fig. 11.4). Another interesting feature in the spatial distribution was that the SRRs in the center of the UB (UB2) were several times higher than those measured in the Japan Basin (JB1) and Yamato Basin (YB1). The spatial distribution of the SRR is consistent with the high primary production (Yamada et al. 2005) and phytoplankton biomass (Yoo and Park 2009) in the water column of the UB compared to any other regions in the East Sea. The SRRs in the UB were higher than those of the Bengulea upwelling system in the Southeast Atlantic in Namibia (Ferdelman et al. 1999; Fossing et al. 2000) and the anoxic continental margin of the Black Sea (Weber et al. 2001). At a similar depth range between 1000 and 2500 m, the SRRs in the UB were even comparable to those

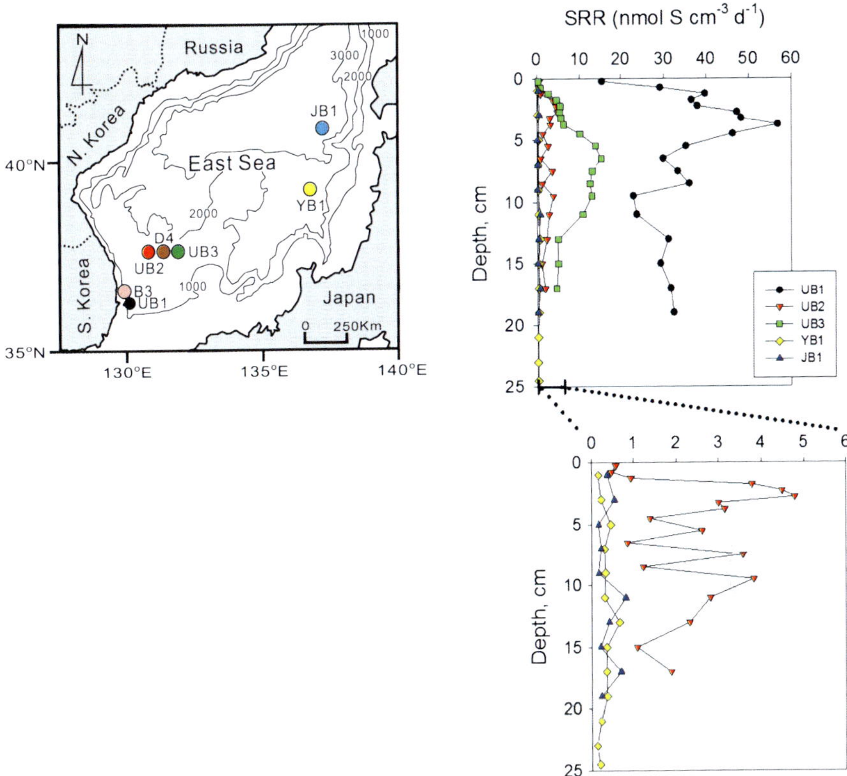

Fig. 11.4 Spatial variations of sulfate reduction rates (SRR) measured in the East Sea. Note that the highest SRR is prominent at the upwelling site (UB1) and at the offshore slope (UB3) (*upper panel*), and the SRR in the center of Ulleung Basin (UB2) was higher than that measured in the Japan Basin (JB) and the Yamato Basin (YB) (*lower panel*)

reported from highly productive Peruvian (5.1 mmol m^{-2} d^{-1}; Fossing 1990) and Chilean (2.7–4.8 mmol m^{-2} d^{-1}; Thamdrup and Canfield 1996) upwelling regions. Consequently, current findings such as the high organic carbon content in the sediment (Lee et al. 2008a), resulting from enhanced primary production associated with coastal upwelling and its subsequent delivery into the basin via the UWE (Hyun et al. 2009a), support exceptionally high carbon oxidation by sulfate reduction in the continental slope and rise of the UB. On the other hand, this high benthic carbon oxidation plays an important role in turnover of organic matter and nutrient regeneration in the UB. From the sulfate reduction rates, Hyun et al. (2010) estimated that total carbon mineralization in the UB may range from 4.2 to 8.4 mmol C m^{-2} d^{-1} (average 6.3 mmol C m^{-2} d^{-1}), which accounts for approximately 30 % of the primary production in overlying waters of the UB and about 60 % of the export carbon.

Table 11.2 Depth integrated (0–10 cm) sulfate reduction rates (SRR) measured at continental slope and basin sites of the Ulleung Basin in spring and summer (modified from Hyun et al. 2010)

Geological region	Station ID	Water depth (m)	Environmental description	Max. [Mn] (μmol cm^{-3})[a]	Max. [Fe] (μmol cm^{-3})[a]	Chl-a (mg m^{-2})[b]	SRR (mmol m^{-2} d^{-1})	Primary production (mmol C m^{-2} d^{-1})
Slope	B3	1051	Summer; upwelling	0.72	12	34	1.22	Spring 2006[c]
	D1	1042	Summer; upwelling	bdl	na	40	2.07	(42.8 ± 22.0)
			Spring; upwelling	bdl	na	156	5.85	
	M1	1453	Spring; upwelling	bdl	na	na	8.07	Summer 2007[d]
Basin	D4	2144	Summer; upwelling	174	53	35–37	0.69–0.89	(23.3 ± 4.3)
	D2	2155	Summer; upwelling	143	82	34–36	1.04–1.22	
			Spring; upwelling	na	na	116	2.52	Upwelling region in shelf[d]
	M2	2050	Spring; upwelling	na	na	na	3.18	(60–145)

[a]Concentration range within 0–4 cm depth of the sediment
[b]Depth integrated down to 150 m depth of the water column
[c]Average from Hyun et al. (2009a)
[d]Average from slope and basin sites in Kim et al. (2009)
bdl indicates below detection level; *na* indicates data not available

11.3.2 Major Carbon Oxidation Pathways

Together with high organic carbon content and the enhanced rate of microbial metabolic activities, another unique sediment property of the UB is that
the surface sediment of the continental rise (i.e. center of the UB) appeared
to be enriched with Mn-oxides (143–174 μmol cm^{-3}) and Fe-oxides (53–
83 μmol cm^{-3}) (Fig. 11.5). The high Mn-oxide concentrations in the UB were
comparable to those in the Panama Basin and Skagerrak (Aller 1990; Canfield
et al. 1993a, b). More surprisingly, at the basin site (D4), electron acceptors such
as O_2, nitrate, Mn- and Fe-oxides were systematically arrayed in a discrete zone
(Fig. 11.5). Thus, high inventories of Mn-oxides and Fe-oxides with discrete
zonation in the surface sediment implied that carbon oxidation pathways such
as Mn- and Fe-reduction are important at the basin site (D4). Actually, despite
the high organic carbon content, sulfate reduction was suppressed at this Mn-
and Fe-oxides-rich basin sites (D2 and D4) (Table 11.2). Recently, Vandieken
et al. (2012) successfully identified acetate-oxidizing bacteria that potentially
reduce manganese in the manganese-rich surface layer at the center of the UB,
which evidenced the microbial manganese reduction as dominating terminal electron accepting processes in the top 2.5 cm. These biogeochemical and

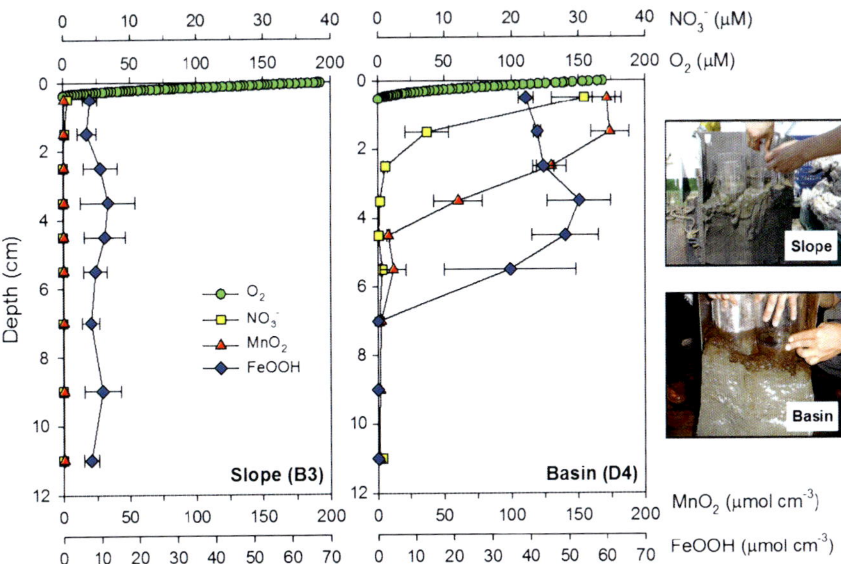

Fig. 11.5 Profiles of major electron acceptors in the continental slope (B3) and center of the
Ulleung Basin (D4). The sites are marked in Fig. 11.4. Note that the surface sediment at D4 is
enriched with manganese-oxides (*reddish color*) and iron-oxides, whereas the slope sediment
exhibited a typical mud color (*grey color*) (Pictures adopted from Hyun et al. 2010)

microbiological findings strongly imply that the carbon oxidation at the surface sediment of the UB is dominated by microbial Mn- and Fe-reduction, yet the actual rates and partitioning of Mn- and Fe-reduction in carbon oxidation have not been determined.

11.4 Composition of Prokaryotes

If we assume that marine microbial ecology is a study of interactions between microorganisms and physico-chemical and biological parameters (Kirchman 2000), information on microbial composition, in combination with biogeochemical and oceanographic parameters, will help us to elucidate unknown specific microbially mediated processes regulating biogeochemical carbon and nutrient cycles in marine environments (Jørgensen 2006). There have been many reports concerning the composition of prokaryotic communities in the East Sea, mostly in the UB. However, most molecular microbiological studies were focused on characterizing phylogenetic diversity and identifying unknown species. Any studies connecting microbial diversity with specific ecological and biogeochemical processes are extremely rare with a few exceptions (Vandieken et al. 2012; Lee et al. 2013). The composition of microorganisms that are identified based on both culture-dependent and culture-independent methods and their habitats are summarized in Tables 11.3, 11.4 and 11.5.

11.4.1 Composition of Culture-Dependent Prokaryotes

So far, most studies on the microbial diversity in the East Sea have been conducted using cultured bacterial strains. Among the 180 species isolated in the East Sea, heterotrophic bacteria from the water column comprised 40 % (Table 11.3), and those from sediment and microbial symbionts with micro- or macrobenthos accounted for 31 and 28 %, respectively. Approximately 73 % of the bacteria isolated from the water column were reported from the UB, and the remaining 27 % were reported in the coastal ocean of Russia. *Bacteriodetes* (19 species), *Alphaproteobactria* (17 species) and *Gammaproteobacteria* (14 species) appeared to be the most abundant bacterial groups isolated in the coastal waters of the UB and the Russian coastal waters of the JB. In contrast, more diverse bacterial groups including *Actinobacteria, Cyanobacteria* and *Firmicutes* were isolated in the off-shore waters (Table 11.3).

In the sediment, 17 and 19 heterotrophic bacterial species were isolated from the coastal and offshore sediment of Russia, respectively, and *Gammaprotebacteria* were the most abundant, comprising 50 % of a 36 total isolates (Table 11.4). Although most bacteria are known as aerobic heterotrophs,

Table 11.3 Inventories of culture-dependent and independent prokaryotes detected in the water column of the East Sea

Habitat	Domain	Phylum	Related to genus (or species)	Ecological and biogeochemical properties	Note	References
Seawater	*Bacteria*	*Actinobacteria*	*Nocardiaceae*	Aerobic heterotroph	UB[a], offshore, cultured	Kim and Khang (2012)
		Actinobacteria	*Nocardioides marinus*	Aerobic heterotroph	UB, offshore, cultured	Choi et al. (2007a)
		Actinobacteria	*Serinicoccus marinus*	Strictly aerobic heterotroph	UB, cultured	Yi et al. (2004)
		Bacteriodetes	*Aquimarina muelleri*	Strictly aerobic heterotroph	JB[b], coast, cultured	Nedashkovskaya et al. (2005g)
		Bacteriodetes	*Cellulophaga pacifica*	Aerobic heterotroph	JB, coast, cultured	Nedashkovskaya et al. (2004e)
		Bacteriodetes	*Cyclobacterium amurskyense*	Aerobic heterotroph	JB, coast, cultured	Nedashkovskaya et al. (2005e)
		Bacteroidetes	*Dokdonia donghaensis*	Aerobic heterotroph	UB, offshore, cultured	Yoon et al. (2005a)
		Bacteroidetes	*Donghaeana dokdonensis*	Aerobic heterotroph	UB, offshore, cultured	Yoon et al. (2006a)
		Bacteroidetes	*Flavobacterium jumunjinense*	Aerobic heterotroph	UB, coast, cultured	Joung et al. (2013)
		Bacteriodetes	*Formosa agariphila*	Aerobic heterotroph, budding	JB, coast, cultured	Nedashkovskaya et al. (2006c)
		Bacteriodetes	*Gillisia mitskevichiae*	Aerobic heterotroph	JB, coast, cultured	Nedashkovskaya et al. (2005d)
		Bacteroidetes	*Hongiella marincola*	Aerobic heterotroph	UB, coast, cultured	Yoon et al. (2004c)
		Bacteroidetes	*Joostella marina*	Aerobic heterotroph	UB, coast, cultured	Quan et al. (2008)

(continued)

Table 11.3 (continued)

Habitat	Domain	Phylum	Related to genus (or species)	Ecological and biogeochemical properties	Note	References
		Bacteriodetes	*Maribacter aquivivus*	Aerobic heterotroph	JB, coast, cultured	Nedashkovskaya et al. (2004b)
		Bacteroidetes	*Maribacter dokdonensis*	Aerobic heterotroph	UB, offshore, cultured	Yoon et al. (2005b)
		Bacteriodetes	*Maribacter orientalis*	Aerobic heterotroph	JB, coast, cultured	Nedashkovskaya et al. (2004b)
		Bacteriodetes	*Maribacter stanieri*	Aerobic heterotroph	JB, coast, cultured	Nedashkovskaya et al. (2010b)
		Bacteriodetes	*Mesonia mobilis*	Aerobic heterotroph	JB, coast, cultured	Nedashkovskaya et al. (2006d)
		Bacteriodetes	*Olleya aquimaris*	Aerobic heterotroph	UB, coast, cultured	Lee et al. (2010b)
		Bacteriodetes	*Polaribacter dokdonensis*	Aerobic heterotroph	UB, offshore, cultured	Yoon et al. (2006c)
		Bacteriodetes	*Reichenbachia agariperforans*	Aerobic heterotroph, agarolytic, gliding	JB, coast, cultured	Nedashkovskaya et al. (2003b)
		Bacteroidetes	*Sufflavibacter maritimus*	Facultatively anaerobic heterotroph	UB, coast, cultured	Kwon et al. (2007a)
		Bacteriodetes	*Tenacibaulum*	Aerobic heterotroph, nitrate reduction	UB, offshore, uncultured	Kim and Khang (2012)
		Bacteroidetes	*Winogradskyella pacifica*	Aerobic heterotroph	UB, coast, cultured	Kim and Nedashkovskaya (2010)
		Bacteriodetes	*Zobellia amurskyensis*	Aerobic heterotroph, gliding	UB, coast, cultured	Nedashkovskaya et al. (2004f)

(continued)

Table 11.3 (continued)

Habitat	Domain	Phylum	Related to genus (or species)	Ecological and biogeochemical properties	Note	References
		Bacteriodetes	*Zobellia laminariae*	Aerobic heterotroph, gliding	UB, coast, cultured	Nedashkovskaya et al. (2004f)
		Bacteriodetes	*Zobellia russellii*	Aerobic heterotroph, gliding	UB, coast, cultured	Nedashkovskaya et al. (2004f)
		Cyanobacteria	*Synechococcus*	Phototrophic oligotrophic autotroph	UB, offshore, cultured	Choi and Noh (2009)
		Firmicutes	*Bacillus hwajinpoensis*	Aerobic heterotroph, nitrate reduction	UB, coast, cultured	Yoon et al. (2004a)
		Firmicutes	*Pelagibacillus goriensis*	Obligately aerobic heterotroph	UB, coast, cultured	Kim et al. (2007c)
		Firmicutes	*Virgibacillus dokdonensis*	Aerobic heterotroph	UB, offshore, cultured	Yoon et al. (2005b)
		Alphaproteobacteria	*Albimonas donghaensis*	Aerobic heterotroph	UB, cultured	Lim et al. (2008)
		Alphaproteobacteria	*Celeribacter baekdonensis*	Anaerobic heterotroph	UB, coast, cultured	Lee et al. (2012)
		Alphaproteobacteria	*Cucumibacter marinus*	Strictly aerobic heterotroph	UB, coast, cultured	Hwang and Cho (2008a)
		Alphaproteobacteria	*Donghicola eburneus*	Aerobic heterotroph	UB, coast, cultured	Yoon et al. (2007b)
		Alphaproteobacteria	*Erythrobacter flavus*	Aerobic heterotroph	UB, coast, cultured	Yoon et al. (2003a)
		Alphaproteobacteria	*Erythrobacter gangjinensis*	Aerobic heterotroph	UB, coast, cultured	Lee et al. (2010a)
		Alphaproteobacteria	*Henriciella marina*	Strictly aerobic heterotroph	UB, coast, cultured	Quan et al. (2009)

(continued)

Table 11.3 (continued)

Habitat	Domain	Phylum	Related to genus (or species)	Ecological and biogeochemical properties	Note	References
		Alphaproteobacteria	Jannaschia donghaensis	Aerobic heterotroph	UB, offshore, cultured	Yoon et al. (2007e)
		Alphaproteobacteria	Kordiimonas aquimaris	Aerobic heterotroph	UB, coast, cultured	Yang et al. (2013)
		Alphaproteobacteria	Lentilitoribacter donghaensis	Aerobic heterotroph	UB, coast, cultured	Park et al. (2013)
		Alphaproteobacteria	Loktanella agnita	Chemoorganotroph	JB, coast, cultured	Ivanova et al. (2005b)
		Alphaproteobacteria	Loktanella rosea	Chemoorganotroph	JB, coast, cultured	Ivanova et al. (2005b)
		Alphaproteobacteria	Marivita cryptomonadis	Strictly aerobic heterotroph	UB, cultured	Hwang et al. (2009a)
		Alphaproteobacteria	Marivita litorea	Strictly aerobic heterotroph	UB, cultured	Hwang et al. (2009a)
		Alphaproteobacteria	Pelagibius litoralis	Strictly aerobic heterotroph, nitrate reduction	UB, coast, cultured	Choi et al. (2009)
		Alphaproteobacteria	Porphyrobacter dokdonensis	Aerobic heterotroph	UB, offshore, cultured	Yoon et al. (2006b)
		Alphaproteobacteria	Porphyrobacter donghaensis	Aerobic heterotroph	UB, offshore, cultured	Yoon et al. (2004b)
		Alphaproteobacteria	Ponticoccus litoralis	Heterotroph, nitrate reduction	UB, coast, cultured	Hwang and Cho (2008b)
		Alphaproteobacteria	Pseudoruegeria aquimaris	Heterotroph	UB, coast, cultured	Yoon et al. (2007f)

(continued)

Table 11.3 (continued)

Habitat	Domain	Phylum	Related to genus (or species)	Ecological and biogeochemical properties	Note	References
		Alphaproteobacteria	*Sulfitobacter donghicola*	Aerobic heterotroph, sulfite oxidation	UB, offshore, cultured	Yoon et al. (2007a)
		Alphaproteobacteria	*Sulfitobacter litoralis*	Aerobic heterotroph, sulfite oxidation	UB, offshore, cultured	Park et al. (2007)
		Alphaproteobacteria	*Sulfitobacter marinus*	Aerobic heterotroph	UB, coast, cultured	Yoon et al. (2007c)
		Alphaproteobacteria	*Roseovarius*	Aerobic heterotroph	UB, offshore, uncultured	Kim and Khang (2012)
		Alphaproteobacteria	*Roseovarius halotolerans*	Aerobic heterotroph	UB, offshore, cultured	Oh et al. (2009)
		Alphaproteobacteria	*Pseudochrobactrum glaciei*	Heterotroph	JB, coast, cultured, sea ice	Romanenko et al. (2008)
		Alphaproteobacteria	*Rhizobium*	Nitrogen fixing	UB, offshore, uncultured	Kim and Khang (2012)
		Alphaproteobacteria	*Sphingopyxis*	Aerobic heterotroph	UB, offshore, uncultured	Kim and Khang (2012)
		Alphaproteobacteria	*Thalassospira*	Facultatively anaerobic heterotroph, nitrate reduction	UB, offshore, uncultured	Kim and Khang (2012)
		Alphaproteobacteria	Candidate division PS1 Clade	Aerobic heterotroph, oligotroph	UB, coast, cultured	Yang et al. (2012)
		Alphaproteobacteria	Candidate division SAR11	Aerobic heterotroph	UB, coast, cultured	Song et al. (2009)
		Gammaproteobacteria	*Alteromonas addita*	Aerobic heterotroph	JB, coast, cultured	Ivanova et al. (2005a)

(continued)

Table 11.3 (continued)

Habitat	Domain	Phylum	Related to genus (or species)	Ecological and biogeochemical properties	Note	References
		Gammaproteobacteria	*Alteromonas marina*	Aerobic heterotroph	UB, coast, cultured	Yoon et al. (2003b)
		Gammaproteobacteria	*Cycloclasticus*	Aerobic heterotroph, PAH degrading	UB, offshore, uncultured	Kim and Khang (2012)
		Gammaproteobacteria	*Halomonas*	Heterotroph	UB, offshore, uncultured	Kim and Khang (2012)
		Gammaproteobacteria	*Litoricola lipolytica*	Facultatively aerobic heterotroph	UB, coast, cultured	Kim et al. (2007c)
		Gammaproteobacteria	*Lysobacter*	Aerobic heterotroph	UB, offshore, uncultured	Kim and Khang (2012)
		Gammaproteobacteria	*Marinobacter*	Heterotroph	UB, offshore, uncultured	Kim and Khang (2012)
		Gammaproteobacteria	*Marinobacter goseongensis*	Aerobic heterotroph	UB, coast, cultured	Roh et al. (2008)
		Gammaproteobacteria	*Marinobacter litoralis*	Aerobic heterotroph, nitrate reduction	UB, coast, cultured	Yoon et al. (2003c)
		Gammaproteobacteria	*Marinicella litoralis*	Aerobic heterotroph, gliding	JB, coast, cultured	Romanenko et al. (2010d)
		Gammaproteobacteria	*Marinomonas primoryensis*	Aerobic heterotroph	JB, coast, cultured, sea ice	Romanenko et al. (2003b)
		Gammaproteobacteria	*Marinomonas dokdonensis*	Aerobic heterotroph	UB, offshore, cultured	Yoon et al. (2005d)
		Gammaproteobacteria	*Paraperlucidibaca baekdonensis*	Aerobic heterotroph	UB, coast, cultured	Oh et al. (2011a)

(continued)

Table 11.3 (continued)

Habitat	Domain	Phylum	Related to genus (or species)	Ecological and biogeochemical properties	Note	References
		Gammaproteobacteria	*Pseudoalteromonas aliena*	Aerobic heterotroph	UB, coast, cultured	Ivanova et al. (2004b)
		Gammaproteobacteria	*Pseudoalteromonas donghaensis*	Aerobic heterotroph	UB, offshore, cultured	Oh et al. (2011b)
		Gammaproteobacteria	*Psychrobacter*	Aerobic heterotroph	UB, offshore, uncultured	Kim and Khang (2012)
		Gammaproteobacteria	*Rheinheimera aquimaris*	Aerobic heterotroph	UB, coast, cultured	Yoon et al. (2007g)
		Gammaproteobacteria	*Salinibacterium amurskyense*	Aerobic heterotroph	JB, coast, cultured	Han et al. (2003)
		Gammaproteobacteria	*Salinisphaera dokdonensis*	Strictly aerobic heterotroph	UB, offshore, cultured	Bae et al. (2010)
		Gammaproteobacteria	*Shewanella dokdonensis*	Facultatively anaerobic heterotroph	UB, offshore, cultured	Sung et al. (2012)
		Gammaproteobacteria	*Shewanella fidelis*	Facultatively anaerobic heterotroph	JB, coast, cultured	Ivanova et al. (2003b)
		Gammaproteobacteria	*Shewanella japonica*	Facultatively anaerobic heterotroph	JB, coast, cultured	Ivanova et al. (2001)
		Gammaproteobacteria	*Shewanella pacifica*	Facultatively anaerobic heterotroph	JB, coast, cultured	Ivanova et al. (2004a)
		Gammaproteobacteria	*Vibrio*	Aerobic heterotroph	UB, offshore, uncultured	Kim and Khang (2012)

Table 11.4 Inventories of culture-dependent and independent prokaryotes detected in the sediments of the East Sea

Habitat	Domain	Phylum	Related to genus (or species)	Ecological and biogeochemical properties	Note	Reference
Sediment	*Bacteria*	*Actinobacteria*	*Kocuria marina*	Aerobic heterotroph, nitrate reduction	JB, coast, cultured	Kim et al. (2004)
		Actinobacteria	*Nocardiopsis*	Heterotroph	JB, offshore, cultured	Romanenko et al. (2013a)
		Actinobacteria	*Paraoerskovia marina*	Facultatively anaerobic heterotroph	YB[c], coast, cultured	Khan et al. (2009)
		Bacteriodetes	*Arenibacter troitsensis*	Aerobic heterotroph	JB, coast, cultured	Nedashkovskaya et al. (2003d)
		Bacteroidetes	*Fulvibacter tottoriensis*	Aerobic heterotroph	YB, coast, cultured	Khan et al. (2008)
		Bacteriodetes	*Maribacter sedimenticola*	Aerobic heterotroph	JB, coast, cultured	Nedashkovskaya et al. (2004b)
		Bacteroidetes	*Salegentibacter flavus*	Microaerophilic chemo-organotroph	JB, coast, cultured	Ivanova et al. (2006)
		Bacteriodetes	*Salinimicrobium marinum*	Facultatively anaerobic heterotroph	JB, offshore, cultured	Nedashkovskaya et al. (2010d)
		Bacteroidetes	*Sediminicola luteus*	Aerobic heterotroph, nitrate reduction	YB, coast, cultured	Khan et al. (2006)
		Bacteriodetes	*Winogradskyella arenosi*	Aerobic heterotroph	JB, coast, cultured	Romanenko et al. (2009d)
		Bacteroidetes	*Winogradskyella pulchriflava*	Strictly aerobic heterotroph	UB, coast, cultured	Kim et al. (2013)
		Chloroflexi	*Dehalococcoidetes*	Dehalogenating polychlorinated aliphatic alkanes	UB, offshore, uncultured, gashydrate	Lee et al. (2013)

(continued)

Table 11.4 (continued)

Habitat	Domain	Phylum	Related to genus (or species)	Ecological and biogeochemical properties	Note	Reference
		Firmicutes	*Paenibacillus donghaensis*	N fixing, Xylan-degradation	UB, offshore, cultured	Choi et al. (2008)
		Firmicutes	*Planococcus donghaensis*	Starch-degradation	UB, offshore, cultured	Choi et al. (2007b)
		Firmicutes	*Bacillus*	Aerobic heterotroph	JB, offshore, cultured	Romanenko et al. (2013a)
		Firmicutes	*Paenibacillus*	Facultatively anaerobic heterotroph	JB, offshore, cultured	Romanenko et al. (2013a)
		Firmicutes	*Paenisporosarcina*	Aerobic heterotroph	JB, offshore, cultured	Romanenko et al. (2013a)
		Alphaproteobacteria	*Bartonella chomelii*	Ammonia oxidation	UB, offshore, uncultured	Park et al. (2010)
		Alphaproteobacteria	*Devosia*	Dinitrogen-fixing root-nodule symbiosis	JB, offshore, cultured	Romanenko et al. (2013a)
		Alphaproteobacteria	*Devosia submarina*	Aerobic heterotroph	JB, offshore, cultured	Yoon et al. (2007d)
		Alphaproteobacteria	*Hoeflea halophila*	Aerobic heterotroph	UB, cultured	Jung et al. (2013)
		Alphaproteobacteria	*Hyphomicrobium*	Aerobic heterotroph	UB, offshore, uncultured, gashydrate	Lee et al. (2013)
		Alphaproteobacteria	*Jannaschia pohangensis*	Strictly aerobic heterotroph	UB, coast, cultured	Kim et al. (2008)
		Alphaproteobacteria	*Kordiimonas gwangyangensis*	Ammonia oxidation	UB, offshore, uncultured	Park et al. (2010)

(continued)

Table 11.4 (continued)

Habitat	Domain	Phylum	Related to genus (or species)	Ecological and biogeochemical properties	Note	Reference
		Alphaproteobacteria	*Litoreibacter janthinus*	Aerobic heterotroph	JB, coast, cultured	Romanenko et al. (2011a)
		Alphaproteobacteria	*Maricaulis maris*	Ammonia oxidation	UB, offshore, uncultured	Park et al. (2010)
		Alphaproteobacteria	*Pacificibacter maritimus*	Strictly aerobic heterotroph	JB, coast, cultured	Romanenko et al. (2011b)
		Alphaproteobacteria	*Pontibaca methylaminivorans*	Facultatively anaerobe, nitrate reduction, trimethylamine-degradation	UB, coast, cultured	Kim et al. (2010c)
		Alphaproteobacteria	*Primorskyibacter sedentarius*	Aerobic heterotroph	JB, coast, cultured	Romanenko et al. (2011c)
		Alphaproteobacteria	*Rhodospirillaceae*	Acetate oxidizing bacteria	UB, offshore, uncultured	Vandieken et al. (2012)
		Alphaproteobacteria	*Roseobacter*	Ammonia oxidation	UB, offshore, uncultured	Park et al. (2010)
		Alphaproteobacteria	*Roseovarius mucosus*	Ammonia oxidation	UB, offshore, uncultured	Park et al. (2010)
		Alphaproteobacteria	*Silicibacter pomeroyi*	Ammonia oxidation	UB, offshore, uncultured	Park et al. (2010)
		Alphaproteobacteria	*Streptomyces*	Produce antibiotics	JB, offshore, cultured	Romanenko et al. (2013a)
		Alphaproteobacteria	*Sulfitobacter*	Aerobic heterotroph, sulfite oxidation	JB, offshore, cultured	Romanenko et al. (2013a)
		Alphaproteobacteria	*Sulfitobacter ponitacus*	Ammonia oxidation	UB, offshore, uncultured	Park et al. (2010)
		Deltaproteobacteria	*Desulfobulbus japonicus*	Sulfate reduction, propionate-oxidation	YB, coast, cultured	Suzuki et al. (2007a)

(continued)

Table 11.4 (continued)

Habitat	Domain	Phylum	Related to genus (or species)	Ecological and biogeochemical properties	Note	Reference
		Deltaproteobacteria	*Desulfoluna butyratoxydans*	Sulfate reduction, butyrate-oxidation	YB, coast, cultured	Suzuki et al. (2008)
		Deltaproteobacteria	*Desulfopila aestuarii*	Sulfate reduction	YB, coast, cultured	Suzuki et al. (2007b)
		Deltaproteobacteria	*Desulfovibrio profundus*	Sulfate reduction	YB, offshore, cultured	Bale et al. (1997)
		Deltaproteobacteria	*Desulfuromonadales*	Acetate dependent Mn reduction	UB, offshore, uncultured	Vandieken et al. (2012)
		Deltaproteobacteria	*Geobacteraceae*	Acetate dependent Mn reduction	UB, offshore, uncultured	Vandieken et al. (2012)
		Deltaproteobacteria	*Nitrospina gracilis*	Ammonia oxidation	UB, offshore, uncultured	Park et al. (2010)
		Gammaproteobacteria	*Aestuariibacter litoralis*	Aerobic heterotroph	JB, coast, cultured	Tanaka et al. (2010)
		Gammaproteobacteria	*Alcanivorax venustensis*	Ammonia oxidation	UB, offshore, uncultured	Park et al. (2010)
		Gammaproteobacteria	*Alishewanella*	Facultatively anaerobe, utilizing various electron acceptors (trimethylamine oxide, nitrate, nitrite and thiosulphate)	UB, offshore, uncultured, gashydrate	Lee et al. (2013)
		Gammaproteobacteria	*Alteromonas*	Aerobic heterotroph	JB, offshore, cultured	Romanenko et al. (2013a)
		Gammaproteobacteria	*Arenicella xantha*	Aerobic heterotroph	JB, coast, cultured	Romanenko et al. (2010b)
		Gammaproteobacteria	*Arenimonas donghaensis*	Aerobic heterotroph	UB, coast, cultured	Kwon et al. (2007b)

(continued)

Table 11.4 (continued)

Habitat	Domain	Phylum	Related to genus (or species)	Ecological and biogeochemical properties	Note	Reference
		Gammaproteobacteria	*Cobetia litoralis*	Aerobic heterotroph	UB, coast, cultured	Romanenko et al. (2013b)
		Gammaproteobacteria	*Cobetia pacifica*	Aerobic heterotroph	UB, coast, cultured	Romanenko et al. (2013b)
		Gammaproteobacteria	*Colwellia*	Acetate dependent Mn reduction	UB, offshore, uncultured	Vandieken et al. (2012)
		Gammaproteobacteria	*Enhydrobacter*	Facultatively anaerobic heterotroph	UB, offshore, uncultured, gashydrate	Lee et al. (2013)
		Gammaproteobacteria	*Graciecola*	Aerobic heterotroph	JB, offshore, cultured, bacterial mat site	Arakawa et al. (2006)
		Gammaproteobacteria	*Glaciecola agarilytica*	Aerobic heterotroph, agar-digesting marine bacterium,	UB, cultured	Yong et al. (2007)
		Gammaproteobacteria	*Halomonas*	Aerobic heterotroph	JB, offshore, cultured	Romanenko et al. (2013a)
		Gammaproteobacteria	*Kangiella japonica*	Aerobic heterotroph	JB, coast, cultured	Romanenko et al. (2010c)
		Gammaproteobacteria	*Luteimonas vadosa*	Aerobic heterotroph	JB, coast, cultured	Romanenko et al. (2012)
		Gammaproteobacteria	*Marinobacter*	Aerobic heterotroph	JB, offshore, cultured	Romanenko et al. (2013a)
		Gammaproteobacteria	*Marinobacter*	Ammonia oxidation	UB, offshore, uncultured	Park et al. (2010)
		Gammaproteobacteria	*Marinobacter excellens*	Aerobic heterotroph, fermentation	JB, coast, cultured	Gorshkova et al. (2003)

(continued)

Table 11.4 (continued)

Habitat	Domain	Phylum	Related to genus (or species)	Ecological and biogeochemical properties	Note	Reference
		Gammaproteobacteria	*Marinobacter koreensis*	Aerobic, nitrate reduction, glucose fermentation	UB, coast, cultured	Kim et al. (2006)
		Gammaproteobacteria	*Marinobacter sediminum*	Aerobic heterotroph	JB, coast, cultured	Romanenko et al. (2005)
		Gammaproteobacteria	*Marinobacterium georgiense*	Ammonia oxidation	UB, offshore, uncultured	Park et al. (2010)
		Gammaproteobacteria	*Marinomonas arenicola*	Aerobic heterotroph	JB, coast, cultured	Romanenko et al. (2009a)
		Gammaproteobacteria	*Methylophaga*	Aerobic methane oxidation	UB, offshore, uncultured, gashydrate	Lee et al. (2013)
		Gammaproteobacteria	*Moritella*	Facultatively anaerobic heterotroph	JB, offshore, cultured, bacterial mat site	Arakawa et al. (2006)
		Gammaproteobacteria	*Oceanisphaera donghaensis*	Manganese-oxidation, nitrate reduction, heterotroph	UB, offshore, cultured	Park et al. (2006)
		Gammaproteobacteria	*Oceanisphaera litoralis*	Aerobic heterotroph	JB, coast, cultured	Romanenko et al. (2003a)
		Gammaproteobacteria	*Oceanobacillus profundus*	Facultatively alkaliphilic bacterium	UB, offshore, cultured	Kim et al. (2007b)
		Gammaproteobacteria	*Oceanospirillaceae*	Acetate dependent Mn reduction	UB, offshore, uncultured	Vandieken et al. (2012)
		Gammaproteobacteria	*Pseudoalteromonas*	Aerobic heterotroph	JB, offshore, cultured	Romanenko et al. (2013a)
		Gammaproteobacteria	*Psychromonas*	Aerobic heterotroph	JB, offshore, cultured, bacterial mat site	Arakawa et al. (2006)

(continued)

Table 11.4 (continued)

Habitat	Domain	Phylum	Related to genus (or species)	Ecological and biogeochemical properties	Note	Reference
		Gammaproteobacteria	*Reinekea marinisedimentorum*	Aerobic heterotroph	JB, coast, cultured	Romanenko et al. (2004)
		Gammaproteobacteria	*Salinicola*	Aerobic heterotroph	JB, offshore, cultured	Romanenko et al. (2013a)
		Gammaproteobacteria	*Shewanella*	Heterotroph	JB, offshore, cultured, bacterial mat site	Arakawa et al. (2006)
		Gammaproteobacteria	*Shewanella donghaensis*	Facultatively anaerobic heterotroph	JB, offshore, cultured	Yang et al. (2007)
		Gammaproteobacteria	*Thalassolituus oleivorans*	Ammonia oxidizer	UB, offshore, uncultured	Park et al. (2010)
		Epsilonproteobacteria	*Sulfurovum lithotrophicum*	Ammonia oxidizer	UB, offshore, uncultured	Park et al. (2010)
		Epsilonproteobacteria	*Arcobacter*	Acetate dependent Mn reduction	UB, offshore, uncultured	Vandieken et al. (2012)
		Candidate division JS1	Environmental clone	Anaerobic heterotroph	UB, offshore, uncultured, gashydrate	Lee et al. (2013)
		Planctomycetes	Environmental clone	Anaerobic heterotroph	UB, offshore, uncultured, gashydrate	Lee et al. (2013)
	Archaea	*Crenarchaeota*	MBGB[d]	Anaerobic heterotroph	UB, offshore, uncultured	Park et al. (2008), Lee et al. (2013)
			MBGC[e]	Anaerobic heterotroph	UB, offshore, uncultured	Park et al. (2008), Kim et al. (2010a)

(continued)

Table 11.4 (continued)

Habitat	Domain	Phylum	Related to genus (or species)	Ecological and biogeochemical properties	Note	Reference
			MCG[f]	Anaerobic heterotroph (putative)	UB, offshore, uncultured	Park et al. (2008), Kim et al. (2010a) Lee et al. (2013)
		Euryarchaeota	MBGD[g]	Anaerobic heterotroph	UB, offshore, uncultured	Park et al. (2008), Kim et al. (2010a) Lee et al. (2013)
			ANME[h] − 1	Anaerobic methane oxidation	UB, offshore, uncultured, gashydrate	Lee et al. (2013)
		Thaumarchaeaota	Candidate *Nitrososphaera*	Ammonia oxidation	UB, offshore, uncultured	Lee et al. (2013)
			Nitrosopumilus maritimus	Ammonia oxidation	UB, offshore, uncultured	Park et al. (2008), (2010), Kim et al. (2010a)
		Candidate division	SAGMEG[i]	Anaerobic methanogenic heterotroph	UB, offshore, uncultured, gashydrate	Lee et al. (2013)
			TMEG[j]	(Unknown)	UB, offshore, uncultured	Park et al. (2008), Kim et al. (2010a)
			MHVG[k]	(Unknown)	UB, offshore, uncultured, Subsurface sediment	Kim et al. (2010a)

[a]*UB* Ulleung Basin; [b]*JB* Japan Basin; [c]*YB* Yamato Basin; [d]*MBGB* Marine Benthic Group B; [e]*MBGC* Marine Benthing Group C; [f]*MCG* Miscellaneous Crenarchaeotic Group; [g]*MBGD* Marine Benthing Group D; [h]*ANME* ANaerobic MEthanotroph; [i]*SAGMEG* South African Gold Mine Euryarchaeotal Group; [j]*TMEG*, Terrestrial Miscellaneous Euryarchaeotal Group; [k]*MHVG* Marine Hydrothermal Vent Group

Table 11.5 Inventories of cultivated bacteria associated with the micro- and macrobenthos in the East Sea

Habitat	Domain	Phylum	Related to genus (or species)	Ecological and biogeochemical properties	Note	Reference
Symbiont	*Bacteria*	*Bacteriodetes*	*Algibacter mikhailovii*	Aerobic heterotroph	JB, sea urchin, cultured	Nedashkovskaya et al. (2007b)
		Bacteriodetes	*Arenibacter echinorum*	Aerobic heterotroph	JB, sea urchin, cultured	Nedashkovskaya et al. (2007a)
		Bacteriodetes	*Bizionia echini*	Aerobic heterotroph	JB, sea urchin, cultured	Nedashkovskaya et al. (2010c)
		Bacteriodetes	*Echinicola pacifica*	Heterotroph, gliding agarolytic	JB, sea urchin, cultured	Nedashkovskaya et al. (2006b)
		Bacteriodetes	*Gramella echinicola*	Strictly aerobic heterotroph	JB, sea urchin, cultured	Nedashkovskaya et al. (2005f)
		Bacteriodetes	*Gramella marina*	Aerobic heterotroph	JB, sea urchin, cultured	Nedashkovskaya et al. (2010a)
		Bacteriodetes	*Maribacter ulvicola*	Aerobic heterotrophic	JB, green alga, cultured	Nedashkovskaya et al. (2004b)
		Bacteriodetes	*Mariniflexile gromovii*	Aerobic heterotroph	JB, sea urchin, cultured	Nedashkovskaya et al. (2006a)
		Bacteriodetes	*Roseivirga echinicomitans*	Strictly aerobic heterotroph	JB, sea urchin, cultured	Nedashkovskaya et al. (2005i)
		Bacteriodetes	*Salegentibacter mishustinae*	Strictly aerobic heterotroph	JB, sea urchin, cultured	Nedashkovskaya et al. (2005h)
		Bacteriodetes	*Winogradskyella echinorum*	Heterotroph	JB, sea urchin, cultured	Nedashkovskaya et al. (2009)
		Bacteriodetes	*Algibacter lectus*	Facultatively anaerobic heterotroph	JB, green alga, cultured	Nedashkovskaya et al. (2004d)

(continued)

Table 11.5 (continued)

Habitat	Domain	Phylum	Related to genus (or species)	Ecological and biogeochemical properties	Note	Reference
		Bacteriodetes	*Arenibacter certesii*	Aerobic heterotroph	JB, green alga, cultured	Nedashkovskaya et al. (2004a)
		Bacteriodetes	*Arenibacter palladensis*	Aerobic heterotroph	JB, green alga, cultured	Nedashkovskaya et al. (2006e)
		Bacteriodetes	*Kriegella aquimaris*	Aerobic heterotroph	JB, green alga, cultured	Nedashkovskaya et al. (2008)
		Bacteriodetes	*Maribacter polysiphoniae*	Aerobic heterotroph	JB, red alga, cultured	Nedashkovskaya et al. (2007c)
		Bacteriodetes	*Mesonia algae*	Aerobic heterotroph	JB, green alga, cultured	Nedashkovskaya et al. (2003a)
		Bacteriodetes	*Pibocella ponti*	Aerobic heterotroph	JB, green alga, cultured	Nedashkovskaya et al. (2005c)
		Bacteriodetes	*Pseudozobellia thermophila*	Heterotroph	JB, green alga, cultured	Nedashkovskaya et al. (2009)
		Bacteriodetes	*Roseivirga ehrenbergii*	Strictly aerobic heterotroph	JB, green alga, cultured	Nedashkovskaya et al. (2005b)
		Bacteriodetes	*Ulvibacter litoralis*	Strictly aerobic heterotroph	JB, green alga, cultured	Nedashkovskaya et al. (2004c)
		Bacteriodetes	*Winogradskyella thalassocola*	Aerobic heterotroph	JB, brown alga, cultured	Nedashkovskaya et al. (2005a)
		Bacteriodetes	*Winogradskyella epiphytica*	Aerobic heterotroph	JB, green alga, cultured	Nedashkovskaya et al. (2005a)
		Bacteriodetes	*Winogradskyella eximia*	Aerobic heterotroph	JB, brown alga, cultured	Nedashkovskaya et al. (2005a)

(continued)

Table 11.5 (continued)

Habitat	Domain	Phylum	Related to genus (or species)	Ecological and biogeochemical properties	Note	Reference
		Bacteriodetes	*Winogradskyella ulvae*	Aerobic heterotroph	JB, green alga, cultured	Nedashkovskaya et al. (2012)
		Bacteriodetes	*Salegentibacter holothuriorum*	Microaerophilic chemo-organotroph	JB, holothurian, cultured	Nedashkovskaya et al. (2004g)
		Bacteriodetes	*Vitellibacter vladivostokensis*	Strictly aerobic, asporogenic heterotroph	JB, holothurian, cultured	Nedashkovskaya et al. (2003c)
		Bacteriodetes	*Pontibacter actiniarum*	Heterotroph	JB, actinian, cultured	Nedashkovskaya et al. (2005j)
		Bacteroidetes	*Croceitalea eckloniae*	Aerobic heterotroph	UB, rhizosphere of the brown alga, cultured	Lee et al. (2008a)
		Bacteroidetes	*Croceitalea dokdonensis*	Aerobic heterotroph	UB, rhizosphere of the brown alga, cultured	Lee et al. (2008a)
		Bacteroidetes	*Costertonia aggregata*	Heterotroph, nitrate reduction	UB, biofilm, rock-bed, cultured	Kwon et al. (2006)
		Alphaproteobacteria	*Altererythrobacter troitsensis*	Aerobic heterotroph	JB, sea urchin, cultured	Nedashkovskaya et al. (2013)
		Alphaproteobacteria	*Hasllibacter halocynthiae*	Aerobic heterotroph	UB, ascidian, cultured	Kim et al. (2012)
		Alphaproteobacteria	*Litoreibacter albidus*	Aerobic heterotroph	JB, marine snail, cultured	Romanenko et al. (2011a)
		Alphaproteobacteria	*Maritalea myrionectae*	Strictly aerobic heterotroph	UB, isolated from culture of cilliate, cultured	Hwang et al. (2009b)

(continued)

Table 11.5 (continued)

Habitat	Domain	Phylum	Related to genus (or species)	Ecological and biogeochemical properties	Note	Reference
		Alphaproteobacteria	*Nitratireductor aquimarinus*	Aerobic, isolated from diatom culture	UB, isolated from culture of diatiom, cultured	Jang et al. (2011)
		Alphaproteobacteria	*Sphingomonas japonica*	Aerobic heterotroph	JB, crustacean, cultured	Romanenko et al. (2009c)
		Alphaproteobacteria	*Sphingomonas molluscorum*	Heterotroph	JB, bivalve, cultured	Romanenko et al. (2007)
		Epsilonproteobacteria	*Arcobacter marinus*	Aerobic or microaerobic heterotroph, nitrate reduction	UB, seaweeds and starfish, cultured	Kim et al. (2010b)
		Gammaproteobacteria	*Alteromonas nigrifaciens*	Aerobic heterotroph	JB, mussels, cultured	Ivanova et al. (1996)
		Gammaproteobacteria	*Colwellia asteriadis*	Facultatively anaerobic heterotroph	UB, starfish, cultured	Choi et al. (2010a)
		Gammaproteobacteria	*Granulosicoccus coccoides*	Aerobic heterotroph	JB, seagrass, cultured	Kurilenko et al. (2010)
		Gammaproteobacteria	*Halomonas halocynthiae*	Aerobic heterotroph	UB, ascidian, cultured	Romanenko et al. (2002)
		Gammaproteobacteria	*Kistimonas asteriae*	Aerobic heterotroph	UB, starfish, cultured	Choi et al. (2010b)
		Gammaproteobacteria	*Pseudoalteromonas citrea*	Aerobic agar degradation	JB, mussels, cultured	Ivanova et al. (1998)
		Gammaproteobacteria	*Pseudoalteromonas ruthenica*	Strictly aerobic chemorganotroph	JB, mussels and scallops, cultured	Ivanova et al. 2002
		Gammaproteobacteria	*Psychrobacter fulvigenes*	Aerobic heterotroph	JB, crustacean, cultured	Romanenko et al. (2009b)

(continued)

Table 11.5 (continued)

Habitat	Domain	Phylum	Related to genus (or species)	Ecological and biogeochemical properties	Note	Reference
		Gammaproteobacteria	*Shewanella affinis*	Facultatively anaerobic heterotroph, nitrate reduction	JB, sipuncula, cultured	Ivanova et al. (2004c)
		Gammaproteobacteria	*Shewanella spongiae*	Psychrophilic, piezosensitive bacterium	UB, sponge, cultured	Yang et al. (2006)
		Gammaproteobacteria	*Shewanella waksmanii*	Facultatively anaerobic heterotroph	JB, sipuncula, cultured	Ivanova et al. (2003a)
		Gammaproteobacteria	*Umboniibacter marinipuniceus*	Aerobic heterotroph	JB, mollusc *Umbonium costatum*, cultured	Romanenko et al. (2010a)

a few facultative anaerobic bacteria such as *Marinobacter, Moritella,* and *Shewanella* were isolated as well (Arakawa et al. 2006; Kim et al. 2006; Yang et al. 2007). In addition, most bacteria from the UB were isolated from the coastal sediment. The bacterial strain closely related to *Oceanisphaera donghaensis* has been known to be a manganese oxidizer using nitrate as an electron acceptor (Park et al. 2006). A total of 7 bacterial species were isolated in the sediment of the Yamato Basin (YB), including *Deltaproteobacteria* that are known to reduce sulfate in anoxic sediments. A total 51 symbiontic bacteria were isolated from the micro- or macrobenthos and their habitats in the Russian coastal area of the JB and the UB (Table 11.5).

11.4.2 Composition of Culture-Independent Prokaryotes

Microbial autecological studies via cultured strains provide thorough information on the physiology and ecological function of specific bacteria. However, since less than 0.1–1 % of the environmental microorganisms are cultivated in defined media (Amann et al. 1995), it was not until recently that culture-independent molecular approaches have revealed the microbial diversity in marine environments. Microbial diversity based on analysis of the bacterial 16S rRNA gene sequence was mostly reported in the water column and sediment of the UB where 38 *Bacteria* and 10 *Archaea* types have been identified (Tables 11.3 and 11.4).

In the water column of the UB, *Gammaproteobacteria* and *Alphaproteobacteria,* which are mostly aerobic heterotrophs, appeared to be the most abundant bacterial group (Table 11.3). On the other hand, the bacterial groups closely related to ammonia oxidation were frequently identified in the surface sediment (Table 11.4). In the manganese-oxide-rich sediment at the center of the UB, bacterial groups related to *Desulfuromonadales, Geobacteraceae, Colwellia, Oceanospirillaceae,* and *Arcobacter* that are potentially affiliated with manganese reduction were also reported (Vandieken et al. 2012). This indicated that microbial manganese reduction should be an important carbon oxidation pathway in the Mn-oxide-rich sediment of the UB (Hyun et al. 2010). Meanwhile, bacterial groups detected in gas hydrate bearing sediment of the UB were related to anaerobic heterotrophs, such as the candidate division JS1, *Chloroflexi,* and *Planctomycetes,* that have not been cultured (Lee et al. 2013). Archaeal groups identified in the sediment of the UB consist of MBGB, MBGC and MCG in the *Crenarchaeota,* and MBGD in the *Euryarchaeota* (Table 11.4). Those archaeal groups are ubiquitously distributed from the surface sediment to gas hydrate bearing deep sediment (Park et al. 2008; Kim et al. 2010a; Lee et al. 2013). The MGI in *Thaumarchaeota* that is associated with ammonium oxidation was identified in the surface sediment (Lee et al. 2013), and ANME-1 that is responsible for the anaerobic ammonium oxidation was also reported in the sulfate-methane transition zone of the gas hydrate bearing sediment of the UB (Lee et al. 2013)

Acknowledgments I would like to acknowledge the support of the EAST-1 (East Asian Seas Time Series) and LTMER (Long-Term Marine Ecological Research) programs funded by the Korean Ministry of Ocean and Fisheries. Many thanks also go to Sung-Han Kim for conducting geochemical analyses and sulfate reduction experiments and Hyeyoun Cho for constructing the inventories of microorganisms reported in the East Sea. Finally, I would like to acknowledge that the comments by Drs. Ju SJ, Noh JH and Choi DH have improved an earlier version of the manuscript.

References

Aller RC (1990) Bioturbation and manganese cycling in hemipelagic sediments. Phil Trans R Soc Lond A 331:51–68

Amann RI, Ludwig W, Schleifer KH (1995) Phylogenetic identification and in situ detection of individual microbial cells without cultivation. Microbiol Rev 59:143–169

Antoine D, André JM, Morel A (1996) Oceanic primary production. 2. Estimation at global scale from satellite (coastal zone color scanner) chlorophyll. Global Biogeochem Cycles 10:57–69

Arakawa S, Nogi Y, Sato T, Yoshida Y, Usami R, Kato C (2006) Diversity of piezophilic microorganisms in the closed ocean Japan Sea. Biosci Biotechnol Biochem 70:749–752

Azam F, Fenchel T, Field JG, Gray JS, Meyer-Reil LA, Thingstad F (1983) The ecological role of water-column microbes in the sea. Mar Ecol Prog Ser 10:257–263

Azam F, Worden AZ (2004) Microbes, molecules, and marine ecosystems. Science 303:1622–1624

Bae GD, Hwang CY, Kim HM, Cho BC (2010) *Salinisphaera dokdonensis* sp. nov., isolated from surface seawater. Int J Syst Evol Microbiol 60:680–685

Bale SJ, Goodman K, Rochelle PA, Marchesi JR, Fry JC, Weightman AJ, Parkes RJ (1997) *Desulforvibrio profundus* sp. nov., a novel barophilic sulfate-reducing bacterium from deep sediment layers in the Japan Sea. Int J Syst Bacteriol 47:515–521

Belkin IM (2009) Rapid warming of large marine ecosystems. Prog Oceanogr 81:207–213

Bidle KD, Azam F (1999) Accelerated dissolution of diatom silica by marine bacterial assemblages. Nature 397:508–512

Bidle KD, Manganelli M, Azam F (2002) Regulation of oceanic silicon and carbon preservation by temperature control on bacteria. Science 298:1980–1983

Böning P, Cuypers S, Grunwald M, Schnetger B, Brumsack HJ (2005) Geochemical characteristics of Chilean upwelling sediments at ~36°S. Mar Geol 220:1–21

Broecker WS (1991) The great ocean conveyor. Oceanography (Wash DC) 4:79–89

Canfield DE, Jørgensen BB, Fossing H, Glud R et al. (1993a) Pathways of organic carbon oxidation in three continental margin sediments. Mar Geol 113:27–40

Canfield DE, Thamdrup B, Hansen JW (1993b) The anaerobic degradation of organic matter in Danish coastal sediments: iron reduction, manganese reduction, and sulfate reduction. Geochim Cosmochim Acta 57:3867–3883

Carlson CA, Ducklow HW (1995) Dissolved organic carbon in the upper ocean of the central equatorial Pacific ocean, 1992: daily and finescale vertical variations. Deep-Sea Res II 42:639–656

Caron DA, Dam HG, Kremer P, Lessard EJ et al. (1995) The contribution of microorganisms to particulate carbon and nitrogen in surface waters of the Sargasso Sea near Bermuda. Deep-Sea ResI 42:943–972

Chang KI, Teague WJ, Lyu SJ, Perkins HT et al. (2004) Circulation and currents in the southwestern East/Japan Sea: overview and review. Prog Oceanogr 61:105–156

Cho BC, Azam F (1990) Biogeochemical significance of bacterial biomass in the ocean's euphonic zone. Mar Ecol Prog Ser 63:253–259

Cho BC, Choi JK, Chung CS, Hong GH (1994) Uncoupling of bacteria and phytoplankton during a spring diatom bloom in the mouth of the Yellow Sea. Mar Ecol Prog Ser 115:181–190

Cho BC, Na SC, Choi DH (2000) Active ingestion of fluorescently labeled bacteria by mesopelagic heterotrophic nanoflagellates in the East Sea, Korea. Mar Ecol Prog Ser. 206:23–32

Choi DH, Hwang CY, Cho BC (2009) *Pelagibius litoralis* gen. nov., sp. nov., a marine bacterium in the family *Rhodospirillaceae* isolated from coastal seawater. Int J Syst Evol Microbiol 59:818–823

Choi DH, Kim HM, Noh JH, Cho BC (2007a) *Nocardioides marinus* sp. nov. Int J Syst Evol Microbiol 57:775–779

Choi DH, Noh JH (2009) Phylogenetic diversity of *Synechococcus* strains isolated from the East China Sea and East Sea. FEMS Microbiol Ecol 69:439–448

Choi DH, Yang SY, Hong GH, Chung CS, Kim SH, Park JS, Cho BC (2005) Different interrelationships among phytoplankton, bacteria and environmental variables in dumping and reference areas in the East Sea. Aquat Microb Ecol 41:171–180

Choi EJ, Kwon HC, Koh HY, Kim YS, Yang HO (2010a) *Colwellia asteriadis* sp. nov., a marine bacterium isolated from the starfish Asterias amurensis. Int J Syst Evol Microbiol 60:1952–1957

Choi EJ, Kwon HC, Sohn YC, Yang HO (2010b) *Kistimonas asteriae* gen. nov., sp. nov., a gammaproteobacterium isolated from Asterias amurensis. Int J Syst Evol Microbiol 60:938–943

Choi JH, Im WT, Yoo JS, Lee SM, Moon DS, Kim HJ, Rhee SK, Roh DH (2008) *Paenibacillus donghaensis* sp. nov., a xylan-degrading and nitrogen-fixing bacterium isolated from East Sea sediment. J Microbiol Biotechnol 18:189–193

Choi JW, Im WT, Liu QM, Yoo JS, Shin JH, Rhee SK, Roh DH (2007b) *Planococcus donghaensis* sp. nov., a starch-degrading bacterium isolated from the East Sea, South Korea. Int J Syst Evol Microbiol 57:2645–2650

Cociasu A, Dorogan L, Humborg C, Popa L (1996) Long-term ecological changes in the Romanian coastal waters of the Black Sea. Mar Pollut Bull 32:32–38

Cole JJ, Honjo S, Erez J (1987) Benthic decomposition of organic matter at a deep-water site in the Panama basin. Nature 327:703–704

Cole JJ, Pace ML, Findlay S (1988) Bacterial production in fresh and saltwater ecosystems: a cross-system overview. Mar Ecol Prog Ser 43:1–10

Cotner JB, Biddanda BA (2002) Small players, large role: microbial influence on biogeochemical processes in pelagic aquatic ecosystems. Ecosystems 5:105–121

Ducklow HW (2000) Bacterial production and biomass in the oceans. In: Kirchman DL (ed) Microbial ecology of the oceans. Wiley-Liss, New York, pp 85–120

Ducklow HW, Purdie DA, Williams PJ leB, Davies JH (1986) Bacterioplankton: a sink for carbon in a coastal marine plankton community. Science 232:865–867

Fenchel T, King GM, Blackburn TH (1998) Bacterial biogeochemistry: the ecophysiology of mineral cycling. Academic Press, p 307

Ferdelman TG, Fossing H, Neumann K, Schulz HD (1999) Sulfate reduction in surface sediments of the southeast Atlantic continental margin between 15°38_S and 27°57_S (Angola and Namibia). Limnol Oceanogr 44:650–661

Fossing H (1990) Sulfate reduction in shelf sediments in the upwelling region off Central Peru. Cont Shelf Res 10:355–367

Fossing H, Ferdelman TG, Berg P (2000) Sulfate reduction and methane oxidation in continental margin sediments influenced by irrigation (South-East Atlantic off Namibia). Geochim Cosmochim Acta 64:897–910

Fouilland E, Mostajir B (2010) Revisited phytoplanktonic carbon dependency of heterotrophic bacteria in freshwaters, transitional, coastal and oceanic waters. FEMS Microb Ecol 73:419–429

Froelich PN, Klinkhammer GP, Bender ML, Luedtke NA, Heath GR, Cullen D, Dauphin P, Hammond D, Hartman B, Maynard V (1979) Early oxidation of organic matter in pelagic sediments of the eastern equatorial Atlantic: suboxic diagenesis. Geochim Cosmochim Acta 43:1075–1090

Fuhrman JA (1999) Marine viruses and their biogeochemical and ecological effects. Nature 399:541–548

Glud RN (2008) Oxygen dynamics of marine sediments. Mar Biol Res 4:243–289

Gorshkova NM, Ivanova EP, Sergeev AF, Zhukova NV, Alexeeva Y, Wright JP, Nicolau DV, Mikhailov VV, Christen R (2003) *Marinobacter excellens* sp. nov., isolated from sediments of the Sea of Japan. Int J Syst Evol Microbiol 53:2073–2078

Han SK, Nedashkovskaya OI, Mikhailov VV, Kim SB, Bae KS (2003) *Salinibacterium amurskyense* gen. nov., sp. nov., a novel genus of the family *Microbacteriaceae* from the marine environment. Int J Syst Evol Microbiol 53:2061–2066

Hansell DA, Carlson CA (1998) Net community production of dissolved organic carbon. Global Biogeochem Cycles 12:443–453

Hong GH, Kim SH, Chung CS, Kang DJ, Shin DH, Lee HJ, Han SJ (1997) ^{210}Pb-derived sediment accumulation rates in the southwestern East Sea (Sea of Japan). Geo Mar Lett 17:126–132

Hoppe HG, Breithaupt P, Walther K, Koppe R, Bleck S, Sommer U, Jürgens K (2008) Climate warming in winter affects the coupling between phytoplankton and bacteria during the spring bloom: a mesocosm study. Aquat Microb Ecol 51:105–115

Hoppe HG, Gocke K, Koppe R, Begler C (2002) Bacterial growth and primary production along a north-south transect of the Atlantic Ocean. Nature 416:168–171

Hwang CY, Bae GD, Yih WH, Cho BC (2009a) *Marivita cryptomonadis* gen. nov., sp. nov., and *Marivita litorea* sp. nov., of the family *Rhodobacteraceae*, isolated from marine habitats. Int J Syst Evol Microbiol 59:1568–1575

Hwang CY, Cho BC (2002) Virus-infected bacteria in oligotrophic open waters of the East Sea, Korea. Aquat Microb Ecol 30:1–9

Hwang CY, Cho BC (2008a) *Cucumibacter marinus* gen. nov., sp. nov., a marine bacterium in the family *Hyphomicrobiaceae*. Int J Syst Evol Microbiol 58:1591–1597

Hwang CY, Cho BC (2008b) *Ponticoccus litoralis* gen. nov., sp. nov., a marine bacterium in the family *Rhodobacteraceae*. Int J Syst Evol Microbiol 58:1332–1338

Hwang CY, Cho KD, Yih W, Cho BC (2009b) *Maritalea myrionectae* gen, nov., sp. nov., isolated from a culture of the marine ciliate *Myrionecta rubra*. Int J Syst Evol Microbiol 59:609–614

Hyun JH, Kim KH (2003) Bacterial abundance and production during the unique spring phytoplankton bloom in the central Yellow Sea. Mar Ecol Prog Ser 252:77–88

Hyun JH, Kim D, Shin CW, Noh JH, Yang EJ, Mok JS, Kim SH, Kim HC, Yoo S (2009a) Enhanced phytoplankton and bacterioplankton production coupled to coastal upwelling and an anticyclonic eddy in the Ulleung basin, East Sea. Aquat Microb Ecol 54:45–54

Hyun JH, Mok JS, Cho HY, Kim SH, Kostka JE (2009b) Rapid organic matter mineralization coupled to iron cycling in intertidal mud flats of the Han River estuary, Yellow Sea. Biogeochem 92:231–245

Hyun JH, Mok JS, You OR, Kim D. Choi DL (2010) Variations and controls of sulfate reduction in the continental slope and rise of the Ulleung basin off the southeast Korean upwelling system in the East Sea. Geomicrobiol J. 27:1–11

Hyun JH, Smith AC, Kostka JE (2007) Relative contributions of sulfate- and iron(III) reduction to organic matter mineralization and process controls in contrasting habitats of the Georgia saltmarsh. Appl Geochem 22:2637–2651

Hyun JH, Yang EJ (2005) Meso-scale spatial variation in the bacterial abundance and production associated with surface water convergence and divergence in the northeast equatorial Pacific. Aquat Microb Ecol 41:1–13

Ivanova EP, Bowman JP, Christen R, Zhukova NV, Lysenko AM, Gorshkova NM, Mitik-Dineva N, Sergeev AF, Mikhailov VV (2006) *Salegentibacter flavus* sp. nov. Int J Syst Evol Microbiol 56:583–586

Ivanova EP, Bowman JP, Lysenko AM, Zhukova NV, Gorshkova NM, Sergeev AF, Mikhailov VV (2005a) *Alteromonas addita* sp. nov. Int J Syst Evol Microbiol 55:1065–1068

Ivanova EP, Gorshkova NM, Bowman JP, Lysenko AM, Zhukova NV, Sergeev AF, Mikhailov VV, Nicolau DV (2004a) *Shewanella pacifica* sp. nov., a polyunsaturated fatty acid-producing bacterium isolated from seawater. Int J Syst Evol Microbiol 54:1083–1087

Ivanova EP, Gorshkova NM, Zhukova NV, Lysenko AM, Zelepuga EA, Prokof'eva NG, Mikhailov VV, Nicolau DV, Christen R (2004b) Characterization of *Pseudoalteromonas distincta*-like sea-water isolates and description of *Pseudoalteromonas aliena* sp. nov. Int J Syst Evol Microbiol 54:1431–1437

Ivanova EP, Kiprianova EA, Mikhailov VV, Levanova GF, Garagulya AD, Gorshkova NM, Yumoto N, Yoshikawa S (1996) Characterization and Identification of Marine *Alteromonas nigrifaciens* strains and emendation of the description. Int J Syst Bacteriol 46:223–238

Ivanova EP, Kiprianova EA, Mikhailov VV, Levanova GF, Garagulya AD, Gorshkova NM, Vysotskii MV, Nicolau DV, Yumoto N, Taguchi T, Yoshikawa S (1998) Phenotypic diversity of *Pseudoalteromonas citrea* from different marine habitats and emendation of the description. Int J Syst Evol Microbiol 48:247–256

Ivanova EP, Nedashkovskaya OI, Sawabe T, Zhukova NV, Frolova GM, Nicolau DV, Mikhailov VV, Bowman JP (2004c) *Shewanella affinis* sp. nov., isolated from marine invertebrates. Int J Syst Evol Microbiol 54:1089–1093

Ivanova EP, Nedashkovskaya OI, Zhukova NV, Nicolau DV, Christen R, Mikhailov VV (2003a) *Shewanella waksmanii* sp. nov., isolated from a sipuncula (*Phascolosoma japonicum*). Int J Syst Evol Microbiol 53:1471–1477

Ivanova EP, Sawabe T, Gorshkova NM, Svetashev VI, Mikhailov VV, Nicolau DV, Christen R (2001) *Shewanella japonica* sp. nov. Int J Syst Evol Microbiol 51:1027–1033

Ivanova EP, Sawabe T, Hayashi K, Gorshkova NM, Zhukova NV, Nedashkovskaya OI, Mikhailov VV, Nicolau DV, Christen R (2003b) *Shewanella fidelis* sp. nov., isolated from sediments and sea water. Int J Syst Evol Microbiol 53:577–582

Ivanova EP, Sawabe T, Lysenko AM, Gorshkova NM, Svetashev VI, Nicolau DV, Yumoto N, Taguchi T, Yoshikawa S, Christen R, Mikhailov VV (2002) *Pseudoalteromonas ruthenica* sp. nov., isolated from marine invertebrates. Int J Syst Evol Microbiol 52:235–240

Ivanova EP, Zhukova NV, Lysenko AM, Gorshkova NM, Sergeev AF, Mikhailov VV, Bowman JP (2005b) *Loktanella agnita* sp. nov. and *Loktanella rosea* sp. nov., from the north-west Pacific Ocean. Int J Syst Evol Microbiol 55:2203–2207

Jahnke RA, Jahnke DB (2000) Rates of C, N, P and Si recycling and denitrification at the US mid-Atlantic continental slope depocenter. Deep-Sea Res I 47:1405–1428

Jahnke RA, Reimers CE, Craven DB (1990) Intensification of recycling of organic matter at the sea floor near ocean margins. Nature 348:50–54

Jang GI, Hwang CY, Cho BC (2011) *Nitratireductor aquimarinus* sp. nov., isolated from a culture of the diatom *Skeletonema costatum*, and emended description of the genus *Nitratireductor*. Int J Syst Evol Microbiol 61:2676–2681

Jensen MM, Thamdrup B, Rysgaard S, Holmer M, Fossing H (2003) Rates and regulation of microbial iron reduction in sediments of the Baltic-North Sea transition. Biogeochem 65:295–317

Jiao N, Herndl GH, Hansell DA, Benner R, Kattner G, Wilhelm SW, Kirchman DL, Weinbauer MG, Luo T, Chen F, Azam F (2010) Microbial production of recalcitrant dissolved organic matter: long-term carbon storage in the global ocean. Nature Rev 8:593–599

Jiao N, Zheng Q (2011) The microbial carbon pump: from genes to ecosystems. Appl Environ Microbiol 77:7439–7444

Jørgensen BB (1982) Mineralization of organic matter in the sea bed-the role of sulphate reduction. Nature 296:643–645

Jørgensen BB (2006) Bacteria and marine biogeochemistry. In: Schulz HD, Zabel M (eds) Marine geochemistry. Springer, Berlin, pp 169–206

Jørgensen BB, Kasten S (2006) Sulfur cycling and methane oxidation. In: Schulz HD, Zabel M (eds) Marine geochemistry. Springer, Berlin, pp 271–309

Joung YC, Kim HE, Joh KS (2013) *Flavobacterium jumunjinense* sp. nov., isolated from a lagoon of East Seaand emended descriptions of *Flavobacterium cheniae*, *Flavobacterium dongtanense* and *Flavobacterium gelidilacus*. Int J Syst Evol Microbiol. doi:10.1099/ijs.0.045286-0

Jung MY, Shin KS, Kim SY, Kim SJ, Park SJ, Kim JG, Cha IT, Kim MN, Rhee SK (2013) *Hoeflea halophila* sp. nov., a novel bacterium isolated from marine sediment of the East Sea. Korea. Anton Leeuw Int J G 103:971–978

Karl DM (1999) A sea of change: biogeochemical variability in the North Pacific subtropical gyre. Ecosystems 2:181–214

Kennett J (1982) Marine geology. Prentice-Hall, Engel Cliffs, p 813

Khan ST, Harayama S, Tamura T, Ando K, Takagi M, Kazuo S (2009) *Paraoerskovia marina* gen. nov., sp. nov., an actinobacterium isolated from marine sediment. Int J Syst Evol Microbiol 59:2094–2098

Khan ST, Nakagawa Y, Harayama S (2006) *Sediminicola luteus* gen. nov., sp. nov., a novel member of the family *Flavbacteriaceae*. Int J Syst Evol Microbiol 56:841–845

Khan ST, Nakagawa Y, Harayama S (2008) *Fulvibacter tottoriensis* gen. nov., sp. nov., a member of the family *Flavobacteriaceae* isolated from marine sediment. Int J Syst Evol Microbiol 58:1670–1674

Kim BY, Weon HY, Yoo SH, Kim JS, Kwon SW, Stackebrandt E, Go SJ (2006) *Marinobacter koreensis* sp. nov., isolated from sea sand in Korea. Int J Syst Evol Microbiol 56:2653–2656

Kim BY, Yoo SH, Weon HY, Jeon YA, Hong SB, Go SJ, Stackebrandt E, Kwon SW (2008) *Jannaschia pohangensis* sp. nov., isolated from seashore sand in Korea. Int J Syst Evol Microbiol 58:496–499

Kim D, Choi MS, Oh HY, Kim KH, Noh JH (2009) Estimate of particulate organiccarbonexport flux using $^{234}Th/^{238}U$ disequilibrium in the south western East Sea during summer. (The Sea) J Korean Soc Oceanogr 14:1–9

Kim HN, Choo YJ, Cho JC (2007a) *Litoricolaceae* fam. nov., to include *Litoricola lipolytica* gen. nov., sp. nov., a marine bacterium belonging to the order *Oceanospirillales*. Int J Syst Evol Microbiol 57:1793–1978

Kim K, Kim KR, Chung JY, Choi BH et al. (1996) New findings from CREAMS observations: water masses and eddies in the East Sea. J Korean Soc Oceanogr 31:155–163

Kim KR, Kim K (1996) What is happening in the East Sea (Japan Sea)? Recent chemical observations during CREAMS 93–96. J Korean Soc Oceanogr 31:164–172

Kim K, Kim KR, Min D, Volkov YN, Yoon JH, Takematsu M (2001) Warming and structural changes in the East Sea (Japan Sea): a clue to the future changes in global oceans? Geophys Res Lett 28:3293–3296

Kim MK, Khang Y (2012) Marine prokaryotic diversity of the deep sea waters at the depth of 1500 m off the coast of the Ulleung Island in the East Sea (Korea). Korean J Microbiol 48:328–331

Kim SB, Nedashkovskaya OI (2010) *Winogradskyella pacifica* sp. nov., a marine bacterium of the family *Flavobacteriaceae*. Int J Syst Evol Microbiol 60:1948–1951

Kim SB, Nedashkovskaya OI, Mikhailov VV, Han SK, Kim KO, Rhee MS, Bae KS (2004) *Kocuria marina* sp. nov., a novel actinobacterium isolated from marine sediment. Int J Syst Evol Microbiol 54:1617–1620

Kim BB, Cho HY, Hyun JH (2010a) Community structure, diversity and vertical distribution of *Archaea* revealed by 16S rRNA gene analysis in the deep sea sediment of the Ulleung Basin, East Sea. Ocean Polar Res 32:309–319

Kim HM, Hwang CY, Cho BC (2010b) *Arcobacter marinus* sp. nov. Int J Syst Evol Microbiol 60:531–536

Kim KK, Lee JS, L KC, Oh HM, Kim SG (2010c) *Pontibaca methylaminivorans* gen. nov., sp. nov., a member of the family *Rhodobacteraceae*. Int J Syst Evol Microbiol 60:2170–2175

Kim SH, Yang HO, Shin YK, Kwon HC (2012) *Hasllibacter halocynthiae* gen. nov., sp. nov., a nutriacholic acid-producing bacterium isolated from the marine ascidian *Halocynthia roretzi*. Int J Syst Evol Microbiol 62:624–631

Kim SJ, Choi YR, Park SJ, Kim JG, Shin KS, Roh DH, Rhee SK (2013) *Winogradskyella pulchriflava* sp. nov., isolated from marine sediment. Int J Syst Evol Microbiol 63:3062–3068

Kim TH, Kim G (2012) Important role of colloids in the cycling of ^{210}Po and ^{210}Pb in the ocean: results from the East/Japan Sea. Geochim Cosmochim Acta 95:134–142

Kim YG, Choi DH, Hyun S, Cho BC (2007b) *Oceanobacillus profundus* sp. nov., isolated from a deep-sea sediment core. Int J Syst Evol Microbiol 57:409–413

Kim YG, Hwang CY, Yoo KW, Moon HT, Yoon JH, Cho BC (2007c) *Pelagibacillus goriensis* gen. nov., sp. nov., a moderately halotolerant bacterium isolated from coastal water off the east coast of Korea. Int J Syst Evol Microbiol 57:1554–1560

Kiørboe T, Kaas H, Kruse B, Mohlenberg F, Tiselius P, Ærtebjerg G (1990) The structure of the pelagic food web in relation to water column structure in the Skagerrak. Mar Ecol Prog Ser 59:19–32

Kirchman DL (2000) Uptake and regeneration of inorganic nutrients by marine heterotrophic bacteria. In: Kirchman DL (ed) Microbial ecology of the oceans. Wiley-Liss, New York, pp 261–288

Kristensen KD, Kristensen E, Jensen MH (2002) The influence of water column hypoxia on the behavior of manganese and iron in sandy coastal marine sediment. Estuar Coast Shelf Sci 55:645–654

Kurilenko VV, Christen R, Zhukova NV, Kalinovskaya NI, Mikhailov VV, Crawford RJ, Ivanova EP (2010) *Granulosicoccus coccoides* sp. nov., isolated from leaves of seagrass (*Zostera marina*). Int J Syst Evol Microbiol 60:97–976

Kwak JH, Lee SH, Park HJ, Choy EJ, Jeong HD, Kim KR, Kang CK (2013) Monthly measured primary and new productivities in the Ulleung Basin as a biological "hot spot" in the East/Japan Sea. Biogeosciences 10:4405–4417

Kwon KK, Lee YK, Lee HK (2006) *Costertonia aggregata* gen. nov., sp. nov., a mesophilic marine bacterium of the family *Flavobacteriaceae*, isolated from a mature biofilm. Int J Syst Evol Microbiol 56:1349–1353

Kwon KK, Yang SJ, Lee HS, Cho JC, Kim SJ (2007a) *Sufflavibacter maritimus* gen. nov., sp. nov., novel *Flavobacteriaceae* bacteria isolated from marine environments. J Microbiol Biotechnol 17:1379–1384

Kwon SW, Kim BY, Weon HY, Baek YK, Go SJ (2007b) *Arenimonas donghaensis* gen. nov., sp. nov., isolated from seashore sand. Int J Syst Evol Microbiol 57:954–958

Lee HS, Kwon KK, Yang S-H, Bae SS, Park CH, Kim SJ, Lee JH (2008a) Description of *Croceitalea* gen. nov. in the family *Flavobacteriaceae* with two species, *Croceitalea eckloniae* sp. nov. and *Croceitalea dokdonensis* sp. nov., isolated from the rhizosphere of the marine alga *Ecklonia kurome*. Int J Syst Evol Microbiol 58:2505–2510

Lee JC (1983) Variation of sea level and sea surface temperature associated with wind-induced upwelling in the southeast coast of Korea in summer. J Korean Soc Oceanogr 18:149–160

Lee JC, Na JY (1985) Structure of upwelling off the southeast coast of Korea. J Korean Soc Oceanogr 20:6–19

Lee JW, Kwon KK, Azizi A, Oh HM, Kim W, Bahk JJ, Lee DH, Lee JH (2013) Microbial community structures of methane hydrate-bearing sediments in the Ulleung Basin, East Sea of Korea. Mar Petrol Geol 47:136–146

Lee SY, Park S, Oh TK, Yoon HH (2012) *Celeribacter baekdonensis* sp. nov., isolated from seawater, and emended description of the genus *Celeribacter Ivanova* et al. 2010. Int J Syst Evol Microbiol 62:1359–1364

Lee SY, Park S, Oh TK, Yoon JH (2010b) Description of *Olleya aquimaris* sp. nov., isolated from seawater, and emended description of the genus *Olleya mancuso* Nichols et al. 2005. Int J Syst Evol Microbiol 60:887–891

Lee T, Hyun JH, Mok JS, Kim D (2008b) Organic carbon accumulation and sulfate reduction rates in slope and basin sediments of the Ulleung basin, East Sea/Japan Sea. Geo-Mar Lett 28:153–159

Lee YS, Lee DH, Kahng HY, Kim EM, Jung JS (2010a) *Erythrobacter gangjinensis* sp. nov., a marine bacterium isolated from seawater. Int J Syst Evol Microbiol 60:1413–1417

Levitus S, Antonov JI, Boyer TP, Stephens C (2000) Warming of the world ocean. Science 287:2225–2229

Li WKW, Dickie PM, Irwin BD, Wood AM (1992) Biomass of bacteria, cyanobacteria, prochlorophytes and photosynthetic eukaryotes in the Sargasso Sea. Deep-Sea Res 39:501–519

Lie HJ, Byun SK, Bang I, Cho CH (1995) Physical structure of eddies in the southwestern East Sea. J Korean Soc Oceanogr 30:170–183

Lim JM, Jeon CO, Jang HH, Park DJ, Shin YK, Yeo SH, Kim CJ (2008) *Albimonas donghaensis* gen. nov., sp. nov., a nonphotosynthetic member of the class *Alphaproteobacteria* isolated from seawater. Int J Syst Evol Microbiol 58:282–285

Liu KK, Atkinson L, Quinones RA, Talaue-McManus L (2010) Biogeochemistry of continental margins in a global context. In: Liu KK, Atkinson L, Quinones RA, Talaue-McManus L (eds) Carbon and nutrient fluxes in continental margins. Springer, Berlin, pp 3–24

Liu KK, Iseki K, Chao SY (2000) Continental margin carbon fluxes. In: Hanson RB, Ducklow HW, Field JG (eds) The changing ocean carbon cycle, pp 187–239

Magen C, Mucci A, Sundby B (2011) Reduction rates of sedimentary Mn and Fe oxides: an incubation experiment with Arctic ocean sediments. Aquatic Biogeochemistry 17:629–643

Nagata T (2000) Production mechanisms of dissolved organic matter. In: Kirchman DL (ed) Microbial ecology of the oceans. Wiley-Liss, New York, pp 121–152

Nedashkovskaya OI, Cho S-H, Joung Y, Joh K, Kim MN, Shin KS, Oh HW, Bae KS, Mikhailov VV, Kim SB (2013) *Altererythrobacter troitsensis* sp, nov., isolated from the sea urchin *Strongylocentrotus intermedius*. Int J Syst Evol Microbiol 63:93–97

Nedashkovskaya OI, Kim SB, Bae KS (2010a) *Gramella marina* sp. nov., isolated from the sea urchin *Strongylocentrotus intermedius*. Int J Syst Evol Microbiol 60:2799–2802

Nedashkovskaya OI, Kim SB, Han SK, Lysenko AM, Mikhailov VV, Bae KS (2004a) *Arenibacter certesii* sp. nov., a novel marine bacterium isolated from the green alga *Ulva fenestrata*. Int J Syst Evol Microbiol 54:1173–1176

Nedashkovskaya OI, Kim SB, Han SK, Lysenko AM, Rohde M, Rhee MS, Frolova GM, Falsen E, Mikhailov VV, Bae KS (2004b) *Maribacter* gen. nov., a new member of the family *Flavobacteriaceae*, isolated from marine habitats, containing the species *Maribacter sedimenticola* sp. nov., *Maribacter aquivivus* sp. nov. *Maribacter orientalis* sp. nov. and *Maribacter ulvicola* sp. nov. Int J Syst Evol Microbiol 54:1017–1023

Nedashkovskaya OI, Kim SB, Han SK, Lysenko AM, Rohde M, Zhukova NV, Falsen E, Frolova GM, Mikhailov VV, Bae KS (2003a) *Mesonia algae* gen. nov., sp. nov., a novel marine bacterium of the family *Flavobacteriaceae* isolated from the green alga *Acrosiphonia sonderi* (Kütz) Kornm. Int J Syst Evol Microbiol 53:1967–1971

Nedashkovskaya OI, Kim SB, Han SK, Rhee MS, Lysenko AM, Falsen E, Frolova GM, Mikhailov VV, Bae KS (2004c) *Ulvibacter litoralis* gen. nov., sp. nov., a novel member of the family *Flavobacteriaceae* isolated from the green alga *Ulva fenestrata*. Int J Syst Evol Microbiol 54:119–123

Nedashkovskaya OI, Kim SB, Han SK, Rhee MS, Lysenko AM, Rohde M, Zhukova NV, Frolova GM, Mikhailov VV, Bae KS (2004d) *Algibacter lectus* gen. nov., sp. nov., a novel member of the family *Flavobacteriaceae* isolated from green algae. Int J Syst Evol Microbiol 54:1257–1261

Nedashkovskaya OI, Kim SB, Han SK, Snauwaert C, Vancanneyt M, Swings J, Kim KO, Lysenko AM, Rohde M, Frolova GM, Mikhailov VV, Bae KS (2005a) *Winogradskyella thalassocola* gen. nov., sp. nov., *Winogradskyella epiphytica* sp. nov. and *Winogradskyella eximia* sp. nov., marine bacteria of the family *Flavobacteriaceae*. Int J Syst Evol Microbiol 55:49–55

Nedashkovskaya OI, Kim SB, Kwak JR, Mikhailov VV, Bae KS (2006a) *Mariniflexile gromovii* gen. nov., sp. nov., a gliding bacterium isolated from the sea urchin *Strongylocentrotus intermedius*. Int J Syst Evol Microbiol 56:1635–1638

Nedashkovskaya OI, Kim SB, Lee DH, Lysenko AM, Shevchenko LS, Frolova GM, Mikhailov VV, Lee KH, Bae KS (2005b) *Roseivirga ehrenbergii* gen. nov., sp. nov., a novel marine bacterium of the phylum 'Bacteroidetes', isolated from the green alga *Ulva fenestrata*. Int J Syst Evol Microbiol 55:231–234

Nedashkovskaya OI, Kim SB, Lee KH, Bae KS, Frolova GM, Mikhailov VV, Kim IS (2005c) *Pibocella ponti* gen. nov., sp. nov., a novel marine bacterium of the family *Flavobacteriaceae* isolated from the green alga *Acrosiphonia sonderi*. Int J Syst Evol Microbiol 55:177–181

Nedashkovskaya OI, Kim SB, Lee KH, Mikhailov VV, Bae KS (2005d) *Gillisia mitskevichiae* sp. nov., a novel bacterium of the family *Flavobacteriaceae*, isolated from sea water. Int J Syst Evol Microbiol 55:321–323

Nedashkovskaya OI, Kim SB, Lee MS, Park MS, Lee KH, Lysenko AM, Oh HW, Mikhailov VV, Bae KS (2005e) *Cyclobacterium amurskyense* sp. nov., a novel marine bacterium isolated from sea water. Int J Syst Evol Microbiol 55:2391–2394

Nedashkovskaya OI, Kim SB, Lysenko AM, Frolova GM, Mikhailov VV, Bae KS, Lee DH, Kim IS (2005f) *Gramella echinicola* gen. nov., sp. nov., a novel halophilic bacterium of the family *Flavobacteriaceae* isolated from the sea urchin *Strongylocentrotus intermedius*. Int J Syst Evol Microbiol 55:391–394

Nedashkovskaya OI, Kim SB, Lysenko AM, Frolova GM, Mikhailov VV, Lee KH, Bae KS (2005g) Description of *Aquimarina muelleri* gen. nov., sp. nov., and proposal of the reclassification of [*Cytophaga*] latercula Lewin 1969 as *Stanierella latercula* gen. nov., comb. nov. Int J Syst Evol Microbiol 55:225–229

Nedashkovskaya OI, Kim SB, Lysenko AM, Lee KH, Bae KS, Mikhailov VV (2007a) *Arenibactere chinorum* sp. nov., isolated from the sea urchin *Strongylocentrotus intermedius*. Int J Syst Evol Microbiol 57:2655–2659

Nedashkovskaya OI, Kim SB, Lysenko AM, Mikhailov VV, Bae KS, Kim IS (2005h) *Salegentibacter mishustinae* sp. nov., isolated from the sea urchin *Strongylocentrotus intermedius*. Int J Syst Evol Microbiol 55:235–238

Nedashkovskaya OI, Kim SB, Lysenko AM, Park MS, Mikhailov VV, Bae KS, Park HY (2005i) *Roseivirga echinicomitans* sp. nov., a novel marine bacterium isolated from the sea urchin *Strongylocentrotus intermedius*, and emended description of the genus *Roseivirga*. Int J Syst Evol Microbiol 55:1797–1800

Nedashkovskaya OI, Kim SB, MiKhailov VV (2010b) *Maribacter stanieri* sp. nov., a marine bacterium of the family *Flavobacteriaceae*. Int J Syst Evol Microbiol 60:214–218

Nedashkovskaya OI, Kim SB, Suzuki M, Shevchenko LS, Lee MS, Lee KH, Park MS, Frolova GM, Oh HW, Bae KS, Park HY, Mikhailov VV (2005j) *Pontibacter actiniarum* gen. nov., sp. nov., a novel member of the phylum "*Bacteroidetes*" and proposal of *Reichenbachiella* gen. nov. as a replacement for the illegitimate prokaryotic generic name *Reichenbachia* Nedashkovskaya et al. 2003. Int J Syst Evol Microbiol 55:2583–2588

Nedashkovskaya OI, Kim SB, Vancanneyt M, Lysenko AM, Shin DS, Park MS, Lee KH, Jung WJ, Kalinovskaya NI, Mikhailov VV, Bae KS, Swings J (2006b) *Echinicola pacifica* gen. nov., sp. nov., a novel flexibacterium isolated from the sea urchin *Strongylocentrotus intermedius*. Int J Syst Evol Microbiol 56:953–958

Nedashkovskaya OI, Kim SB, Vancanneyt M, Snauwaert C, Lysenko AM, Rohde M, Frolova GM, Zhukova NV, Mikhailov VV, Bae KS, Oh HW, Swings J (2006c) *Formosa agariphila* sp. nov., a budding bacterium of the family *Flavobacteriaceae* isolated from marine environments, and emended description of the genus *Formosa*. Int J Syst Evol Microbiol 56:161–167

Nedashkovskaya OI, Kim SB, Zhukova NV, Kwak J, Mikhailov VV, Bae KS (2006d) *Mesonia mobilis* sp. nov., isolated from seawater, and emended description of the genus *Mesonia*. Int J Syst Evol Microbiol 56:2433–2436

Nedashkovskaya OI, Kukhlevskiy AD, Zhukova NV (2012) *Winogradskyella ulvae* sp. nov., an epiphyte of a Pacific seaweed, and emended descriptions of the genus *Winogradskyella* and *Winogradskyella thalassocola*, *Winogradskyella echinorum*, *Winogradskyella exilis* and *Winogradskyella eximia*. Int J Syst Evol Microbiol 62:1450–1456

Nedashkovskaya OI, Suzuki M, Kim SB, Mikhailov VV (2008) *Kriegella aquimaris* gen. nov., sp. nov., isolated from marine environments. Int J Syst Evol Microbiol 58:2624–2628

Nedashkovskaya OI, Suzuki M, Lee JS, Lee KC, Shevchenko LS, Mikhailov VV (2009) *Pseudozobellia thermophila* gen. nov., sp. nov., a bacterium of the family *Flavobacteriaceae*, isolated from the green alga *Ulva fenestrata*. Int J Syst Evol Microbiol 59:806–810

Nedashkovskaya OI, Suzuki M, Lysenko AM, Snauwaert C, Vancanneyt M, Swings J, Vysotskii MV, Mikhailov VV (2004e) *Cellulophaga pacifica* sp. nov. Int J Syst Evol Microbiol 54:609–613

Nedashkovskaya OI, Suzuki M, Vancanneyt M, Cleenwerck I, Lysenko AM, Mikhailov VV, Swings J (2004f) *Zobellia amurskyensis* sp. nov., *Zobellia laminariae* sp. nov. and *Zobellia russellii* sp. nov., novel marine bacteria of the family *Flavobacteriaceae*. Int J Syst Evol Microbiol 54:1643–1648

Nedashkovskaya OI, Suzuki M, Vancanneyt M, Cleenwerck I, Zhukova NV, Vysotskii MV, Mikhailov VV, Swings J (2004g) *Salegentibacter holothuriorum* sp. nov., isolated from the edible holothurian *Apostichopus japonicus*. Int J Syst Evol Microbiol 54:1107–1110

Nedashkovskaya OI, Suzuki M, Vysotskii MV, Mikhailov VV (2003b) *Reichenbachia agariperforans* gen. nov., sp. nov., a novel marine bacterium in the phylum *Cytophaga-Flavobacterium-Bacteroidetes*. Int J Syst Evol Microbiol 53:81–85

Nedashkovskaya OI, Suzuki M, Vysotskii MV, Mikhailov VV (2003c) *Vitellibacter vladivostokensis* gen. nov., sp., nov., a new member of the phylum *Cytophaga-Flavobacterium-Bacteroides*. Int J Syst Evol Microbiol 53:1281–1286

Nedashkovskaya OI, Suzuki M, Vysotskii MV, Mikhailov VV (2003d) *Arenibacter troitsensis* sp. nov., isolated from marine bottom sediment. Int J Syst Evol Microbiol 53:1287–1290

Nedashkovskaya OI, Vancanneyt M, Cleenwerck I, Snauwaert C, Kim SB, Lysenko AM, Shevchenko LS, Lee KH, Park MS, Frolova GM, Mikhailov VV, Bae KS, Swings J (2006e) *Arenibacter palladensis* sp. nov., a novel marine bacterium isolated from the green alga *Ulva fenestrata*, and emended description of the genus *Arenibacter*. Int J Syst Evol Microbiol 56:155–160

Nedashkovskaya OI, Vancanneyt M, Kim SB (2010c) *Bizionia echini* sp. nov., isolated from a sea urchin. Int J Syst Evol Microbiol 60:928–931

Nedashkovskaya OI, Vancanneyt M, Kim SB, Han J, Zhukova NV, Shevchenko LS (2010d) *Salinimicrobium marinum* sp. nov., a halophilic bacterium of the family *Flavobacteriaceae*, and emended descriptions of the genus *Salinimicrobium* and *Salinimicrobium catena*. Int J Syst Evol Microbiol 60:2303–2306

Nedashkovskaya OI, Vancanneyt M, Kim SB, Hoste B, Bae KS (2007b) *Algibacter mikhailovii* sp. nov., a novel marine bacterium of the family *Flavobacteriaceae*, and emended description of the genus *Algibacter*. Int J Syst Evol Microbiol 57:2147–2150

Nedashkovskaya OI, Vancanneyt M, Vos PD, Kim SB, Lee MS, Mikhailov VV (2007c) *Maribacter polysiphoniae* sp. nov., isolated from a red alga. Int J Syst Evol Microbiol 57:2840–2843

Oh KH, Lee SY, Lee MH, Oh TK, Yoon JH (2011a) *Paraperlucidibaca baekdonensis* gen. nov., sp. nov., isolated from seawater. Int J Syst Evol Microbiol 61:1382–1385

Oh YS, Lim HJ, Cha IT, Im WT, Yoo JS, Kang UG, Rhee SK, Roh DH (2009) *Roseovarius halotolerans* sp. nov., isolated from deep seawater. Int J Syst Evol Microbiol 59:2718–2723

Oh YS, Park AR, Lee JK, Lim CS, Yoo JS, Roh DH (2011b) *Pseudoalteromonas donghaensis* sp. nov., isolated from seawater. Int J Syst Evol Microbiol 61:351–355

Pakhomova S, Hall POJ, Kononets M, Rozanov AG, Tenberg A, Vershinin AV (2007) Fluxes of iron and manganese across the sediment—water interface under various redox conditions. Mar Chem 107:319–331

Park BJ, Park SJ, Yoon DN, Schouten S, Damsté JSS, Rhee SK (2010) Cultivation of autotrophic ammonia-oxidizing archaea from marine sediments in coculture with sulfur-oxidizing bacteria. Appl Environ Microbiol 76:7575–7578

Park JR, Bae JW, Nam YD, Chang HW, Kwon HY, Quan ZX, Park YH (2007) *Sulfitobacter litoralis* sp. nov., a marine bacterium isolated from the East Sea, Korea. Int J Syst Evol Microbiol 57:692–695

Park SJ, Kang CH, Nam YD, Bae JW, Park YH, Quan ZX, Moon DS, Kim HJ, Roh DH, Rhee SK (2006) *Oceanisphaera donghaensis* sp. nov., a halophilic bacterium from the East Sea, Korea. Int J Syst Evol Microbiol 56:895–898

Park SJ, Park BJ, Rhee SK (2008) Comparative analysis of archaeal 16S rRNA and *amoA* genes to estimate the abundance and diversity of ammonia-oxidizing archaea in marine sediments. Extremophiles 12:605–615

Park SY, Lee JS, Lee KC, Yoon JH (2013) *Lentilitoribacter donghaensis* gen. nov., sp. nov., a slowly-growing alphaproteobacterium isolated from coastal seawater. Anton Leeuw Int J G 103:457–464

Quan ZX, Xiao YP, Roh SW, Nam YD, Chang HW, Shin KS, Rhee SK, Park YH, Bae JW (2008) *Joostella marina* gen. nov., sp. nov., a novel member of the family *Flavobacteriaceae* isolated from the East Sea. Int J Syst Evol Microbiol 58:1388–1392

Quan ZX, Zeng DN, Xiao YP, Roh SW, Nam YD, Chang HW, Yoon JH, Oh HM, Bae JW (2009) *Henriciella marina* gen. nov., sp. nov., a novel member of the family *Hyphomonadaceae* isolated from the East Sea. J Microbiol 47:156–161

Reimers CE, Jahnke RA, McCorkle DC (1992) Carbon fluxes and burial rates over the continental slope and rise off central California with implications for the global carbon cycle. Global Biogeochem Cycles 6:201–224

Reschke S, Ittekkot V, Panin N (2002) The nature of organic matter in the Danube river particles and north-western Black Sea sediments. Estuar Coast Shelf Sci 54:563–574

Rivkin RB, Legendre L (2001) Biogenic carbon cycling in the upper ocean: Effects of microbial respiration. Science 291:2398–2400

Robinson C (2008) Heterotrophic bacterial respiration. In: Kirchman DL (ed) Microbial ecology of the oceans. Wiley-Liss, New York, pp 299–334

Roh SW, Quan ZX, Nam YD, Chang HW, Kim KH, Rhee SK, Oh HM, Jeon CO, Yoon JH, Bae JW (2008) *Marinobacter goseongensis* sp. nov., from seawater. Int J Syst Evol Microbiol 58:2866–2870

Romanenko LA, Schumann P, Rohde M, Mikhailov VV, Stackebrandt E (2002) *Halomonas halocynthiae* sp. nov., isolated from the marine ascidian *Halocynthia aurantium*. Int J Syst Evol Microbiol 52:1767–1772

Romanenko LA, Schumann P, Rohde M, Mikhailov VV, Stackebrandt E (2004) *Reinekea marinisedimentorum* gen. nov., sp., nov., a novel gammaproteobacterium from marine coastal sediments. Int J Syst Evol Microbiol 54:669–673

Romanenko LA, Schumann P, Rohde M, Zhukova NV, Mikhailov VV, Stackebrandt E (2005) *Marinobacter bryozoorum* sp. nov. and *Marinobacter sediminum* sp. nov., novel bacteria from the marine environment. Int J Syst Evol Microbiol 55:143–148

Romanenko LA, Schumann P, Zhukova NV, Rohde M, Mikhailov VV, Stackebrandt E (2003a) *Oceanisphaera litoralis* gen. nov., sp. nov., a novel halophilic bacterium from marine bottom sediments. Int J Syst Evol Microbiol 53:1885–1888

Romanenko LA, Tanaka N, Frolova GM (2009a) *Marinomonas arenicola* sp. nov., isolated from marine sediment. Int J Syst Evol Microbiol 59:2834–2838

Romanenko LA, Tanaka N, Frolova GM (2010a) *Umboniibacter marinipuniceus* gen. nov., sp. nov., a marine gammaproteobacterium isolated from the molusc *Umbonium costatum* from the Sea of Japan. Int J Syst Evol Microbiol 60:603–609

Romanenko LA, Tanaka N, Frolova GM, Mikhailov VV (2008) *Pseudochrobactrum glaciei* sp. nov., isolated from sea ice collected from Peter the Great Bay of the Sea of Japan. Int J Syst Evol Microbiol 58:2454–2458

Romanenko LA, Tanaka N, Frolova GM, Mikhailov VV (2009b) *Psychrobacter fulvigenes* sp. nov., isolated from a marine crustacean from the Sea of Japan. Int J Syst Evol Microbiol 59:1480–1486

Romanenko LA, Tanaka N, Frolova GM, Mikhailov VV (2009c) *Sphingomonas japonica* sp. nov., isolated from the marine crustacean *Paralithodes camtschatica*. Int J Syst Evol Microbiol 59:1179–1182

Romanenko LA, Tanaka N, Frolova GM, Mikhailov VV (2009d) *Winogradskyella arenosi* sp. nov., a member of the family *Flavobacteriaceae* isolated from marine sediments from the Sea of Japan. Int J Syst Evol Microbiol 59:1443–1446

Romanenko LA, Tanaka N, Frolova GM, Mikhailov VV (2010b) *Arenicella xantha* gen. nov., sp. nov., a gammaproteobacterium isolated from a marine sandy sediment. Int J Syst Evol Microbiol 60:1832–1836

Romanenko LA, Tanaka N, Frolova GM, Mikhailov VV (2010c) *Kangiella japonica* sp. nov., isolated from a marine environment. Int J Syst Evol Microbiol 60:2583–2586

Romanenko LA, Tanaka N, Frolova GM, Mikhailov VV (2010d) *Marinicella litoralis* gen. nov., sp. nov., a gammaproteobacterium isolated from coastal seawater. Int J Syst Evol Microbiol 60:1613–1619

Romanenko LA, Tanaka N, Frolova GM, Svetashev VI, Mikhailov VV (2011a) *Litoreibacter albidus* gen. nov., sp. nov. and *Litoreibacter janthinus* sp. nov., members of the class *Alphaproteobacteria* isolated from the seashore. Int J Syst Evol Microbiol 61:148–154

Romanenko LA, Tanaka N, Kalinovskaya NI, Mikhailov VV (2013a) Antimicrobial potential of deep surface sediment associated bacteria from the Sea of Japan. World J Microbiol Biotechnol 29:1169–1177

Romanenko LA, Tanaka N, Svetashev VI, Falsen E (2013b) Description of *Cobetia amphilecti* sp. nov., *Cobetia litoralis* sp. nov. and *Cobetia pacifica* sp. nov., classification of *Halomonas halodurans* as a later heterotypic synonym of *Cobetia marina* and emended descriptions of the genus *Cobetia* and *Cobetia marina*. Int J Syst Evol Microbiol 63:288–297

Romanenko LA, Tanaka N, Svetashev VI, Kalinovskaya NI (2011b) *Pacificibacter maritimus* gen. nov., sp. nov., isolated from shallow marine sediment. Int J Syst Evol Microbiol 61:1375–1381

Romanenko LA, Tanaka N, Svetashev VI, Kurilenko VV, Mikhailov VV (2012) *Luteimonas vadosa* sp. nov., isolated from seashore sediment. Int J Syst Evol Microbiol 63:1261–1266

Romanenko LA, Tanaka N, Svetashev VI, Mikhailov VV (2011c) *Primorskyibacter sedentarius* gen. nov., sp. nov., a novel member of the class *Alphaproteobacteria* from shallow marine sediments. Int J Syst Evol Microbiol 61:1572–1578

Romanenko LA, Uchino M, Frolova GM, Tanaka N, Kalinovskaya NI, Latyshev N, Mikhailov VV (2007) *Sphingomonas molluscorum* sp. nov., a novel marine isolate with antimicrobial activity. Int J Syst Evol Microbiol 57:358–363

Romanenko LA, Uchino M, Mikhailov VV, Zhukova NV, Uchimura T (2003b) *Marinomonas primoryensis* sp. nov., a novel psychrophile isolated from coastal sea-ice in the Sea of Japan. Int J Syst Evol Microbiol 53:829–832

Romankevich EA (1984) Geochemistry of organicmatter in the ocean. (Springer, Berlin, Heidelberg, New York, Tokyo) p 334

Sherr EB, Sherr BF (1988) Role of microbes in pelagic food webs: a revised concept. Limnol Oceanogr 33:1225–1227

Sherr EB, Sherr BF, Paffenhöfer GA (1986) Phagotrophicprotozoa as food for metazoans: a 'missing' trophic link inmarine pelagic food webs? Mar Microb Food Webs 1:61–80

Shin HR, Shin CW, Kim C, Byun SK, Hwang SC (2005) Movement and structural variation of warm eddy WE92 for three years in the western East/Japan Sea. Deep-Sea Res II 52:1742–1762

Shubert CJ, Ferdelman TG, Strotmann B (2000) Organic matter composition and sulfate reduction rates in sediments off Chile. Organic Geochim 31:351–361

Simon M, Grossart H-P, Schweitzer B, Ploug H (2002) Microbial ecology of organic aggregates in aquatic ecosystems. Aquat Microb Ecol 28:175–211

Slomp CP, Malschaert JFP, Lohse L, van Raaphorst W (1997) Iron and manganese cycling in different sedimentary environments on the North Sea continental margin. Cont Shelf Res 17:1083–1117

Song J, Oh HM, Cho JC (2009) Improved culturability of SAR11 strains in dilution-to-extinction culturing from the East Sea, West Pacific Ocean. FEMS Microbiol Ecol 295:141–147

Sung HR, Yoon JH, Ghim SY (2012) *Shewanella dokdonensis* sp. nov., isolated from seawater. Int J Syst Evol Microbiol 62:1636–1643

Suzuki D, Ueki A, Amaishi A, Ueki K (2007a) *Desulfobulbus japonicus* sp. nov., a novel Gram-negative propionate-oxidizing, sulfate-reducing bacterium isolated from an estuarine sediment in Japan. Int J Syst Evol Microbiol 57:849–855

Suzuki D, Ueki A, Amaishi A, Ueki K (2007b) *Desulfopila aestuarii* gen. nov., sp. nov., a Gram-nagative, rod-like, sulfate-reducing bacterium isolated from an estuarine sediment in Japan. Int J Syst Evol Microbiol 57:520–526

Suzuki D, Ueki A, Amaishi A, Ueki K (2008) *Desulfoluna butyratoxydans* gen. nov., sp. nov., a novel gram-negative, butyrate-oxidizing, sulfate-reducing bacterium isolated from an estuarine sediment in Japan. Int J Syst Evol Microbiol 58:826–832

Tanaka N, Romanenko LA, Frolova GM, Mikhailov VV (2010) *Aestuariibacter litoralis* sp. nov., isolated from a sandy sediment of the Sea of Japan. Int J Syst Evol Microbiol 60:317–320

Thamdrup B, Canfield DE (1996) Pathways of carbon oxidation in continental margin sediments off central Chile. Limnol Oceanogr 41:1629–1650

Thamdrup B, Canfield DE (2000) Benthic respiration in aquatic sediments. In: Sala O, Jackson E, Mooney BE, Howarth RW (eds) Methods in ecosystem science. Springer, New York, pp 86–103

Tyler PA (1970) *Hyphomicrobia* and the oxidation of manganesse in aquatic ecosystems. Antonie Van Leeuwenhoek 36:567–578

Vandieken V, Nickel M, Jørgensen BB (2006) Carbon mineralization in Arctic sediments northeast of Svalbard: Mn(IV) and Fe(III) reduction as principal anaerobic respiratory pathways. Mar Ecol Prog Ser 322:15–27

Vandieken V, Pester M, Finke N, Hyun JH, Friedrich MW, Loy A, Thamdrup B (2012) Three manganese oxide-rich marine sediments harbor similar communities of acetate-oxidizing manganese-reducing bacteria. ISMEJ 6:2078–2090

Walsh JJ (1991) Importance of continental margins in the marine biogeochemical cycling of carbon and nitrogen. Nature 350:53–55

Weber A, Riess W, Wenzhoefer F, Jørgensen BB (2001) Sulfate reduction in Black Sea sediments: in situ and laboratory radiotracer measurements from the shelf to 2000 m depth. Deep-Sea Res I 48:2073–2096

Wollast R (1991) The coastal organic carbon cycle: fluxes, sources and sinks. In: Mantoura RFC, Martin J-M, Wollast R (eds). Ocean marine processes in global change. Wiley, pp 366–381

Yamada K, Ishizaka J, Nagata H (2005) Spatial and temporal variability of satellite primary production in the Japan Sea from 1998 to 2002. J Oceanogr 61:857–869

Yang EJ, Kang HK, Yoo S, Hyun JH (2009) Contribution of auto- and heterotrophic protozoa to the diet of copepods in the Ulleung Basin, East/Japan Sea. J Plankton Res 31:647–659

Yang SH, Kim MR, Seo HS, Lee SH, Lee JH, Kim SJ, Kwon KK (2013) Description of *Kordiimonas aquimaris* sp. nov., isolated from seawater, and emended descriptions of the genus *Kordiimonas* Kwon et al. 2005 emend. Xu et al. 2011 and of its existing species. Int J Syst Evol Microbiol 63:298–302

Yang SH, Kwon KK, Lee H-S, Kim SJ (2006) *Shewanella spongiae* sp. nov., isolated from a marine sponge. Int J Syst Evol Microbiol 56:2879–2882

Yang SH, Lee JH, Ryu JS, Kato C, Kim SJ (2007) *Shewanella donghaensis* sp. nov., a psychrophilic, piezosensitive bacterium producing high levels of polyunsaturated fatty acid, isolated from deep-sea sediments. Int J Syst Evol Microbiol 57:208–212

Yang SJ, Kan I, Cho JC (2012) Genome sequence of strain IMCC14465, isolated from the East Sea, belonging to the PS1 clade of *Alphaproteobacteria*. J Bacteriol 194:6952–6953

Yi H, Schumann P, Sohn K, Chun J (2004) *Serinicoccus marinus* gen. nov., sp. nov., a novel actinomycete with L-ornithine and L-serine in the peptidoglycan. Int J Syst Evol Microbiol 54:1585–1589

Yong JJ, Park SJ, Kim HJ, Rhee SK (2007) *Glaciecola agarilytica* sp. nov., an agar-digesting marine bacterium from the East Sea, Korea. Int J Syst Evol Microbiol 57:951–953

Yoo S, Kim HC (2003) Primary productivity in the East Sea. In: Choi JK. (ed.) The plankton ecology of Korean coastal waters. Dongwha Press, pp 96–111 (in Korean)

Yoo S, Park JS (2009) Why is the southwest the most productive region of the East Sea/Sea of Japan? J Mar Syst 78:301–315

Yoon JH, Kang SJ, Lee CH, Oh TK (2005a) *Dokdonia donghaensis* gen. nov., sp. nov., isolated from sea water. Int J Syst Evol Microbiol 55:2323–2328

Yoon JH, Kang SJ, Lee CH, Oh TK (2006a) *Donghaeana dokdonensis* gen. nov., sp. nov., isolated from sea water. Int J Syst Evol Microbiol 56:187–191

Yoon JH, Kang SJ, Lee MH, Oh HW, Oh TK (2006b) *Porphyrobacter dokdonensis* sp. nov., isolated from sea water. Int J Syst Evol Microbiol 56:1079–1083

Yoon JH, Kang SJ, Lee MH, Oh TK (2007a) Description of *Sulfitobacter donghicola* sp. nov., isolated from seawater of the East Sea in Korea, transfer of Staleya guttiformis Labrenz et al. 2000 to the genus *Sulfitobacter* as *Sulfitobater guttiformis* comb. nov. and emended description of the genus *Sulfitobacter*. Int J Syst Evol Microbiol 57:1788–1792

Yoon JH, Kang SJ, Lee SY, Lee CH, Oh TK (2005b) *Maribacter dokdonensis* sp. nov., isolated from sea water off a Korean island, Dokdo. Int J Syst Evol Microbiol 55:2051–2055

Yoon JH, Kang SJ, Lee SY, Lee MH, Oh TK (2005c) *Virgibacillus dokdonensis* sp. nov., isolated from a Korean island, Dokdo, located at the edge of the East Sea in Korea. Int J Syst Evol Microbiol 55:1833–1837

Yoon JH, Kang SJ, Oh TK (2005d) *Marinomonas dokdonensis* sp. nov., isolated from sea water. Int J Syst Evol Microbiol 55:2303–2307

Yoon JH, Kang SJ, Oh TK (2006c) *Polaribacter dokdonensis* sp. nov., isolated from seawater. Int J Syst Evol Microbiol 56:1251–1255

Yoon JH, Kang SJ, Oh TK (2007b) *Donghicola eburneus* gen. nov., sp. nov., isolated from seawater of the East Sea in Korea. Int J Syst Evol Microbiol 57:73–76

Yoon JH, Kang SJ, Oh TK (2007c) *Sulfitobacter marinus* sp. nov., isolated from seawater of the East Sea in Korea. Int J Syst Evol Microbiol 57:302–305

Yoon JH, Kang SJ, Park S, Oh TK (2007d) *Devosia insulae* sp. nov., isolated from soil, and emended description of the genus *Devosia*. Int J Syst Evol Microbiol 57:1310–1314

Yoon JH, Kang SJ, Park SY, Oh TK (2007e) *Jannaschia donghaensis* sp. nov., isolated from seawater of the East Sea, Korea. Int J Syst Evol Microbiol 57:2132–2136

Yoon JH, Kim HG, Kim IG, Kang KH, Park YH (2003a) *Erythrobacter flavus* sp. Nov., a slight halophile from the East Sea in Korea. Int J Syst Evol Microbiol 53:1169–1174

Yoon JH, Kim IG, Kang KH, Oh TK, Park YH (2004a) *Bacillus hwajinpoensis* sp. nov. and an unnamed *Bacillus genomospecies*, novel members of *Bacillus* rRNA group 6 isolated from sea water of the East Sea and the Yellow Sea in Korea. Int J Syst Evol Microbiol 54:803–808

Yoon JH, Kim IK, Kang KH, Oh TK, Park YH (2003b) *Alteromonas marina* sp. nov., isolated from sea water of the East Sea in Korea. Int J Syst Evol MicrobiolInt J Syst Evol Microbiol 53:1625–1630

Yoon JH, Lee MH, Oh TK (2004b) *Porphyrobacter donghaensis* sp. nov., isolated from sea water of the East Sea in Korea. Int J Syst Evol Microbiol 54:2231–2235

Yoon JH, Lee SY, Kang SJ, Lee C-H, Oh TK (2007f) *Pseudoruegeria aquimaris* gen. nov., sp. nov., isolated from seawater of the East Sea in Korea. Int J Syst Evol Microbiol 57:542–547

Yoon JH, Park SE, Kang SJ, Oh TK (2007g) *Rheinheimera aquimaris* sp. nov., isolated from seawater of the East Sea in Korea. Int J Syst Evol Microbiol 57:1386–1390

Yoon JH, Shin DY, Kim IG, Kang KH, Park YH (2003c) *Marinobacter litoralis* sp. nov., a moderately halophilic bacterium isolated from sea water from the East Sea in Korea. Int J Syst Evol Microbiol 53:563–568

Yoon JH, Yeo SH, Oh TK (2004c) *Hongiella marincola* sp. nov., isolated from sea water of the East Sea in Korea. Int J Syst Evol Microbiol 54:1845–1848

Chapter 12
Zooplankton

Chul Park, Hae-Lip Suh, Young-Shil Kang, Se-Jong Ju and Eun-Jin Yang

Abstract Zooplankton plays a key role in marine ecosystems as an energy link between primary producers and higher trophic levels in the pelagic food web. In the East Sea (Japan Sea), some studies of zooplankton in terms of their species and distribution have been conducted in selected regions, but there is a lack of reviews and/or summaries of zooplankton species composition and distribution covering the entire area of the East Sea. Therefore, we here summarize the species composition and abundance of zooplankton, where possible covering the entire area of the East Sea. We report several interesting findings (spatial distribution, the trophic role of microzooplankton in pelagic food web, etc.) based on the currently available information. We believe that a survey, covering the entire region of the East Sea, should be conducted in the near future to better understand the species composition and distribution of mesozooplankton, including microzooplankton,

C. Park (✉)
Department of Oceanography and Ocean Environmental Sciences,
Chungnam National University, Daejeon 34134, Republic of Korea
e-mail: chulpark@cnu.ac.kr

H.-L. Suh
Department of Oceanography, Chonnam National University,
Gwangju 61186, Republic of Korea
e-mail: suhhl@chonnam.ac.kr

Y.-S. Kang
Korea Fisheries Resources Agency, Busan 48058, Republic of Korea
e-mail: kys6207@fira.or.kr

S.-J. Ju
Deep-Sea and Seabed Resources Research Center, Korea Institute of Ocean Science
and Technology, Ansan 15627, Republic of Korea
e-mail: sjju@kiost.ac.kr

E.-J. Yang
Korea Polar Research Institute, Korea Institute of Ocean Science and Technology,
Incheon 21990, Republic of Korea
e-mail: ejyang@kiost.ac.kr; ejyang@kopri.re.kr

© Springer International Publishing Switzerland 2016
K.-I. Chang et al. (eds.), *Oceanography of the East Sea (Japan Sea)*,
DOI 10.1007/978-3-319-22720-7_12

which will eventually help us to properly monitor and predict the functional and structural changes of the East Sea ecosystem due to the influences of climate change as well as human impacts such as intensive fishing.

Keywords East Sea (Japan Sea) · Zooplankton · Species list · Abundance distribution · Microbial food web

12.1 Introduction

Biological energy comes from primary production by phytoplankton in the pelagic ecosystem. It is consumed by zooplankton and subsequently this zooplankton is consumed by larger animals in a sequence known as 'the classical food web' in pelagic ecosystems. Since publication of Azam et al. (1983), recognition of the important role of microzooplankton (heterotrophic and mixotrophic organisms and metazoan: generally in the size range from 20 to 200 μm) as nutritionally packaging and improving low quality prey (microbes) and effectively transferring energy to mesozooplankton (planktonic animals: generally in the size range from 200 to 2000 μm) in oligotrphic environments. In spite of their important role in the transfer of energy in marine ecosystems, studies of zooplankton in the East Sea were not only rare but also limited to the mesozooplankton. Previous reports on zooplankton in this area were mainly done by Japanese and Russian scientists before World War II. Park's (1956) study on the seasonal change of the plankton in the Korea Strait was the first work on zooplankton in the seas around the Korean peninsula by a Korean scientist. Unfortunately, there has not yet been a survey covering the entire East Sea.

Here we report on the species composition and abundance distribution of mesozooplankton in the Korean waters of the East Sea where they have been primarily and intensively surveyed by Korean Scientists. Some local patterns and the trophic role of microzooplankton in the pelagic food web are also included. We believe that a survey, covering the whole area of the East Sea, should be conducted in the near future to better understand the species composition and distribution of mesozooplankton as well as microzooplankton. Concurrently, in situ experiments and incubations are also strongly recommended to determine the vital rates of key species of zooplankton and their trophic interactions with multiple environmental forcing factors, which could significantly improve model predictions. These results will help us to properly monitor and predict the functional and structural changes of the East Sea ecosystem caused by human activities such as global warming, ocean acidification, overfishing, eutrophication, etc.

12.2 Species Composition of Mesozooplankton

The great diversity of mesozooplankton in the southern East Sea has been well recognized (Chihara and Murano 1997; Park et al. 1997). A general description of the major taxa is provided with some indication of horizontal distribution (neritic or oceanic) and water masses (warm or cold). Zooplankton in the southern East Sea are commonly divided into four communities in the upper pelagic layer: neritic warm-water, neritic cold-water, oceanic warm-water, and oceanic cold-water communities. The neritic community is distributed in a narrow range along the coast in the southern East Sea down to *ca.* 50 m depth. It has been well known that the meroplankton, including polychaetes, benthic crustaceans and mollusk larvae, usually form a large portion of the neritic community. However, almost all meroplankton and juveniles were not included in the zooplankton list of the southern East Sea. Based on the literature reported previously and Suh's unpublished data, a total of 175 species was recorded in the zooplankton list. According to the species list (Table 12.1), the neritic warm-water, neritic cold-water, oceanic warm-water, and oceanic cold-water communities consisted of 21, 7, 131, and 16 species, respectively. The zooplankton diversity in the southern East Sea was strongly affected by the Tsushima Warm Current, a branch of the Kuroshio.

12.2.1 Cnidaria and Mollusca

The Cnidaria, known as jellyfish or gelatinous zooplankton, form a number of groups which are important members of the marine plankton. Reports on jellyfish blooms have increased substantially in recent decades (Condon et al. 2012), and their ecological and economic impacts on marine ecosystem have also increased (Purcell 2009).

In the southern East Sea, 18 species of Cnidaria, 10 species of Hydrozoa and 8 species of Scyphozoa, were recorded. Of these, 3 species including *Diphyes chamissonis, Nemopilema nomurai* and *Dactylometra quinquecirrha* belonged to the neritic community. Of the remaining 15 oceanic species, 12 species were distributed in warm water, whereas 3 species, *Muggiaea bargmannae, Dipleurosoma typicum* and *Aglantha digitale*, were reported in cold water.

In the phylum Mollusca three taxa of gastropods including Heteropoda, Thecosomata and Gymnosomata have adopted a holoplanktonic life. In the southern East Sea, *Desmopterus papilio* belonging to Thecosomata was recorded.

12.2.2 Arthropoda

12.2.2.1 Branchiopoda and Ostracoda

Of branchiopod crustaceans, the Cladocera are usually a more important group in fresh water than in marine. There are three genera in the marine cladocerans: *Penilia, Podon* and *Evadne*. In the southern East Sea, four species of cladocerans, including *Penilia avirostris, Evadne nordmanni, Evadne spinifera* and *Evadne tergestina*, were recorded. In the Ostracoda, a single species, *Euconchoecia pacifica*, was reported.

Table 12.1 List of the zooplankton taxa in the southern East Sea

Phylum	(Sub-) Class	Order	Family	Species	Type	Reference
Cnidaria	Hydrozoa	Anthoathecata	Porpitidae	*Porpita porpita*	OW	8
		Siphonophorae	Abylidae	*Abylopsis eschscholtzii*	OW	1, 8
				Bassia bassensis	OW	1
			Diphyidae	*Diphyes chamissonis*	NW	8
				Muggiaea bargmannae	OC	8
			Physaliidae	*Physalia physalis*	OW	8
		Leptothecata	Dipleurosomatidae	*Dipleurosoma typicum*	OC	8
			Cirrholoveniidae	*Cirrholovenia tetranema*	OW	1
			Eirenidae	*Eutima japonica*	OW	1
		Narcomedusae	Aeginidae	*Solmundella bitentaculata*	OW	1, 8
		Trachymedusae	Geryoniidae	*Liriope tetraphylla*	OW	8
			Rhopalonematidae	*Aglantha digitale*	OC	1, 2, 8
	Scyphozoa	Rhizostomeae	Rhizostomatidae	*Nemopilema nomurai*	NW	8, 9
		Semaeostomeae	Cyaneidae	*Cyanea nozakii*	OW	8
			Pelagiidae	*Chrysaora quinquecirrha*	NW	8
				Pelagia noctiluca	OW	8
			Ulmaridae	*Aurelia aurita*	OW	1, 8
Mollusca	Gastropoda	Thecosomata	Desmopteridae	*Desmopterus papilio*	OW	1
Arthropoda	Branchiopoda	Cladocera	Sididae	*Penilia avirostris*	NW	9
			Pononidae	*Evadne nordmanni*	NC	9
				E. spinifera	OW	9
				Pseudevadne tergestina	NW	9
	Ostracoda	Halocyprida	Halocyprididae	*Euconchoecia pacifica*	OW	1
	Copepoda	Calanoida	Acartiidae	*Acartia danae*	OW	1, 9

(continued)

Table 12.1 (continued)

Phylum	(Sub-) Class	Order	Family	Species	Type	Reference
				A. negligens	OW	1, 9
				A. omorii	NC	1, 9
				A. pacifica	NW	1, 9
			Aetidadae	Aetideus acutus	OW	1, 2, 9
				Chridius gracilis	OW	1, 2, 9
			Calanidae	Calanus sinicus	NC	1, 2, 9
				Canthocalanus pauper	OW	1, 2, 9
				Cosmocalanus dawinii	OW	1, 2, 9
				Mesocalanus tenuicornis	OW	1, 2, 9
				Nannocalanus minor	OW	1, 9
				Neocalanus cristatus	OC	2, 9
				N. plumchrus	OC	2, 9
				Undinula vulgaris	OW	1, 2, 9
			Candaciidae	Candacia bipinnata	OW	1, 2, 9
				C. catula	OW	1, 2, 9
				C. curta	OW	3, 1, 9
				C. discaudata	OW	1, 9
				C. tuberculata	OW	1, 9
				C. truncata	OW	1, 9
			Centropagidae	Centropages abdominalis	NC	9
				C. bradyi	OW	1, 2, 9
				C. furcatus	OW	1, 2, 9
				C. gracilis	OW	1, 9
				C. tenuiremis	NW	1, 9

(continued)

Table 12.1 (continued)

Phylum	(Sub-) Class	Order	Family	Species	Type	Reference
			Clausocalanidae	*Clausocalanus arcuicornis*	OW	1, 5, 9
				C. farrani	OW	1, 5
				C. furcatus	OW	1, 2, 5, 9
				C. pergens	OW	1, 2, 5, 9
				Ctenocalanus vanus	NW	1, 5, 9
				Pseudocalanus minutus	OC	1, 9
				P. newmani	OC	5, 9
			Eucalanidae	*Eucalanus bungii*	OC	9
				E. hyalinus	OW	1, 9
				E. elongatus	OW	1, 9
				Pareucalanus attenuatus	OW	1, 9
				Subeucalanus crassus	OW	1, 2, 9
				S. mucronatus	OW	1, 9
				S. subcrassus	OW	1, 9
				S. subtenuis	OW	1, 2, 9
			Euchaetidae	*Euchaeta concinna*	OW	1, 2, 7, 9
				E. longicornis	OW	7, 9
				E. plana	OW	1, 2, 7, 9
				E. rimana	OW	1, 2, 7, 9
				Paraeuchaeta elongata	OC	1, 2, 7, 9
				P. russelli	OW	1, 2, 7, 9
			Lucicutidae	*Lucicutia flavicornis*	OW	1, 9
				L. clausi	OW	1, 9
			Metridinidae	*Metridia pacifica*	OC	1, 2, 9

(continued)

Table 12.1 (continued)

Phylum	(Sub-) Class	Order	Family	Species	Type	Reference
				Pleuromamma gracilis	OW	1, 9
				P. scutullata	OW	9
			Paracalanidae	*Acrocalanus gibber*	OW	1, 2, 9
				A. gracilis	OW	1, 3, 9
				A. longicornis	OW	1, 9
				Paracalanus aculeatus	OW	1, 2, 9
				P. parvus s. l.	NW	1, 2, 9
				Calocalanus pavo	OW	1, 2, 9
				C. plumulosus	OW	1, 9
			Pontellidae	*Calanopia minor*	OW	1
				Labidocera acuta	OW	1, 2, 9
				L. euchaeta	NW	1, 2, 9
				L. japonica	OW	1, 2, 9
			Rhincalanidae	*Rhincalanus cornutus*	OW	1, 2
				R. nasutus	OW	1, 2, 9
			Scolecitrichidae	*Scolecithrix danae*	OW	1, 2, 9
				Scolecithricella longispinosa	OW	2, 9
				S. minor	OC	1, 2, 9
				S. nicobarica	OW	9
			Temoridae	*Temora discaudata*	NW	1, 2, 9
				T. stylifera	NW	2, 9
		Cyclopoida	Corycaeidae	*Agetus flaccus*	OW	9
				A. limbatus	OW	9
				A. typicus	OW	9

(continued)

Table 12.1 (continued)

Phylum	(Sub-) Class	Order	Family	Species	Type	Reference
				Corycaeus clausi	OW	1, 9
				C. crassiusculus	OW	1, 9
				C. speciosus	OW	1, 2, 9
				Ditrichocorycaeus affinis	NC	1, 2, 9
				D. andrewsi	OW	1, 9
				D. asiaticus	OW	1, 9
				D. dahli	OW	1, 9
				D. erythraeus	OW	1, 9
				D. lubbocki	OW	9
				D. subtilis	OW	9
				Farranula carinata	OW	9
				F. concinna	OW	9
				F. gibbula	OW	1, 9
				Onychocorycaeus agilis	OW	1, 9
				O. catus	OW	1, 2, 9
				O. pacificus	OW	1, 2, 9
				O. pumilus	OW	1, 9
				Urocorycaeus furcifer	OW	1, 9
			Oithonidae	*Oithona atlantica*	NC	1
				O. plumifera	OW	1, 2, 9
				O. setigera	OW	1, 2, 9
				O. similis	OC	1, 2, 9

(continued)

Table 12.1 (continued)

Phylum	(Sub-) Class	Order	Family	Species	Type	Reference
				O. tenuis	OW	1, 9
			Oncaeidae	*Oncaea clevei*	OW	1, 4, 9
				O. media	OW	1, 4, 9
				O. mediterranea	OW	1, 4, 9
				O. scottodicarloi	OW	1, 4, 9
				O. venella	OW	1, 3, 9
				O. venusta	OW	1, 2, 3, 9
				O. waldemari	OW	4, 9
				Triconia borealis	OC	6, 9
				T. conifera	OW	1, 2, 6, 9
				T. hirsuta	OW	6, 9
				T. pararedacta	OW	9
			Sapphirinidae	*Copilia mirabilis*	OW	1, 9
				C. quadrata	OW	1, 9
				Sapphirina stellata	OW	1, 9
				S. gemma	OW	1, 9
				S. ovatolanceolata	OW	1, 9
		Harpacticoida	Peltidiidae	*Goniopsyllus rostratus*	OW	1, 2, 9
				Clytemnestra scutellata	OW	1, 9
			Ectinosomatidae	*Microsetella norvegica*	OW	1, 9
			Miraciidae	*Macrosetella gracilis*	OW	1, 9
			Euterpinidae	*Euterpina acutifrons*	OW	1, 2, 9
	Malacostraca	Amphipoda	Hyperiidae	*Themisto japonica*	OC	1, 2, 9
		Cumacea	Bodotriidae	*Heterocuma sarsi*	OW	1

(continued)

Table 12.1 (continued)

Phylum	(Sub-) Class	Order	Family	Species	Type	Reference
				Bodotria serrulata	OW	1
			Diastylidae	*Dimorphostylis asiatica*	OW	1
		Mysidacea	Mysidae	*Orientomysis tenuicauda*	OW	1
		Dacopoda	Luciferidae	*Lucifer chacei*	OW	9
				L. hanseni	OW	9
			Sergestidae	*Acetes japonicus*	NW	2
		Euphausicea	Euphausiidae	*Euphausia pacifica*	OC	1, 9
Chaetognatha	Sagittoidea	Aphragmophora	Sagittidae	*Aidanosagitta crassa*	NC	1, 9
				A. neglecta	OW	1
				A. regularis	OW	1
				Ferosagitta ferox	OW	1
				F. robusta	OW	1
				Flaccisagitta enflata	OW	1, 2,9
				Krohnitta pacifica	OW	1
				Parasagitta elegans	OC	1, 2, 9
				Sagitta bipunctata	OW	1, 2
				Serratosagitta pacifica	OW	1
				Zonosagitta bedoti	NW	1, 2, 9
				Z. nagae	OW	1
				Z. pulchra	OW	1, 2
Chordata	Appendiculari	Copelata	Fritillariidae	*Fritillaria charybdae*	OW	1
				F. formica	OW	1
				F. messanensis	OW	1

(continued)

Table 12.1   (continued)

Phylum	(Sub-) Class	Order	Family	Species	Type	Reference
			Oikopleuridae	*Oikopleura (Vexillaria) cophocerca*	NW	1, 2, 9
				O. (Vexillaria) dioica	NW	1, 2
				O. (Vexillaria) parva	NW	1, 2
				O. (Vexillaria) rufescens	NW	1, 2
				O. (Coecaria) fusiformis	NW	1, 2
				O. (Coecaria) fusiformis cornutogastra	NW	1, 2
				O. (Coecaria) gracilis	NW	1, 2
	Thaliacea	Doliolida	Doiliolidae	*Doliolum nationalis*	OW	1, 2
		Salpida	Salpidae	*Thalia democratica*	OW	2, 9

NC neritic cold-water species, *NW* neritic warm-water species, *OC* oceanic cold-water species, *OW* oceanic warm-water species
References *1* Chihara and Murano (1997), *2* Park et al. (1997), *3* Wi et al. (2008), *4* Wi et al. (2009), *5* Im (2010), *6* Wi et al. (2010), *7* Jeong et al. (2011), *8* Yoon et al. (unpublished data), *9* Suh (unpublished data)

12.2.2.2 Copepoda

The Copepoda, the most dominant taxon in the marine zooplankton, play an important role in marine ecosystems. Their abundance and distribution are strongly influenced by hydrographic conditions of their habitat such as water temperature and currents, suggesting them as biological indicators for water masses (Beaugrand et al. 2002).

In the southern East Sea, copepods were the most dominant taxon, comprising 108 species. The neritic species, such as *Calanus sinicus*, *Paracalanus parvus* s. l. and *Ditrichocorycaeus affinis*, were most abundant in the coastal region, while oceanic species, including *Metridia pacifica* and *Oithona plumifera*, predominated in the offshore region. The seasonal change of water masses affects the community structure of copepods. The oceanic warm-water species, *Oithona plumifera*, *Clausocalanus furcatus*, *Mesocalanus tenuicornis* and *Euchaeta plana*, were the dominant species from summer to fall, while cold-water species, such as *Neocalanus plumchrus*, *Scolecithricella minor* and *Paraeuchaeta elongata*, predominated from winter to spring.

12.2.2.3 Malacostraca

The superorder Peracarida is a large group of malacostracan crustaceans, having members in marine, freshwater, and terrestrial habitats. They are chiefly defined by the presence of a brood pouch, or marsupium. In the Peracarida there are five species in three orders. Each single species of Amphipoda and Mysida, and three species of Cumacea were recorded in the southern East Sea. The amphipod *Themisto japonica* is an oceanic cold-water species and a dominant carnivore in the epipelagic region, suggesting an important dietary source for whales, birds and fishes (Nemoto 1962; Vermeer and Devito 1988).

The orders Euphausiacea and Decapoda belong to the superorder Eucarida morphologically characterized by having the carapace fused to all thoracic segments, and by the presence of stalked eyes. In the southern East Sea, the euphausiid *Euphausia pacifica* was reported as an oceanic cold-water species (Ponomareva 1963), while in spawning season krill swarms were usually found in coastal waters (Suh et al. 1998). In the Decapoda, two species of the genus *Lucifer* (i.e., *L. chacei* and *L. hanseni*) and *Acetes japonicus* were reported in the southern East Sea. *Acetes japonicus* is usually known as a neritic warm-water species, and both *L. chacei* and *L. hanseni* as oceanic warm-water species.

12.2.3 Chaetognatha

The Chaetognatha is a small and rather isolated phylum, confined to the marine environment. This group is a relatively abundant carnivore, often ranking second or third in frequency after the ubiquitous copepods. Generic names of chaetognaths proposed by Bieri (1991) have been accepted. In the southern East Sea, a total of 13 species has been reported: *Aidanosagitta crassa*, *A. neglecta*, *A. regularis*, *Ferosagitta ferox*, *F. robusta*, *Flaccisagitta enflata*, *Krohnitta pacifica*, *Parasagitta elegans*, *Sagitta bipunctata*, *Serratosagitta pacifica*, *Zonosagitta bedoti*, *Z. nagae* and *Z. pulchra*. Most species of chaetognaths belong to the oceanic community, except two neritic species, *Aidanosagitta crassa* and *Zonosagitta bedoti*. In addition, most of them were warm-water species, but one was a cold-water species, *Parasagitta elegans*.

12.2.4 Chordata—Tunicata

Tunicata of the phylum Chordata includes two classes of holoplankton: Appendicularia and Thaliacea. In the southern East Sea, ten species of appendicularian have been recorded: *Oikopleura (Vexillaria) dioica*, *O. fusiformis*, *O. gracilis*, *O. cornutogastra*, *O. refescens*, *O. parva*, *O. cophocerca*, *Fritillaria formica*, *F. charybdae*, and *F. messanensis*. Species of *Oikopleura* and *Fritillaria* belong to the neritic and oceanic warm-water communities, respectively. Moreover, two species of Thaliacea, *Doliolum nationalis* and *Thalia democratica*, were reported from summer to fall, and they were well known as oceanic warm-water species.

12.3 Abundance Distribution

There is no regular survey or monitoring of zooplankton in the whole East Sea, so information on abundance distribution is limited. Hirota and Hasegawa (1999) synthesized the data of regional monitoring from the Maizuru Marine Observatory, Meteorological Agency of Japan (1966–1991) and the National Fisheries Research and Development Institute, Republic of Korea (1968–1992). They reported that the total zooplankton biomass was 9×10^6 tons in daytime and 16×10^6 tons at night.

In this section, we synthesize the information from regional scale research and review the spatio-temporal variations in zooplankton abundance with emphasis on seasonal patterns. Moreover, we describe spatio-temporal distributions of the major zooplankton groups and species.

12.3.1 Spatio-Temporal Distributions of Zooplankton Abundance

Zooplankton abundance/biomass generally peaked in spring (April–May), and was lowest in winter across the whole East Sea (Morioka 1985). However, there were some regional differences in seasonality (Table 12.2). Hirota and Hasegawa (1999) reported that peaks occurred in both spring and autumn. The spring peak was larger than that in autumn, appearing in May to June in the coastal area (shallower than 100 m depth) and in April in the offshore area (deeper than 100 m depth). On the other hand, in the northwestern shelf and subpolar front zones, the zooplankton peak occurred in summer, and the lowest abundance occurred in the autumn or winter (Dolganova 2000, 2001). In contrast, on the southeastern shelf, the peak appeared earlier than in the northwestern shelf: zooplankton exhibited the largest peak in winter to early spring (February–April) and a smaller peak in autumn (Kang 2008).

Zooplankton abundance/biomass was generally high in the coastal and northern areas (Table 12.3). Morioka (1985) had simply classified the water masses of the East Sea into the three groups as cold, warm and coastal water, and noted the annual mean of zooplankton biomass in each separate water mass as 95, 53 and 57 mg m^{-3}, respectively. On the other hand, Hirota and Hasegawa (1999) divided data of zooplankton biomass collected in the whole East Sea into three time periods, daytime, nighttime and twilight-time and analyzed the distributional patterns (Table 12.3 and Fig. 12.1). In daytime, zooplankton biomass was more than 50 mg m^{-3} in the area north of 39°N; in particular, the zooplankton biomass was highest, with more than 150 mg m^{-3} in the northwestern area, whereas zooplankton biomass was less than 50 mg m^{-3} in the region south of 39°N and also off Hokkaido and off Tohoku, but was higher, more than 100 mg m^{-3}, in the region from the Korea Strait to the coast of Shimane.

Considering the regional small scale, zooplankton abundance was 55-916 ind. m^{-3} in the southwestern region (34–38°N and 129–131°E, Table 12.3; Kang 2008; Rho et al. 2010). Zooplankton abundance was higher in the inshore than in the offshore area, 241 and 409 ind. m^{-3}, respectively. In the subpolar front zone (37–38.5°N and 129.5–132°E), zooplankton abundance was 214.6 ind. m^{-3} in the cold water (the north part of the frontal zone) and 469.1 ind. m^{-3} in the warm water (the south part of the frontal zone) (Park and Choi 1997). As previously stated, zooplankton biomass/abundance was generally higher in the northern and colder

Table 12.2 Seasonality of zooplankton biomass/abundance in sub-regions of the East Sea

Sub-region	Seasonality		Reference
	Maximum	Minimum	
East Sea	Spring (April–May)	Winter	Morioka (1985)
Coastal area (<100 m depth) in day	Spring (May–June) > Autumn (October)	Winter (December–February)	Hirota and Hasegawa (1999)
Offshore area (>100 m depth) in day	Spring (April) > August–November	Winter (December–January)	
Northwestern shelf	Summer	Autumn	Dolganova (2000 and 2001)
Subpolar front zone	Summer	Winter	
Southwestern shelf	Winter-spring (February–April) > Autumn (October)	Winter (December)	Kang (2008)

Table 12.3 Zooplankton biomass/abundance in sub-regions of East Sea

Sub-region		Zooplankton biomass/abundance	Reference
The East Sea	Cold water	95 mg m^{-3}	Morioka (1985)
	Warm water	53 mg m^{-3}	
	Coastal water	57 mg m^{-3}	
The East Sea (33–48°N and 128–141°E)	Most north area of 39°N	More than 50 mg m^{-3} (in day)	Hirota and Hasegawa (1999)
	- Northwestern area	>150 mg m^{-3} (in day)	
	Most south of 39°N (including off Hokkaido and off Tohoku)	<50 mg m^{-3} (in day)	
	- Korea Strait-off Shimane	>100 mg m^{-3} (in day)	
The southwestern region (34–38°N and 129–131°E)	The whole region	55–916 ind. m^{-3}, 87.6 ind. m^{-3}	Rho et al. (2010), Kang (2008)
	Offshore (>1000 m depth)	241 ind. m^{-3}	
	Inshore (<150 m depth)	409 ind. m^{-3}	
Subpolar Front zone (37–38.5°N and 129.5–132°E)	Warm water (south area from front)	469.1 ind. m^{-3}	Park and Choi (1997)
	Cold water (north area from front)	214.6 ind. m^{-3}	

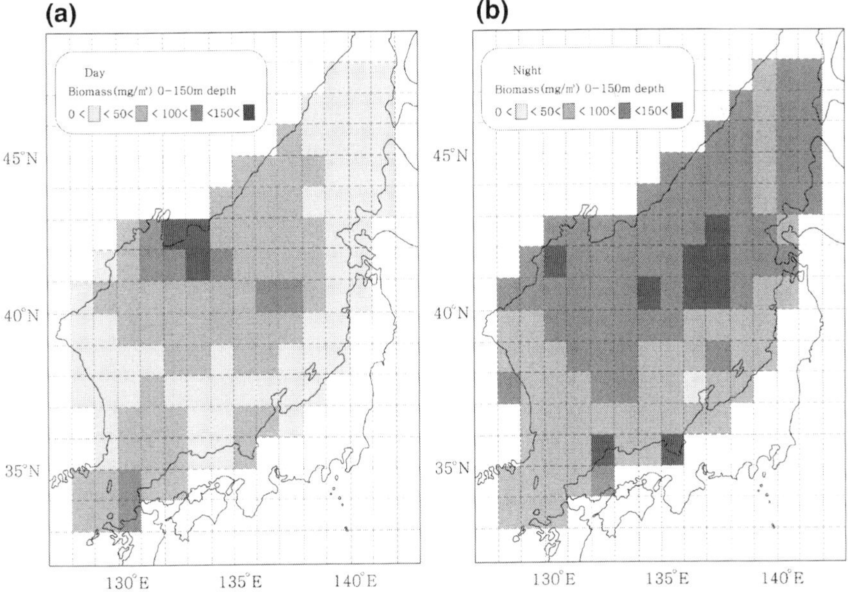

Fig. 12.1 Horizontal distribution of annual zooplankton biomass in the East Sea (**a**) in the daytime; (**b**) at night (Redrawn from Hirota and Hasegawa 1999)

regions than in the southern and warmer regions, and more abundant in the inshore area than in the offshore area.

12.3.2 Abundance Distributions of Major Zooplankton Taxa

Vinogradov and Sazhin (1978) reported the abundance of the major zooplankton taxa in three representative regions (the north, south and subpolar front areas) of the East Sea (Table 12.4). Among the major zooplankton taxa, copepods predominated, with abundances of chaetognaths, amphipods, euphausiaceas and ostracods decreasing in order. In the northern area, copepods, euphausiaceas, chaetognaths, amphipods and ostracods were 71.8, 1.5, 8.1, 1.8 and 5.3 mg m^{-3} in abundance, respectively. In the subpolar front, copepods and chaetognaths were overwhelmingly greater in abundances than in the northern area. Among the three areas, the south area had the lowest abundances for all the zooplankton taxa mentioned above.

Ashjian et al. (2005) surveyed zooplankton in the East Sea during the summer of 1999 with high- and low-magnification cameras, and net tows (Fig. 12.2a). From the net sampling, they determined that copepods were remarkably dominant, with greater abundance around the Ulleung Basin region (Fig. 12.2b). Copepods were mostly abundant near the subpolar front and in the northern area. Unidentified copepods and *Oithona* spp. that were observed by the high-magnification camera had the greatest abundance around the Ulleung Basin region, but not in the Tsushima Warm Current (TWC), the North Korean Cold Current (NKCC) or the East Korean Warm Current (EKWC) (Fig. 12.3). Observations by the low-magnification camera indicated that distributions of copepods and *Calanus* spp. were similar to distributions of copepods and *Oithona* spp. observed by the high-magnification camera in some regions, but not in the NKCC. In summary, Ashjian et al. (2005) found that copepods, *Oithona* spp and *Calanus* spp. were more abundant in the northern portion of the East Sea than in the southern portion. They considered the distributions to be closely associated with hydrographically distinct regions, with the southern portion being more tropical/oligotrophic and the northern portion more boreal/eutrophic.

12.3.3 Distribution of Major Zooplankton Taxa in Local Areas

Abundance of zooplankton species and taxa was studied for five local areas (Fig. 12.4), (a) the vicinity of the Korea Strait, (b) the southwestern shelf around the Ulleung Basin, (c) the front zone in the southwestern shelf, (d) the southeastern area (PM line), (e) and Toyama Bay.

Table 12.4   Abundance of major zooplankton taxa in the sub-regions of East Sea

	Copepoda	Euphausiacea	Chaetognatha	Amphipoda	Ostracoda	Remarks
Northern part (41° 59′N and 133° 37′E)	71.8 mg m^{-3}	1.5 mg m^{-3}	8.1 mg m^{-3}	1.8 mg m^{-3}	5.3 mg m^{-3}	Vinogradov and Sazhin (1978)
In the subpolar front zone (41° 28′N and 130° 06′E)	299.9 mg m^{-3}	1.4 mg m^{-3}	65.6 mg m^{-3}	1.3 mg m^{-3}	0.06 mg m^{-3}	
Southern part (38° 25′N and 134° 43′E)	5.5 mg m^{-3}	0.3 mg m^{-3}	1.3 mg m^{-3}	1.3 mg m^{-3}	0.009 mg m^{-3}	
Southeastern shelf	188.7 ind. m^{-3}	2.6 ind. m^{-3}	6.1 ind. m^{-3}	5.4 ind. m^{-3}	–	Kang (unpublished data)

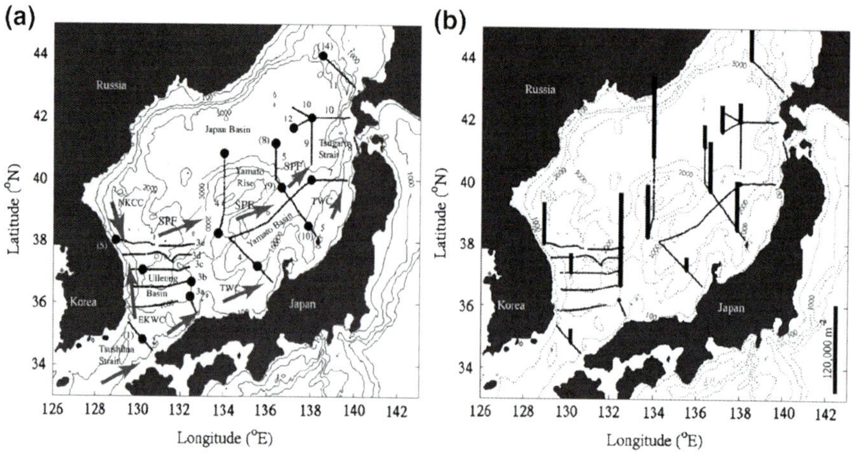

Fig. 12.2 Cruise track in the East Sea (**a**) and integrated water-column abundance (0–80 m depth) of copepods from the net tows (**b**). *TWC* Tsushima Warm Current; *NKCC* North Korean Cold Current; *EKWC* East Korean Warm Current; *SPF* Subpolar Front (from Ashjian et al. 2005)

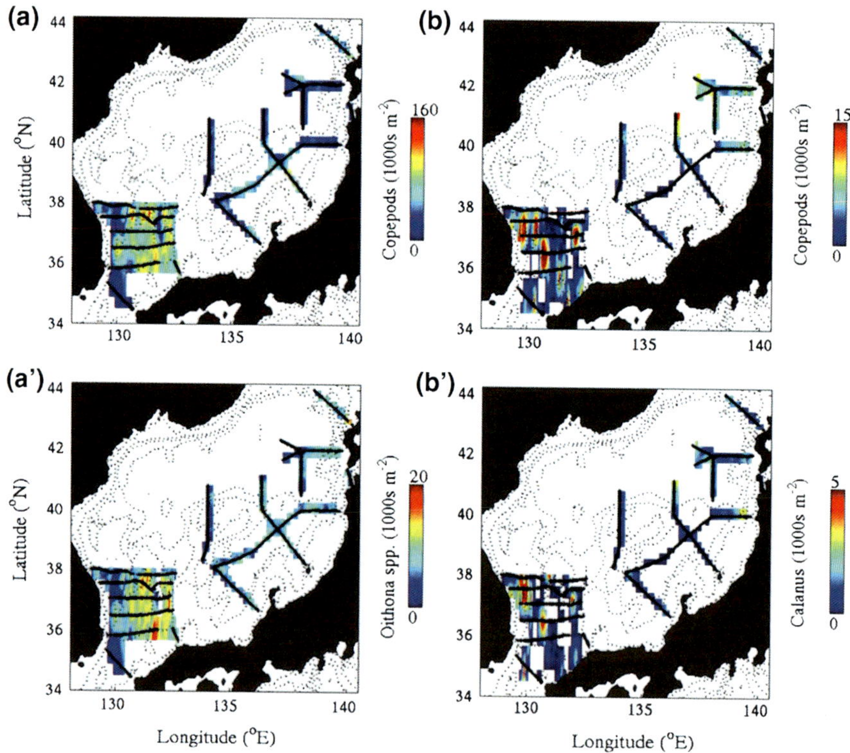

Fig. 12.3 Total water-column abundance (0–80 m depth) of copepods (**a**) and *Oithona* spp. (**a´**) observed using the high-magnification camera, and copepods (**c**) and *Calanus* spp. (**b´**) observed using the low-magnification camera (from Ashjian et al. 2005)

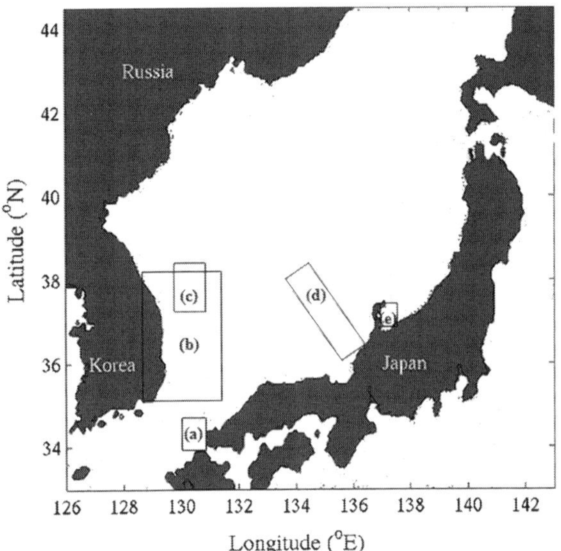

Fig. 12.4 Map showing the local areas to summarize the spatio-temporal distribution of zooplankton species and taxa. The vicinity of the Korea Strait (**a**), the southwestern shelf around Ulleung Basin (**b**), the front zone in the southwestern shelf (**c**), the southeastern area (PM line) (**d**), and Toyama Bay (**e**)

Of the local areas mentioned above, three areas (the vicinity of the Korea Strait, the southwestern shelf around the Ulleung Basin and the subpolar front zone in the southwestern shelf) were located in the southwestern part of the East Sea. In the vicinity of the Korea Strait and the subpolar front zone in the southwestern shelf, dominant copepods in November were listed using percent composition (Table 12.5; Fig. 12.5). Among the copepods, *Paracalanus parvus* and *Calanus sinicus* were the major species at 15.8 and 4.8 % composition, respectively, in the front zone on the southwestern shelf. In contrast, *Paracalaus aculeatus* and

Table 12.5 Abundance of dominant species or taxa in the subpolar front zone of the southwestern shelf and vicinity of the Korea Strait in November (modified from Hirakawa et al. 1995; Park and Choi 1997)

Subpolar front zone of the southwestern shelf		Vicinity of the Korea Strait	
Species or taxon	Ind. m^{-3} (%)	Species	100 Ind. m^{-3} (%)
Paracalanus parvus s. l.	124.4 (15.2)	*Paracalanus aculeatus*	35.1 (13.5)
Oikopleura spp.	109.9 (13.4)	*Euchaeta plana*	31.7 (12.2)
Noctiluca scintillans	93.9 (11.5)	*Oithona plumifera*	23.7 (9.1)
Calanus sinicus	39.4 (4.8)	*Euchaeta flava*	22.4 (8.6)
Foraminifera	32.2 (3.9)	*Paracalanus parvus*	18.2 (7.0)
Oncaea venusta	24.9 (3.0)	*Clausocalanus arcuicornis*	13.5 (5.2)
Unidentified copepodites	24.8 (3.0)	*Clausocalanus furcatus*	12.7 (4.9)
Eucalanus copepodite	22.7 (2.8)	*Oncaea venusta*	12.7 (4.9)
Parathemisto japonica	18.5 (2.3)	*Ctenocalanus vanus*	11.4 (4.4)
Oithona setigera	18.0 (2.2)	*Calanus minor*	8.8 (3.4)

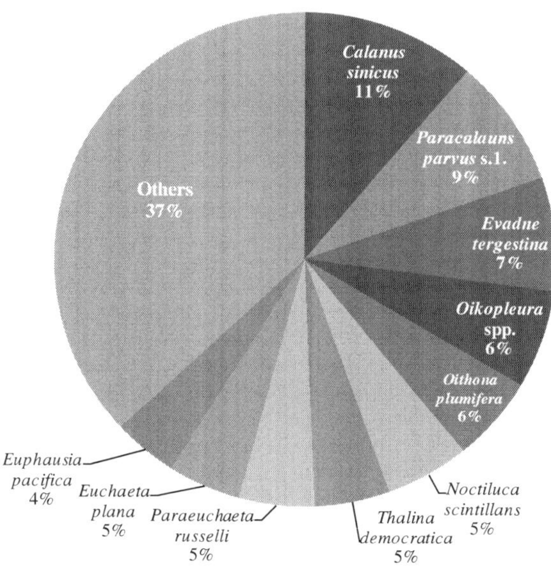

Fig. 12.5 Relative abundances of the 10 dominant zooplankton in the southwestern shelf around Ulleung Basin (from Rho et al. 2010)

Euchaeta plana were dominant in the vicinity of the Korea Strait with percent compositions of 13.5 and 12.2 %, respectively. Similarly, in the southwestern shelf around the Ulleung Basin (Fig. 12.5), *C. sinicus* was again predominant, and *Paracalaus parvus* s. 1, less so at 8.4 % composition. *Evadne tergestina, Oikopleura* spp. *Oithona plumifera, Noctiluca scintillans, Thalina democratica, Paraeuchaeta russelli, Euchaeta plana* and *Euphausia pacifica* were equally dominant with *ca.* 4–8 % composition. In the subpolar front zone on the southwestern shelf, *Paracalanus parvus* s. 1. showed the greatest abundance, followed by *Oikopleura* spp, *Noctiluca scintillans* and *Calanus sinicus* in order (Table 12.5). Their abundances ranged from 39.4 to 124.4 ind. m^{-3}.

The spatio-temporal variations of dominant zooplankton species and taxa were studied in two local areas: the southeastern area (PM line) and Toyama Bay. The composition of dominant zooplankton species was very different from those of the southwestern part. In the southeastern area (PM line), the cold-water copepods, *Metridia pacifica* and *Oithona atlanita* were dominant in all seasons, but not in the southwestern part (Table 12.6). Cold-water copepods were dominant as well as diverse in winter and spring with relative abundance values of 46.2 % (winter) and 75.7 % (spring) of the mean total abundance, while they were less than 20 % of the total in summer and autumn. In particular, *Pseudocalanus newmanni* showed the greatest abundance in spring with 267.8 ind. m^{-3}. On the other hand, gelatinous plankton (appendicularians and doliolids), cladoceras and chaetognaths increased greatly in summer and autumn.

Similar to the patterns seen for the southeastern area (PM line), cold-water copepods were dominant in winter and spring in Toyama Bay (Table 12.7).

Table 12.6 Abundance of dominant species and taxa for each season in the southeastern shelf (PM line) of the East Sea (modified from Chiba and Saino 2003)

Winter		Spring		Summer		Autumn	
Species or taxon	Ind. m^{-3} (%)	Species or taxon	Ind. m^{-3} (%)	Species or taxon	Ind. m^{-3} (%)	Species or taxon	Ind. m^{-3} (%)
Metridia pacifica	33.6 (25.8)	*Pseudocalanus newmani*	267.8 (38.4)	*Metridia pacifica*	19.9 (11.2)	*Metridia pacifica*	8.5 (4.1)
Pseudocalanus newmani	14.0 (10.7)	*Metridia pacifica*	168.2 (24.1)	*Oithona atlantica*	10.3 (5.8)	*Oithona atlantica*	6.4 (3.1)
Oithona atlantica	7.9 (6.0)	*Oithona atlantica*	49.8 (7.1)	*Calanus sinicus*	6.7 (3.8)	*Mesocalanus tenuicorinis*	9.7 (4.7)
Paracalanus parvus s.l.	4.6 (3.6)	*Neocalanus plumchrus*	17.0 (2.4)	*Corycaeus affinis*	3.9 (2.2)	*Clausocalanus furcatus*	5.5 (2.6)
Mesocalanus tenuicorinis	12.8 (9.8)	*Paracalanus parvus* s.l.	15.1 (2.2)	*Ctenocalanus vanus*	5.5 (3.1)	*Clausocalanus minor*	5.0 (2.4)
Clausocalanus pergens	2.8 (2.1)	*Neocalanus cristatus*	10.7 (1.5)	Appendicularia	34.5 (19.4)	*Scolecithricella minor*	5.3 (2.6)
Ctenocalanus vanus	5.3 (4.0)	*Mesocalanus tenuicorinis*	31.6 (4.5)	Cladocera	22.2 (12.5)	Chaetognatha	26.4 (12.6)
Ostracoda	4.8 (3.7)	*Corycaeus affinis*	27.9 (4.0)	Chaetognatha	10.0 (5.6)	Appendicularia	17.8 (8.5)
Appendicularia	10.6 (8.1)	*Clausocalanus pergens*	18.0 (2.6)	Amphipoda	9.9 (5.6)	Amphipoda	8.0 (3.8)
Amphipoda	7.8 (6.0)	Euphausiacea	13.7 (2.0)	Doliolida	5.1 (2.9)	Cladocera	7.7 (3.7)

Table 12.7 Abundance of dominant copepods in the Toyama Bay (modified from Hirakawa et al. 1992)

Winter		Spring	
Species	100 Ind. m^{-3} (%)	Species	100 Ind. m^{-3} (%)
Metridia pacifica	1570.0 (38.1)	*Oithona atlantica*	14858.5 (75.5)
Oithona atlantica	677.0 (16.4)	*Metridia pacifica*	1822.0 (9.3)
Paracalanus sp.	333.3 (8.1)	*Corycaeus affinis*	645.0 (3.3)
Pseudocalanus minutus	263.3 (6.4)	*Calanus pacificus* s.l.	622.5 (3.2)
Scolecithricella minor	234.0 (5.7)	*Scolecithricella minor*	430.0 (2.2)
Pareuchaeta japonica	154.0 (3.7)	*Mesocalanus tenuicornis*	391.0 (2.0)
Mesocalanus tenuicornis	143.7 (3.5)	*Pareuchaeta japonica*	191.0 (1.0)
Ctenocalanus vanus	94.7 (2.3)	*Paracalanus* sp.	190.5 (1.0)
Clausocalanus pergens	91.0 (2.2)	*Pseudocalanus newmani*	180.5 (0.9)
Pleuromanna gracilis	91.0 (2.2)	*Gaidius variabilis*	73.5 (0.4)

Particularly, *Metridia pacifica* and *Oithona atlantica* were greatest in abundance showing 1570.0 and 677.0 ind. m^{-3} in winter and 1822.0 and 14,858.5 ind. m^{-3} in spring, respectively. They dramatically increased in spring, in particular *O. atlantica*. On the other hand, the warm-water species, *Mesocalanus tenuicornis*, *Clausocalanus pergens* and *Corycaeus affinis* appeared less abundantly. Their abundances varied with season, but they were more abundant in the southeastern part than in the southwestern part.

12.4 Patterns of Interest and Trophic Role of Microzooplankton

As mentioned in previous chapters, the processes and phenomena in the East Sea are similar to those occurring in large-scale basins. Warm and cold waters flow in from the south and north at surface, respectively, and result in a 'subpolar front' and several cyclonic and anticyclonic gyre systems often called 'meso-scale eddies' in the central part. Relatively strong flows at the surface and a deep homogeneous water mass are other features similar to large-scale basins. Upwelling often occurs in the coastal area depending upon the direction and fetch of the wind.

In this section, we deal with the zooplankton community characteristics in the diverse oceanographic conditions, which include the subpolar front, the coastal upwelling area, and the warm water entrance zone of the Korea Strait. Diel vertical migration patterns and trophic role of microzooplankton are also addressed.

12.4.1 Subpolar Front

Patterns of mesozooplankton distribution in the subpolar front zones were studied by Park and Choi (1997) and Park et al. (1998). They determined the target area by Satellite images of surface seawater temperatures prior to field sampling. Park and Choi (1997) collected zooplankton samples in area A of Fig. 12.6 in November 1994 and 1995 using a 'Bongo' net and a Multiple Plankton Sampler of five nets while Park et al. (1998) collected in the area B of Fig. 12.6 in November 1996 using a 1 m^2 MOCNESS (Multiple Opening/Closing Nets and Environmental Sensing System; Wiebe et al. 1976). All three types of nets were fitted with 333 μm mesh.

Taxonomic diversity in the frontal zone was very high as shown in the appendix of Park and Choi (1997) with more than 130 taxa. It was more diverse than that found in the Yellow Sea or in the northern part of the East China Sea. Total abundances (in terms of individuals per cubic meter) at the subpolar front zone were less than those in the nearby cold water zone and more than those in the warm water zone. Abundance varied within the range of several hundreds to two to three thousand individuals per cubic meter. Based on clustering of co-occurrence, Park and Choi (1997) reported that warm water species showed high probability of co-occurrence with low abundances while cold water species showed the opposite pattern. They suggested that taxonomic diversity of the zooplankton community at the subpolar front zone was supported by warm water while high zooplankton biomass was mostly supported by cold water. The relationship between water temperature and abundance also showed different patterns depending on the range of seawater temperatures. Abundance was positively correlated with temperature in the warm water zone, negatively correlated in the cold water zone and showed no significant relationship in the mixing zone.

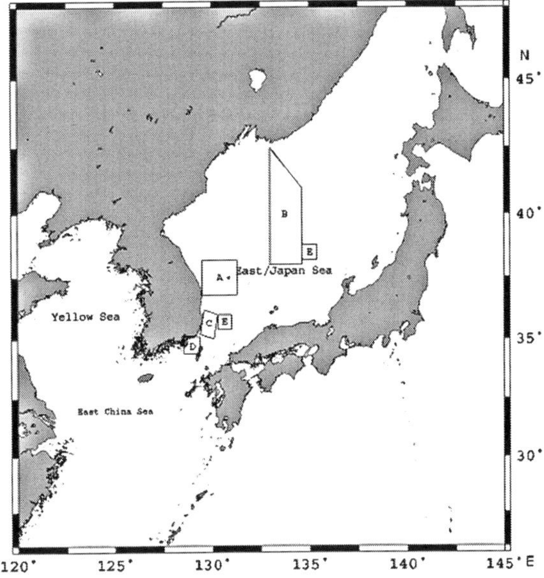

Fig. 12.6 Map showing study areas of zooplankton distribution patterns with respect to the subpolar front (*A* and *B*), the upwelling area (*C*), vertical distribution in the Korea Strait (*D*) and diel vertical migrations (*E*)

12.4.2 Upwelling Area

It has been reported that upwelling occurs when the wind blows from the south along the southeastern coast of Korea in summer (Lee and Na 1985; Byun 1989). The effect of this upwelling on zooplankton distribution was studied by Lee et al. (2004) at area C of Fig. 12.6. Vertically stratified samples at four sites near the coast and four sites in offshore waters were collected three times from mid July to early August, 2001, with an interval of about two weeks, using a 1 m^2 MOCNESS fitted with 333 μm mesh. Cold waters near the coast were detected which were about 7–8 °C colder than offshore waters. Zooplankton diversity and biomass showed significant differences between coastal sites of upwelling and offshore sites.

Taxonomic diversity was low but abundance was high in the area of upwelling. The correlation between the numbers of species found and seawater temperature was significantly positive. However, the negative correlation between abundance and seawater temperature was not statistically significant. Lee et al. (2004) postulated that the increase of zooplankton abundance was caused by the increase of phytoplankton resulted from the high nutrient supply through upwelling.

12.4.3 Korea Strait

Lee et al. (1999) surveyed the vertical distribution of Copepod zooplankton in the western channel of the Korea Strait (area D in Fig. 12.6) using a 1 m^2 MOCNESS fitted with 333 μm mesh over several seasons in 1996–97. They reported that more than 50 copepod species drifted into the East Sea seasonally. The number of copepod species identified was the highest in fall with 86 taxa. It was about 55 in other seasons. In contrast, abundance was the highest in spring with 163.8 ind. m^{-3}. Copepod abundance was reduced in summer to the level of 25.0 ind. m^{-3} and then bounced back in fall (41.3) and winter (54.2). The depth layer having maximum abundance was seasonally varied. Maximum abundance was found at the mid depth in spring and at the surface in summer and fall. In winter, the vertical difference of abundance was negligible.

Warm water flows into the East Sea through the Korea Strait throughout the year. Occasional intrusion of Cold Water into the Korea Strait in the deeper layer was reported in winter (Lee et al. 1998) or in summer (Cho and Kim 1998). Cold water species were found in the deep layer (such as *Metridia pacifica*) in summer with less than 1 ind. m^{-3}. Obviously this area was the southern limit for the cold water species. The three most abundant warm water copepod species entering the East Sea through this strait included *Nannocalanus minor*, *Ditricocorycaeus affinis*, and *Paracalanus parvus* s. l. in spring, *Nannocalanus minor*, *Acrocalanus gracilis* and *Undinula vulgaris* in summer, *Nannocalanus minor*, *Clausocalanus furcatus* and *Oncaea venusta* in fall and *Paracalanus parvus* s. l., *Nannocalanus minor*, and *Ditricocorycaeus affinis*, in winter. Copepodites were relatively abundant in winter. *Nannocalanus sinicus*, one of the important species in the Yellow Sea did not appear at all in this strait. Only *Acartia* spp, the most abundant species in the Yellow Sea and northern East China Sea, was found with very low abundance.

12.4.4 Diel Vertical Migration

Diel vertical migration (DVM) of zooplankton was investigated in area E (Fig. 12.6) by Park et al. (1997). They collected zooplankton twice, first in November 1995 in the northern E area with a Multiple Plankton Sampler (Hydro Bios, 5 nets) and a second time in May 1996 at the southern section of the E area of Fig. 12.6 with a 1 m^2 MOCNESS (9 nets). Both samplers were fitted with 333 μm mesh. At each time period the sampling consisted of seven consecutive vertically stratified net tows at 4 h intervals resulting in 35 and 63 samples, respectively. Stratified net sampling was carried out from the surface to 250 m at depth intervals of 50 m during the November cruise, and from the surface down to 270 m with 30 m intervals during the May cruise. To ensure the nets towed in the same water parcel, a buoy with two drogues at 50 and 150 m depths was deployed at the beginning of the tow series.

The water parcel sampled was thermally stratified both in November and May with thermoclines at 50–100 m depths. The most abundant species *Metridia pacifica* showed DVM in this thermally stratified water. The depth range of movement was from 60–90 m to 240–270 m and the copepods remained in the deeper layer during the daytime. Another copepod species *Scolecithricella minor* also showed DVM but the pattern and depth range were different. The depth range was rather narrow from about 100 to 200 m and this species occupied the shallow depth during nighttime. Most other less abundant taxa of zooplankton did not show any significant patterns of DVM. They preferred the warm surface layers where food was probably more abundant.

12.4.5 Trophic Role of Microzooplankton

Research over the past three decades has significantly shifted views of the pelagic food-web structure in marine ecosystems. The role of microbes has been recognized and added to a suite of new trophic levels, called "the microbial food web" (or sometimes "the microbial loop"), in the classic food-web "phytoplankton-zooplankton-fish" (Pomeroy 1974). The major components of the microbial food web are heterotrophic/autotrophic bacteria and heterotrophic protists (such as cililates, dinoflagellates, etc.), considered as primary prey and its grazers, respectively. Moreover, large heterotrophic protists are capable of feeding on phytoplankton and they consume a significant proportion of the primary production (more than half of the total primary production) (Schmoker et al. 2013).

In the East Sea, the microbial loop is also a significant support for higher trophic production as recognized in recent studies by Yang et al. (2009) and Hyun et al. (2009). They found that ciliates and heterotrophic dinoflagellates (>10 μm in size) were preferentially ingested over phytoplankton by major copepods (*Calanus sinicus* and *Neocalanus plumchrus*) and their contributions to the copepod diets were significant, ranging from 47.2 % under oligotrophic conditions (i.e., phytoplankton <40 μg/L) to 27.4 % in productive environments (i.e., phytoplankton >40 μg/L) in the Ulleung Basin of the East Sea, but still only equal to or less than the phytoplankton contribution. The simplest scenario explaining these observations is that copepods were feeding on a diverse assemblage of heterotrophic protists (i.e., larger ciliates and heterotrophic dinoflagellates), and the larger heterotrophic protists were feeding on the phytoplankton and small flagellates. Therefore, these microzooplankton assemblages (nanoplanktonic and microplanktonic consumers: size range from 2 to 200 μm in Fig. 12.7) are a significant factor in controlling the phytoplankton biomass and they may play a pivotal role in the transfer of organic carbon to higher trophic levels in the East Sea (Table 12.8). As a result, a large part of the primary production capacity might have indirectly reached the copepods through their consumption of ciliates and heterotrophic dinoflagellates, which are significant components of microzooplankton in the East Sea (KORDI 2007). Thus, they directly and indirectly link primary producers and copepods. In conclusion, copepods ingested heterotrophic protozoa

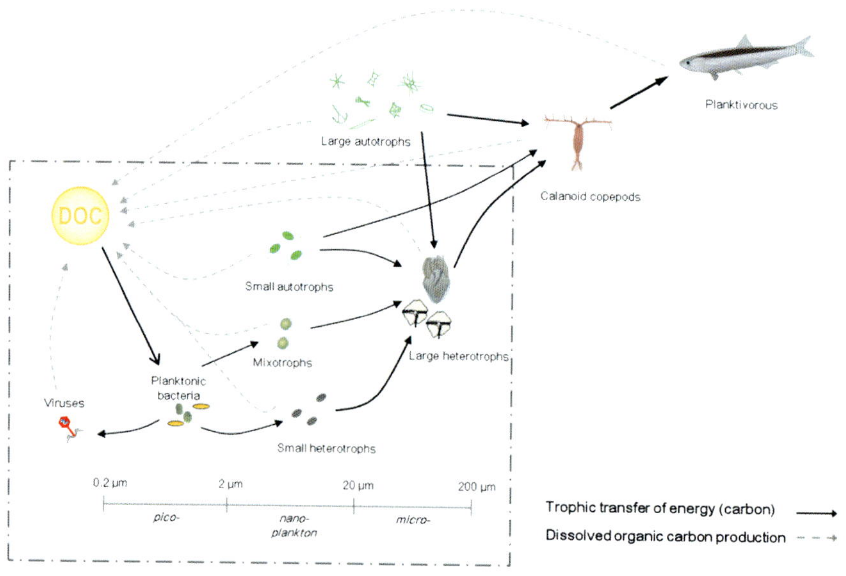

Fig. 12.7 Simplified diagram of the pelagic food-web with emphasis on the microbial loop. Organisms inside the *dotted-line* box, whose sizes are indicated roughly by the adjacent scale bar, are responsible for the microbial loop (Redrawn from Jumars and Hay 1999)

Table 12.8 Relative contribution of phytoplankton[a] and microzooplankton for copepod diets in oligotrophic and eutrophic regions

	Word ocean		East Sea	
	oligo-trophic	eutrophic	oligo-trophic	eutrophic
Chl-a con. (ugC/L)	<1	>3	<1	>2
% phytoplankton production grazed by microzooplankton	69.6[b]	59.9[b]	78.2	49.7
% microzooplankton in total carbon ration ingested by copepod[(2)]	48.0[c]	19.0[c]	47.2	27.4
% phytoplankton in total carbon ration ingested by copepod[(2)]	5.0[a]	12.3[a]	52.8	73.6

[a]Including only diatoms

Values are taken from the following references: [b]Calbet and Landry (2004), [c]Saiz and Calbet (2011)

in amounts that were sufficient to provide a substantive part of copepod nutrition and to impact heterotrophic protozoan populations. More significantly, copepods did not indiscriminately ingest all sizes of particles but instead primarily consumed the large particles and thereby modified the size composition and structure of the microplankton community, indirectly altering the carbon flow of the pelagic ecosystem by a trophic cascade (Fig. 12.7). Therefore, the microbial food web in the East Sea could be strongly coupled with the first consumer (copepods) in the pelagic ecosystem, making the production of the pelagic ecosystem in the East Sea more sustainable and efficient.

The East Sea is experiencing climate change resulting from increasing atmospheric CO_2. The changes include both accelerated warming and acidification (Kim et al. 2001; Park et al. 2006). These changes may affect protist dynamics directly (e.g., through changing temperature, O_2, pH, pCO_2) or indirectly (e.g., through changing food sources and predators) (Caron and Hutchins 2013). Thus, to better understand and predict the response of microzooplankton assemblages to climate change, long-term observational and experimental studies examining the influence of multiple environmental forcing factors are required rather than studies focused only on short-term acclimation to single factors. At the same time, modeling efforts that incorporate existing information on the effects of climate change on ecosystem structure and function are necessary to improve our prediction of future changes to the function and structure of the pelagic ecosystem. It is especially important to study regions like the East Sea, where microzooplankton play a pivotal role, and where knowledge of their fate is essential to understanding the overall effects of climate change on ocean biology and biogeochemistry.

References

Ashjian CJ, Davis CS, Gallager SM et al (2005) Characterization of the zooplankton community, size composition, and distribution in relation to hydrography in the Japan/East Sea. Deep-Sea Res II 52:1363–1392

Azam F, Fenchel T, Field JG, Meyer-Reil LA, Thingstad F (1983) The ecological role of water-column microbes in the sea. Mar Ecol Prog Ser 10:257–263

Beaugrand G, Ibanez F, Lindley JA (2002) Diversity of calanoid copepods in the North Atlantic and adjacent seas: species associations and biogeography. Mar Ecol Prog Ser 232:179–195

Bieri R (1991) Systematics of Chaetognatha. In: Bone Q, Kapp H, Pierrot-Bults AC (eds) The biology of chaetognaths. Oxford University Press, London, pp 122–136

Byun SK (1989) Sea surface cold water near the southeastern coast of Korea: wind effect. J Korean Soc Oceanogr 24(3):121–131

Calbet A, Landry MR (2004) Phytoplankton growth, microzooplankton grazing, and carbon cycling in marine systems. Limnol Oceanogr 40:51–57

Caron DA, Hutchins DA (2013) The effects of changing climate on microzooplankton grazing and community structure: drivers, predictions and knowledge gaps. J Plankton Res 35:235–252

Chiba S, Saino T (2003) Variation in mesozooplankton community structure in the Japan/East Sea (1991–1999) with possible influence of the ENSO scale climatic variability. Prog Oceangr 57:317–339

Chihara M, Murano M (1997) An illustrated guide to marine plankton in Japan. Tokai University Press, Tokyo, p 1574

Cho YK, Kim K (1998) Structure of the Korea Strait Bottom Cold water and its seasonal variation in 1991. Cont Shelf Res 15:763–777

Condon RH, Graham WM, Duarte CM et al (2012) Questioning the rise of gelatinous zooplankton in the world's oceans. Bioscience 62:160–169

Dolganova NT (2000) Composition, seasonal and interannual dynamics of plankton in the northwestern part of Sea of Japan. Dissertation, TINRO-Center (in Russian)

Dolganova NT (2001) Composition, seasonal and interannual dynamics of plankton in the northwestern part of Sea of Japan. Ixvestiya TINRO (TINRO Transaction) 128:810–889 (in Russian)

Hirakawa K, Imamura A, Ikeda T (1992) Seasonal variability in abundance and composition of zooplankton in Toyama Bay, Southern Japan Sea. Bull Japan Sea Natl Fish Res Inst 42:1–15

Hirakawa K, Kawano M, Nishihama S et al (1995) Seasonal variability in abundance and composition of zooplankton in the vicinity of the Tsushima Straits, southwestern Japan Sea. Bull Japan Sea Natl Fish Res Inst 45:25–38

Hirota Y, Hasegawa S (1999) The zooplankton biomass in the Sea of Japan. Fish Oceanogr 8(4):274–283

Hyun JH, Kim D, Shin CW et al (2009) Enhanced phytoplankton and bacterioplankton production coupled to coastal upwelling and an anticyclonic eddy in the UB, East Sea. Aquat Microb Ecol 54:45–54

Im DH (2010) Taxonomical study of clausocalanid copepods from the Korean waters. Dissertation, Chonnam National University

Jeong MK, Suh HL, Soh HY (2011) Taxonomy and zoogeography of euchaetid copepods (Calanoida, Clausocalanoida) from Korean waters, with notes on their female genital structure. Ocean Sci J 46:117–132

Jumars P, Hay M (1999) Ocean ecology: understanding and vision for research. In: Report of the OEUVRE Workshop. Keystone Resort, Colorado, 1–6 Mar 1998

Kang YS (2008) Seasonal variation in zooplankton related to north Pacific Regime Shift in Korea Sea. J Korean Fish Soc 41(6):493–504 (in Korean)

Kim K, Kim KR, Min DH et al (2001) Warming and structural changes in the East (Japan) Sea: a clue to future changes in global oceans? Geophys Res Lett 28(17):3293–3296

KORDI (2007) Carbon cycle in the East Sea I. The Ulleung Basin, Ansan, p 324 KORDI

Lee JC, Na JY (1985) Structure of upwelling off the southeast coast of Korea. J Korean Soc Oceanogr 20(3):6–19 (in Korean)

Lee JC, Lee SR, Byun SK et al (1998) Variability of current and sea level difference in the western channel of the Korea Strait in winter 1995–96. J Fish Sci Tech 1:276–282

Lee CR, Lee PG, Park C (1999) Seasonal and vertical distribution of planktonic copepods in the Korea Strait. J Korean Fish Soc 32(4):525–533 (in Korean)

Lee CR, Park C, Moon CH (2004) Appearance of cold water and distribution of zooplankton off Ulsan-Gampo area, Eastern coastal area of Korea. J Korean Soc Oceanogr 9(2):51–63 (in Korean)

Morioka Y (1985) Zooplankton biomass in the Japan Sea with reference to the southwestern region. Bull Japan Soc Fish Oceanogr 47(48):63–66 (In Japanese)

Nemoto T (1962) Food of baleen whales collected in recent Japanese Antarctic whaling expeditions. Scient Rep Whales Res Inst 16:89–103

Park T (1956) On the seasonal change of the plankton at Korean channel. Bull Pusan Fish Coll 1(1):1–12 (in Korean)

Park C, Choi JK (1997) Zooplankton community in the front zone of the East Sea of Korea (the Sea of Japan): 1 species list, distribution of dominant taxa, and species association. J Korean Fish Soc 30(2):225–238 (in Korean)

Park C, Lee CR, Hong SY (1997) Patterns of vertical distribution and diel vertical migration of zooplankton in the East Sea of Korea (Sea of Japan). J Korean Soc Oceanogr 32(1):38–45

Park C, Lee CR, Kim JC (1998) Zooplankton community in the front zone of the East Sea of Korea (the Sea of Japan): 2 relationship between abundance distribution and seawater temperature. J Korean Fish Soc 31(5):749–759 (in Korean)

Park GH, Lee K, Tishchenko P, Min DH et al (2006) Large accumulation of anthropogenic CO_2 in the East (Japan) Sea andits significant impact on carbonate chemistry. Glob Biogeochem Cycles 20:GB4013. doi:10.1029/2005GB002676

Pomeroy LR (1974) The ocean's food web, a changing paradigm. Bioscience 24:499–504

Ponomareva LA (1963) Euphausiids of the North Pacific, their distribution and ecology. Israel Program for Scientific Translations, Israel

Purcell JE (2009) Extension of methods for jellyfish andctenophore trophic ecology to large-scale research. Hydrobiologia 616:23–50

Rho T, Kim Y, Park JI et al (2010) Plankton community response to physico-chemical forcing in the Ulleung Basin, East Sea during summer 2008. Ocean Polar Res 32(3):269–289 (in Korean)

Saiz E, Calbet A (2011) Copepod feeding in the ocean: scaling patterns, composition of their diet and the bias of estimates due to microzooplankton grazing during incubations. Hydrobiologia 666:181–196

Schmoker C, Hernández-LeónS Calbet A (2013) Microzooplankton grazing in the oceans: impacts, data variability, knowledge gaps and future directions. J Plankton Res 35(4):691–706

Suh HL, Lim JW, Soh HY (1998) Population structure of surface swams of the euphausiid *Euphausia pacifica* caught by drum screens of Uljin Nuclear Power Plant in the east coast of Korea. J Korean Soc Oceanogr 33:35–40

Vermeer K, Devito K (1988) The importance of *Paracallisoma coecus* and myctophid fishes to nesting folk-tailed and Leach's storm-petrels in the Queen Charlotte Islands, British Columbia. J Plankton Res 10:63–75

Vinogradov ME, Sazhin AF (1978) Vertical distribution of the major groups of zooplankton in the northern part of the Sea of Japan. Oceanology 18:205–209

Wi JH, Suh HL, Yang HS et al (2008) Two species of the genus *Oncaea* (Copepoda, Poecilostomatoida, Oncaeidae) from the East Sea, Korea. Ocean Sci J 43:183–193

Wi JH, Yoon YH, Soh HY (2009) Five *Oncaea* species (Copepoda, Poecilostomatoida, Oncaeidae) from the Korean waters, with notes on the spatio-temporal distribution of Korean *Oncaeid* species. Ocean Sci J 44:95–115

Wi JH, Böttger-Schnack R, Soh HY (2010) Species of *Triconia* of the conifera-subgroup (Copepoda, Cyclopoida, Oncaeidae) from Korean waters, including new species. J Crustacean Biol 30:673–691

Wiebe PH, Hurt KH, Boyd SH et al (1976) A multiple opening/closing net and environmental sensing system for sampling zooplankton. J Mar Res 34:313–326

Yang EJ, Kang CK, Yoo S et al (2009) Contribution of auto- and heterotrophic protozoa to the diet of copepods in the Ulleung Basin, East Sea/Japan Sea. J Plankton Res 31(6):647–659

Chapter 13
Fish and Fisheries

Suam Kim and Chang-Ik Zhang

Abstract The East Sea (Japan Sea) is a physically dynamic regional sea, and fish communities have adapted to the variety of habitats that characterize the Sea. There are various opinions about the number of fish species in the East Sea. Although the recent tendency for increasing occurrence of subtropical species is likely due to ocean warming, it has been generally understood that about 400–500 fish species, including some brackish water or anadromous species, reside in the East Sea. Warm water pelagic species such as anchovy and chub mackerel occupy the surface layer in the south, while coldwater species such as chum salmon reside in the surface layer of the northern East Sea and subsurface water column in the whole East Sea. The annual fish catch statistics from the entire East Sea have not been available because of the political obstacles currently existing among neighboring nations. However, historic catch records indicated that pollock was a dominant species exceeding about 3 million metric tons (MT) from the whole East Sea during the 1980s. Pollock biomass seems to be much reduced since the 1990s. The Japanese fish catch in the East Sea indicated that fish production increased greatly after 1970, and reached a peak of 1.76 million MT in 1989. Then, it decreased abruptly with the sardine collapse, and remained at about 500,000 MT in the 2000s. In Japanese waters, the fish community structure changed evidently around the late 1980s with warm-water species increasing and cold-water species decreasing. The annual catches of Korean East Sea fisheries showed a peak of 275,000 MT in 1982 and then gradually declined until the early 1990s due to the decline of pollock stock. After that, the annual catches started increasing with the increase in common squid, which comprised 45 and 55 % of

S. Kim (✉)
Department of Marine Biology, Pukyong National University, Busan 608-737,
Republic of Korea
e-mail: suamkim@pknu.ac.kr

C.-I. Zhang
Division of Marine Production System Management, Pukyong National University,
Busan 608-737, Republic of Korea
e-mail: cizhang@pknu.ac.kr

© Springer International Publishing Switzerland 2016
K.-I. Chang et al. (eds.), *Oceanography of the East Sea (Japan Sea)*,
DOI 10.1007/978-3-319-22720-7_13

the Korean East Sea fisheries in the 1990s and 2000s, respectively. This shift of dominant species caused a continuous decreasing trend in trophic level. Recently, Korean catches fluctuated around 200,000 MT annually. The climatic regime shifts (CRS), which occurred in the mid 1970s and the late 1980s, caused changes in the structure of the East Sea ecosystem. In Korean waters, the 1976 CRS was manifested by a decrease of saury population, but an increased biomass and catch production of sardine and filefish. Correlation studies have found that high saury and sandfish catches occurred during periods when the Southern Oscillation Index was generally positive (i.e. La Niña conditions), when the mixed layer depth was shallower, and when the catches of sardine and pollock were low. Results from a mass-balance model (ECOPATH) indicated that the mean trophic level increased from 3.09 in the pre-1976 CRS period (1970–1975) to 3.28 in the post-1976 CRS period (1978–1984) due to the increase of high trophic level species such as pollock. Also, the biomass and production of fisheries resources in the southwestern East Sea changed before and after the 1988 CRS. Due to overfishing or seawater warming in the East Sea, the fish biomass in the East Sea seems to be reduced in recent years. Implementation of a proper management scheme using ecosystem-based fisheries assessment and international cooperation in data exchange and scientific surveys would increase the likelihood of sustainable utilization of fish resources in the East Sea.

Keywords East Sea (Japan Sea) · Fish · Commercial fisheries · Aquaculture · Climate change impacts · Ecosystem-based fisheries assessment and management

13.1 Introduction

The East Sea is semi-enclosed sea, sometimes called a miniature ocean. Two different current systems collide in the middle of the East Sea forming the meandering subpolar front which moves toward the east. High biotic as well as abiotic variability has been observed frequently in the frontal system, and the formation of eddies around the frontal system create biologically productive areas. Owing to the presence of a subpolar front, the East Sea is divided into cold and warm water masses in the surface layer of the ocean, which results in different spatial biomes in the fish community (i.e. coldwater ecosystem in the north and warm-water one in the south). A thermocline beneath the surface layer also differentiates ecosystems vertically, so that pelagic and demersal fish communities occupy the surface and subsurface layers in the water column, respectively. A very diverse group of resident and migratory fish and fishery species is distributed according to physiological preference for various environmental conditions. Each fish community usually selects a unique habitat, and the timing and routes of fish migration in the East Sea depend on the variability of climate and water mass.

The East Sea is regarded as the southern boundary of coldwater species in the North Pacific. For example, natal rivers for chum salmon (*Oncorhynchus keta*) spawning are located mostly in the north and middle coasts of the Korean Peninsula, and salmon migration to the southern Korean Peninsula is limited (Kim et al. 2007). Spawning areas of semi-demersal walleye pollock (*Gadus chalcogrammus*, pollock hereafter) are in the coastal area of North Korea (Kim and Kang 1998). The distribution of pollock, however, has expanded to the south and the catch in South Korea increased during the cool phase. On the other hand, small pelagic warm-water species such as common squid (*Todarodes pacificus*, or Japanese flying squid, common squid hereafter) are in general much more abundant in the south, and show a strong seasonal migration pattern latitudinally. Most pelagic fishes in Korean waters spawn in the southern coastal areas during the spring, move for feeding to the north during the summer, and return to the south for overwintering.

Warming of seawater would be a critical factor for fish behavior. The warming rate of the East Sea is one of the highest in the world (Belkin 2009), and such a high rate of warming (0.96–1.35 °C/1982–2006 period) must be a stressor for marine fish and fishery species. Aside from the continuous warming trend, the abrupt basin-scale climate change, called a climate regime shift (CRS), also influences marine biota in Korean waters. Seasonal and interannual ecological variability (i.e. the changes in ocean productivity, species composition, recruitment and growth of fish species) in the East Sea seems to be closely linked with regional or basin scale climate and environmental changes in the North Pacific (Seo et al. 2006; Zhang et al. 2000). In this chapter, we review fish species and their ecological characteristics in the East Sea. A number of scientific issues regarding the fluctuation of fish biomass under a changing climate and the relationship between environmental variability and fish populations are investigated, and possible explanations for the observed patterns are discussed. Finally, the East Sea is surrounded by different political regimes which limit cooperative research activity, and hinder scientific achievement. Some suggestions for promoting East Sea ecosystem science are introduced.

13.2 Fish Species and Habitat

13.2.1 Species

There have been various estimates of the number of fish species in the East Sea. A series of pioneering works conducted by Russian scientists in the mid 20th century indicated at least 376 species plus some systematically primitive species as the resident fish species in the East Sea (Lindberg and Legeza 1959, 1965; Lindberg and Krasyukova 1969). Estimates from North Korean scientists reported similar numbers of fish species in the East Sea off the Korean Peninsula. Kim and Kang (1998) cited and re-summarized Son's (1980) research indicating totals of 335 species as marine fish, 32 species as brackish water fish, and 9 species as anadromous fish (376 species, 257 genera, 126 families, and 21 orders) in the East Sea.

Son (1986) revised the numbers to 389 species in the East Sea: 345 marine species (Order Perciformes—143 sp), 32 brackish water species, and 12 diadromous species.

Totals of 439 fish species were identified from the southwestern part of the East Sea (Kim and Nam 2003): Perciformes occupied about 44 % of the total species, Scorpaeniformes 19 %, Pleuronectiformes 7 %, and Tetraodontiformes 6 %. They showed that approximately 50 % of total marine fish from Korean waters were found in the East Sea (i.e. 439 species out of 824 species). Kim (2009) reviewed an investigation on fish fauna of the East Sea using various types of fishing gear in coastal areas (depth <50 m) of South Korea, and demonstrated that 99 fish species were found only in the East Sea among three regional seas. Japanese scientists reported higher numbers of fish species compared to other areas: Katayama (1940) reported 411 species from Toyama Bay, and Mori (1956) listed fishes of the Oki Island and adjacent waters, indicating 542 marine and freshwater fishes. Kawano et al. (2011), however, identified a total of 1336 fish species in the East Sea during the 1930s–2000s with a recent trend for increasing numbers of tropical/subtropical species.

13.2.2 Distribution and Habitat

In general, demersal species are dominant in the north, while pelagic species dominate the south. In the East Sea off the Korean Peninsula, Son (1986) found that more species resided in the central (343 species) than the northern (187 species) and the southern (275 species) areas. He also examined the proportion of pelagic and demersal groups in each geographical area: Pelagic fish populations are greater in the south (31 %) than in the north (19 %), and demersal species represent 75 % of all species in the northern area. This ratio decreases to 45 % in the waters off the southern Korean Peninsula (Table 13.1). Also, anadromous fish such as chum salmon were more common in the central and northern areas than in the southern area. Kim (1986) identified a total of 65 fish species in the offshore area: 59 species resided in both coastal and offshore areas, but there were only 6 species that were only in the offshore area. The study of Shinohara et al. (2011) revealed from voucher specimens of fish in several organizations that there were about 160 deep-sea fish species belonging to 68 families and 21 orders in the East Sea.

Table 13.1 Areal difference in number and proportion of fish species found in the East Sea off the Korean Peninsula

	Habitat area			Depth zone occupied		
	Marine	Brackish	Anadromous	Pelagic	Intermediate	Demersal
Northern area	152 (81 %)	27 (14 %)	8 (4 %)	36 (19 %)	11 (6 %)	150 (75 %)
Middle area	305 (89 %)	29 (9 %)	9 (2 %)	87 (25 %)	57 (17 %)	199 (58 %)
Southern area	252 (92 %)	21 (8 %)	2 (≒0 %)	86 (31 %)	64 (23 %)	125 (45 %)

Warm water species such as Japanese anchovy (*Engraulis japonica*) and chub mackerel (*Scomber japonicus*) occupy the surface layer of the Tsushima Warm Current (TWC), while coldwater species reside either in the surface layer of the northern East Sea (e.g. chum salmon) or in the subsurface water column over the entire East Sea (e.g. Pacific cod (*Gadus macrocephalus*, cod hereafter)). Due to overfishing or seawater warming, the biomass of cold water species has decreased in South Korean waters in recent years, so that an enhancement program by the Korean Government was started for stock rebuilding in the 1980s. Warm water species such as common squid, Pacific sardine (*Sardinops melanostica*, sardine hereafter), and Pacific saury (*Colorabis saira*, saury hereafter) are relatively abundant, but not all at the same time. A strong interdecadal fluctuation in their abundances has been observed. Some were abundant and widely distributed in specific periods, while others showed reduced biomass in limited areas.

13.3 Fisheries

Common squid, saury, chub mackerel, horse mackerel (*Trachurus japonica*), yellowtail (*Seriola quinqueradiata*) and sandfish (*Arctoscopus japonicus*) are economically valuable species in the East Sea, especially in Korean and Japanese waters. The annual fish catch statistics from the entire East Sea have not been available, but it likely exceeded about 3 million MT in the 1980s–1990s due to high abundance of pollock and sardine. The total annual catch of pollock was about 1 million MT from the East Sea in the 1970s–1980s (Kim and Kang 1998), but the catch has dramatically decreased since the 2000s (Kang et al. 2013). The North Korean Government did not provide the UN Food and Agriculture Organization (FAO) reliable information on catch statistics, but it was likely that the fish catch from the East Sea might be double of that from the Yellow Sea in North Korea. In particular, the Order Gadiformes consists of only 2 families, 4 genera, and 4 species, but fisheries on pollock contributed to the national economy of North Korea enormously in the 1980s (Kim and Kang 1998). The largest fishery for pollock occurs near East Korea Bay (or Won-San Bay) in North Korean waters. Because of difficulties in collecting fishery statistics from Russia and North Korea, we briefly summarize only Korean and Japanese fisheries in the East Sea.

13.3.1 Korean Commercial Fisheries

13.3.1.1 Yields

The Korean East Sea fisheries account for about 16 % (range: 10–23 %) of the total Korean coastal and offshore fisheries in terms of annual production. The mean annual catch from the East Sea was about 200,000 MT, ranging from

141,000–275,000 MT during the last four decades (1971–2011), and comprised 19 % (230,000 MT) of the National total in 2009. The annual catches of the Korean East Sea fisheries peaked at 275,000 MT in 1982 and then gradually declined until the 1990s due to the decrease in the pollock stock. After that, the annual catches began increasing due to the higher catch of common squid and then fluctuated around the 200,000 MT level until recently (MIFAFF 2012) (Fig. 13.1). Currently, common squid is the major species in commercial fisheries.

Table 13.2 summarizes fish species caught by Korean fisherman in the East Sea during the last five decades. Common squid made up the largest proportion of catch (33.7 %), followed by pollock (16.6 %) and red snow crab (13.4 %). The average annual production of three pelagic fishes (common squid, saury, and anchovy) during the last five decades was 85,442 MT or 43.8 % of the total fishery production of 195,022 MT in Korean waters of the East Sea, while pollock and

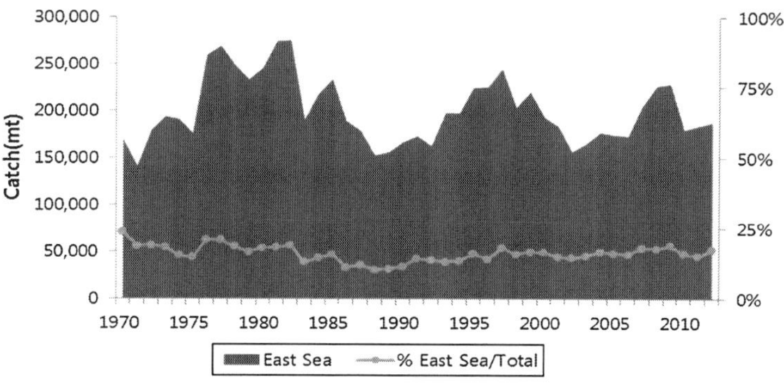

Fig. 13.1 Annual catches by Korean East Sea fisheries and their proportions of the total Korean catches for 1971–2011 (data from MIFAFF 2012)

Table 13.2 Catches of major species from the East Sea fisheries, 1961–2011

Species	Mean catch (MT)	Proportion to total (%)
Common squid (Japanese flying squid)	65,776	33.7
Pollock (Walleye or Alaska pollock)	32,458	16.6
Red snow crab (Queen crab)	26,081	13.4
Saury (Pacific saury)	13,226	6.8
Filefish	8648	4.4
Anchovy	6420	3.3
Sandfish	4557	2.3
Herring	4317	2.2
Flatfishes	3562	1.8
Other minor species	29,977	25.0
Total	195,022	100.0

red snow crab which are the major semi-demersal and demersal species, occupied about 30 % of total East Sea yields.

13.3.1.2 Types of Fisheries

About 40 fishing methods are used for the exploitation of fish and shellfish resources in the East Sea. The mid-sized trawl fishery is the major fishery in the offshore area of the East Sea, followed by jigging fishery, pot fishery, Danish seine fishery, gill-net fishery and long-line fishery. The main target species of the mid-sized trawl fishery, jigging fishery and small-sized purse seine fishery is common squid, which accounts for more than 90 % of the catches by the three fisheries. The pot fishery targets red snow crab and octopus, while the Danish seine fishery targets multiple species such as flatfishes, sand fish, herring and shrimp (MIFAFF 2012). In the coastal area, however, gill-net fishery is the major fishery, followed by set-net fishery, coastal purse-seine fishery and pot fishery in the East Sea. The major target species of the gillnet fishery is sandfish, flatfishes, sand lance and herring. The set-net fishery targets common squid, chub mackerel, yellowtail, herring and horse mackerel. The major target species of the coastal purse-seine fishery are herring and anchovy, while those of the pot fishery are octopus, sea snails and conger eel (MIFAFF 2012).

13.3.1.3 Species Composition

It has been known that the species composition of major small pelagic fisheries catches has changed remarkably during the last 50 years (Fig. 13.2). The ecological regime shift between small pelagic fish and demersal fish is especially evident in relationship to the CRS in the mid 1970s and in the late 1980s. In the 1960s,

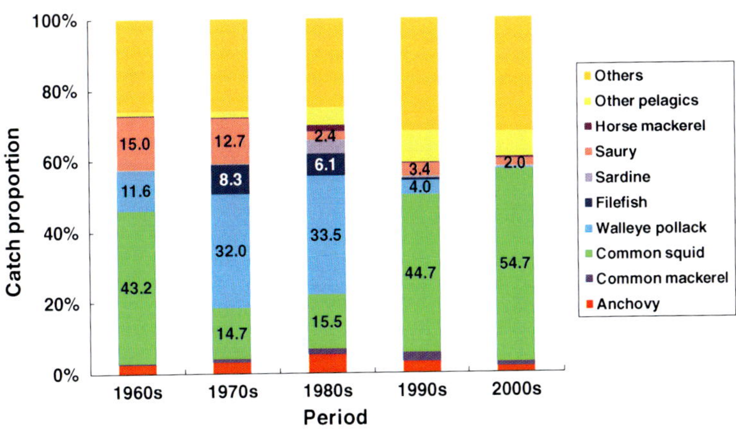

Fig. 13.2 Korean catch proportions of major species in the East Sea (1961–2008)

common squid catch occupied 43 % of the catch in the East Sea, followed by saury (15 %) and pollock (12 %). Pollock increased, occupying about 33 % from the 1970s to the 1980s. Concurrently, the proportions of common squid and saury greatly decreased in the 1970s and the 1980s. The dominant fisheries catch also shifted from pollock in the 1980s, to squid in the 1990s (45 %) and 2000s (55 %), and pollock and saury occupied less than 5 % in recent years.

13.3.1.4 Mean Trophic Level

The trophic levels of resource organisms in the catches from the East Sea showed a significant decreasing trend from 1967 to 2011 (Fig. 13.3). Mean trophic levels ranged from 3.38 to 3.65 during the last four-decades. The mean trophic level increased during the 1976 CRS period mainly due to the increase of pollock catch and then declined substantially in the mid-1980s probably due to the collapse of pollock (Zhang et al. 2004). Since the late 1990s the mean trophic level started increasing again due to the increase of the squid catch (Zhang and Lee 2013). However, it decreased again since the early 2000s, and reached a historical low (3.38) in 2011 due to the sharp decrease of the squid catch.

13.3.1.5 Important Fisheries

The most common commercial fish species in the East Sea was pollock and common squid. The catch records of pollock in South Korea have changed dramatically since commercial fisheries began in the early 20th century: the highest catch in the 1930s, a sudden decrease during the 1940s–1960s due to low fishing activity, another boom in the 1970s–1980s, and a continuous decrease in the 1990s before they collapsed completely in the 2000s. On the other hand, common squid catches were at low levels (around 50,000 MT) until the late 1980s, but increased

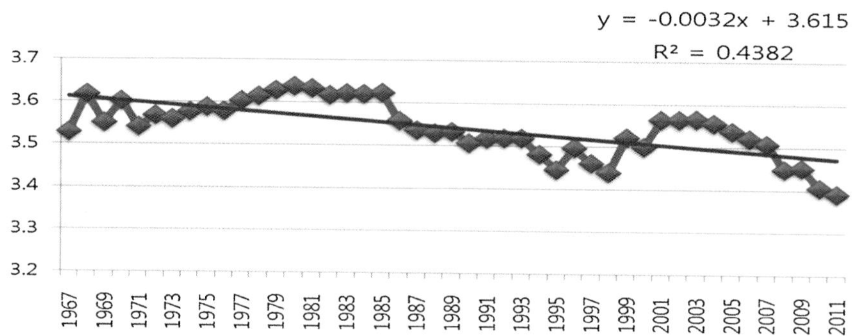

Fig. 13.3 Mean trophic levels of resource organisms in the catches from Korean East Sea fisheries, showing a significant decreasing trend from 1967 to 2011 (from Zhang and Lee 2013)

rapidly after the early 1990s. Annual catch ranged from 100,000 to 250,000 MT during the 1990–2002 period. The main fishing grounds of the squid angling fishery begin moving to the north in the East Sea in April and to the south in September. It is concluded that the high water temperatures and the weak strength of cold water from the north are the major controlling factors for good fishing conditions for common squid. Assuming that catches are correlated with abundance, there is an indication of an inverse relationship between common squid and sardine catches in the 1980s, while sardine and pollock catch patterns were roughly similar coincident in the 1930s and 1980s.

13.3.2 Japanese Commercial Fisheries

The Japanese fish catch in the East Sea ranged from about 500,000 MT in the early 2000s to 1,500,000 MT in the mid 1980s. Tian et al. (2008) examined the Japanese catch for 54 taxa (26 pelagic and 28 demersal species), finding that fish production in the East Sea increased greatly after 1970, and reached a peak (1.76 million MT) in 1989. Then, it decreased abruptly with the collapse of the sardine catch in Japanese waters. The long-term trend in fisheries production was largely dependent on the sardine fishery which contributed up to 60 % of the total production in the Japanese waters of the East Sea during the late 1980s suggesting that dominant species have a large effect on the structure of the fish community. However, their appearance is sporadic, and showed remarkable fluctuation historically. The sardine catch showed an increasing trend around 1920 and attained its first peak during the 1930s. Fisheries almost ceased during World War II, and resumed in 1946. Catches reached a second peak in 1956, but abruptly decreased to their lowest level from 1962 to 1971. After 1974 catches recovered and attained their historically highest level of 1,170,000 MT in 1989 (Terazaki 1999). This long-term fluctuation has been accompanied by noticeable geographical shifts in their distribution including spawning and nursery grounds. When the population size stayed at its peak level during the 1930s, the spawning ground was located to the south of Kyushu, the southernmost sector of the distributional range. In recent years, the major spawning and nursery grounds have been located in the western coastal waters off Honshu and Kyushu.

The production of other small pelagic species resulted in higher yields in the 1960s and 1990s, but lower yields in the 1970s and 1980s. As we can see in Korean waters, this pattern of variation generally corresponds with the changing trend in water temperature: higher catch in the early 1960s and after the 1990s (i.e. warmer periods), but lower ones during cool periods in the 1970s and 1980s (Tian et al. 2006). Common squid were particularly abundant around the main Japanese Islands in the warm periods (Sakurai et al. 2000, 2002), although the biomass and catch during cool periods were still quite large. The migration routes and spawning areas of common squid in the East Sea vary with abundance. In autumn, common squid usually undergo a southward spawning migration. In the 1970s, adult common squid usually migrated westward along the northern edge of the subpolar front to an area east of Korea, and then migrated southward to spawn in the East China Sea. But in the 1980s, they often migrated southward to the coast of Honshu Island, crossing the subpolar front. In the 1990s the migration route returned to the pattern observed in the 1970s. The cold water demersal species show opposite responses to the water temperature in the TWC region.

In Japanese waters, the catch proportion of demersal species of the total catch fluctuated from over 30 % in the late 1960s to about 7 % in 1989 and to over 20 % in the 2000s (Tian et al. 2011). The fish community structure evidently changed around the late 1980s with warm-water species increasing and cold-water ones decreasing (Tian et al. 2006, 2008, 2011). However, the large predatory and demersal fishes seem to be facing stronger fishing pressure with the collapse of the sardine since the late 1980s (Tian et al. 2006).

Purse seine, bottom trawl, and set-net are the three main fisheries in the region of the TWC, where more than 10 % of total Japanese landings are caught (Tian et al. 2006). The proportions of warm-water species increased in trawl catches after the late 1980s regime shift in spite of the declining trend in the catch. In addition, the fishing grounds had shifted: coldwater species, such as pollock and cod, extended their distribution to the southwestern region in the 1980s, but the area of distribution retreated to the north after the regime shift to warming; in contrast, warm-water species such as pointhead flounder (*Hippoglossoides pinetorum*) and shotted halibut (*Eopsetta grigorjewi*) extended their presence into the southern region of the East Sea after the late 1980s. The different responses between warm- and cold-water species to the climate shift reflected their different thermal preferences and life history characteristics.

13.3.3 Climate Change and Its Impacts on Fish and Fisheries

13.3.3.1 Capture Fishery in Korean Waters

Climatic regime shifts (CRS) caused changes in the structure of the ecosystem and the role of major species, as well as large variations in biomass and production of fisheries resources. There are indicators of CRSs that occurred in 1976 and around 1988 in the East Sea ecosystem. The CRS over the East Sea may have affected the dynamics of the marine ecosystem and fisheries resources in the waters. The 1976 CRS was manifested by a decreased biomass and catch production of saury, but an increased biomass and catch production of sardine and filefish in these waters. Correlation studies have found that saury and sandfish catches occurred during periods when the Southern Oscillation Index was generally positive (La Niña conditions), when spring chlorophyll in the East Sea was high, and when the air temperatures were cooler, the mixed layer depth more shallow, when the Northeast Pacific Pressure Index was low, and when sardine and pollock were low in catch (Fig. 13.4) (PICES 2004; Kang et al. 2000).

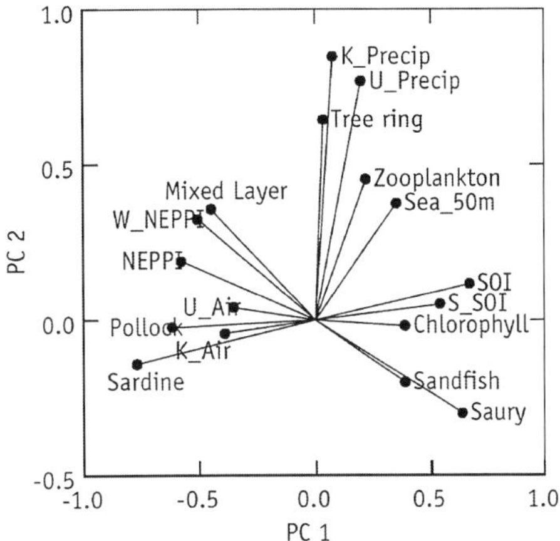

Fig. 13.4 Principal component ordination of correlations between fish catches and ocean/climate variables in Korean waters from 1960–1990 (from PICES 2004) North Pacific Marine Science Organization (PICES)

The species composition of the catch for major small pelagics has changed remarkably during the period from the 1960s–1990s, which might be attributed to the influence of the 1976 CRS (Zhang et al. 2000). Common squid was most dominant, followed by saury in the 1960s through the early 1970s. After the mid-1970s, saury drastically decreased in the East Sea, because the mismatch between early seasonal migration and poor spring blooms of phytoplankton combined with a large increase of fishing effort in Korea (Zhang and Gong 2005). From the 1970s, the catch of filefish increased while saury decreased in the 1980s. Filefish almost disappeared after the late 1980s. Common squid was replaced by pollock in the 1970s and 1980s. Based on a mass-balance model (ECOPATH), the weighted mean trophic level was estimated. The mean trophic level increased from 3.09 in the pre-1976 CRS period (1970–1975) to 3.28 in the post-1976 CRS period (1978–1984). This was the result of the increase of species or species groups that had relatively high trophic levels. The most significant difference between the pre-1976 CRS and the post-1976 CRS models occurred at the mid-trophic levels, occupied by fishes and cephalopods (Zhang et al. 2004). Total biomass of species groups in the East Sea ecosystem increased by 15 % and total catch production increased by 48 % due to the 1976 CRS.

Based on comparison of the structure of the ecosystem and the role of major species, it was concluded that the biomass and production of fisheries resources in the southwestern East Sea changed before and after the 1988 CRS. Total biomass of all species groups in the ecosystem increased by 59 % after the CRS. These results indicate that there were substantial changes in the role of major species in the southwestern East Sea ecosystem. As shown in Fig. 13.5, the relative contribution of pollock, at trophic level III, to the total flow of energy decreased drastically from 33.0 % in the pre-1988 CRS period to 4.3 % in the post-1988 CRS period, while that of common squid, at the same trophic level, doubled from 34.2 % to 72.2 % during these periods (Zhang et al. 2007).

When catches were low in the 1980s, the proportion of autumn spawning (i.e. August and September) common squid was the highest among spawning

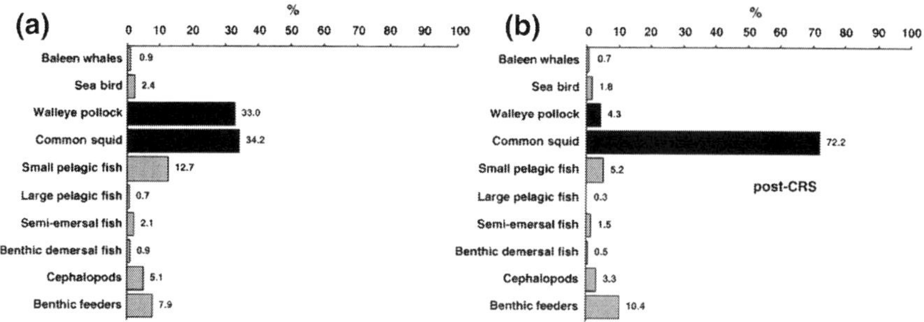

Fig. 13.5 Relative contribution (%) of species to the total flow of energy (throughout) at trophic level III in the **a** pre-1988 and **b** post-1988 CRS model of the southwestern East Sea ecosystem (from Zhang et al. 2007)

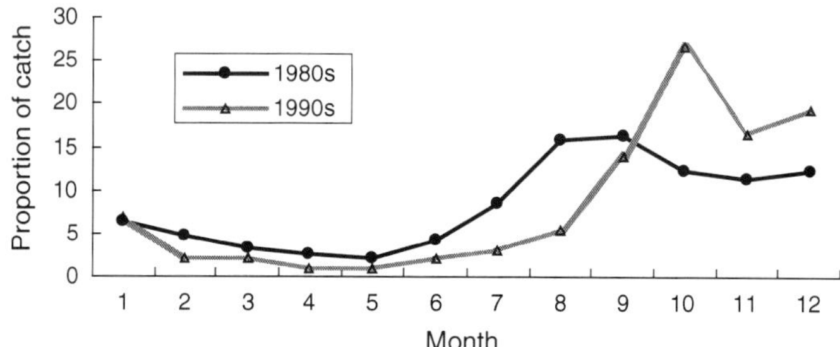

Fig. 13.6 Mean proportions of the monthly catch of common squid in Korean waters during the 1980s and 1990s

populations. However, as catch increased in the 1990s, the highest proportion of spawners shifted to October and November (Fig. 13.6). The increased catch from late autumn and early winter caused difficulties in distinguishing autumn and winter spawners. During the early 1990s, there was a rapid increase in winter catch. Seawater temperature might be the most important parameter for controlling stock abundance. When squid catches increased in the early 1990s, the SST in Korean waters increased concurrently. In particular, seawater temperature during winter seemed to be more important than in other seasons. When the winter temperature rose in the 1990s, the winter population of common squid also increased rapidly compared to the autumn population.

13.3.3.2 Capture Fishery in Japanese Waters

Ecological characteristics of the fish community on the Japanese side of the East Sea were very similar to those observed on the Korean side. Based on a principal component analysis for 58 time-series data sets of fisheries catches,

Tian et al. (2006) concluded that the fish community on the Japanese side of the East Sea varied on interannual to interdecadal scales, and the abrupt changes in the mid 1970s and late 1980s seemed to correspond closely with the CRSs in the North Pacific. Also, they argued that the structure of the fish community was largely affected by climatic and oceanic regime shifts rather than by fishing, because there was no evidence of 'fishing down food webs' in the East Sea.

Tian et al. (2008) demonstrated that a climatic regime shift occurred in the East Sea during the late 1980s. Warm-water pelagic and coldwater demersal species responded to this late 1980s CRS in the East Sea. Besides the fluctuation of sardines in Japanese waters, catches of other small pelagic species such as anchovy and common squid, and several higher-trophic fishes, such as yellowtail and tuna increased markedly in the 1990s compared to the early-mid 1980s. Step changes were detected in these pelagic species during 1989–1992. Catch of demersal species (crab, pink shrimp, cod and pollock) was high during most of the 1970–1980s, but declined in the late 1980s to generally low catches in the 1990s. Cold-water species (e.g. pollock and cod) decreased in abundance and the regions in which their abundances remained high became greatly reduced in extent. Conversely, warm-water species (e.g. pointhead flounder, spotted halibut) increased in abundance and/or extended their spatial range during the warm 1990s.

13.3.3.3 Aquaculture

The coastal area of the East Sea is characterized by rocky intertidal zones with a very small tidal range and strong wave action, so that aquaculture on the east coast of the Korean Peninsula is poorly developed relative to the western and southern coasts. Since the mid 1990s, small-scale scallop (*Patinopecten yessoensis*) farming has been developed in the east using a long-line suspension culture system. According to the report by the Intergovernmental Panel on Climate Change (IPCC), global warming could cause changes in the coastal marine environment including sea level rise and subsequent shoreline erosion, sea surface water temperature increase, increase in ocean acidity and changes in rainfall pattern (IPCC 2001). It is likely that these impacts of climate change are more intense for shellfish and seaweed aquaculture than for land-based fish farming. Drier and warmer winters could increase the incidence of pathogenic disease outbreak and the incidence of mass mortalities is expected to increase in clam culture. Increase in the SST may retard growth and reproduction of *Undaria* and *Laminaria* spp., which are known to be cold temperate species. No clear impacts of acidification of seawater on survival and growth of aquaculture species have been reported yet in Korean waters, although acidification may affect survival and growth of any calcitic organisms such as shellfish larvae and juveniles in the long run.

13.3.3.4 Invasive Species

Kim (2009) examined recent studies of unreported fish species published in the Korean Journal of Ichthyology between 1999 and 2008, and found a mean of 4.7 unrecorded species every year. Unrecorded species were prevalent in the TWC region: 20 species (43 %) from Jejudo (Jeju Island), 12 species (26 %) from Busan and its vicinity, and 8 species (17 %) from the southwestern East Sea. Recently, the appearance of subtropical species has been reported often in society meetings and news media of Korea. Pacific bluefin tuna (*Thunnus orientalis*) and yellow skate (*Dasyatis* sp.) are typical subtropical fish species, and some invertebrates such as the common paper nautilus (*Argonauta argo*), giant jellyfish (*Nemopilema nomurai*), shrimp species (*Plesionika orientalis*, *Alpheus compressus*, etc.) also appeared frequently in the southwestern East Sea.

In Japanese waters, pelagic warm-water predators (e.g. yellowtail, Japanese-Spanish mackerel, and tuna) in the ecosystem increased greatly after the late 1980s (Tian et al. 2008). Recent records of fish species on the Japanese side of the East Sea showed that the number of fish species varied with time, but a higher diversity was seen in the 2000s compared to the early and mid 20th century (Kawano et al. 2011).

13.3.4 Ecosystem-Based Fisheries Assessment and Management Issues

Fisheries resources for the Korean East Sea fisheries have been managed by a variety of tools, such as input and output controls and technical measures. Current fisheries management in Korea include the establishment of precautionary total allowable catch (TAC)-based fishery management and some technical measures such as closed fishing seasons/areas, fish size- and sex-controls, and fishing gear restrictions. A number of enhancement projects are also underway to recover depleted fisheries resources and the associated habitats for Korean East Sea fisheries.

Based on the limited knowledge of the status of fish species harvested in commercial fisheries in the East Sea, stocks are generally considered to be declining in quality as well as in quantity. However, only a few studies have been carried out on the scientific assessment of these resources. Only a few species out of more than 100 have been currently managed by scientifically determined TACs in the East Sea fisheries, and thus the stocks for the remaining production are not efficiently managed. There have been insufficient budget and human resources for further fisheries resource surveys and assessment in Korea. In addition, TACs are determined by a traditional population-based assessment, which does not consider habitat quality, biodiversity and socio-economic status of marine ecosystems.

Several difficulties are associated with traditional control devices. For example, mesh and minimum landing size regulations were adopted to avoid the dangers of recruitment overfishing; however, such measures did not prevent fishing effort from increasing because the number and power of vessels were not limited. Some areas were closed to allow stocks to reproduce and grow undisturbed by reducing fishing mortality; however, adequate funds for monitoring and enforcement were not available.

In addition, because the width of the East Sea is less than 400-nautical miles between Korea and Japan, negotiation of a boundary for fishing grounds between the neighboring countries was inevitable based on United Nations Convention on the Law of the Sea (UNCLOS). Fisheries agreements with neighbors were required for the management of overlapping fish stocks (Zhang and Lee 2002). The most commercially important fishery resources in the East Sea are migratory species, moving seasonally across national boundaries. Competition between countries has resulted in heavy exploitation of the most economically valuable species. Although suggestions made to improve management of fishery resources are important steps, realization of their full benefits will require addressing the trans-boundary issue. The production potential of the fishery resources shared by Japan and Korea is large. Effective management will be required if this potential is to be fully realized. However, one of the difficulties in managing fisheries resources of the East Sea is the lack of international cooperation among neighboring countries. Effective management will be possible by the creation of a regional fisheries body among coastal countries of the East Sea. Joint studies on the assessment and management of fisheries resources are needed among the neighboring countries of the East Sea to maintain and to sustainably utilize fisheries resources in the Sea. These studies should include exchanges of fishery data and other information such distribution, biomass, ecological parameters of fish stocks and ecosystem characteristics as well (Zhang and Lee 2002).

The commercially exploited fish stocks are parts of a marine ecosystem, with many species feeding on one another, and thus sustainable fisheries require considering a holistic view. There has been considerable recent interest in ecosystem-based management in fisheries. Several factors have contributed to the current relevance and awareness of this issue, including the debate over the limitations of single-species management and use of this perspective as an alternative tool to justify any position. Consideration of factors that impact marine resource populations in a context beyond just the species level has a long and notable history in fisheries science. Therefore, it is necessary to study and explore impacts of fishing and other human-induced changes on the East Sea ecosystem and their living resources (Zhang et al. 2006).

An ecosystem-based fisheries assessment (EBFA) approach was developed for the assessment of fisheries resources in Korea and detailed methods and data used for scoring indicators and nested risk indices are explained in MOMAF (2007) and Zhang et al. (2009, 2010, 2011). A comprehensive ecosystem-based integrated fisheries management framework has been developed in Korea (MOMAF 2007). Eventually this management framework will be designed for the enhancement and efficient management of fisheries in their use of resources, and thus it requires an in-depth understanding of the ecological interactions of major species in their relationships with predators, competitors and prey species, the effects of climate on fish ecology, the complex interactions between fishes and their habitats, and the effects of fishing on fish stocks and their ecosystems. In addition, the EBFA requires extensive data and information, and so data collection and monitoring of fisheries are essential tasks for this approach. There are many factors that can act to constrain the ecosystem-based fisheries management efforts and their effectiveness. These factors are insufficient resources of money, time and manpower, lack of common vision and goals, inter- and intra-agency conflict, lack of ecological knowledge and lack of public participation. A significant enhancement in resources in terms of funding, research manpower and fishery research vessels will be required if these research themes are to be fulfilled. At the same time strong governmental actions to stop overfishing, restore overfished stocks, protect biodiversity and habitat, and support expanded research and monitoring programs will also be necessary to improve an ecosystem-based approach for proper management of fisheries resources.

13.4 Suggestions

The East Sea shows the most rapid warming rate in the world's ocean. Species composition of dominant fish species has varied across each decadal period in the East Sea, either due to environmental changes or anthropogenic activities. Under warming condition of the East Sea, we can notice the changes in trophic level, growth and survival, recruitment, distribution and migration, new fishing and spawning areas. For proper management of commercial fish resources in the region, scientific advancement of knowledge of fish biology and recruitment processes should be supported. Due to the political obstacles among neighboring nations (i.e. South and North Korea, Japan, and Russia), however, activity in the area of data exchange as well as international scientific cooperation has been very limited.

Cooperative interdisciplinary research on the physical and biological aspects of the East Sea is necessary to understand the mechanisms of oceanographic phenomena, especially to understand the mechanisms of climate change and variations in biological productivity. To ensure the conservation of fishery resources, the creation of regional fisheries bodies among coastal countries should be given serious consideration. The roles of the proposed regional body could be, for example, (1) to collect, analyze, and exchange statistical data on the exploited stocks; (2) to estimate stock biomass using commercial catch information and scientific surveys and furthermore to diagnose the current stock condition; (3) to establish conservation measures such as TAC in order to maintain a healthy state of the stocks of mutual concern; and (4) to disseminate scientific information to the public.

Besides the foundation of an international management body, each nation must establish long-term plans in science and fisheries policy for adapting to climate change. Future research directions in fisheries include the investigation of the relationship between major climate/ocean indices and fisheries, bio-physical coupling processes in the ocean, modeling efforts to predict the future condition of ecosystems, and ecosystem-based management for efficient control of future fishery yields. Of particular interest are findings showing impacts on recruitment, abundance, and distribution of commercial fishes, which suggest that early recognition of such events could be useful for fishery management (Wooster and Zhang 2004).

Acknowledgments   The authors would like to thank the laboratory staffs and scientists in Pukyong National University and the National Fisheries Research and Development Institute for their information collection, and preparation of figures and tables. Special thanks go to Prof. J.K. Kim for consultation on fish species in the East Sea.

References

Belkin IM (2009) Rapid warming of large marine ecosystems. Prog Oceanogr 81:207–213

IPCC (2001) Climate change 2001 synthesis report: contribution of working groups I, II, and III to the third assessment report of the intergovernmental panel on climate change. Cambridge University Press, Cambridge

Kang S, Kim S, Bae SW (2000) Changes in ecosystem components induced by climate variability off the eastern coast of the Korean Peninsula during 1960–1990. Prog Oceanogr 47:205–222

Kang S, Park JH, Kim S (2013) Size-class estimation of the number of walleye pollock *Theragra chalcogramma* caught in the Southwestern East Sea during the 1970s–1990s. Korean J Fish Aquat Sci 46:445–453 (in Korean)

Katayama M (1940) A catalogue of fishes of Toyama Bay. Toyama Gakkaishi 3:1028 (in Japanese)

Kawano M, Doi H, Hori S (2011) List of the fishes in the Japan Sea (Preliminary report). Bull Yamaguchi Prefectural Fisher Res Centre, No. 9, pp 65–94

Kim JK (2009) Diversity and conservation of Korean marine fishes. Korean J Ichthyol 21 (supplement):52–62

Kim MS (1986) Production and species composition in the deep East Sea. In: Proceedings in fishery science (1). Industrial Pub Co Pyung-Yang, pp 132–150 (in Korean)

Kim S, Kang S (1998) The status and research direction for fishery resources in the East Sea/Sea of Japan. J Korean Soc Fish Res 1:44–58 (in Korean)

Kim S, Lee CS, Kang S (2007) Present status and future prospect in salmon research in Korea. Sea J Korean Soc Oceanogr 12:57–60 (in Korean)

Kim YS, Nam MM (2003) Fish fauna in the East Sea. In: Son YM (ed) Current and preservation of Korean fishes, Korean Ichthyological Society, pp 5–36 (in Korean)

Lindberg GU, Krasyukova ZV (1969) Fishes of the Sea of Japan and the adjacent areas of the Sea of Okhotsk and the Yellow Sea, part 3: Percoidei (XC. Serranidae–CXLIV. Champsodontidae). Nauka SSSR, Leningrad, 480 pp (in Russian, but translated into English in 1971)

Lindberg GU, Legeza MI (1959) Fishes of the Sea of Japan and the adjacent areas of the Sea of Okhotsk and the Yellow Sea, part 1: Amphioxi Petromyzones Myxini Elasmobranchii Holocephali. Izdatel'stvo Akademii Nauk SSSR, Moscow, 208 pp (in Russian, but translated into English in 1967)

Lindberg GU, Legeza MI (1965) Fishes of the Sea of Japan and the adjacent areas of the Sea of Okhotsk and the Yellow Sea, part 2: Teleostomi XII. Acipenseriformes–XXVIII. Polynemiformes. 392 pp. Izdatel'stvo Akademii Nauk SSSR, Moscow (in Russian, but translated into English in 1969.)

MIFAFF (2012) Statistical yearbook of agriculture and fisheries. In: Ministry for food, agriculture, forestry and fisheries, Korea, Seoul (in Korean)

MOMAF (2007) Studies on the ecosystem-based resource management system. In: Ministry of maritime affairs and fisheries, Seoul, Korea, 341 pp (in Korean with English abstract)

Mori T (1956) Fishes of San-in district including Oki Islands and its adjacent waters (southern Japan Sea). Memoirs Hyogo Univ Agric Biol Ser 2(2):1–62 (in Japanese)

PICES (2004) Marine ecosystems of the North Pacific, vol 1. PICES Special Publication, 280 p

Sakurai Y, Kiyofuji H, Saitoh S, Goto T, Hiyama Y (2000) Change in inferred spawning areas of *Todarodes pacificus* (Cephalopoda : Ommastrephidae) due to changing environmental conditions. ICES J Mar Sci 57:24–30

Sakurai Y, Kiyofuji H, Saitoh S, Yamamoto J, Goto Y, Mori K (2002) Stock fluctuations of the Japanese common squid, *Todarodes pacificus*, related to recent climate changes. Fish Sci 68(Suppl. I):226–229

Seo H, Kim S, Seong K, Kang S (2006) Variability in scale growth rates of chum salmon (*Oncorhynchus keta*) in relation to climate changes in the late 1980s. Prog Oceanogr 68:205–216

Shinohara G, Shirai SM, Yabe MV (2011) Preliminary list of the deep-sea fishes of the Sea of Japan. Bull Natl Mus Nat Sci Ser A 37:35–62

Son YH (1980) Marine fishes in Korean waters. Science and Encyclopedia Pub., 464 p (in Korean. Author name, title, publishing company were translated)

Son YH (1986) Distribution pattern of fishes in the East Sea off coastal areas of the Korean Peninsula. In: Proceedings in fishery science (1). Industrial Pub. Co. Pyung-Yang, pp 132–150 (in Korean. Author name, title, publishing company were translated)

Terazaki M (1999) The Sea of Japan large marine ecosystem. In: Sherman, Tang (eds) Large marine ecosystems of the Pacific rim assessment, sustainability and management. Blackwell, Oxford, 465 pp

Tian Y, Kidokoro H, Fujino T (2011) Interannual-decadal variability of demersal fish assemblages in the Tsushima warm current region of the Japan Sea: impacts of climate regime shifts and trawl fisheries with implications for ecosystem-based management. Fish Res 112:140–153

Tian Y, Kidokoro H, Watanabe T (2006) Long-term changes in the fish community structure from the Tsushima warm current region of the East Sea with an emphasis on the impacts of fishing and climate regime shift over the last four decades. Prog Oceanogr 68:217–237

Tian Y, Kidokoro H, Watanabe T, Iguchi N (2008) The late 1980s regime shift in the ecosystem of Tsushima warm current in the Japan/East Sea: evidence from historical data and possible mechanisms. Prog Oceanogr 77:127–145

Wooster WS, Zhang CI (2004) Regime shifts in the North Pacific: early indicators of the 1976–1977 event. Prog Oceanogr 60:183–200

Zhang CI, Gong Y (2005) Effect of ocean climate changes on the Korean stock of Pacific saury, *Cololabis saira* (BREVOORT). J Oceanogr 61:313–325

Zhang CI, Hollowed AB, Lee JB, Kim DH (2011) An IFRAME approach for assessing impacts of climate change on fisheries. ICES J Mar Sci 68(6):1318-1328

Zhang CI, Kim S, Gunderson D, Marasco R, Lee JB, Park HW, Lee JH (2009) An ecosystem-based fisheries assessment approach for Korean fisheries. Fish Res 100:26-41

Zhang CI, Lee SG (2002) Fisheries management. Sejong Press, Sejong, 500 pp (in Korean)

Zhang CI, Lee SK (2013) Trophic levels and fishing intensities in Korean marine ecosystems. Unpublished manuscript, 26 p (in Korean)

Zhang CI, Lee JB, Kim S, Oh JH (2000) Climatic regime shifts and their impacts on marine ecosystem and fisheries resources in Korean waters. Prog Oceanogr 47:171–190

Zhang CI, Lee JB, Seo YI, Yoon SC, Kim S (2004) Variations in the abundance of fisheries resources and ecosystem structure in the Japan/East Sea. Prog Oceanogr 61:245–265

Zhang CI, Park HW, Lim JH, Kwon HC, Kim DH (2010) A study on indicators and reference point for the ecosystem-based resource assessment. J Korean Soc Fish Technol 46(1):32-49

Zhang CI, Radchenko V, Sugimoto T, Hyun S (2006) Interdisciplinary physical and biological processes of the Sea of Okhotsk and the Japan/East Sea. In: Robinson AR, Brink KH (eds) The Sea: the global coastal ocean, interdisciplinary regional studies and syntheses. Harvard University Press, Cambridge, 815 pp

Zhang CI, Yoon SC, Lee JB (2007) Effects of the 1988/89 climatic regime shift on the structure and function of the southwestern Japan/East Sea ecosystem. J Mar Syst 67:225–235

Chapter 14
Benthic Animals

Jin-Woo Choi

Abstract The species composition and distribution of macrobenthic fauna in the East Sea (Japan Sea) are summarized and briefly introduced in this chapter from many investigations in both small scale local areas including coastal bays and shallow subtidal areas and large scale areas extending from shelf to bathyal and abyssal depths. There have been few studies on benthic organisms in the deep bottoms of the East Sea, and some recent investigations have not been completed or the target benthic fauna were restricted to a particular faunal group such as bivalves, polychaetes, sipunculids and pericarid crustaceans. Because both the faunal composition and the distribution of the benthic fauna are closely related to the sedimentary facies they inhabit, there were complex benthic faunal assemblages in the southwestern shelf area of the East Sea. The East Sea contains various benthic environments from shallow coastal zones to bathyal and abyssal depths and the distribution of macrobenthic fauna clearly represents the diversity of environmental factors related to the depth gradient. It is expected that when the water depth increased, the abundance and diversity of macrobenthic fauna decreased, although this conclusion is based on only a few fragmental fauna data. From the recent research by Russian investigators on bathyal and abyssal fauna, it can be concluded that there was no real abyssal fauna in the East Sea because they are different from fauna commonly found at near abyssal depths in the North Pacific Ocean. The macrobenthic fauna communities in the ocean dumping areas responded to the organic enrichment of sediments, and thus a few opportunistic species like *Capitella capitata* occurred in a temporally closed region of the dumping area for several years as well as in the present dumping area in the southwest of the East Sea.

Keywords Macrobenthic fauna · Community structure · Shelf and slope · Bathyal and abyssal · Ocean dumping · East Sea (Japan Sea)

J.-W. Choi (✉)
South Sea Research Institute, Korea Institute of Ocean Science and Technology, Geoje 53201, Republic of Korea
e-mail: jwchoi@kiost.ac.kr

© Springer International Publishing Switzerland 2016
K.-I. Chang et al. (eds.), *Oceanography of the East Sea (Japan Sea)*,
DOI 10.1007/978-3-319-22720-7_14

14.1 Introduction

Soft bottom benthic animals are an important component of the marine ecosystem. Benthic organisms are a food source for benthic fishes and large invertebrates, and they play important roles in ecosystems such as nutrient cycling, pollutant metabolism, and secondary production (Snelgrove 1998). Marine benthic organisms are mainly sessile or have no mobility and they have a relatively long life span integrating the effect of continuous environmental disturbances on them, thus they have been used as environmental indicators in many studies (Bilyard 1987; Dauer 1993; Borja et al. 2000). It is known that the body size of macrobenthic fauna become smaller with increasing water depth (Rex et al. 2000; Van Hoey et al. 2004). This decrease in body size in deep waters has been proposed to be related to primary productivity at the surface layer, total organic carbon (TOC), and particulate organic matter (POM), but the direct and important factors have not been clearly defined until the present time. In shallow benthic habitats, however, the species richness and abundance of macrobenthos has been closely related to the stability of sediments and the availability of food to the macrobenthos (Gray 1981). The distribution of macrobenthic fauna is likely determined mainly by sedimentary properties such as grain size and composition and organic content (Sanders 1958). The food supply to benthic environments will determine the composition of feeding types of macrobenthic fauna (Levinton 1972; Jumars and Fauchald 1977).

The benthic fauna described in this chapter only includes the soft-bottom animals inhabiting the continental shelf and slope and deep basin of the East Sea. Thus the sessile fauna on the hard bottoms and macroalgae are not included here. Among the three size categories of benthic fauna [macro- (>1 mm), meio- (0.062–1 mm), and microbenthos (<0.062 mm)], only macrobenthic fauna are dealt with here.

Most previous studies on the macrofauna in the East Sea were mainly focused on faunal composition and spatial distributions. Due to the limited research facilities, including research vessels and sampling gear operating for the deep water, there is no repeated time series monitoring data, and also limited faunal data from the deep basins of the East Sea. Thus most faunal data were collected from the shallow coastal regions.

There were some small and local scale investigations on the macrobenthic fauna or a specific fauna group along the coasts of the Korean Peninsula and far eastern Russia and an isolated ecosystem around a small island, Dokdo (Dok Island), in the middle of the East Sea. However, a series of large scale surveys in the Korean waters of the East Sea was started in 1982 and 5 succeeding surveys were conducted along the shelf and upper slope of the East Sea from 35°N to 37° 30′N until 1987 (Fig. 14.1). The final results of these surveys on the fauna groups of polychaetes (Choi and Koh 1988, 1990), mollusks (Je 1993), and ophiuroids (Shin and Koh 1993) were reported. There was a research report on the spatial distribution of macrobenthos in the slope area and a deep basin (Ulleung Basin) of the East Sea

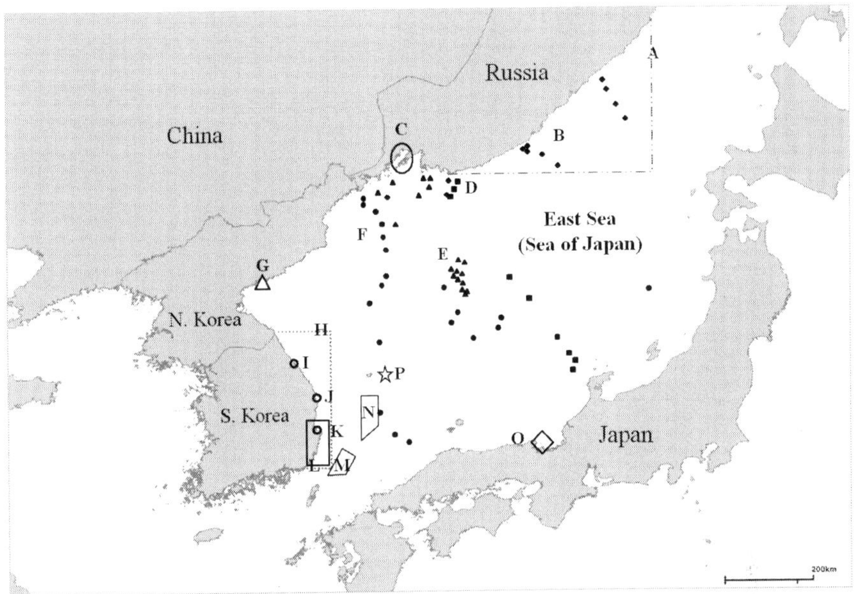

Fig. 14.1 Map showing the study areas of macrobenthic fauna in the East Sea. *A* Northern slope (Kharlamenko et al. 2013), *B* northern bathyal and abyssal (Kamenev 2013), *C* Amursky Bay (Belan 2003), *D* northern bathyal and abyssal (Kamenev 2013), *E* Japan Basin (Kamenev 2013), *F* northern shelf and slope (Kamenev 2013), *G* northern coast (KORDI 1999), *H* southern shelf and slope (Shin and Koh 1993), *I* Youngdong PP (Choi et al. 2000), *J* Uljin NPP (Yu et al. 2011; Shin per. comm.), *K* Youngil Bay (Shin et al. 1992), *L* southern shelf (Choi and Koh 1988), *M* dumping area (Dong-haejeong) (KORDI 2008), *N* dumping area (Donghaebyung) (KORDI 2009), *O* Wakasa Bay (Antonio et al. 2010), *P* Dokdo (Choi et al. 2002; Park et al. 2002; Ryu et al. 2012)

(KORDI 1997). Also there was a small scale survey including the shelf area and the slope area around Dokdo in the East Sea (Choi et al. 2000).

Recently Russian benthologists carried out several benthic surveys during the Russian-German joint expedition Sea of Japan Biodiversity Studies (SoJaBio) on the spatial distribution and trophic relations of macro- and megabenthic fauna (>5 cm) in the northwestern part of the East Sea in August–September 2010 (Belan 2003; Alalykina 2013; Brandt et al. 2013; Kamenev 2013; Golovan et al. 2013; Maiorova and Adrianov 2013). These results will be used to compare with the benthic fauna occurring in the southwestern part of the East Sea. However, there are few studies on the trophic relationships among macrofauna in the soft bottom sediments of the East Sea (Antonio et al. 2010; Kharlamenko 2013).

There are two ocean dumping areas located at the deep central part of the Ulleung Basin (Donghaebyung dumping area) and at the southwestern shallow shelf (Donghaejeong dumping area) of the East Sea designated by the Korean government in 1992. The dumped materials were composed of feces of animals, and wastes from homes and industrial companies which contained high concentrations of nitrogen and phosphate, heavy metals, and polycyclic aromatic hydrocarbons

(PAHs) (Hong and Shin 2009). The amount of ocean dumping has increased during the period from 1992 to 2006. Thus the southern part of the Donghaebyung dumping area in the Ulleung Basin (UB) has been temporarily closed since 2006 due to the high organic enrichment and heavy metal contamination. Recently the Korean government decided to decrease gradually the amount of waste dumping and to cease completely ocean dumping in 2014. The intensive monitoring of benthic environmental conditions including the macrobenthos in the Donghaejeong and Donghaebyung dumping areas started in 2007 and 2009, respectively. Here the impacts of ocean dumping on the macrobenthic fauna were briefly described by focusing on changes in benthic community structures found at different water depths and sedimentary facies.

14.2 Local Scale Macrofauna Distributions

14.2.1 Southwestern Coasts

There were a few studies on the macrobenthic faunal communities in the southwestern coasts of the East Sea, that is, along the east coast of the Korean Peninsula (Fig. 14.1). These studies have focused on the distribution of macrobenthos around the discharge outfalls of power plants constructed along the coastal line in order to determine the influence of the discharged warm water on the benthic fauna and flora. However, there was an investigation on the macrobenthic faunal communities in a semi-enclosed bay, Youngil Bay, and the harbor area within the bay. An electric generating power plant, the Youngdong Power Plant, using coal as the fuel materials, is located near Gangneung city (37° 45′N) on the east coast of Korea. A seasonal investigation on the macrobenthic fauna was conducted at 14 stations in the coastal area including the discharge outfall of the power plant from April 1993 to February 1994 (Choi et al. 2000). The study area is located on the east coast of the Korean Peninsula where there are well developed sandy beaches along the coast, and thus the study area consists of sandy sediments out to a water depth of 50 m.

A total of 109 and 70 macrobenthic fauna species were collected in summer and winter, respectively. The mean density was in the range of 1995 and 631 ind. m^{-2} in summer and winter, respectively. Polychaete worms accounted for the majority of individuals and mollusks contributed to the biomass of the macrobenthic communities. Dominant species were *Spiophanes bombyx*, *Prionospio* sp., *Alvenius ojianus*, *Wecomedon* sp., *Urothoe* spp.

Based on cluster analysis of the abundance of macrobenthos, the study area was divided into two station groups by water depth (Fig. 14.2). The typical dominant species in the offshore area were *S. bombyx* and *A. ojianus*. They occurred during all seasons. However, the dominant species of the inner shallow area were *Scoloplos armiger*, *Mactra chinensis*, *Corophium* sp., and *Urothoe* spp. Their abundance changed from season to season, especially in autumn and winter possibly due to river discharge and warm water discharge from the power plant.

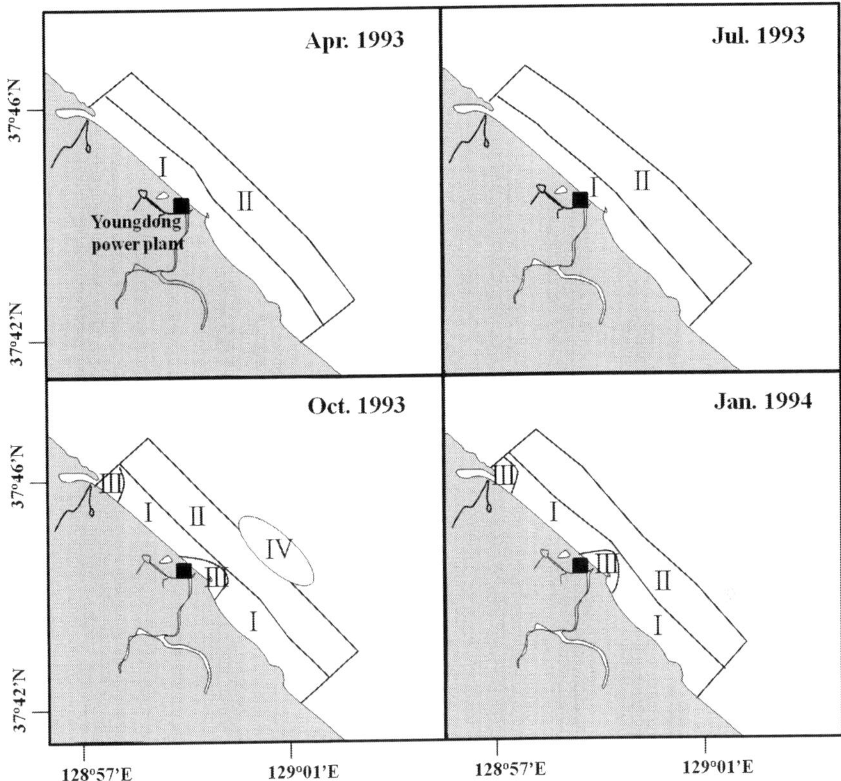

Fig. 14.2 Map showing the distribution of macrobenthic communities in the sandy sediments along the coast line of the east coast of the Korean peninsula. Station Group I: *Scoloplos-Mactra* community, Station Group II: *Spiophanes* community during spring and summer (from Choi et al. 2000)

The second case study of the small scale surveys conducted along the south-western coasts of the East Sea is from the subtidal region around the outfall of the warm water discharge of the nuclear power plant in Uljin province of Korea. There was an ecological survey of the macrobenthic polychaete communities on the subtidal soft-bottom around the discharge outfall of the Uljin Nuclear Power Plant (UNPP) in 2008 (Shin, personal communication).

A total of 84 polychaete species occurred and the average density was 2378 ind. m^{-2}. The typical dominant species were *Spiophanes bombyx*, *Magelona* sp., *Praxillella affinis*, and *Lumbrineris longifolia*. The high macrobenthic density was primarily due to the very high population density of *Spiophaes bombyx* which is known as a typical polychaete worm in sandy sediments. There were three poly-chaete assemblages in the study area (Fig. 14.3). While there was no single species characteristic of the shallower coastal area at less than 5 m depth, a *Spiophanes* assemblage occupied the coastal area where water depths ranged from 5 to 30 m, and a *Magelona* assemblage existed in the offshore area deeper than 30 m.

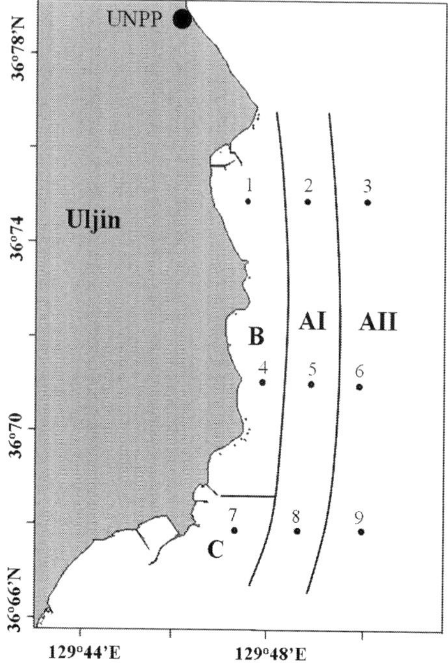

Fig. 14.3 The distribution of macrobenthic polychaete worms around the discharge outfall of the Uljin nuclear power plant (AI: *Spiophanes* community, AII: *Magelona* community) (created by J.-W. Choi)

This distribution pattern of polychaete assemblages indicates that there is little disturbance in the benthic environment around the UNPP except for the near shore coastal area.

A seasonal ecological survey was conducted from August 2007 to May 2008 to determine the effects of the warm water discharge from the UNPP (Yu et al. 2011). The macrobenthic fauna around the UNPP were divided into two regional assemblages with similar faunal composition according to water depth shallower and deeper than 30 m. The shallow coastal fauna assemblage was composed of some opportunistic fauna such as *Urothoe convexa, Mandibulophoxus mai, Felaniella sowerbyi*, and *Rhynchospio* sp. near the warm water discharge outfall. The deeper offshore fauna assemblage was characterized by polychaetes like *Magelona japonica, Spiophanes bombyx, Scoletoma longifolia*, and *Chaetozone setosa* which have been present since 1987, before the construction of the UNPP.

The third example of a small scale benthic survey was from a unique sheltered bay, Youngil Bay, along the east coast of Korea. An ecological study was carried out to determine the distribution patterns of the benthic polychaete worms in 1991 (Shin et al. 1992). A total of 72 polychaete species was found and the mean density was 1485 ind. m^{-2}. *Spiophanes bombyx, Pseudopolydora* sp. *Lumbrineris longifolia, Maldane cristata*, and *Polydora* sp. were the major contributors to the abundance of the polychaete worms in Youngil Bay. *Capitella capitata*, a well-known organic pollution indicator, occurred in the harbor region and the estuarine region of the bay. Based on a multivariate analysis, three station

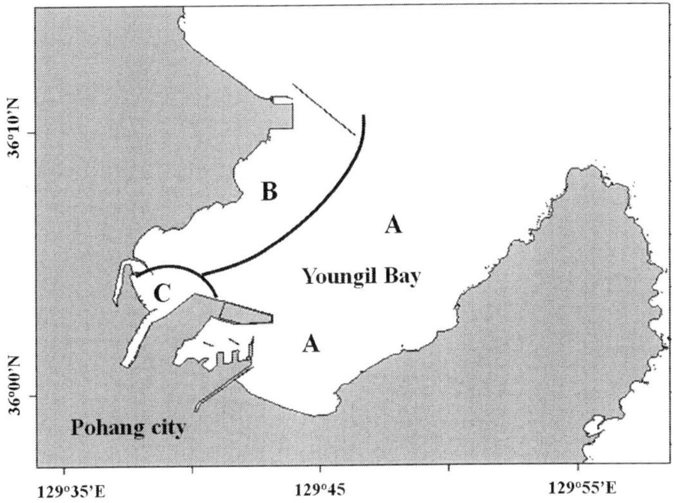

Fig. 14.4 The spatial distribution of polychaete communities in Youngil Bay during 1991. **a** *Maldane-Praxillella* assemblage, **b** *Spiophanes-Nephtys* assemblage, **c** *Pseudopolydora-Polydora-Capitella* assemblage (from Shin et al. 1992)

groups with similar species composition of polychaete worms were identified (Fig. 14.4). A *Maldane-Paxillella* assemblage occupied the southern half of the bay (A), a *Spiophanes-Nephtys* assemblage existed in the northern half (B), and a *Pseudopolydora-Polydora-Capitella* assemblage was located in the harbor and estuarine area (C). This distribution pattern of polychaete communities seemed to be influenced by the sedimentary facies under the current flow pattern in Youngil Bay and organic loads from Pohang City and seaport.

14.2.2 Northwestern Coasts

There was a marine environmental survey report on the macrobenthic fauna around the expected construction site of a Nuclear Power Plant in Shinpo City of North Korea (KORDI 1999). The sampling was conducted seasonally at 13 stations located around the discharge outfall of the power plant in 1998 (Fig. 14.1). A total of 244 species including 79 molluscan species, 75 polychaete species and 61 crustacean species was collected. The mean density and biomass was 2458 ind. m^{-2} and 121 g wet m^{-2}, respectively. The most abundant benthic fauna were *Mryiochele oculata* (23.5 %), *Nuculoma tenuis* (14.6 %), *Axinopsida subquadrat* (13.8 %), *Lumbrineris nipponica* (10.9 %), *Prionospio malmgreni* (4.1 %), and *Alvenius ojianus* (3.2 %). The dominant species among 45 megabenthos collected by Agassiz trawl were *Ophiura sarsi* and *Asterias amurensis*. The maximum biomass of megabenthos was estimated to be 9 kg wet $trawl^{-1}$.

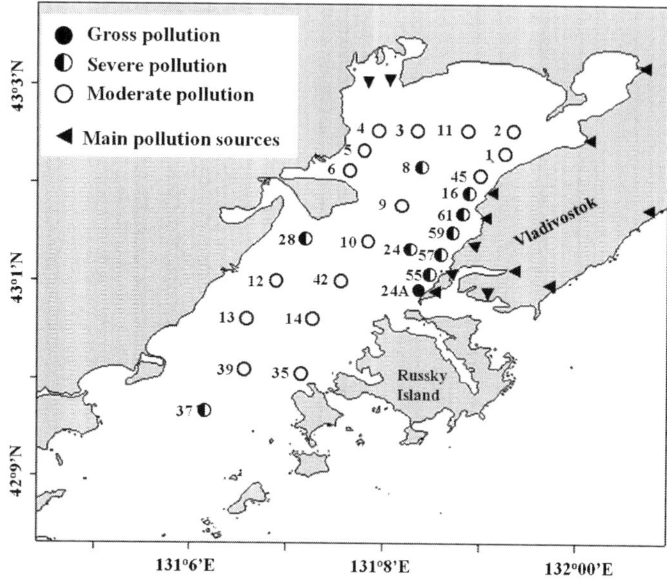

Fig. 14.5 Pollution level of *bottom* sediments and distribution of benthic communities in Amursky Bay during 1986–1989 (from Belan 2003)

There was another important local survey on the macrofauna assemblages at the northwestern coast of the East Sea, that is, in Amursky Bay, Russia. Belan (2003) investigated a contamination gradient, and assessed the effect of pollution on the macrobenthic fauna using quantitative and qualitative benthic parameters in Amursky Bay. He collected macrobenthic fauna at 30 stations from four survey cruises between 1986 and 1989. Significant changes in the number of taxonomic groups, density and ecological indices were observed between severe and moderate contamination levels. *Tharyx pacifica*, *Dorvillea japonica*, *Dipolydora cardalia*, *Capitella capitata*, and a phoronid, *Phoronopsis harmeri* were selected as the most contaminant insensitive species whereas *Maldane sarsi*, *Lumbrinereis* sp., and *Scoloplos armiger* were assigned as contaminant sensitive species. From the results of cluster analysis and n multidimensional scaling (MDS) ordination, two macrobenthic fauna assemblages were identified in Amursky Bay in response to the pollution gradient (Fig. 14.5).

14.2.3 Dokdo (Dok Island)

There was a unique isolated benthic ecosystem around Dokdo in the middle of the East Sea. The macrobenthic faunal assemblages are under the influence of the Tsushima Warm Current entering into the East Sea through the Korea Strait.

The surface sediments of the shelf and upper slope area were composed of sand particles with sponge spicules whereas those in the slope were composed of mud. An ecological survey on the macrobenthic fauna and their habitat conditions was conducted on the soft-bottom around Dokdo in September 1999 and May 2000. The macrobenthic fauna in the shelf and slope area were collected using a van Veen grab and a box corer, respectively.

The macrobenthos occurring on the slope were represented by 15 faunal groups in 8 phyla and the most abundant faunal group consisted of polychaete worms which accounted for 81 % of the slope fauna and 85 % of the shelf fauna. The mean faunal densities of the slope and shelf were 2028 and 456 ind. m^{-2}, respectively. The dominant species on the slope were *Exogone verugera* (41 %), *Cossura longocirrata* (8 %), *Tharyx* sp. (7 %), *Scalibregma inflatum* (5 %), *Aedicira* sp. (5 %), *Arricidea ramosa* (4 %), and *Sigambra tentaculata* (4 %). However, the dominant species in the shelf area were different from those in the slope area. They were *Chone* sp. (49 %), *Tharyx* sp. (18 %), *Ophelina acuminata* (7 %), *Chaetozone setosa* (4 %), *Glycera* sp. (3 %), and *Aedicira* sp. (2 %). The trophic composition of the macrobenthos differed between the slope and shelf areas, that is, surface deposit feeding worms were dominant in the slope area whereas filter feeders were dominant in the shelf area. According to cluster analysis and nMDS ordination (Fig. 14.6), the spatial distribution of macrobenthos in the slope area around Dokdo was related to sediment properties such as grain size and composition and organic content. Most macrobenthos inhabited the top 2 cm sediment layer of the slope area.

There is little published literature dealing with macro- and megabenthic fauna on the rocky hard-bottoms around Dokdo despite many investigations (Park et al. 2002; Ryu et al. 2012). Park et al. (2002) reported on densities and biomass of the subtidal hard-bottom megabenthic fauna (Fig. 14.7). Megabenthos were related to water depth and topographic conditions, and thus in the area shallower than 10 m, a turbo shell *Batillus cornutus* and a mussel *Mytilus coruscus* were dominant species with their distinct vertical distributions (Fig. 14.8). However, hard-bottoms at depths deeper than 10 m showed a diverse fauna distribution pattern depending on bottom topography, and a sea cucumber *Stichopus japonicus* was its dominant fauna. *B. cornutus*, *M. coruscus*, and *S. japonicus* were three commercially useful fauna on the hard-bottoms of Dokdo. The estimated amount of total biomass of these three species were 6.54, 3.89, and 8.92 tones, respectively. Abalones were rare and their biomass was very small.

Ryu et al. (2012) reported on the topographic distributions of macrofauna on the shallow rocky shores around Dokdo during August 2007–June 2008. A total of 98 macrofaunal species, including 21 newly recorded species which are all known from the Korean and Japanese waters, were collected at Dokdo. Among them, mollusks and crustaceans comprised more than 70 % of the total species at each station. Thus by 2008, 403 macrofaunal species were identified on the rocky shores of Dokdo. The fauna composition and distribution of invertebrates were related to the topographical characteristics of each habitat. A North American isopod *Idotea metallica*, first described at Helgoland Island in the German North Sea in 1994, was collected again in this survey. *I. metallica* had already been found at

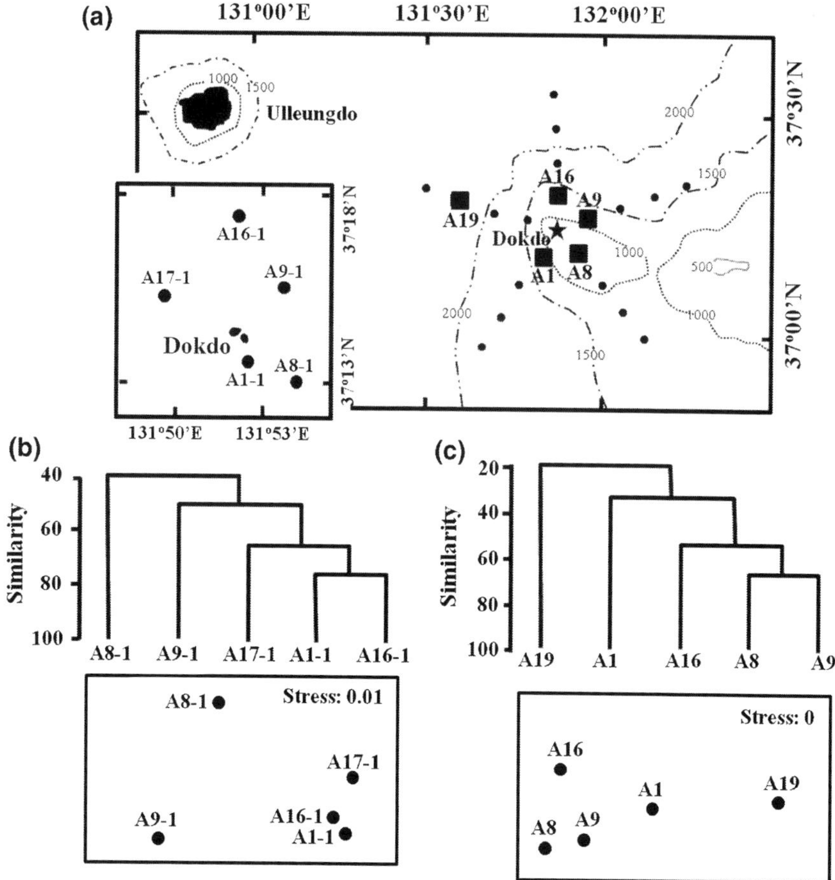

Fig. 14.6 Map showing the sampling stations for macrobenthos around Dokdo (**a**), and the dendrograms and nMDS plots indicating the fauna similarity between stations on the shelf area (**b**) and the slope area (**c**) of the East Sea (from Choi et al. 2002)

Dokdo in 1909. The Dokdo marine ecosystem seems to be highly dynamic in biodiversity, and Dokdo itself may play a role as a good settling site for warm-water invasive species entering the East Sea through the East China Sea.

14.3 Large Scale Macrobenthic Fauna Communities

14.3.1 Southwestern Shelf Area

The macrobenthic polychaete communities on the continental shelf area in the southwestern part of the East Sea were investigated through 4 April survey cruises from 1982 to 1984, and 1987. The surface sediment of this study area was

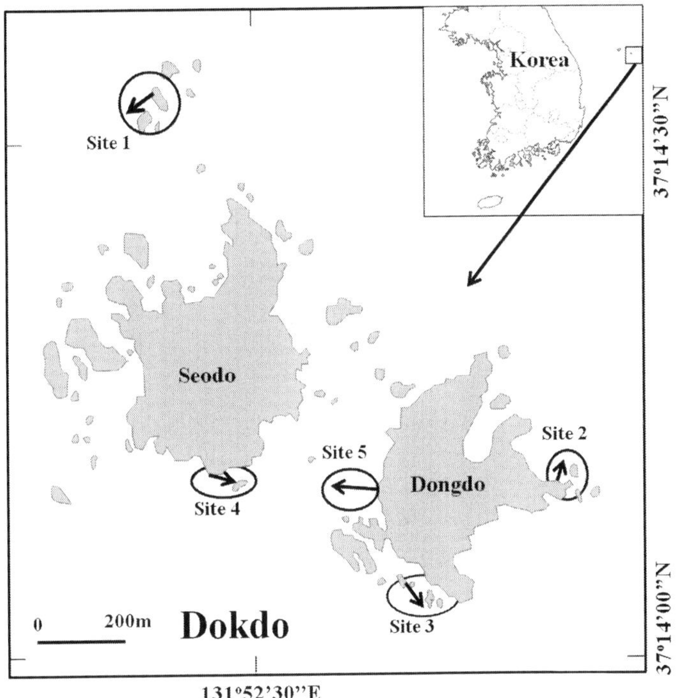

Fig. 14.7 The locations of sampling sites around Dokdo. The *arrows* indicate the direction of sampling from intertidal to subtidal region (from Park et al. 2002)

composed of various sedimentary facies from mud to muddy sand. Thus there existed many different macrobenthic fauna assemblages in the shelf area according to the sedimentary facies (Fig. 14.9). The shallow coastal region sustained 4 polychaete assemblages: a *Nothria* assemblage in the most northern part near Pohang city with fine sediments, a *Magelona-Maldane* assembalge off Gampo, an *Ophelina* assemblage in the middle coastal area off Ulsan city, and a *Nothria* assemblage in the southern section near Busan city. These coastal polychaete assemblages contained fewer than 10 species per station and showed a relatively low species diversity ($H' = 1.22$–1.52). The offshore shelf area also contained 4 assemblages: a *Terebellides-Aglaophamus* assemblage in the deep northern area of very fine sediments, a *Myriochele* assemblage and a *Spiophanes* assemblage in the central area of sandy sediments, and a *Ninoe* assemblage in the southern offshore area composed of a sandy bottom. The offshore polychaete assemblages showed a relatively greater species richness and higher diversity than coastal ones ($H' = 1.90$–2.26). However, the offshore region consisting of sandy sediment showed very low population densities. Some dominant species showed a preference for a specific sediment type, and this phenomenon could be detected by the feeding modes of dominant species. It seemed that the complicated distribution of various polychaete assemblages was determined by both the gradient of sedimentary properties and bottom temperature (Choi and Koh 1988).

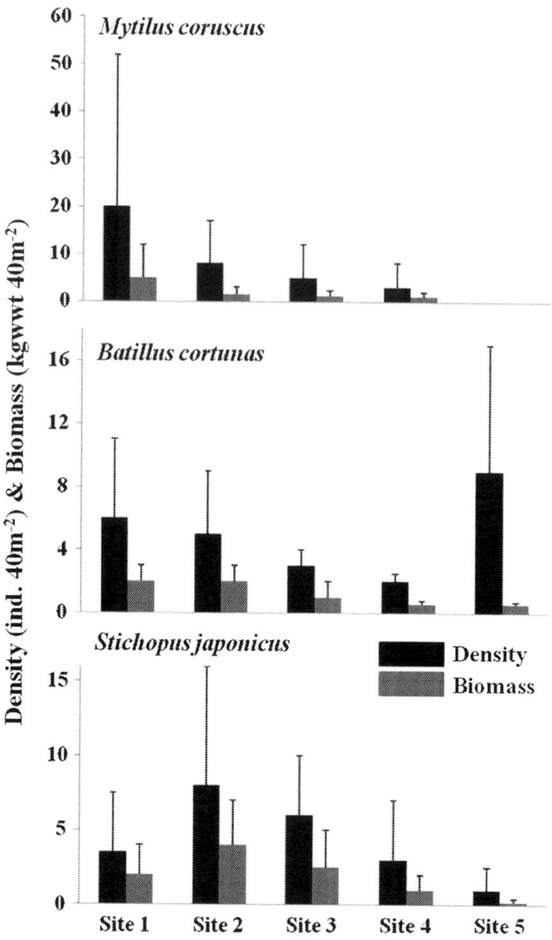

Fig. 14.8 Spatial variation in the density and biomass of three commercial species at each survey site around the Dokdo shown in Fig. 14.7 (from Park et al. 2002)

14.3.2 Southwestern Shelf and Slope Area

In the continental shelf and slope of the southwestern part of the East Sea, there was a study of the species composition and distribution patterns of the polychaete communities (Choi and Koh 1990). The polychaete worms were collected using a van Veen grab at 65 stations covering 15,000 km^2 from 35° 30′N to 37° 50′N on three sampling cruises in April 1985 and April and October 1987. The surface sediment showed five sedimentary facies. Fine sediments such as silt and clay prevailed on the slope area of the East Sea and the Hupo Basin, whereas mixed sediments like muddy sand and sand facies existed on the Hupo Bank and the shelf area off Jeokbyun.

A total of 112 polychaete species from 36 families were collected. The overall mean density of polychaete worms was 300 ind. m^{-2} and the species richness

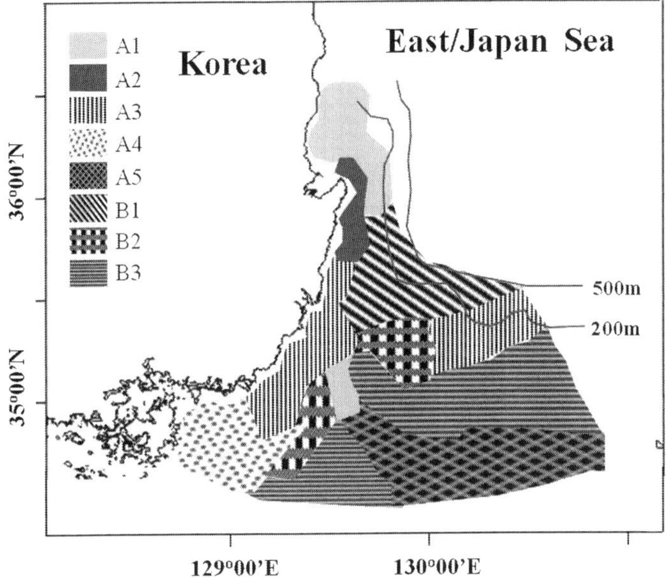

Fig. 14.9 The spatial distribution of macrobenthic polychaete communities in the southwestern shelf area of the East Sea (A1: *Nothria* assemblage, A2: *Magelona-Maldane* assemblage, A3: *Ophelina* assemblage, A4: *Nothria* assemblage, A5: *Ninoe* assemblage, B1: *Terebellides-Ampharete* assemblage, B2: *Myriochele* assemblage, B3: *Spiophanes* assemblage) (Redrawn from Choi and Koh 1988)

was about 15 species 0.2 m^{-2}. Dominant polychaete species were *Chaetozone setosa* (13.3 %), *Aglaophamus malmgreni* (6.7 %), *Ampharete acrtica* (5.9 %), *Terebellides horikoshii* (5.5 %), *Tharyx* sp. (4.8 %), and *Magelona pacifica* (7.2 %). Boreal and cold water polychaetes were major contributors to the polychaete communities in this area. Most dominant species had specific depth ranges from the shelf area to the upper slope and middle slope depth. A significant change in the mean polychaete density and species richness was found at a water depth of approximately 600 m (Fig. 14.10).

The distributional pattern of some dominant brittle stars was investigated in the East Sea (Shin and Koh 1993). They were *Amphiodia craterodometa*, *Amphioplus macraspis*, *Ophiura leptotenia*, and *Ophiura sarsi*. These brittle stars had specific distribution patterns in response to the water depth, that is, *Amphiodia craterodometa* was dominant on the shelf area, and *Ophiura sarsi* was abundant on the upper slope area in the depth range from 200 to 300 m, and *Ophiura leptotenia* was a dominant species on the middle and lower slope area deeper than 300 m (Fig. 14.11). This distribution pattern of brittle stars in the East Sea seemed to be related to the rapid decrease of water temperature (less than 1.0 °C below 200 m) with increasing water depth along the slope area.

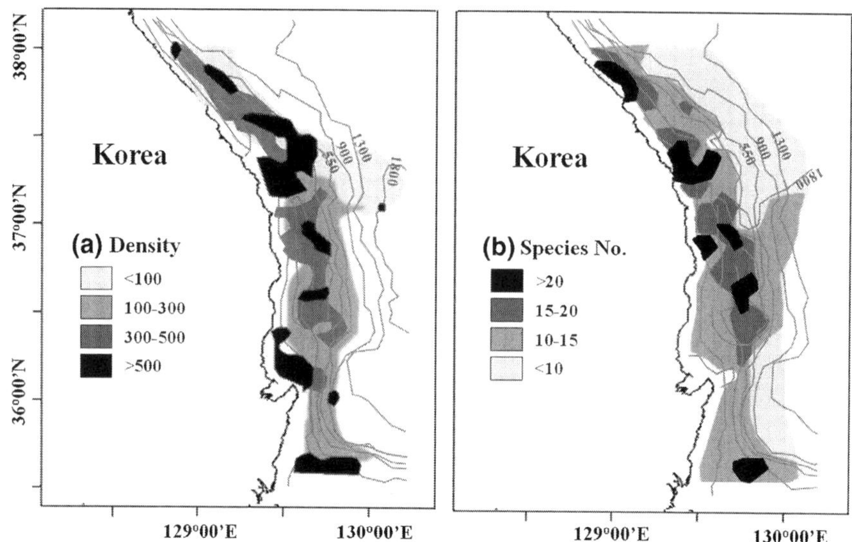

Fig. 14.10 Spatial distribution of density (**a**) and species number (**b**) of macrobenthic fauna in the continental shelf and slope area of the southwestern East Sea (Redrawn from Choi and Koh 1990)

14.3.3 Western Slope Area

Quantitative sampling of the deep sea macrobenthic fauna on the slope and UB in the East Sea was conducted in 1996. A large box corer was used, instead of a grab sampler, to determine the vertical distribution of macrofauna and a mesh screen of 300 µm was used to increase the sampling efficiency (KORDI 1997). The sediments from all sampling stations were composed of fine particles and the sedimentary facies was mud. The benthic animals collected in this deep bottom region were classified into 10 phyla, and polychaetes comprised the greatest abundance of the faunal groups, accounting for 43 % of total individuals. Nematodes (30 %), mollusks (14 %), and crustaceans (10 %) were also major faunal groups in this area. There were few echinoderms in these samples and their proportion was less than 1 % of the total faunal abundance.

Unfortunately, no fauna were identified to the species level. The fauna composition of this deep bottom region was very similar to that of the shallow soft bottom benthic communities except for the nematodes. A high proportion of polychaetes in the fauna of deep bottom communities was previously reported (Blake and Grassle 1994; Blake and Hilbig 1994). In the case of the Atlantic Ocean, polychaetes comprised 58 and 47 % of the total number of species and abundance of individuals, respectively. The overall density of the macrobenthos was 2655 ind. m^{-2}, ranging from 44 ind. m^{-2} in UB to 17,511 ind. m^{-2} at the station near Ulleungdo (Ulleung Island). The density of macrobenthos decreased from stations near the east coast of Korean Peninsula to UB at the latitude of 37°N (Fig. 14.12).

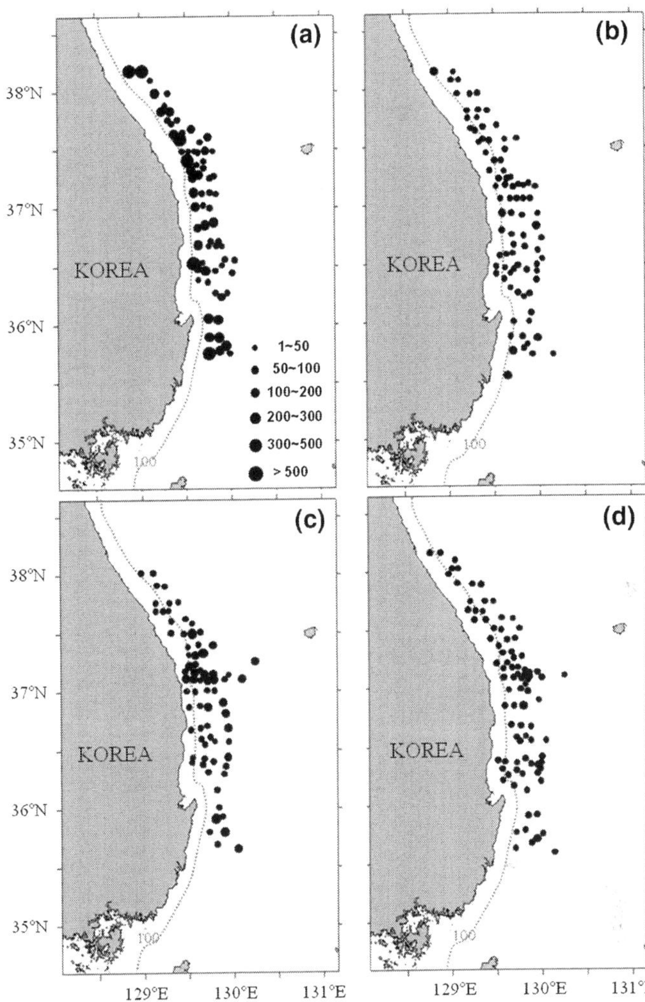

Fig. 14.11 The spatial distribution of four dominant brittle star species in the continental shelf and slope areas of the southeast coast of Korean Peninsula in the East Sea (**a** *Ampiodia craterodometa*, **b** *Amphioplus macraspis*, **c** *Ophiura leptoctenia*, **d** *Ophiura sarsi*) (Redrawn from Shin and Koh 1993)

From the vertical sections of sediment samples collected by a box corer, we can obtain the vertical distribution of macrofauna in the bathyal depths of stations like UB. More than 55 % of the total macrobenthic fauna was distributed in the surface sediment layer (0–2 cm) and more than 80 % of benthic animals were concentrated within 4 cm of the surface sediment layer (Fig. 14.13). At 8 cm below the sediment surface, only long-bodied polychaetes occurred. This vertical distribution pattern was found also on the deep slope benthic communities off Cape Lookout

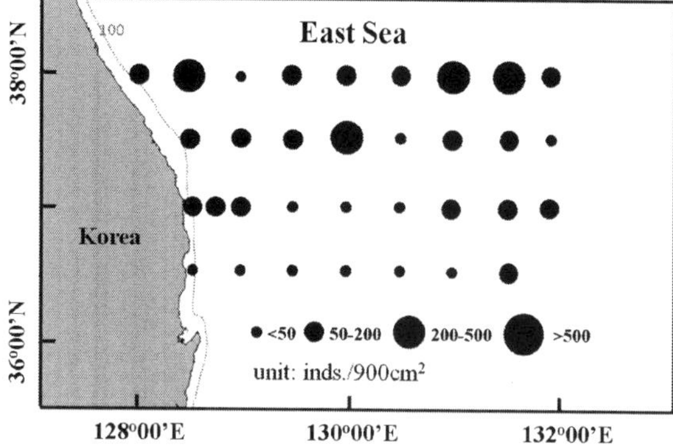

Fig. 14.12 Map showing the sampling stations and density of macrobenthos collected by a box corer (30 cm × 30 cm) in the Southwestern East Sea extended from the southeast coast of Korea to Ulleung Basin

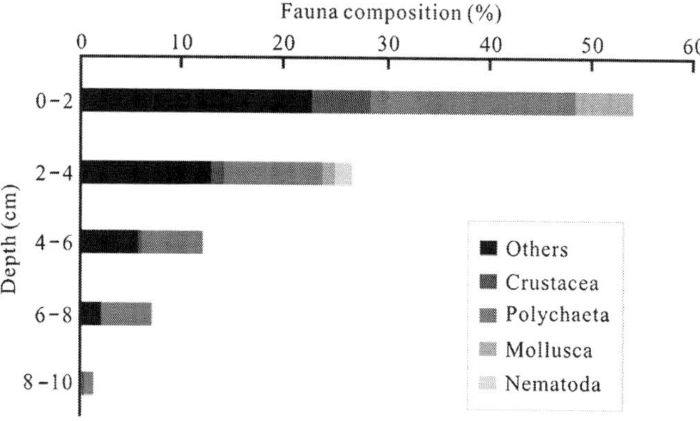

Fig. 14.13 The vertical distribution of macrobenthos occurring in the Ulleung Basin of the East Sea

in the Atlantic Ocean (Blake 1994). Major faunal groups also showed the same trend in vertical distribution as polychaetes, but crustaceans and mollusks were limited to the upper 4 cm layer, whereas polychaetes and nematodes appeared in relatively deeper sediment depths.

Megabenthos in the slope area of the southwestern part of the East Sea and UB were collected at 10 stations using an Agassiz trawl (KORDI 1997). The most abundant species were *Ophiura sarsi*, *Ctenodiscus crispatus*, and *Chinoecetes japonicus*. Thus the faunal composition of deep bottom megabenthos in the East Sea was dominated by echinoderms and decapods. A similar faunal composition of megabenthos was also reported for the slope area of the Atlantic Ocean where echinoderms accounted for 34 % of total individuals and other major contributors were cnidarians (24 %) and demersal fish (21 %) (Hecker 1994). A maximum density of 412 ind. trawl^{-1} and biomass of 4.0 kg trawl^{-1} were found at stations in the slope off Hupo Bank. The maximum biomass of *C. japonicus* was estimated to be 4.0 g m^{-2}. The estimated density of megafauna in the southwestern part of the East Sea collected from the Agassiz trawl survey was 0.4 ind. m^{-2} which is comparable to those found for the slope of the Atlantic Ocean, estimated in the ranges from 0.10 to 5.59 ind. m^{-2}.

14.3.4 North Bathyal and Abyssal Area

Kamenev (2013) studied the species composition and distribution of bivalves in bathyal and abyssal depths (465–3435 m) of the East Sea (Fig. 14.1). Twenty six species were collected during five cruises. The deep-water bivalve fauna of the East Sea were characterized by an impoverished shelf fauna and consisted of eurybathic species that extend from the shelf to the bathyal and abyssal zones. Most of bivalve species have a wide geographic distribution and inhabit cold water regions of the Northern Atlantic, Northern Pacific, and Arctic Ocean. Only five species were endemic to the East Sea. The species diversity of bivalves decreased with the increase of water depth. At depths below 3000 m, only *Dacrydium vitreum*, *Delectopecten vencouverensis*, and *Thyasira* sp. were found. The lack of typical abyssal species of bivalves in the deep East Sea seemed to be connected to the isolation of this water body from the Pacific abyssal depths.

Alalykina (2013) reported that a total of 92 polychaete species in 28 families were collected from the SoJaBio expedition, and dominant species were *Laonice* sp., *Ophelina cylindricaudata*, *Arcteobia spinelytris*, *Ampharete* sp., and *Chaetozone* sp.1. The most abundant polychaete families were Spionidae, Opheliidae, Polynoidae, Ampharetidae, and Cirratulidae. There was a depth related distribution pattern in polychaete assemblages such that the upper bathyal area of the East Sea had higher polychaete diversity than the abyssal area, likely because the fauna assemblages in the bathyal area might be in a zoogeographic zone that overlaps with the shelf area. The dominance of polychaete worms in the fauna of the bathyal and abyssal bottoms of the East Sea was similar with that found in deep bottoms world-wide (Brandt et al. 2013). Maiorova and Adrianov (2013) reported that 8 sipunculid species were recently found in the Russian coastal areas of the East Sea from the SoJaBio expedition. Until now, a total of 31 sipunculid species have been found in the East Sea. Among them, *Phascolosoma agassizii*, *P. scolops*, *Themiste nigra*, and *T. hexadactyla* were distributed in the shallow area (0–50 m), while *Golfingia margaritacea*, *G. vulgaris*, *Nephasoma capilleforme*, *N. wodjanizkii*, and *P. strombus* occurred at depths from 50 to 1000 m, especially *G. margaritacea* which was found at the deepest depth of 1699 m. Identification keys were provided for all sipunculids in the East Sea including Russian, Japanese, and Korean coasts.

Golovan et al. (2013) also investigated the diversity and distribution of peracarid crustaceans in the bathyal and abyssal depths of the East Sea (Fig. 14.1). In total, 146 species from 85 genera and 42 families were collected. The diversity and species richness and abundance of most peracarid species were highest at the shallowest slope stations (450–550 m) and decreased with depth, however, several opportunistic species such as *Eurycope spinifrons* (Isopoda), *Chauliopleona hanknechti*, and *Paratyphlotanais japonicus* (Tanaidacea) increased below the slope. They suggested that these opportunistic species may have recently colonized the geologically young pseudo-abyssal habitats of the East Sea.

There was only one study on trophic relationships among macrobenthic fauna in the bathyal and abyssal depths of the East Sea (Kharlamenko et al. 2013). They used stable isotope techniques (TL = $(\delta^{15}N_{consumer} - \delta^{15}N_{base})/\delta^{15}N + TL_{base}$, Post 2002) to estimate the trophic levels from primary producers to top predators. They selected the selective deposit feeder *Megayoldia* sp. as the baseline species with trophic level of 2. Suspended particulate organic matter (POM) and bottom sediment organic matter (SOM) were assigned to the basement of a food web, and primary consumers were categorized as surface deposit feeders and selective deposit feeders, for example, the foraminiferan *Elphidium* sp. and a protobranch bivalve *Megayoldia* sp. The third trophic group was occupied by the scaphopod *Fustiaria nipponica*, the ophiuroid *Ophiura leptoctenia*, and the crinoid *Heliometra glacialis*. The fourth trophic level of the continental slope benthic community in the East Sea was comprised of the large crab *Chionocetes japonicus*, the shrimp *Eualus biungus*, the septibranch bivalve *Cardiomya beringensis*, and the sea star *Ctenodiscus crispatus*. Antonio et al. (2010) reported the trophic relationships among benthic fauna from the estuarine coastal area to the offshore shelf area of Wakasa Bay in the southeastern part of the East Sea using stable isotopes. They defined the community structure and assigned the trophic levels from these faunal data. Benthic microalgae were an important energy source in the shallow coastal area, and their proportion decreased in the offshore area. The main food sources in the offshore shelf area changed to marine POM and phytoplankton.

14.4 Ocean Dumping Areas

There are two ocean dumping areas in the East Sea, The Donghaebyung dumping area in UB and the Donghaejeong dumping area in the southwestern shelf area. These ocean dumping areas were designated as receiving areas for the disposal of sewage sludge and wastes from land beginning in 1992 and scheduled to be closed by 2014. The amount of disposed wastes was 20,800 m^3 for ten years from 1993 to 2003. Thus some intensified ecological surveys were required in the ocean dumping areas in order to manage and maintain the marine ecosystem properly.

Two benthic fauna surveys were conducted to monitor and assess the ecological impact of ocean dumping such as the abundance and biological diversity of macrobenthos in the Donghaejeong area (1616 km^2) during June 2007 and 2010 (Fig. 14.1).

Macrozoobenthos collected during the 2007 and 2010 surveys comprised a total of 202 and 274 species in 8 animal phyla, and the mean faunal densities were 922 and 1337 ind. m^{-2}, respectively (KORDI 2008, 2010). Polychaete worms were the most dominant faunal group accounting for 51 % of the total abundance in the benthic community, with 92 and 152 species in 2007 and 2010, respectively. Total species number of the major benthic faunal groups in the reference area was very similar to that in the waste dumping area. However, in the case of mean density, the dumping area showed higher density than the reference area, especially for polychaete worms. Some faunal groups like crustaceans and echinoderms showed higher densities in the reference area.

The spionid *Spiophanes kroyeri* and the maldanid *Maldane cristata* were the most abundant fauna in the ocean dumping area with mud and muddy sand facies. These polychaete worms were also known as dominant species in surveys from the 1980s. Before waste dumping, a *Spiophanes kroyeri* assemblage mainly occupied the broad offshore area including the Donghaejeong dumping area, and an *Ophelina acuminata* assemblage was found in a coastal area. A *Myriochele* assemblage and a *Terebellides-Aglaophamus* assemblage occupied the region between the coastal and offshore shelf areas (Choi and Koh 1988). Several pollution indicative species were found at sites in the dumping area during the intensified survey in 2007 (Table 14.1). The indicators of organic enrichment were *Paraprionospio coora*, *Dorvillea* sp., *Polydora ligni*, *Spiochaetopterus* sp., *Euchone analis*, etc. These species were not found in the previous studies conducted during the 1980s before the waste dumping in the shelf area of the East Sea (Choi and Koh 1988). However, the abundance of these indicative species was very low and was not comparable to abundances in the semi-enclosed coastal bays of Korea. The study area sustained three different macrobenthic communities whose mean densities increased twice during the 20 years from the 1980s to 2007. The health of macrobenthic communities was assessed using several benthic indices such as Benthic Pollution Index (Seo et al. 2012) and was estimated to be below normal or indicative of slightly disturbed condition.

There is another marine dumping area (the Donghaebyung area) located in the central area of the Ulleung Basin at a distance of 125 km from Pohang City located at the southeast coast of Korea Peninsula (Fig. 14.1). The total designated area is about 3700 km^2 covering water depths of 200–2000 m. The macrobenthic faunal assemblages in this dumping area have been investigated in order to assess the benthic environmental health status in 2009 (KORDI 2009).

The macrobenthic fauna were collected at 25 stations using a box corer (covering 0.2 m^2) in May 2009, sieved through a 1 mm mesh screen. A total of 35 species occurred: the mean density and biomass were 525 ind. m^{-2} and 3.7 g wet m^{-2}, respectively. The Donghaebyung dumping area showed lower values of species richness, faunal density, and biomass compared with those from the Donghaejeong dumping area located in the southwestern shelf region of the East Sea. Within the dumping area, however, there was an increase in species richness and abundance of the macrobenthic fauna in the previously dumped region of the Donghaebyung dumping area compared with those in the present dumping region as well as the reference area (Fig. 14.14). Dominant species were *Thyasira tokunagai* (50.6 %), *Capitella capitata* (10.1 %), *Chaetozone* sp. (6.3 %), and

Table 14.1 Dominant macrofauna occurring in the dumping area and the reference area of the Donghaejeong dumping area in the East Sea

Dumping area			Reference area		
Species name	Density	%	Species name	Density	%
(a) 2007					
Spiophanes sp.	122	21.7	*Ennucula niponica*	75	21.1
Musculus senhausia	113	20.1	*Maldane cristata*	37	10.3
Maldane cristata	69	12.3	*Ophelina acuminata*	32	9.0
Thyasira tokunagai	17	3.1	*Ophiura* sp.	28	7.8
Ophelina acuminata	16	2.9	*Modiolus margaritaceus*	25	7.0
Ampelisca sp.	16	2.9	*Byblis japonicus*	18	5.1
Nothria sp.1	13	2.3	*Chaetozone setosa*	11	3.2
Ophiura sp.	10	1.8	*Spiophaneskroyeri*	11	3.2
(b) 2010					
Spiophanes kroeyeri	295	22.4	*Aoroides* sp.	190	14.0
Madane cristata	156	11.8	*Yoldiella philippiana*	150	11.0
Aoroides sp.	82	6.3	*Ophiura sarsi*	75	5.5
Ophiura sarsi	75	5.7	*Lumbrineris* sp.	73	5.4
Ampelisca sp.	48	3.6	*Ampelisca* sp.	65	4.8
Spiochaetopterus costarum	25	1.9	*Ampharete arctica*	46	3.4

Yoldiella philippiana (9.4 %) (Table 14.2). *T. tokunagai* was considered to be an opportunistic species with small body size and a short life span and reproductive maturation (Seo et al. 2013), thus this species can occupy any vacant habitat space after a strong disturbance. Although *T. tokunagai* occurred as a dominant species in the entire dumping area, *T. tokunagai* and *C. capitata* showed their high densities in temporarily closed part of the Donghaebyung dumping area since 2006 (Fig. 14.15). In conclusion, the ocean dumping in the East Sea resulted in the organic enrichment of surface sediments and changes in the community structure of macrobenthic fauna due to their responses to this long-term disturbance with the overall increase of faunal density and changed faunal composition by newly adding opportunistic fauna and reducing previously inhabited equilibrium species.

14.5 Summary and Further Study

In this chapter, the macrobenthic fauna assemblages in the East Sea investigated in both small and large scale areas were introduced. However, there are a few studies of the benthic organisms in the deep bottom regions of the East Sea. Moreover, these were incomplete or only limited to a particular target fauna like bivalves, polychaetes, and pericarid crustaceans. Because benthic faunal composition and distribution is closely related to the sedimentary facies where they reside, the benthic fauna assemblages found in the southwestern shelf area of the East Sea

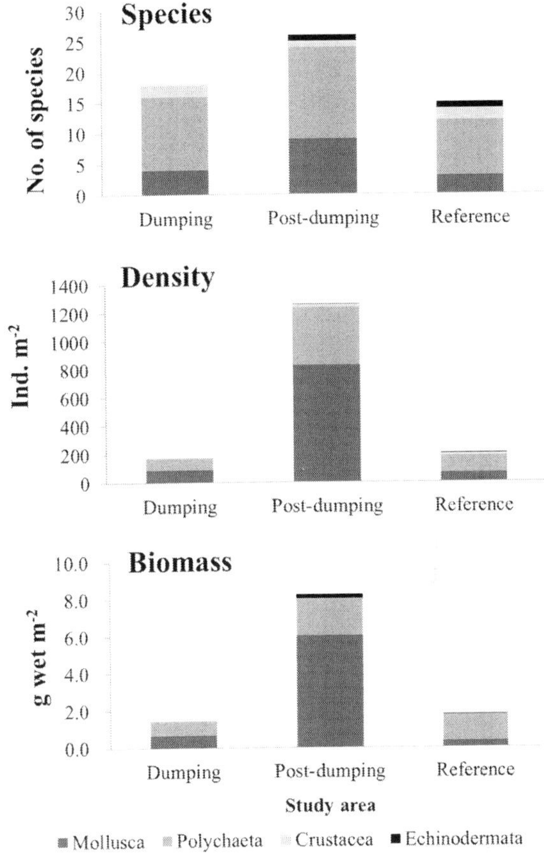

Fig. 14.14 Species number, density, and biomass of macrofauna occurring in the dumping area and the reference area of the Donghaebyung dumping area in the East Sea during May 2009

were complex. The East Sea contains various benthic environments from shallow coastal areas to the bathyal and abyssal depths, for which the distribution of macrobenthic fauna represents the diversity of environmental factors related to the depth gradient. It was clear that when the water depth increased, the abundance and diversity of macrobenthic fauna decreased, despite having only a small fragmented faunal dataset on which to base this conclusion.

The faunal data covering the macrobenthic communities were obtained from investigations conducted more than 20 years ago, making it difficult to compare the recent fauna data and old data sets directly to detect the temporal change in faunal composition for any specific location (Table 14.3). In order to detect changes in macrobenthic faunal assemblages attributable to global warming, such as the increase of sea water temperature, it would be necessary to investigate the faunal composition simultaneously at various local regions at the same latitude. In the East Sea, there was a specific faunal composition reflecting benthic

Table 14.2 Dominant macrofauna occurring in the dumping area and the reference area of the Donghaebyung dumping area in the East Sea during May 2009 (*density* ind. m^{-2})

Sampling area	Species name	Density	%
Dumping area	*Thyasira tokunagai*	130	43.4
	Yoldiella philippiana	65	21.9
	Aglaophamus malmgreni	26	8.7
	Capitella capitata	23	7.7
	Sigambra tentaculata	19	6.4
	Ophelina acuminata	19	6.2
Post-dumping area (dumping ceased in 2006)	*Thyasira tokunagai*	692	58.8
	Chaetozone sp.	169	14.4
	Capitella capitata	146	12.4
	Sigambra tentaculata	51	4.4
	Hyperiid sp.	25	2.1
	Yoldiella philippiana	23	2
Reference area	*Thyasira tokunagai*	1324	50.6
	Capitella capitata	264	10.1
	Yoldiella philippiana	246	9.4
	Chaetozone sp.	244	9.3
	Sigambra tentaculata	164	6.3
	Aglaophamus malmgreni	95	3.6

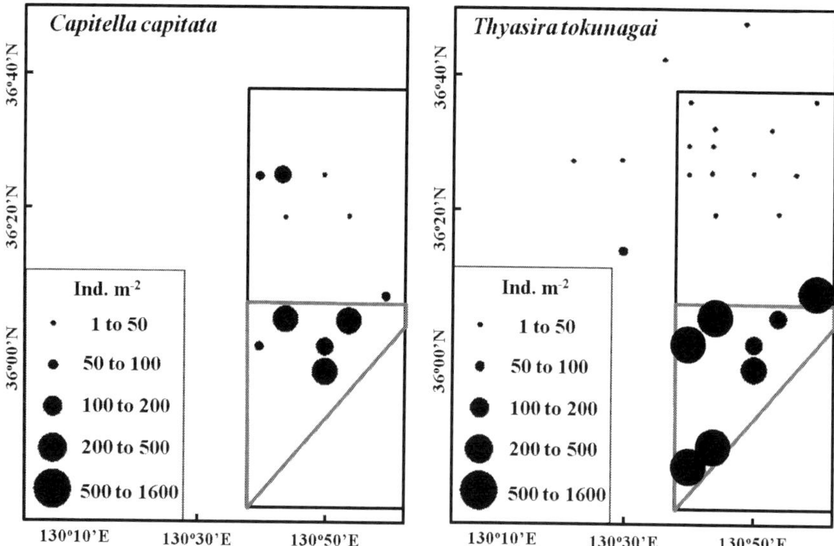

Fig. 14.15 The spatial distribution of dominant macrofauna occurring in the Donghaebyung dumping area in the East Sea during May 2009. The region enclosed with *red* color is the post-dumping region closed in 2006

Table 14.3 The dominant species of macrobenthic faunal communities in the coastal regions and bathyal and abyssal regions of the East Sea (created by J.-W. Choi)

Study area	Coasts	Bathyal and abyssal area	
Southern East Sea	Gangneung	Southwestern slope	
	Spiophanes bombyx	*Chaetozone setosa*	*Amphiodia craterodometa*
	Prionospio sp.	*Aglaophamus malmgreni*	*Amphioplus macraspis*
	Alvenius ojianus	*Ampharete acrtica*	*Ophiura leptotenia*
	Wecomedon sp.	*Terebellides horikoshii*	*Ophiura sarsi*
		(Polychaeta)	(Ophiuroidea)
	Uljin	Dokdo	
	Spiophanes bombyx	*Exogone verugera*	
	Magelona sp.	*Cossura longocirrata*	
	Praxillella affinis	*Tharyx* sp.	
	Lumbrineris longifolia	*Scalibregma inflatum*	
	Youngil Bay	Ulleung Basin	
	Spiophanes bombyx	*Ophiura sarsi*	
	Pseudopolydora sp.	*Crenodiscus crispatus*	
	Lumbrineris longifolia	*Chinoecetes japonicus*	
	Maldane cristata	(Ophiuroidea)	
	Polydora sp.		
Northern East Sea	Shinpo	Russian coast	
	Mryiochele oculata	*Dacrydium vitreum*	*Eurycope spinifrons*
	Nuculoma tenuis	*Delectopecten vencouverensis*	*Chauliopleona hanknechti*
	Axinopsida subquadrat	*Thyasira* sp.	*Paratyphlotanais japonicus*
	Lumbrineris nipponica	(Bivalvia)	(Crustacea)
	Amursky Bay		
	Tharyx pacifica	*Golfingia margaritacea*	
	Dorvillea japonica	(Sipuncula)	
	Dipolydora cardalia		
	Capitella capitata		

environmental conditions such as grain size and composition, sediment enrichment, and warm water discharge at small scales, but there was also a large scale distributional pattern of macrobenthic fauna reflecting a water depth gradient in bathyal and abyssal basins.

The macrobenthic fauna are closely associated with large demersal fishes or benthic decapods as important food organisms. However, there are few studies on the trophic structures or food webs of macrobenthic assemblages in the East Sea. Thus, it is necessary to investigate the functional aspects of the benthic macrofauna such as the secondary production, the amount of resources, their life cycles, and the pelagic-benthic coupling as well as the trophic structures. In order to effectively conduct these kinds of studies, international co-operation and communication among scientists are strongly recommended. In the ocean dumping areas, the community structure of macrobenthic fauna was changed by increasing opportunistic species and also excluding the previously existed native species there. Therefore, it has been decided to stop the ocean dumping from 2014 and after that a long-term monitoring will be needed to understand the natural restoration and recovery of benthic environment from ocean dumping.

Acknowledgments The author would like to acknowledge all supports from the Korea Institute of Ocean Science and Technology (KIOST). Also the author would like to thank Prof. H-C Shin at Chonnam National University for providing the faunal data and JY Seo for her help drawing the figures.

References

Alalykina I (2013) Preliminary data on the composition and distribution of polychaetes in the deep-water areas of the north-western part of the Sea of Japan. Deep-Sea Res II 86–87:164–171

Antonio ES, Kasai A, Ueno M, Won N, Ishihi Y, Yokoyama H, Yamashita Y (2010) Spatial variation in organic matter utilization of benthic communities from Yura-River Estuary to offshore of Tango Sea, Japan. Estuar Coast Shelf Sci 86:107–117

Belan TA (2003) Benthos abundance pattern and species composition in conditions of pollution in Amursky Bay (the Peter the Great Bay, the Sea of Japan). Mar Pollut Bull 46:1111–1119

Bilyard G (1987) The value of benthic infauna in pollution monitoring studies. Mar Pollut Bull 18:581–585

Blake JA (1994) Vertical distribution of benthic infauna in continental slope sediments off Cape Lookout, North Carolina. Deep-Sea Res II 41:919–927

Blake JA, Grassle JF (1994) Benthic community structure on the U.S. South Atlantic slope off the Carolinas: spatial heterogeneity in a current-dominated system. Deep-Sea Res II 41:835–874

Blake JA, Hilbig B (1994) Dense infaunal assemblages on the continental slope off Cape Hatteras, North Carolina. Deep-Sea Res II 41:875–899

Borja A, Franco J, Perez V (2000) A marine biotic index to establish the ecological quality of soft-bottom benthos within European estuarine and coastal environments. Mar Pollut Bull 40:1100–1114

Brandt A, Elsner N, Brenke N, Golovan O, Malyutina MV, Riehl T, Schwabe E, Würzberg L (2013) Epifauna of the Sea of Japan collected via a new epibenthic sledge equipped with camera and environmental sensor systems. Deep-Sea Res II 86–87:43–55

Choi JW, Koh CH (1988) The polychaete assemblages on the continental shelf off the southeastern coast of Korea. J Korean Soc Oceanogr 23:169–183

Choi JW, Koh CH (1990) Distribution pattern of polychaete worms on the continental shelf and slope of the East Sea (Sea of Japan), Korea. J Korean Soc Oceanogr 25:36–48

Choi JW, Je JG, Lee JH, Lim HS (2000) Distributional pattern of macrobenthic invertebrates on the subtidal sandy bottom near Gangrung, east coast of Korea. Sea J Korean Soc Oceanogr 5:346–356 (in Korean)

Choi JW, Hyun S, Kim D, Kim WS (2002) Macrobenthic assemblages on the soft bottom around Dokdo in the East Sea, Korea. Ocean Polar Res 24:429–442 (in Korean)

Dauer DM (1993) Biological criteria, environmental health and estuarine macrobenthic community structure. Mar Pollut Bull 26:249–257

Golovan OA, Blazewicz-Paszkowycz M, Brandt A, Budnikova LL, Elsner NO, Ivin VV, Lavrenteva AV, Malyutina MV, Petryashov VV, Tzareva LA (2013) Diversity and distribution of peracarid crustaceans (Malacostraca) from the continental slope and the deep-sea basin of the Sea of Japan. Deep-Sea Res II 86–87:66–78

Gray JS (1981) The ecology of marine sediments. Cambridge University Press, London (185 pp)

Hecker B (1994) Unusual megafaunal assemblages on the continental slope off Cape Hatteras. Deep-Sea Res II 41:809–834

Hong S, Shin KH (2009) Alkylphenols in the core sediment of a waste dumpsite in the East Sea (Sea of Japan), Korea. Mar Pollut Bull 58:1566–1587

Je JG (1993) Distribution of molluscs in soft bottom in Korean seas. PhD. Thesis Dissertation, Seoul National University, pp 296

Jumars PA, Fauchald K (1977) Between-community contrasts in successful polychaete feeding strategies. In: Coull BC (ed) Ecology of marine benthos. University of South Carolina Press, USA, pp 1–20

Kamenev GM (2013) Species composition and distribution of bivalves in bathyal and abyssal depths in the Sea of Japan. Deep-Sea Res II 86–87:124–139

Kharlamenko VI, Brandt A, Kiyashko SI, Würzberg L (2013) Trophic relationship of benthic invertebrate fauna from the continental slope of the Sea of Japan. Deep-Sea Res II 86–87:34–42

KORDI (Korea Ocean Research & Development Institute) (1997) A study on the oceanographic atlas in the adjacent seas to Korea—Southwest of the East Sea. BSPN 00316–963–1 (in Korean)

KORDI (Korea Ocean Research & Development Institute) (1999) Report on the marine ecological survey in the adjacent seas for the overseas construction of nuclear power plant in North Korea, pp 27–40 (in Korean)

KORDI (Korea Ocean Research & Development Institute) (2008) Annual report on the establishment of management system for the waste sea dumping (IV). BSPM 44200–1934–4 (in Korean)

KORDI (Korea Ocean Research & Development Institute) (2009) Annual report on the establishment of management system for the waste sea dumping (VI). BSPM 54871–2135–4 (in Korean)

KORDI (Korea Ocean Research & Development Institute) (2010) Annual report on the establishment of management system for the waste sea dumping (VII). BSPM 55601–2246–4 (in Korean)

Levinton J (1972) Stability and trophic structure in deposit-feeding and suspension-feeding communities. Am Nat 106:472–486

Maiorova AS, Adrianov AV (2013) Peanut worms of the Phylum Sipuncula from the Sea of Japan with a key to species. Deep-Sea Res II 86–87:140–147

Park HS, Kang R-S, Myoung JG (2002) Vertical distribution of mega-invertebrate and calculation to the stock assessment of commercial species inhabiting shallow hard-bottom in Dokdo, Korea. Ocean Polar Res 24(4):457–464 (in Korean)

Post DM (2002) Using stable isotopes to estimate trophic position: models, methods, and assumptions. Ecology 83:703–718

Rex MA, Stuart CT, Coyne G (2000) Latitudinal gradients of species richness in the deep-sea benthos of the North Atlantic. Proc Natl Acad Sci USA (PNAS) 97(8):4082–4085

Ryu SH, Jang K-H, Choi EH, Kim SK, Song SJ, Cho HJ, Ryu JS, Kim YM, Sagong J, Lee JH, Yeo MY, Bahn SY, Kim HM, Lee GS, Lee DH, Choo YS, Pak JH, Park JS, Ryu JS, Khim JS, Hwang UW (2012) Biodiversity of marine invertebrates on rocky shores of Dokdo, Korea. Zool Stud 51(5):710–726 (in Korean)

Sanders HL (1958) Benthic studies in Buzzards Bay I: animal-sediment relationship. Limnol Oceanogr 3:245–258

Seo JY, Park SH, Lee JH, Choi JW (2012) Structural changes in macrozoobenthic communities due to summer hypoxia in Gamak Bay, Korea. Ocean Sci J 47:27–40

Seo JY, Lee JS, Choi JW (2013) Distribution patterns of opportunistic molluscan species in Korean waters. Korean J Malacol 31:1–9 (in Korean)

Shin HC, Koh CH (1993) Distribution and abundance of ophiuroids on the continental shelf and slope of the East Sea (southwestern Sea of Japan), Korea. Mar Biol 115:393–399

Shin HC, Choi SS, Koh CH (1992) Seasonal and spatial variation of polychaetous community in Youngil Bay, southeastern Korea. J Korean Soc Oceanogr 27:46–54 (in Korean)

Snelgrove PVR (1998) The biodiversity of macrofaunal organisms in marine sediments. Biodivers Conserv 7:1123–1132

Van Hoey G, Degraer S, Vincx M (2004) Macrobenthic community structure of soft-bottom sediments at the Belgian continental shelf. Estuar Coast Shelf Sci 59:559–613

Yu OH, Paik SG, Lee HG, Lee JH (2011) Spatiotemporal distribution of macrobenthic communities in the coastal area of Uljin and its relation to environmental variables. Ocean Polar Res 33(4):421–434 (in Korean)

Chapter 15
Marine Mammals

Kyum Joon Park

Abstract The coasts of the East Sea (Japan Sea) of the Korean Peninsula are known to have been populated by whales since prehistoric times. This association has lasted throughout the history of Korea's dynasties without interruption. Traces indicating the presence of whales are written in petroglyphs and in the histories of the dynasties. From the late 19th century, many foreign whaling ships started entering Korean waters, especially in the East Sea. After Japan won the Russo-Japanese War, monopoly whaling by the Japanese funded by their capital lasted until 1945. At that time, statistical data on whaling in the East Sea was first recorded. According to those records, the major species of whales were fin whales and the minke whales, with gray whales, humpback whales and others. Koreans introduced Japanese-style whaling ships and started whaling in postcolonial times. Commercial whaling in Korea continued actively, focused mainly on minke whales, until the International Whaling Commission declared a moratorium on commercial whaling in 1986. According to whaling data, bycatches and stranding, the baleen whales observed in the East Sea are northern right whales, blue whales, fin whales, sei whales, Bryde's whales, minke whales, humpback whales, and gray whales. Thus far, the minke whale has been the dominant species. For toothed whales, there are large species such as sperm whales, Baird's beaked whales, and Stejneger's beaked whales, which are discovered once or twice every year through bycatches or stranding. Small and medium-sized species are extremely diverse, including killer whales, Pacific white-sided dolphins, Risso's dolphins, harbor porpoises, and common dolphins. The most commonly observed species are common dolphins and Pacific white-sided dolphins. In the case of common dolphins, about a hundred of them are bycaught or stranded every year along the coast of the East Sea. These species are distributed throughout the year in the East Sea, but Dall's porpoises and harbor porpoises show changes in latitudinal distributions

K.J. Park (✉)
Cetacean Research Institute, National Fisheries Research and Development Institute,
Ulsan 44780, Republic of Korea
e-mail: mogas@korea.kr

© Springer International Publishing Switzerland 2016
K.-I. Chang et al. (eds.), *Oceanography of the East Sea (Japan Sea)*,
DOI 10.1007/978-3-319-22720-7_15

depending on seasonal changes. Pinnipeds living in the East Sea are the Steller sea lion, the northern fur seal and the spotted seal. Most of them live in the northern East Sea and a few migrate to the south coast of the Korean Peninsula. The Japanese sea lion was once abundant in the East Sea, but it is now extinct.

Keywords Marine mammal · Cetacean · Pinniped · East Sea (Japan Sea) · Whaling

15.1 Introduction

Interest in wild animals and marine mammals has certainly increased not only in the academic world but also among the general public. The media, as well as scientific institutions, societies, government administrators, and environmental activists are all expressing intense interest in wild animals and marine mammals. The general public wants to see for themselves and experience wild animals to satisfy their desire for both education and adventure. At the same time, there is an increasing awareness of the importance of marine mammals in maintaining a healthy state of the marine ecosystem, and the impact of human activities on marine mammals and their habitats has become a worldwide issue.

Most biologists classify marine mammals into five mammalian groups: Cetaceans (whales, dolphins, and porpoises), pinnipeds (seals, sea lions, and walruses), sirenians (manatees, dugongs, and sea cows), marine and sea otters, and the polar bear. The five groups that we call marine mammals have different species of origin and show considerable differences in their life cycles. What they do have in common is that all the animals that we call marine mammals find all of their food (or most of their food) from the sea (and often in fresh water). All marine mammals are adapted to living in water. Cetaceans and sirenians spend their whole life cycle under the water, while other marine mammals live on the beach during specific periods for various reasons (reproduction, molting, or resting). With regard to morphological modifications, cetaceans and sirenians have obsolete legs and developed tails, while pinnipeds have legs that evolved to gain thrust in the water. However, cetaceans, sirenians, and pinnipeds have all evolved a streamlined body to increase their hydrodynamic efficiency. These types of morphological modifications in marine and sea otters and in the polar bear are less adapted for marine life, and most of them are similar to those of land mammals in their surroundings.

Of the seas surrounding the Korean Peninsula, the East Sea shows a remarkable diversity of marine mammals, but there are not many data on them. Accurate statistics and research on species classification in the past were limited to large whales, which were utilized commercially until the 1980s, whereas until the 21st century, almost no statistical information was produced for other species such as small cetaceans, or pinnipeds. To find out more about the ecological features, migration patterns, and statuses of the marine mammals distributed in the East Sea, efforts such as consistent investigations, research, and cooperation with neighbouring countries in the East Sea are necessary.

In this chapter, the historical records of whales and whaling in Korean waters, mainly focusing on the East Sea, is described. And through data on whale fishing and recent research, I introduce information on the distribution, occurrence, and abundance of marine mammals (especially, cetaceans and pinnipeds) with some biological characteristics, if any are known for Korean waters including areas of the East Sea.

15.2 The Historic Records of Whales in Korean Waters

The relationship between Korea and whales became known to the world, several decades ago; the world recognized Korea as a strong whaling nation when Korea argued for continued whaling in order to maintain Korea's food culture of eating whales. Although Korea drew the world's attention to its whaling tradition only a few decades ago, the East Sea area had maintained connections with whales since prehistoric times, and these connections did not break but survived on the Korean Peninsula throughout several dynasties. However, there are relatively few people who are aware of this, not only outside Korea, but also within Korea.

In 1971, an array of petroglyphs, 10 m long and 3 m wide, were discovered at Ulsan, located in the southern East Sea. A total of 200 carvings with 75 types of art depicting land animals and humans hunting them, including whales, tigers, wild boars, and deer, were drawn on this rock. In particular, more than 10 species of whales are depicted, each with clearly visible morphological differences, a number that is larger than that found on the oldest known whale petroglyph found on Røddøy Island in Norway. There is also a scene of a ship carrying several people who have captured a whale using a harpoon with a rope attached. Therefore, it is thought that people were already hunting and using whales at the time the carvings were made, and that the people depicted in this art had considerable knowledge of whales. Although there is some debate among Korean academics about the time period during which these petroglyphs were created, the general consensus is that they originated in the Neolithic era or the early Bronze Age. This series of petroglyphs earned the name "Bangudae petroglyphs" and was designated as national treasure No. 285 in 1995.

The early written history of the Korean Peninsula contains several records of sightings of a large type of fish in the history books of the Three Kingdoms (BC57–AD668) and Unified Shilla (AD668–AD918) periods. Also, according to the records, people ate this fish. This large fish was most likely a whale. Later, during the Goryeo and Joseon dynasties, 918–1392 and 1392–1910, respectively, a history book written about the Goryeo dynasty recorded that an ambassador from China came to Korea and obtained some whale oil from the East Sea coastal areas. There are also several records of whales from the last Korean dynasty, the Joseon dynasty, describing the finless porpoise in detail, as well as records about dolphins, such as killer whales. Hendrik Hamel, a Dutchman who arrived in Joseon in 1653 after a shipwreck, wrote in his book that the people in Joseon harvested whales (Park 1987).

In the late 19th century, the Joseon dynasty began to deteriorate and this coincided with many whaling ships entering the seas surrounding the Korean Peninsula, especially into the East Sea. Although many whaling ships from Western nations arrived in the East Sea, the existing ship journals of American whaling ships are the most numerous and the most revealing. The first records are from 1848, when the journals of several ships recorded sightings of Joseon people, or recorded that the ships had entered the East Sea. There are records of almost 40 ships, and in reality there were probably more. However, the largest number of American whaling ships entered the East Sea in the following year, 1849, when records from the journals of 120 ships stated this. One ship journal written in 1848 from an American whaling ship contains a record of its encounter with both French and German whaling ships. Thus, American whaling ships as well as whaling ships from other Western nations entered the East Sea.

In 1849, encounters with many French whaling ships were recorded in the journals of American whaling ships. In fact, the French whaling ship Liancourt discovered Dokdo (Dok Island). Henceforth, Dokdo is referred to as Liancourt Rocks on Western maps. It appears that German whaling ships fished for whales in the East Sea until the 1860s. From the 1870s, the whaling industry began to deteriorate because the population of right whales, which was the main target of the whaling ships, decreased. This led to a decrease in the number of whaling ships entering Korean waters (Park 1987).

Although the advance of Russian whaling ships into the East Sea was later than other Western nations, they engaged in a substantial amount of whaling in the East Sea and in the northwest Pacific. The whaling operations were based in Vladivostok, a city located in the northern East Sea. The Russians first started whaling in 1889, entering into an agreement with the Joseon, which tried to open its doors to Western nations and become a modern nation. Until the early 1900s, the Russians continued to enter various ports in Vladivostok and Korea to conduct significant whale hunts (Park 1987).

Like Korea, Japan has a long whaling history, achieving as well modernization earlier than Korea in the 19th century. As Japan started whaling with modern equipment, it began to compete with Russia and advanced into various locations in the Joseon area for effective whaling. In 1905, as the Russo-Japanese war ended with Japan's victory, Japan had no more competition in the East Sea as the Russian whaling ships pulled out. Later, as Japan forcefully annexed Joseon and it became a Japanese colony, the Japanese monopoly on whaling in Korean waters, including the East Sea, began as Japanese whaling companies advanced into Joseon. Although the Korean people had harvested whales and used them from ancient times, there was neither systematic development of the practice nor being institutionalized. Therefore, it soon died out with the introduction of modern technology and mass whaling by foreign powers. Japan, however, succeed in modernizing and institutionalizing traditional whaling, and this practice was modernized.

As the Japanese began monopolizing whaling in Korean waters, a system evolved which produced accurate statistics on caught whales. The official number of whales caught of each species was recorded from 1911 onwards. Modern

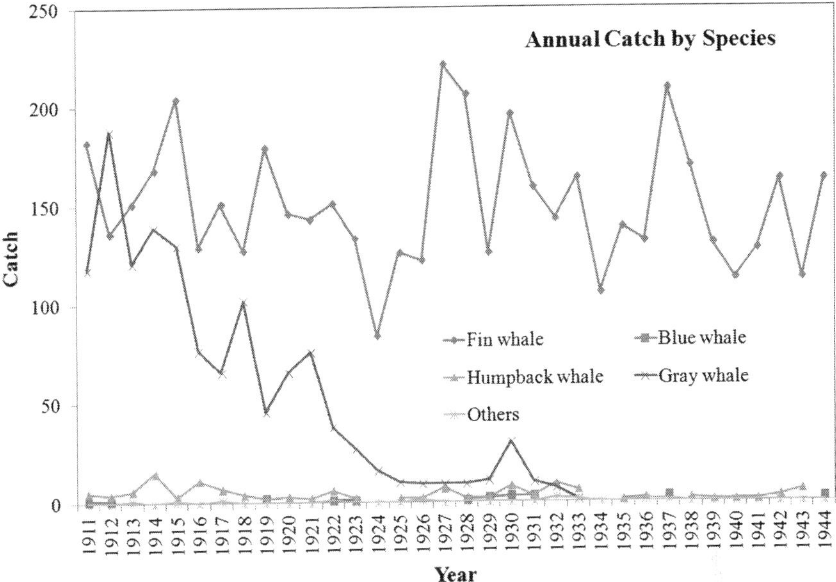

Fig. 15.1 Annual catch of cetaceans in Korean waters by Japanese whaling vessels from 1911 to 1944

whaling in Korean waters can be divided into that occurring before Korean independence and that conducted after independence. Before Japan surrendered and Korea gained its independence in 1945, only whaling companies that had received permission from the Japanese colonial government could catch whales in Korean waters, which meant that whaling was a monopoly for the Japanese people and for Japanese capital.

The types and numbers of whales captured in Korean waters by Japanese whaling ships before Korean independence are shown in Fig. 15.1. Statistical data exist for dates between 1911, which was the year after the forced annexation of Joseon by Japan, and 1944, which was the year before Japan surrendered. The whales that were captured in the largest numbers during this period were fin whales and gray whales. The annual number of fin whales captured ranged from 84 to 221, with a total of 5114 fin whales captured during this period. More than 100 gray whales were captured each year until 1915 and more gray whales than fin whales were captured in 1912. After then, the number of gray whales captured decreased year by year until no more were captured after a single gray whale was taken in 1933 (Fig. 15.1). The total number of gray whales captured during this period was 1306. In addition to these two species, humpback whales were captured almost every year, as were blue whales, sei whales, sperm whales, and right whales (Park 1987).

Immediately after Korean independence in 1945, several Koreans took over Japanese whaling ships to begin a very small-scale whaling industry. The number of whaling ships increased after the Korean War as people renovated fishing

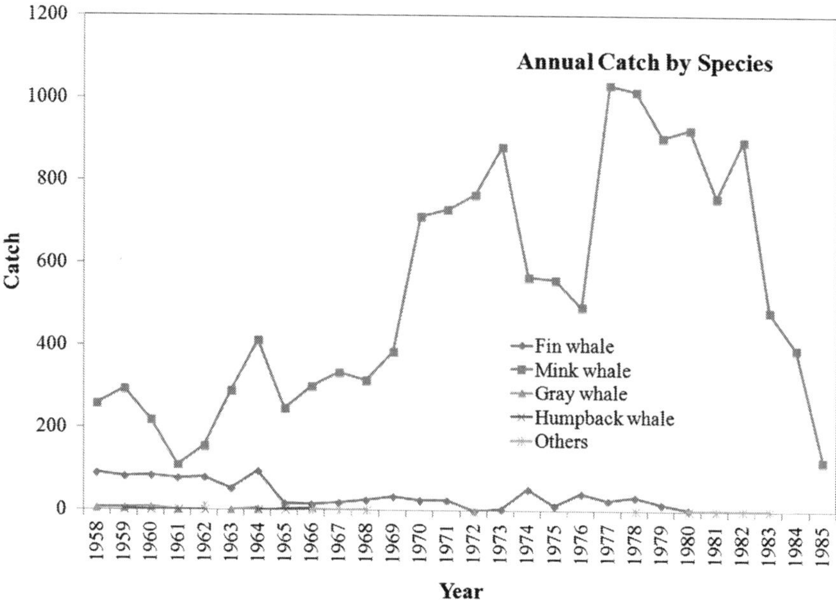

Fig. 15.2 Annual catch of cetaceans in Korean waters by Korean whaling vessels from 1958 to 1985

vessels or purchased whaling ships from Japan; the numbers increased from 17 whaling ships in late 1950 to 20 in the 1960s, and 22 in the 1970s. In 1979, Korea became a member of the IWC (International Whaling Commission), and continued active whaling in Korean waters until the commercial whaling moratorium was enacted in 1986 (Fig. 15.2).

The collection of statistics on caught whales after Korean independence only started after 1958 due to the confusion caused by the U.S. military government and the Korean War immediately after independence. Figure 15.2 shows the number of whales caught in Korean waters starting in 1958 until 1985, when whaling was banned. The main species that were caught were minke whales and fin whales; when the population of fin whales, which were the main target species, decreased, people began hunting minke whales, and minke whales consequently became the most important target species. The number of minke whales captured each year ranged from 110 to 1033, with a total of 14,587 captured during this period. The number of fin whales captured each year decreased to 100 after Korean independence, and continued to decrease, until no more were captured after 4 fin whales were captured in 1980 (Park 1987).

When commercial whaling was banned in 1986, Korea captured 69 minke whales on the grounds of scientific research, but immediately stopped whaling due to the lack of scientific investigations. For the last 20 years, and continuing until the present time, whaling has been banned on the Korean peninsula. The coastal areas of the East Sea have maintained a culture of whale meat consumption, using whales that have become trapped in nets and died, or dead whales that have drifted to shore.

In an effort to continue the long history and food culture of whales on the Korean peninsula, the people in the coastal areas of the East Sea argued for permission to conduct aboriginal subsistence whaling, such as has been allowed in Alaska in the United States, Chukotka in Russia, and in Greenland, but the general consensus is that such whaling is not applicable to Korea because it has achieved high economic development. There are also arguments for whaling for scientific purposes, as occurs in Japan, but this view has not received much support in Korea, much less around the world.

15.3 Marine Mammals

15.3.1 Cetacean

15.3.1.1 Baleen Whales

Northern Right Whale (*Eubalaena japonica*)

In the northern Pacific Ocean, baleens are distributed over temperate climates and subpolar regions, such as the East Sea, East China Sea, the Sea of Okhotsk, and the Bering Sea. They mainly feed on copepods and other small crustaceans (mainly Calanoida), and they feed by slowly swimming with open mouths to filter food that is concentrated near the surface.

Right whales in the northern Pacific Ocean became depleted in the 1800s, and there are records of large-scale whaling by foreign whaling ships between 1848 and 1913. However, only two right whale was captured in Korean waters after 1911, which was when statistics on whaling began to be recorded. Two right whales were captured in 1915 and 1974 respectively. Although there are no reliable population estimates or population trends known for right whales, they are thought to be seriously endangered. Scarcely any research on right whales has been conducted, even on an international scale.

Blue Whale (*Balaenoptera musculus*)

The worldwide population of blue whales is estimated to be between 8,000 and 9,000, but accurate estimates are lacking (Calambokidis and Barlow 2004). There are records of 20 blue whales being captured in Korean waters, including the East Sea, between 1911 and 1944.

Fin Whale (*Balaenoptera physalus*)

Fin whales are distributed around the oceans of the world; they prefer deep seas and stay away from warm areas. Their exact migration is not known because almost no research has been conducted on fin whales of the northern Pacific Ocean. They feed in the East Sea, the Yellow Sea, the East China Sea, and the Sea of Okhotsk during summer, and breed in warm waters during winter (Aguilar 2002). Around the Korean peninsula, fin whales are observed in the East Korea Bay in North Korea during spring and autumn, in the central and southern regions of the East Sea during August and November, and in the Yellow sea from October to May. In the East Sea, they aggregate around the East Korea Bay, along the coasts of North and South Gyeongsang provinces, and around Ulleungdo (Ulleung Island), and they appear to be distributed throughout the Yellow Sea. They tend to group more strongly than other whales, forming schools of 2–7 whales. Their swimming speed is rather fast at 37 km/h. They hardly show their tail flukes when they submerge, and they sometimes emerge on the sea surface (Aguilar 2002).

Fin whales have historically been a target species for commercial whaling. 921 fin whales were captured in Korean waters from 1911 to 1982. For nearly two centuries between the 18th and 20th century, fin whales in Korea were hunted by foreign whaling ships. Almost none have been observed near the Korean coasts recently, although one mature fin whale, 9.8 m long, was stranded at Songdo, Incheon, in 1996.

Minke Whale (*Balaenoptera acutorostrata*)

Minke whales are distributed in all oceans, from equatorial to polar seas. Although different hypotheses have been proposed, it is likely that there are three stocks in the northern Pacific Ocean. Other stocks in the east are the East Sea-Yellow Sea-East China Sea stock and the Sea of Okhotsk-Western Pacific Ocean stock. They are known to feed in the northern East Sea and the Sea of Okhotsk during summer, and they spend the winter in the southern East China Sea and near equatorial regions (Perrin and Brownell 2002). However, they are observed on the Korean coasts all year round. The main stomach content of minke whales captured in Korean coastal waters is recorded to be anchovy.

Because minke whales are relatively small, they were not traditional targets for commercial whaling, but they were hunted from the 1900s onwards and were heavily hunted until whaling was banned altogether. Almost all large whales have disappeared from the East Sea, and the most frequently observed baleen whale there is the minke whale. Around 80 minke whales are bycaught in nets in Korean waters every year (Fig. 15.3a).

Humpback Whale (*Megaptera novaeangliae*)

Humpback whales are distributed throughout the northern Pacific Ocean, and their breeding ground is usually found in the tropical regions. They migrate across the ocean to the glacial zones of subpolar regions, which is their feeding ground. There is one stock of these whales that migrates from the Sea of Okhotsk to the

Fig. 15.3 Observations of whale species in the East Sea. **a** A minke whale sighted by a Korean sighting survey in May 2007, **b** Landed gray whales in Ulsan by a Japanese whaling ship in the early 1900s, **c** Pacific white-sided dolphins, and **d** A northern fur seal found in September 2009

East Sea, the Southern sea of Korea, and the Yellow Sea. They usually migrate in groups of 1–3 but form large groups in their feeding grounds or breeding grounds (Clapham 2002).

Humpback whales also used to be targets of commercial whaling, and their hunting was banned from 1965 on. There are about 6,000–8,000 humpback whales in the northern Pacific Ocean (Calambokidis and Barlow 2004). In Korean coastal waters, there are records of 128 whales captured between 1911 and 1944, and 13 captured between 1959 and 1966.

Gray Whale (*Eschrichtius robustus*)

Gray whale stocks in the Atlantic Ocean have become extinct, and currently there are stocks in the northwestern Pacific Ocean and the eastern Pacific Ocean. The northwestern Pacific Ocean stock, also called the Korean gray whale, feed in the shallow waters of the northern East Sea and the Sea of Okhotsk during summer and move towards the south in late autumn to reproduce in the Yellow Sea and the East China Sea, after moving south along the East Sea coast during November and

December. They then return north between March and May, passing through the East Sea.

After fin whales, gray whales were the most hunted species in the East Sea up until the early 1900s because they are relatively large and move slowly (Fig. 15.3b). 1338 gray whales were captured between 1911 and 1964, but the number captured rapidly decreased thereafter. No more gray whales were captured after 5 were taken in 1964; indeed, they were thought to be extinct for some time. Recently, gray whales were observed to feed in the Sea of Okhotsk, and estimations of individual identifications and stock biomass are being made through photographs. This stock has been estimated to be about 100 in number (Weller et al. 2002).

15.3.1.2 Toothed Whales

Sperm Whale (*Physeter macrocephalus*)

The sperm whale is the most widespread species of all whale species. Sperm whales are distributed in all oceans, from the tropics to the poles. They are rarely observed in the East Sea and the Sea of Okhotsk and are rather concentrated in the southern East Sea and the East China Sea. Sperm whales form a matriarchal society, in which 20–40 whales form a group to care for infants, and from which mature males leave the infant care group to roam the oceans alone or in schools of several whales. During the mating season, which is during summer and autumn, the males return to the breeding grounds and partake in breeding for a short period of time (Whitehead 2002).

It is estimated that at least 44 thousand sperm whales live west of 170° east latitude in the northern Pacific Ocean. Five sperm whales were captured in the East Sea between 1911 and 1944, 9 were observed in the East China Sea in June 1999, and 8 were observed in the East Sea in March 2004.

Baird's Beaked Whale (*Berardius bairdii*)

This species is not very well known. Some are being directly captured in the northwestern Pacific by the Japanese, and some are bycaught or stranded from time to time. A total of 12 Baird's beaked whales were bycaught in the East Sea between 1996 and 2012.

Stejneger's Beaked Whale (*Mesoplodon stejnegeri*)

There is hardly anything known about this species. There are records of them occasionally being captured by Japanese salmon drift nets. In Korea, 24 Stejneger's beaked whales were reportedly stranded or bycaught on the coasts of the East Sea between 1996 and 2012.

Killer Whale (*Orcinus orca*)

There is little known about killer whales in the East Sea. However, 3 killer whales were bycaught near Busan in 2008, and a dolphin was found inside its stomach; from its external features, this killer whale is thought to be a Type A transient killer whale.

False Killer Whale (*Pseudorca crassidens*)

False killer whales are distributed around the tropical and temperate open seas between 50° northern latitude and 50° southern latitude. They can be observed around the Korean coasts, including the East Sea, between May and July, when the temperature of the sea water is above 17° Celsius.

They form schools with numbers ranging from tens into the hundreds, and they have been observed to mingle with bottlenose dolphins off the coasts of Jejudo (Jeju Island). There are many cases in which false killer whales became stranded as a group by following their leader. It is estimated that there are about 3000 killer whales in the eastern part of the East China Sea. Six whales were bycaught or stranded in the East Sea between 1996 and 2012.

Pacific White-Sided Dolphin (*Lagenorhynchus obliquidens*)

This species is only found in the northern Pacific Ocean; they are usually found in the neighboring seas across temperate and polar regions, they also come and go to continental shelves and coastal areas. Due to their strong collective behavior, they form schools of a few hundred to a few thousand. They display various behaviours, such as surfing, jumping, flicking water, and doing somersaults (Fig. 15.3c). Pacific white-sided dolphins are the most numerous species in the East Sea along with common dolphins, and they concentrate in waters with temperatures between 7 and 25 °C. Their estimated number in the East Sea is 5,000. Every year, around 10–40 of them are stranded or bycaught in the coastal areas of the East Sea.

Short-Beaked Common Dolphin (*Delphinus delphis*)

Short-beaked common dolphins tend to concentrate mainly in tropical and temperate seas between 50°N and 50°S. They display strong collectivity and crowding behavior (several thousand dolphins in a group). Their numbers in the Pacific Ocean is not known, but there are thought to be more than 30,000 of them; they are the most commonly observed species in the East Sea.

Dall's Porpoise (*Phocoenoides dalli*)

As an indigenous species of the northern Pacific, they are distributed widely between 30 and 62°N, mainly around the coasts. They are distributed above 35°N in the East Sea. The Dall's porpoises found in the East Sea are dalli-type porpoises; truei-type Dall's porpoises are distributed only along the Pacific coasts of Japan and the Sea of Okhotsk.

Finless Porpoise (*Neophocaena asiaeorientalis*)

Finless porpoises are the most typical marine mammal species around the Korean Peninsula. The body is mainly greyish white in color. A finless porpoise has no dorsal fin. It has a crista 2 cm high that is similar to mastodon bones running from the thorax to the caudal peduncle. Its pectoral fins are relatively large, being one sixth of their body length. The tail fluke is also relatively wide, being a quarter of the body width.

Finless porpoises inhabit large rivers, shallow seas, and coastal zones, especially in shallow waters within 5–6 km of the coast, through the East Sea; the Southern sea of Korea; the Yellow Sea; the Japanese coastal areas; and around Taiwan, China, Borneo, Sumatra, and Singapore. They have also been seen in the Persian Gulf. In recent research, finless porpoises have been divided into finless porpoises that inhabit waters between Persia and southern China (*N. phocaenoides*), and finless porpoises that inhabit central and northern China, Korea, and Japan (*N. asiaeorientalis*) (Wang et al. 2008).

Around the Korean Peninsula, they are frequently observed in the Yellow Sea and Southern sea of Korea coasts, and are observed very close to the coast in the southern East Sea. They are eurythermal animals that can live in temperatures between 5 and 28 °C.

They usually do not form groups; although they may form groups in scores when shoals of anchovy form around the coast, they do not swim close to each other. They do not approach ships, and when followed, they swiftly swim away. They feed on various organisms, such as fish, squid, shrimp, and crustaceans. Although there are known to be about 30,000 of them in the Yellow Sea, there is nothing known about their numbers in the Southern sea of Korea and in the East Sea (Park et al. 2007).

15.3.2 Sea Lions, Seals, Walrus

15.3.2.1 Fur Seals and Sea Lions

Steller Sea Lion (*Eumetopias jubatus*)

Groups of Steller sea lions are distributed around the western, northern and eastern coasts of Hokkaido (Hattori et al. 2009). They are not distributed in groups around the Korean coasts, but are sometimes observed around East Sea coastal areas; two Steller sea lions were discovered, one in 2008 and the other in 2009. One dead Steller sea lion was seen in the Southern sea of Korea, on Jejudo in 2013.

Northern Fur Seal (*Callorhinus ursinus*)

Northern fur seals are widely distributing in the waters of the North Pacific. Eight seals in total were observed between 2009 and 2013, one of which was discovered dead. In 2013, one seal that drifted onto the coast was discovered and rescued, and it was decided that an aquarium would continue to house it considering that the seal was blind in both eyes (Fig. 15.3d).

Japanese Sea Lion (*Zalophus japonicus*)

A large population of Japanese sea lions is known to have inhabited areas around the coast of Hokkaido, including the East Sea; in particular, they were also found around Dokdo. They appear to be completely extinct at the present due to commercial hunting in the 1950s (Reijders et al. 1993).

15.3.2.2 True Seals

Spotted Seal (*Phoca largha*)

Spotted seals are distributed only around the northern Pacific Ocean, including Alaskan coasts, the Bering Sea, coasts of the Kamchatka Peninsula, the Sea of Okhotsk, the coasts of Hokkaido, and the Yellow Sea (Lowry et al. 2000). There is a high possibility that the spotted seals in the Yellow Sea and the East Sea are of different stocks. Satellite tagging was used in 2008 to confirm that spotted seals in Bohai, China, migrated to Baekryeongdo in Korea. In 2013, spotted seals that were captured in the southern East Sea and given satellite tags were found to have moved to Vladivostok in Russia. Spotted seals have not been observed to live in groups in the East Sea; the closest winter habitat is the Sea of Okhotsk, where the population is estimated to be around 100,000–130,000 (Burns 2002).

15.4 Summary

The coasts of the East Sea on the Korean Peninsula have had connections with whales since prehistoric times, and this connection was not broken but continued through the many dynasties which ruled the Korean Peninsula. Beginning in the late 19th century, many foreign whaling ships started entering Korean waters, especially the East Sea. The records that reveal the most about this period of whaling come mostly from the journals of American whaling ships, and they record that it wasn't only American ships that hunted there but also whaling ships from other western nations that entered the East Sea.

The Russians nearly monopolized whaling in the East Sea for a brief period in the early 1900s, but after Japan defeated Russia in the Russo-Japanese War, the Japanese people and Japanese capital monopolized whaling in the East Sea until 1945. At this time, beginning in 1911, statistical data on whaling in the East Sea began to be recorded. The main catch targets were fin whales and minke whales, with gray whales, humpback whales, and others captured as well. Commercial whaling in Korea continued after independence; it was begun by introducing Japanese whaling ships and continued actively, focused mainly on minke whales, until the IWC declared a moratorium on commercial whaling in 1986.

According to whaling data, bycatches and stranding, the baleen whales observed in the East Sea are northern right whales, blue whales, fin whales, sei whales, Bryde's whales, minke whales, humpback whales, and gray whales. Up to the present, the minke whale has been the dominant species. For toothed whales, there are large species, such as sperm whales, Baird's beaked whales, and Stejneger's beaked whales, which are discovered once or twice every year through bycatch or stranding. Small and medium-sized species are extremely diverse, including killer whales, Pacific white-sided dolphins, Risso's dolphins, and common dolphins. The most commonly observed species are common dolphins and Pacific white-sided dolphins; for common dolphins, nearly one hundred are bycaught or stranded every year around the coasts of the East Sea. These species are distributed throughout the year in the East Sea, but Dall's porpoises and harbor porpoises show changes in their latitudinal distributions with changes in the season.

Some of the pinnipeds that are distributed in the East Sea include Steller sea lions, northern fur seals, and spotted seals. Most of them live in the northern East Sea in groups, but they are sometimes found in the southern East Sea. Japanese sea lions used to live in the East Sea, but they are now extinct.

References

Aguilar A (2002) Fin whale *Balaenoptera physalus*. In: Perrin WF, Würsig B, Thewissen JGM (eds) Encyclopedia of marine mammals, Academic Press, pp 435–438

Burn JJ (2002) Harbor seal and spotted seal *Phoca vitulina* and *P. largha*. In: Perrin WF, Würsig B, Thewissen JGM (eds) Encyclopedia of marine mammals. Academic Press, Salt Lake City, pp 552–560

Calambokidis J, Barlow J (2004) Abundance of blue and humpback whales in the eastern North Pacific estimated by capture-recapture and line-transect methods. Mar Mammal Sci 20:63–85

Clapham PJ (2002) Humpback whale *Megaptera novaeangliae*. In: Perrin WF, Würsig B, Thewissen JGM (eds) Encyclopedia of marine mammals. Academic Press, Salt Lake City, pp 589–592

Hattori K, Isono T, Wada A et al (2009) The distribution of Steller sea lions (Eumetopias jubatus) in the sea of Japan off Hokkaido, Japan: a preliminary report. Mar Mammal Sci 25(4):949–954

Lowry LF, Burganov VN, Frost KJ et al (2000) Habitat use and habitat selection by spotted seals (*Phoca largha*) in the Bering Sea. Can J Zoolog 78:1–13

Park GB (1987) The whaling history in Korean waters. Minjok-munwha-sa, Busan, Korea

Park KJ, Kim ZG, Zhang CI (2007) Abundance estimation of the finless porpoise, *Neophocaena phocaenoides*, using models of the detection function in a line transect. J Korean Fish Soc 40(4):201–209

Perrin WF, Bownell RL (2002) Minke whales *Balaenoptera acutorostrata* and *B. bonaerensis*. In: Perrin WF, Würsig B, Thewissen JGM (eds) Encyclopedia of marine mammals. Academic Press, Salt Lake City, pp 750–754

Reijders P, Brasseur S, Toorn J et al (1993) Status survey and conservation action plan—seals, fur seals, sea lions and walrus. IUCN/SSC Seal Specialist Group. Gland

Wang JY, Frasier TR, Wang SC et al (2008) Detecting recent speciation events: the case of the finless porpoise (genus Neophocaena). Heredity 101:145–156

Weller DW, Burdin AM, Würsig B et al (2002) The western gray whale: a review of past exploitation, current status and potential threats. J Cetacean Res Manag 4:7–12

Whitehead H (2002) Sperm whale *Physeter macrocephalus*. In: Perrin WF, Würsig B, Thewissen JGM (eds) Encyclopedia of marine mammals. Academic Press, Salt Lake City, pp 1165–1172

Chapter 16
Physiography and Late Quaternary Sedimentation

Sang Hoon Lee, Jang Jun Bahk, Seong-Pil Kim and Jun-Yong Park

Abstract The East Sea (Japan Sea) is a semi-enclosed back-arc basin or marginal sea connecting with the Pacific Ocean through four shallow straits. There are three deep basins (the Ulleung, Japan and Yamato basins) separated by submarine topographic highs (the Korea Plateau, Oki Bank and Yamato Rise). The deepwater seafloor of the East Sea beyond the coastal and shelf regions is generally covered by fine-grained sediments which have been deposited by various pelagic and hemipelagic processes, such as pelagic settling, aeolian transport, and lateral transport by surface currents, river plumes, and mid- or bottom-water nepheloid layers. The oceanographic condition and pelagic and hemipelagic sedimentation pattern in the East Sea have undergone considerable variations due to Quaternary global climate changes, resulting in characteristic cyclic alternations of dark and light layers in the deep-water sediments. Along the margins of the East Sea, slope failures, mass movements and mass-flow deposits occur extensively in the uppermost (late Quaternary) sedimentary sequences. The slope failures and mass movements have been generated by various factors, such as earthquakes, sea-level changes, gas-hydrate dissociation and high sedimentation rate. The complex geomorphology of the eastern coast of Korea seems to have originated from the interaction of tectonic history of uplift, bedrock of panoramic lithologies and ages, and

S.H. Lee (✉) · J.-Y. Park
Korean Seas Geosystem Research Unit, Korea Institute of Ocean Science and Technology, Ansan 426-744, Republic of Korea
e-mail: sanglee@kiost.ac.kr

J.-Y. Park
e-mail: jypark@kiost.ac.kr

J.J. Bahk
Petroleum and Marine Research Division, Korea Institute of Geoscience and Mineral Resources, Daejeon 305-350, Republic of Korea
e-mail: jjbahk@kigam.re.kr

S.-P. Kim
Korea Institute of Geoscience and Mineral Resources, Pohang Branch, Pohang, Republic of Korea
e-mail: spkim@kigam.re.kr

© Springer International Publishing Switzerland 2016
K.-I. Chang et al. (eds.), *Oceanography of the East Sea (Japan Sea)*,
DOI 10.1007/978-3-319-22720-7_16

wave-dominated, micro-tidal hydrodynamic conditions. A number of sedimentary environments, such as beaches, lagoons, marine terraces, coastal dunes, river (or stream) mouth spits, rocky headlands and cliffs, can be found in the coastal land area. In the nearshore zone, submerged sand bars, channels (buried or active), tombolos and sea stacks are notable geomorphic features.

Keywords Bathymetry · Late quaternary sedimentation · Paleoceanography · Submarine slope failures and mass flows · Coastal morphology · Ulleung Basin · East Sea (Japan Sea)

16.1 Physiography

The East Sea is a semi-enclosed back-arc basin or marginal sea surrounded by the eastern Asian continent and the Japanese Islands (Fig. 16.1). The East Sea is connected with the Pacific Ocean through four shallow straits; the Korea (140 m deep), Tsugaru (130 m deep), Soya (55 m deep), and Tartarsky (12 m deep) straits. According to ETOPO1 dataset (http://www.ngdc.noaa.gov/mgg/global/global.html), average depth of the East Sea is about 1650 m and the maximum depth is about 3800 m. There are three deep basins (the Ulleung, Japan, and Yamato basins) separated by submarine topographic highs (the Korea Plateau, Oki Bank, and Yamato Rise) that rise to within about 500 m of the sea surface (Fig. 16.1).

In the southwestern part of the East Sea, the Ulleung Basin is a deep, bowl-shaped depression, bounded by continental slopes of the Korean Peninsula and the southwestern Japanese Islands on the west and south, respectively, and by the submarine topographic highs of the Korea Plateau and Oki Bank on the north and east, respectively (Fig. 16.2; Yoon and Chough 1995). The northern and western margins of the basin are relatively steep with gradients of up to 10°. The western slope of the basin is characterized by a few small-scale scoop-shaped failure scars and gullies without large-scale, distinctive canyons and channels (Fig. 16.2; Chough and Lee 1987). Along the eastern and southern margins, the basin is bordered by a gentle (less than 3°) slope, in which numerous large-scale (more than 70–80 m deep and several kilometers wide) gullies and scars are present on the upper to middle slope sectors (Fig. 16.2; Lee et al. 2014). The northeastern part of the basin is punctuated by two volcanic Ulleung Island (Ulleungdo) and Dok Island (Dokdo), and several volcanic seamounts. The basin floor lies at water depths of 2000–2500 m, and gradually deepens toward the north and northeast, connecting to the Japan Basin through the Ulleung Interplain Gap (UIG) (Fig. 16.2).

The UIG is bounded in the southeast by the slopes of the Oki Bank and Dokdo, and in the northwest by an ENE–WSW-trending escarpment of the South Korea Plateau (Fig. 16.2). In the UIG, a deep-sea channel system (Ulleung Interplain Channel, UIC) occurs along the base-of-slope of the South Korea Plateau (Fig. 16.3; Lee et al. 2004b). The UIC is 1.5–13.6 km wide and 15–85 m deep. It is asymmetric in cross section: steep on the northwestern side (the Korea Plateau) and gentle on the southeastern side (the Oki Bank and Dokdo) (Fig. 16.3). Near the North Ulleung Interplain Seamount (NUIS), the UIC is relatively shallow, less than 20 m deep. To the south of the NUIS, it deepens southwestward, up to 85 m deep (Fig. 16.3). Near the Ulleung Seamount, it abruptly narrows and splits into several branches immediately south of the Ulleung Seamount (Fig. 16.3). These branches gradually shallow southward, but the easternmost branch extends to 36° 50′N along the western base-of-slope of Dokdo.

The Korea Plateau, north of the Ulleung Basin, is a topographically rugged feature with 1000–1500 m of relief. The Japan Basin indenting the plateau divides it into the North and South components (Lee et al. 2002). The plateau ends abruptly to the south against an ENE-WSW-trending escarpment (Fig. 16.2). It is bounded in the north and east by the Wonsan Trough and the Japan Basin, respectively.

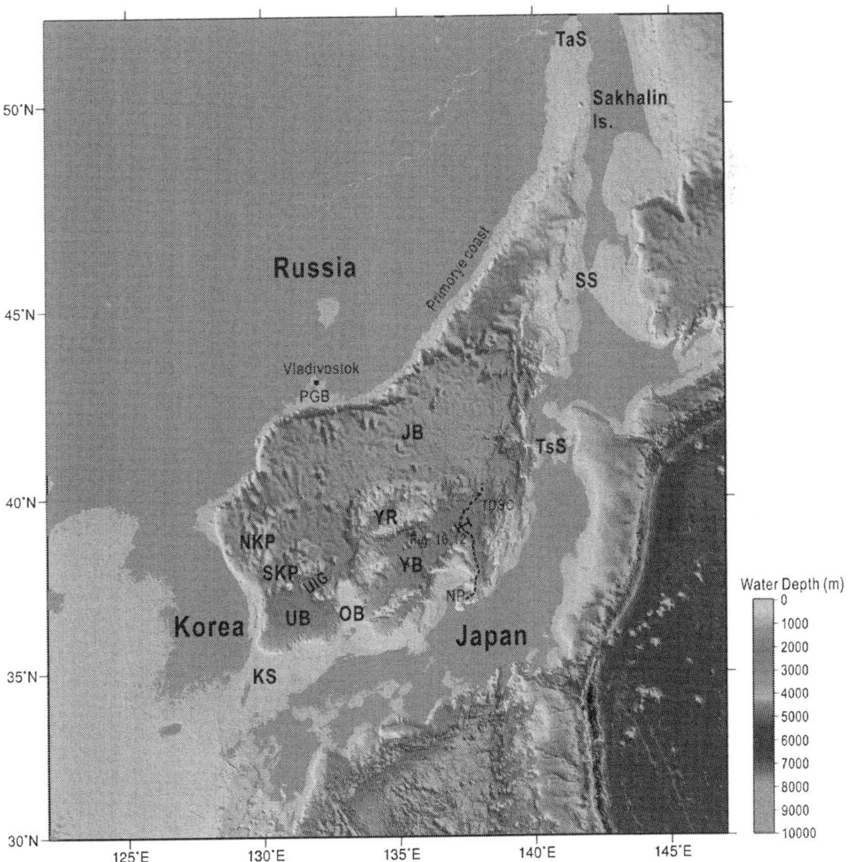

Fig. 16.1 Major physiographic features of the East Sea (Japan Sea). *JB* Japan Basin; *KS* Korea Strait; *NKP* North Korea Plateau; *NP* Noto Peninsula; *OB* Oki Bank; *PGB* Peter the Great Bay; *SKP* South Korea Plateau; *SS* Soya Strait; *TaS* Tatarsky Strait; *TDSC* (a *dotted line*) Toyama Deep-Sea Channel; *TsS* Tsugaru Strait; *UB* Ulleung Basin; *UIG* Ulleung Interplain Gap; *YB* Yamato Basin; *YR* Yamato Rise. For a color version of the figure, see Fig. 1.1

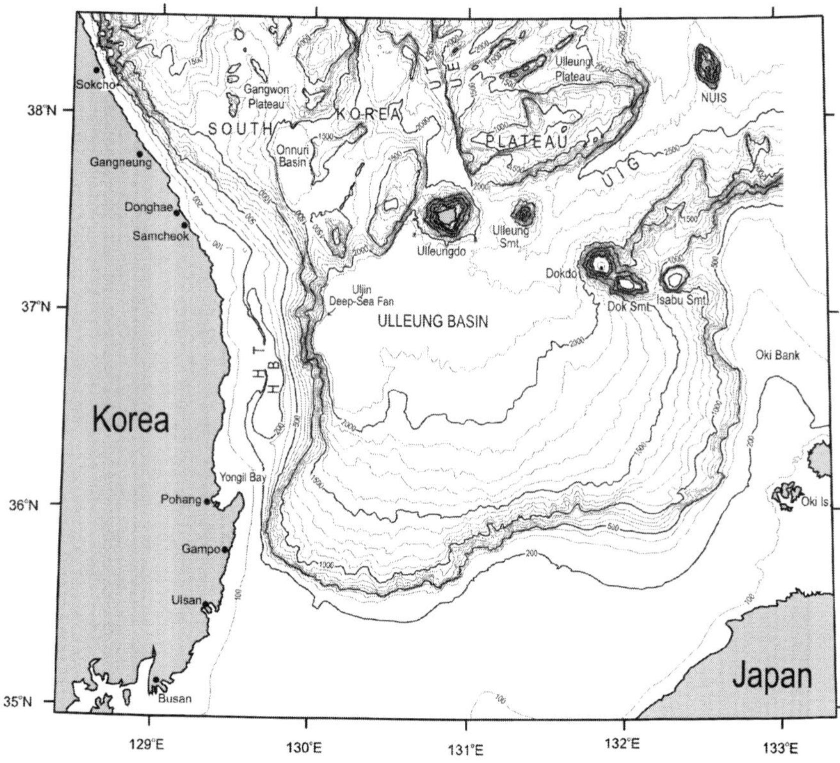

Fig. 16.2 Detailed physiographic and bathymetric features of the southwestern part of the East Sea. Bathymetry in meters. *HB* Hupo Bank; *HT* Hopo Trough; *NUIS* North Ulleung Interplain Seamount; *Smt.* Seamount; *UE* Usan Escarpment; *UT* Usan Trough

The southern component of the plateau (the South Korea Plateau) consists of numerous ridges, seamount chains, and intervening troughs/basins (Fig. 16.2; Lee et al. 2002). The ridges and seamount chains occur at water depths of about 600–1500 m. The ridges are relatively broad (more than 30 km wide) and rarely exceed 1° in slope gradient. In contrast, the seamount chains are mostly steep and narrow (less than 10 km wide). The South Korea Plateau is divided into a western part (called the Gangwon Plateau) and an eastern part (called the Ulleung Plateau) by the Usan Trough which deepens northward and extends to the Japan Basin (Fig. 16.2). In the western part of the plateau, the ridges and troughs dominantly trend N-S or NNE-SSW. On the other hand, the eastern part shows ENE-WSW-trending ridges, seamount chains and troughs.

The northern part of the East Sea is occupied by the Japan Basin, which is about 200–300 km wide and about 700 km long, trending NE-SW. The Japan Basin is bounded on the north by a relatively straight, steep slope along Russia (Fig. 16.1). The eastern margin of the basin shows a relatively steep slope that has experienced recent seismic activities (Satake 1985; Nakajima and Kanai 2000).

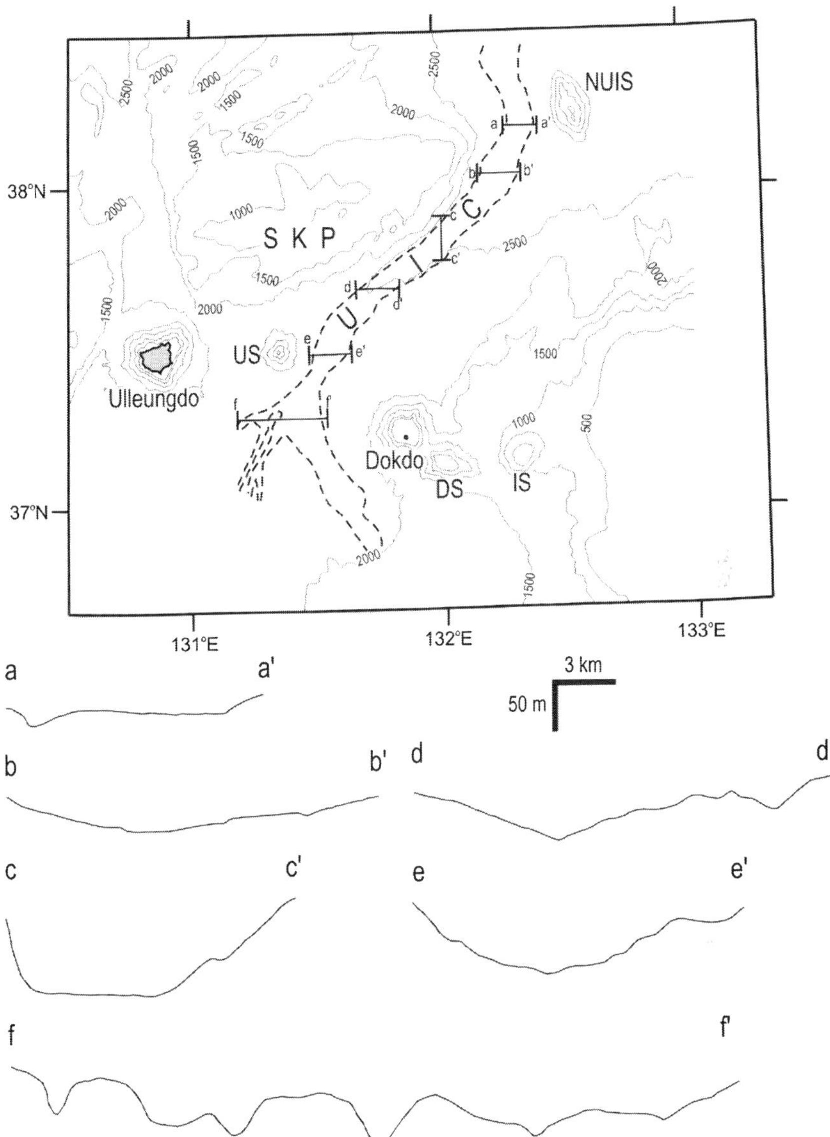

Fig. 16.3 Distribution (*top*) and cross-sectional geometry (*bottom*) of the Ulleung Interplain Channel (*UIC*) in the Ulleung Interplain Gap (from Lee et al. 2004a). Bathymetry in meters. *DS* Dok Seomount; *IS* Isabu Seamount; *SKP* South Korea Plateau

The basin is bounded on the south and west by the Yamato Rise and Korea Plateau, respectively. Its basin floor shows 3500–3800 m water depth, and the deepest area is located between Sikhote-Alin and the southwestern part of Hokkaido. The basin floor is rather smooth and flat except for several seamounts and hills which rise up to 2000 m above the surrounding seafloor.

The Yamato Basin in the southeastern part of the sea is a NE-SW-trending depression bounded by the Yamato Rise and Honshu of Japan (Fig. 16.1). The basin is shallower than the Japan Basin with an average water depth of 2500–2700 m and maximum water depth of 2970 m. Its basin floor is rather smooth and flat except for a few seamounts and hills. The basin is connected to the Japan Basin in the northeastern part. At the southeastern margin of the basin, the Toyama Deep-Sea Channel descends about 750 km from several river mouths along Toyama Bay to the abyssal Yamato and Japan basins (Fig. 16.1; Nakajima 2006). North of the Yamato Basin, a volcanic topographic high (the Yamato Rise) is parallel to the long axis of the Yamato Basin. The Yamato Rise consists of a few topographic highs, separated from each other by transverse depressions. The prominent Kita-Yamato Trough, a NE-SW-trending longitudinal depression, divides the Yamato Rise into the northwestern component (the Kita-Yamato Bank) and the southeastern component (the Yamato and Takuyo banks).

16.2 Late Quaternary Sedimentation

16.2.1 Deep-Water Pelagic and Hemipelagic Sedimentation and Its Relation to Paleoenvironmental Changes

16.2.1.1 Characteristics of Surface Sediments and Modern Sedimentation Pattern

The deep-water seafloor of the East Sea beyond the coastal and shelf regions is generally covered by silt, silty clay, and clay (Fig. 16.4; Chough et al. 2000), which were deposited by various pelagic and hemipelagic processes such as pelagic settling, aeolian transport, and lateral transport by surface currents, river plumes, and mid- or bottom-water nepheloid layers. Annual mean mass flux determined by sediment trap experiments at 1 km water depth was 16.6 mg/cm^2/yr in the western Japan Basin, 9.2 mg/cm^2/yr in the eastern Japan Basin and 5.4 mg/cm^2/yr in the Yamato Basin (Otosaka et al. 2004). Recent sedimentation rates estimated from excess ^{210}Pb activity profiles in the deep southwestern East Sea are 18–136 mg/cm^2/yr, being highest on the western slope of the Ulleung Basin off Gampo (water depth 1160 m) and lowest on the Yamato Rise (water depth 1650 m) (Hong et al. 1997). Cha et al. (2007) also reported that sedimentation rates in the surface sediments of the Ulleung Basin decreased rapidly in a radial pattern from offshore of Gampo (more than 0.3 cm/yr) to the Ulleung Interplain Gap (less than 0.05 cm/yr).

Sediment trap experiments by Otosaka et al. (2004) during the period 1999–2002 revealed that more than 60 % of settling particles consisted of lithogenic aluminosilicates and biogenic opal. They suggested three sources of lithogenic materials to the East Sea, based on La/Yb and Mn/Al ratios; (1) atmospheric input

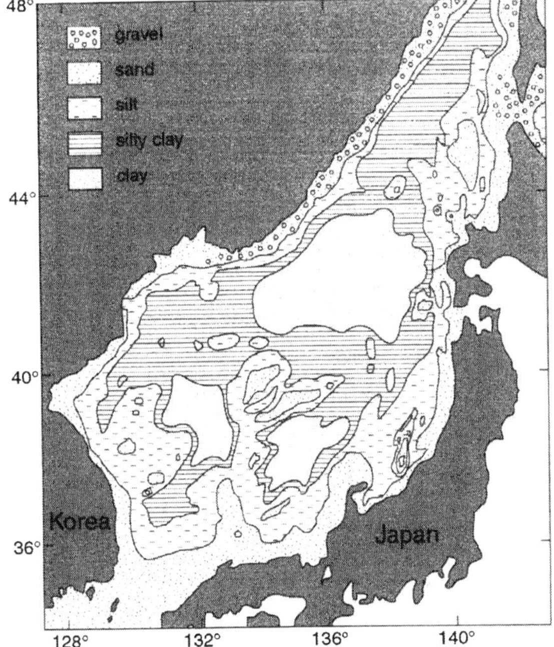

Fig. 16.4 Grain size distribution of surface sediments in the East Sea (from Chough et al. 2000)

of "fresh" Asian loess characterized by relatively high La/Yb and low Mn/Al, (2) lateral transport of "old" Asian loess from the East China Sea by the Tsushima Warm Current, high La/Yb and high Mn/Al, and (3) lateral transport from the island arcs of the Sakhalin and Japan Islands, low La/Yb. In the southwestern East Sea, sediments from the Nakdong and Changjiang rivers were also suggested as two major sources of lithogenic materials in the coastal and upper slope areas based on metal/Al ratios of surface sediments (Cha et al. 2007).

Biogenic components of the surface sediments are dominated by biogenic opal which constitutes more than 15 % and up to about 60 % of settling particles (Otosaka et al. 2004). Opal contents of core top sediments from the southwestern East Sea are usually about 10–30 % (Hyun et al. 2007) and mainly composed of diatoms and silicoflagellates (Chough et al. 2000). The present Carbonate Compensation Depth (CCD) in the East Sea lies between 1500 and 2000 m in water depth (average water depth: about 1900 m) (Ujiie and Ichikura 1973) and carbonate contents of the surface sediments below the CCD are usually less than a few percent (Cha et al. 2007; Hyun et al. 2007). In the Ulleung Basin, the surface carbonate contents rapidly increase above the CCD and reach their highest value, more than 30 %, on the outer shelves in the Korea Strait (Cha et al. 2007). Organic carbon contents of the surface sediments in the East Sea are generally less than 1 % in the basin floor (Chough et al. 2000), except in the deep southwestern part where they are more than 1.5 % and reach up to 2.5 % in the Korea Plateau and the Ulleung Interplain Gap (Cha et al. 2007).

Surface sediments in the East Sea deeper than about 2000 m are characterized by reddish brown oxidized layers, reflecting the presence of highly oxygenated deep waters (Niino et al. 1969). In the southwestern East Sea, they are relatively enriched in Mn oxides and exhibit maximum concentrations of Mn, more than 15,000 μg/g, in the Ulleung Interplain Gap (Cha et al. 2007).

16.2.1.2 Late Quaternary Paleoceanographic Changes as Recorded in the Pelagic and Hemipelagic Sediments

The oceanographic conditions and pelagic/hemipelagic sedimentation pattern in the East Sea have experienced considerable variations because of the Quaternary global climate changes, and such variations resulted in characteristic cyclic alternations of dark and light layers in the deep-water sediments. Since the East Sea is connected to adjacent seas through four shallow and narrow straits with sill depths of 12–140 m, significant changes in oceanographic conditions have been expected in response to the glacio-eustatic sea-level fluctuations which resulted in a sea-level drop of as much as about 120 m during the last glacial maximum. Tada et al. (1999) suggested that the modulation of the volume and character of the surface water inflow through the Korea Strait associated with glacio-eustatic sea level changes played a key role in the changes in deep-water ventilation and surface productivity (Fig. 16.5). According to their hypothesis, the East Sea was nearly isolated by a sea-level drop of more than 90 m during the glacial maximum periods. During such periods, low-salinity surface water developed because of excess precipitation in relation to evaporation, and caused strong density stratification of the water column, resulting in the euxinic deep water and low surface productivity. They also postulated that the millennial-scale climate changes of Dansgaard-Oeschger (D-O) cycles were also responsible for significant changes in paleoceanographic conditions in the East Sea. For example, during the D-O interstadials (warm phases) the influx of low-saline, high-nutrient East China Sea Coastal Water was increased by enhanced Asian summer monsoons, which caused high surface productivity and enhanced density stratification (Fig. 16.5).

Numerous studies on variations of paleoceanographic proxies recorded in the late Quaternary sediments of the East Sea have confirmed the hypothesis suggested by Tada et al. (1999), and revealed specific changes in sea surface salinity (SSS), sea surface temperature (SST), deep-water ventilation, surface productivity, and dust transport. Freshening of surface water during the last glacial maximum in the East Sea was manifested by lighter oxygen isotope values in planktonic foraminifera than today, opposite to those from the open ocean at the time (e.g. Oba et al. 1991; Gorbarenko and Southon 2000; Khim et al. 2007), and the lower estimated SSS (12 ‰) than today (Lee 2007). Reconstruction of SST based on alkenone temperature records indicates that the SST changes were generally consistent with orbital-scale global climate records since the penultimate glacial period of Marine Isotope Stage (MIS) 6 and were associated with volume

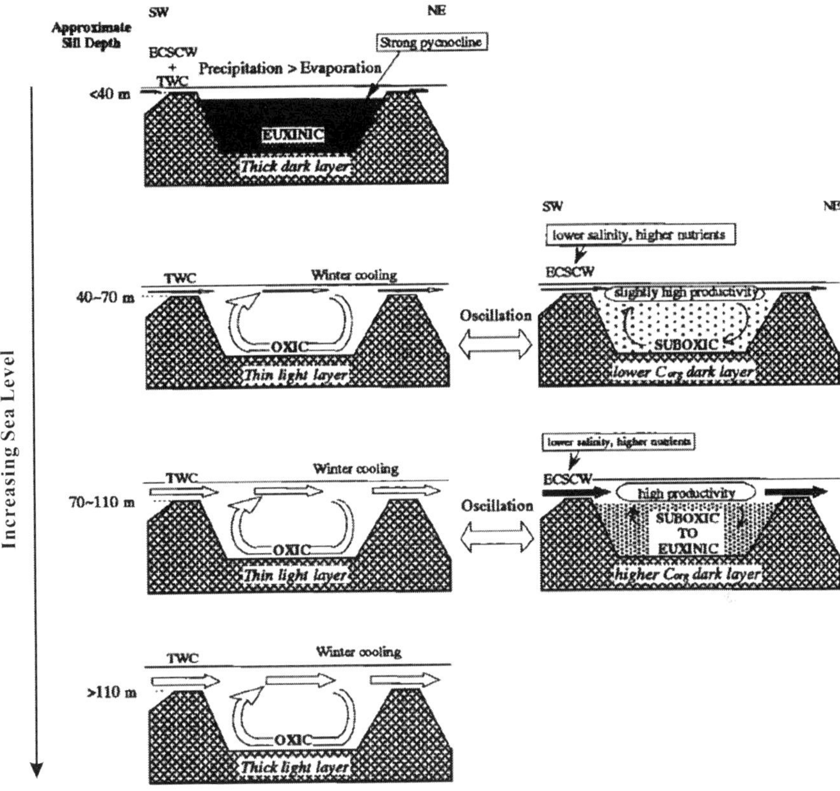

Fig. 16.5 Models for different modes of ocean circulation in the East Sea proposed to have occurred in response to different eustatic sea levels (from Tada et al. 1999)

transport change of the Tsushima Warm Current and north-south movement of the subpolar front (Lee et al. 2008; Fujine et al. 2009; Choi et al. 2012a).

Studies on the elemental compositions of core sediments revealed that the thick dark laminated mud (or TL: thinly laminated) layers formed in the glacial periods of MIS 2 and 6 are characterized by relatively low organic carbon and high total sulfur contents and enrichment of chalcophile elements such as Mo, demonstrating the development of euxinic deep-water condition and low surface productivity during these periods (e.g. Crusius et al. 1999; Kido et al. 2007; Khim et al. 2007, 2012; Lim et al. 2011; Zou et al. 2012). In addition to the orbital-scale changes, millennial-scale variations of deep-water ventilation and surface productivity are also indicated by the presence of relatively thin TL layers correlative with D-O cycles, which are often characterized by relatively high organic and sulfur contents (e.g. Tada et al. 1999; Kido et al. 2007; Khim et al. 2012). Examination of the detrital silt fraction in the hemipelagic sediments identified two different source areas of aeolian dust (the Taklimakan Desert-Loess Plateau and Siberia-Northeast China), and suggested controls of north-south shifts of westerly jet axis on the dust size and flux at both orbital- and millennial-scales (Nagashima et al. 2007).

16.2.2 Slope Failures, Mass Movements, and Mass Flows

Along the margins of the East Sea, the uppermost (about 20–60 m thick) sedimentary sequences show abundant slope failures, mass movements and mass-flow deposits (Chough et al. 2000; Nakajima and Kanai 2000; Lee et al. 2004a). The slope failures and mass movements have been generated by various factors, such as earthquakes, sea-level changes, gas-hydrate dissolution, high sedimentation rate, and so on (Lee et al. 1996, 2004a, 2010). These slope failures and mass movements have been recently considered as an important geohazard because of tsunami generation and underwater-facility destruction (Locat and Lee 2002). Therefore, the exact distribution and morphology of slope failures and mass movements, and their triggers, mechanisms and dynamics are important issues in the East Sea. In this section, the slope failures, mass movements and mass-flow deposits on the eastern, western and southern margins are introduced. Those on the northern margin along the Sikhote-Alin Mountains of Russia have been rarely studied.

16.2.2.1 Ulleung Basin and South Korea Plateau

In the Ulleung Basin, slope failures, mass movements and mass-flow deposits occur extensively along the entire basin margin (Fig. 16.6; Chough et al. 1997; Lee et al. 2004a). On the western margin, mass movements (slides and slumps) have relatively small dimensions with a few small, scoop-shaped scars and gullies deeper than 700–800 m water depth (Figs. 16.6 and 16.7; Chough and Lee 1987; Lee et al. 2014). The downslope mass-flow deposits (debrites and turbidites) occur as small or subdued, solitary lobes restricted at the base-of-slope (Fig. 16.8). In contrast, the southern margin is characterized by large gullied scars with huge slides and slumps deeper than 250 m water depth (Figs. 16.6 and 16.9; Lee et al. 1993; Lee et al. 2014). These catastrophic failures evolved into extensive mass flows (debris flows and turbidity currents), which travelled downslope for more than several tens of kilometers (Figs. 16.6 and 16.10; Lee et al. 1999, 1999, 2004a). On the slopes north of Dokdo, mass movements with large-scale gullied scars occur deeper than 500–1000 m in water depth (Fig. 16.6; Lee et al. 2004a). The mass movements are transitional downslope to mass-flow chutes/channels, debrites and turbiditic sediment waves (Fig. 16.6; Lee et al. 2004b). The different styles of slope failures, mass movements and mass-flow deposits between the western and southern margins could be ascribed to variations in shelf morphology and sediment supply to the shelf within the basin (Lee et al. 2014). The regional (i.e. basinwide) downslope change from mass movements to mass-flow deposits (debrites and turbidites) is suggestive of successive downslope evolution from slide/slump to debris flows and turbidity currents (Chough et al. 1997; Lee et al. 1999). In the Ulleung Basin, various thin- to very thick-bedded turbidites accumulated (an

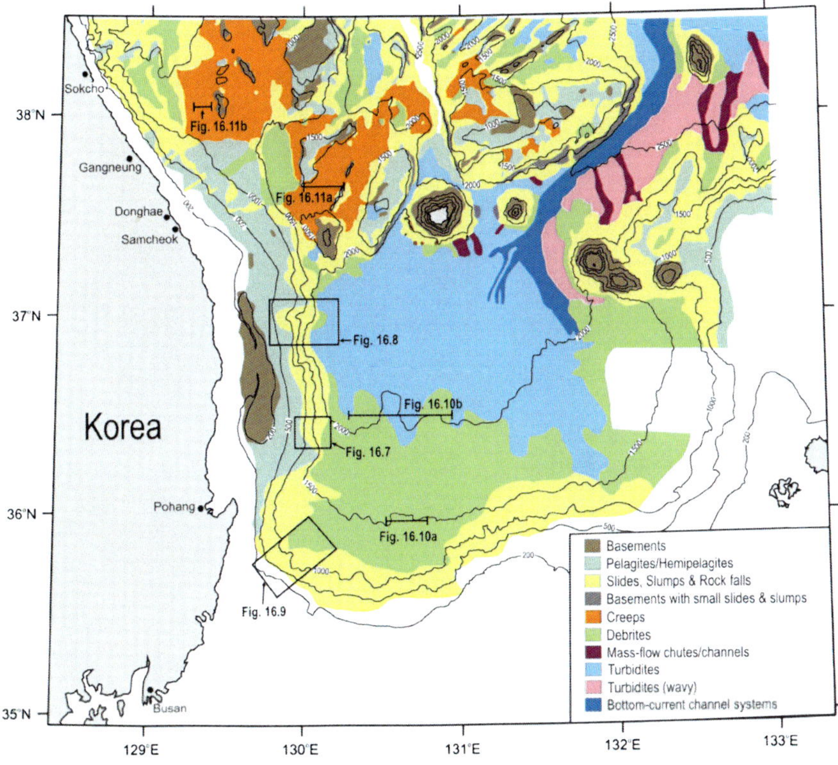

Fig. 16.6 Distribution of mass-movement and mass-flow deposits in the southwestern part of the East Sea, identified from Chirp (2–7 kHz) subbottom profiles (from Lee et al. 2004a)

average recurrence interval of ca. 605 years) between 29.4 and 19.1 cal. ka BP (Lee et al. 2010). After 19.1 cal. ka BP, turbidites were deposited with an average recurrence interval of 3183 years, and their thickness abruptly decreased upward (Lee et al. 2010). These features suggest that various-scale slope failures occurred frequently during the eustatic lowering of sea level, and the frequency and relative volumes of the slope failures suddenly decreased after sea level began to rise. The frequent slope failures and mass movements between 29.4 and 19.1 cal. ka BP were plausibly triggered by earthquakes, in combination with reduced hydrostatic pressure that promoted gas-hydrate dissociation during the eustatic lowering of sea level (Lee et al. 1996, 2004a, 2010).

On the South Korea Plateau, slides and slumps with scarps or scars occur predominantly on the relatively steep slopes of the seamount chains and submarine ridges (Fig. 16.6; Lee et al. 2002). In contrast, the broad, gently sloping areas on the submarine ridges are dominated by submarine creep deposits (Figs. 16.6 and 16.11; Lee and Chough 2001). The creep deposits are transitional downslope to slide and slump masses with scars or scarps on the upper to middle slopes of the ridges (Fig. 16.6). These voluminous mass movements with failure scars or

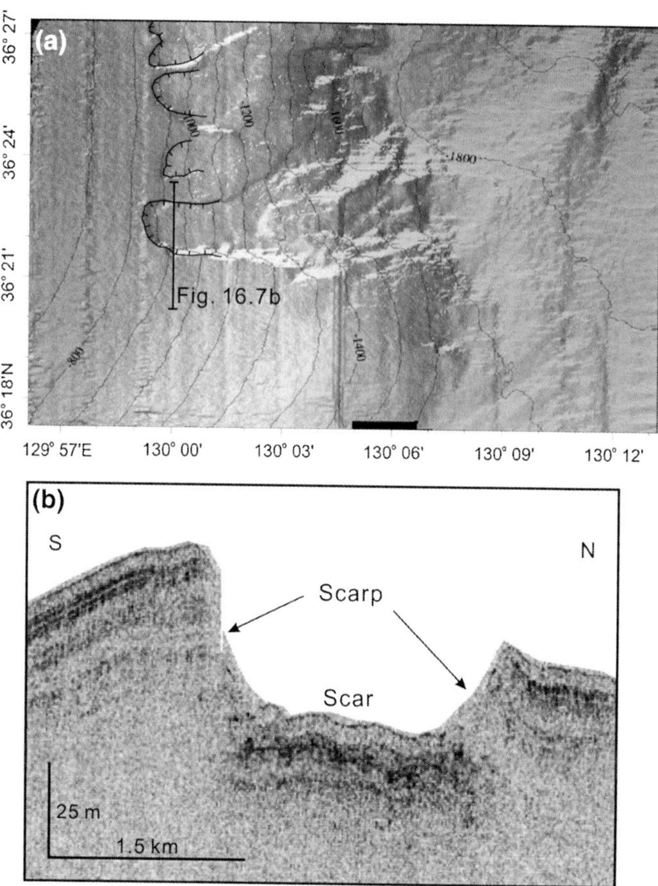

Fig. 16.7 **a** Scoop-shaped slope-failure scars on the western slope of the Ulleung Basin. Water depth in meters. For location of (**a**), see Fig. 16.6. **b** A chirp (2–7 kHz) subbottom profile showing sedimentary features of the scoop-shaped scars in cross section (from Lee et al. 2014)

scarps, deeper than 500–1000 m in water depth, were most likely generated by earthquakes that have frequently occurred in the South Korea Plateau (Lee et al. 2002). The slide/slump masses change downslope to mass-flow deposits (debrites and turbidites) (Fig. 16.6). Debrites are dominant in the intervening troughs and basins which are bounded by large areas of slides and slumps (Fig. 16.6). In contrast, turbidites occur predominantly in the intervening troughs which are bordered by small areas of slope failures (Fig. 16.6). These distribution patterns imply that the amounts of sediment involved in sliding and slumping have an important role in controlling the downslope evolution of sediment gravity flows (Lee et al. 2002).

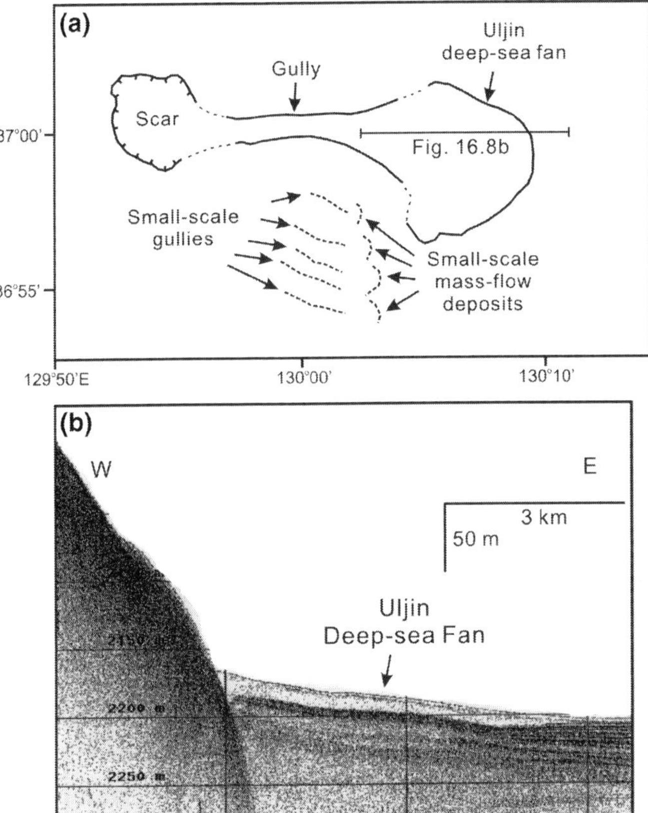

Fig. 16.8 **a** Line drawings of scoop-shaped scars, gullies, and downslope associated mass-transport deposits (MTDs) from sonar images on the western slope of the Ulleung Basin. For location of (**a**), see Fig. 16.6. **b** A chirp (2–7 kHz) subbottom profile showing a distinct fan-shaped mass-flow deposits (Uljin deep-sea fan) at the base-of-slope (from Lee et al. 2014)

16.2.2.2 Eastern Margin of the Japan Basin

Along the eastern margin of the Japan Basin, many scoop-shaped failure scars occur on the upper slope at water depths of 300–600 m (Nakajima and Kanai 2000). Slide and slump masses on the middle to lower slope, downslope of the failure scars, merge into debrites and turbidites in the basin plain (Nakajima and Kanai 2000). The slope failures and mass movements along the eastern margin of the Japan Basin have been mostly generated by earthquakes which have frequently occurred in this area (Satake 1985; Tanioka et al. 1995; Nakajima and Kanai 2000). Nakajima and Kanai (2000) recognized several mass-failure deposits and seismo-turbidites generated by the 1983 Japan Sea earthquake and other historic earthquakes.

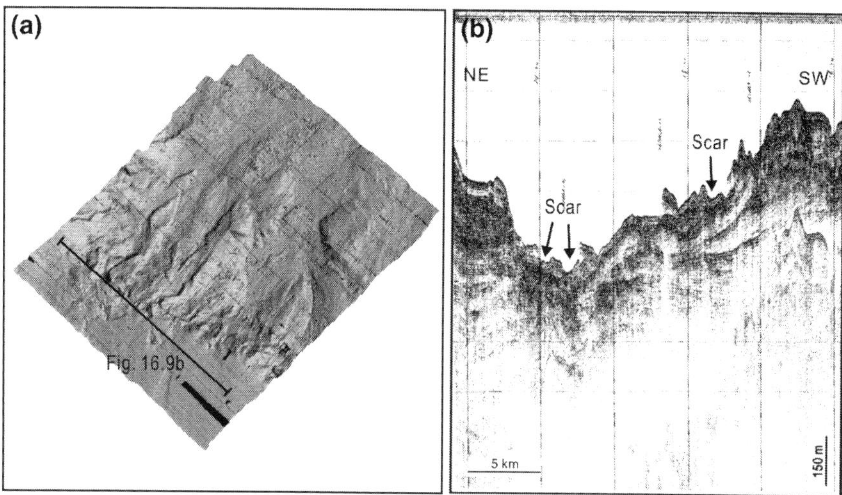

Fig. 16.9 Shaded relief image (**a**) and a single-channel air-gun seismic profile (**b**) showing gullied slope-failure scars on the upper slope of the southern Ulleung Basin. For location of (**a**), see Fig. 16.6 (from Lee et al. 2014)

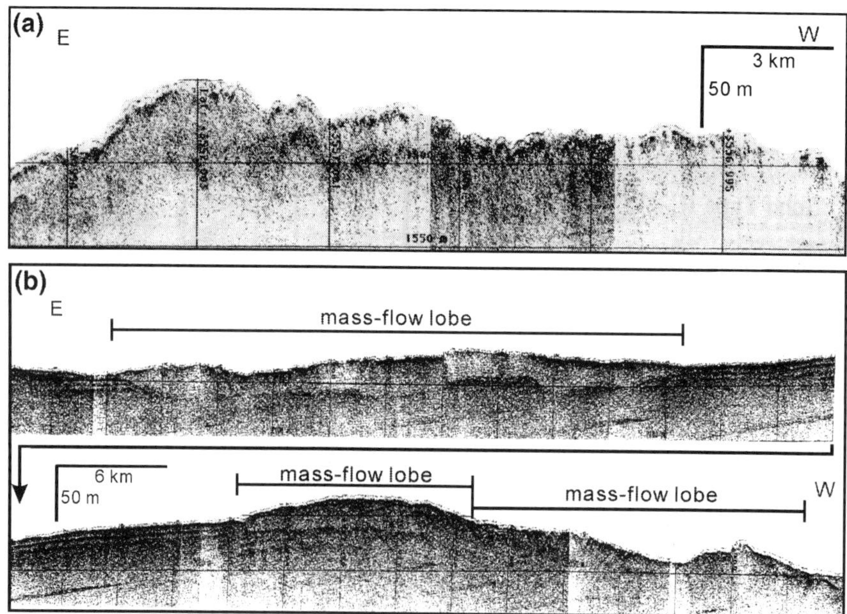

Fig. 16.10 Chirp (2–7 kHz) subbottom profiles showing hyperbolic surfaced mass-flow deposits on the southern middle slope (**a**) and sedimentary features of mass-flow lobes on the southern lower slope (**b**). For location of profiles, see Fig. 16.6 (from Lee et al. 2014)

Fig. 16.11 Submarine creep deposits showing irregular wavy bottom echoes with several distinct to indistinct subsurface reflectors. Note basal mingled or transparent reflections in (**a**). A subbottomward increase in wave amplitude (*open arrow* in **b**). *CIF* compressional internal fault; *NIF* normal internal fault. For location of profiles, see Fig. 16.6 (from Lee et al. 2004a)

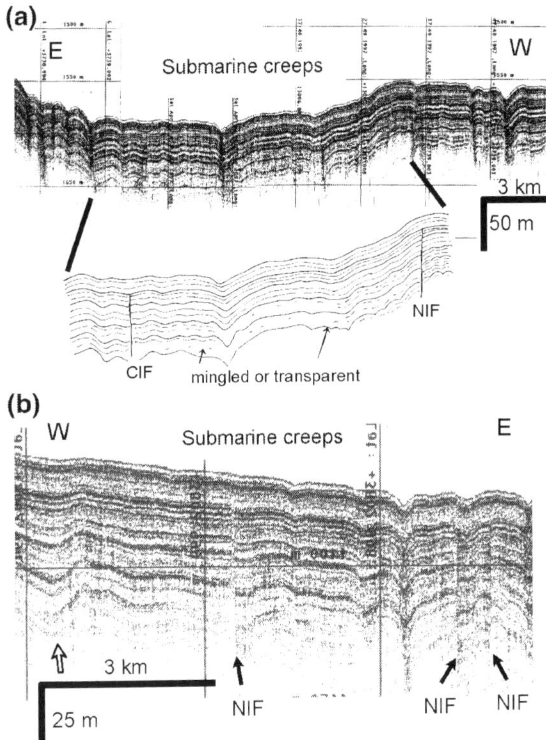

16.2.2.3 Toyama Deep-Sea Channel

In the East Sea, the Toyama deep-sea channel (TDSC) is a prominent submarine channel, about 750 km long, extending from the central Japan (Toyama Bay) to the abyssal Yamato and Japan basins (Fig. 16.1; Nakajima et al. 1998). The TDSC has been supplied with sediment derived by rivers from the Northern Japan Alps through tributary canyons in the narrow shelf of Toyama Bay (Nakajima 2006). The canyons in Toyama Bay merge downslope into the TSDC in the Toyama Trough (Nakajima and Satoh 2001). In the Yamato and Japan basins, prominent levees have developed along the TSDC (Nakajima et al. 1998; Nakajima 2006). Sediment waves (or mudwaves) formed by overflowing turbidity currents occur extensively on the levees (Fig. 16.12; Nakajima and Satoh 2001). In the Japan Basin, the TDSC splits into smaller distributary channels that have formed a series of small-scale depositional lobes on the terminal Toyama Fan (Nakajima 2006).

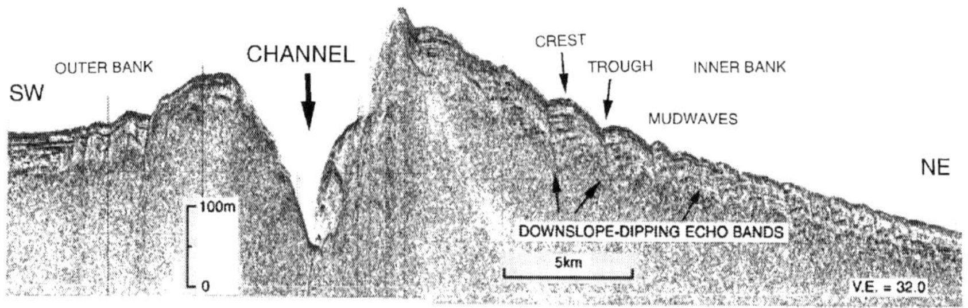

Fig. 16.12 A 3.5-kHz subbottom seismic reflection profile across the Toyama deep-sea channel (from Nakajima and Satoh 2001). Vertical exaggeration (V.E.) = 32. For location of a profile, see Fig. 16.1

16.2.3 Coastal Geomorphology of the Eastern Korean Peninsula

At first glance, it appears that the East Sea meets the Korean peninsula along gently curved and smooth shorelines without conspicuous interference by river systems. However, on closer examination, the coastal geomorphology is rather complex, including many types of sedimentary environments. This complexity seems to have originated from the tectonic history of the uplifted eastern part of the peninsula (Chough et al. 2000), the wave-dominated, micro-tidal hydrodynamic conditions of the coastal waters, and the panoramic lithologies and ages of the regional bedrock (Fig. 16.13; Lee et al. 2011).

Compared to the western and southern coasts of Korea, the eastern coast has a small areal portion of recent sediments. Due to a narrow and steep coastal plain, the sediments are mostly sands and gravels from nearby mountains and shores. The sediments in the coastal sea are transported along the shore and fed to nearby beaches by wind-generated coastal currents (Fig. 16.14). However, on a regional scale, no conspicuous longshore sorting in grain size or composition has been recognized (Lee and Kim 2006).

Although the coastal land area has experienced numerous civil engineering works a number of natural sedimentary environments can be found. Beaches, lagoons, marine terraces, coastal dunes, river (or stream) mouth spits, rocky headlands and cliffs are the most representative features found from the backshore to the shoreface. In the nearshore zone, submerged sand bars, channels (buried or active), tombolos and sea stacks are major conspicuous morphological features.

16.2.3.1 Beaches

Beaches are one of the most representative environments in the eastern coast of Korea (Fig. 16.14). The beach sediments generally consist of either sands (generally coarse and medium sands) or gravels (shingle) and probably originated either

Fig. 16.13 Pictures showing comparatively different lithologies of basement rocks and adjacent beaches in the eastern coast of Korea. Clockwise from the *upper left* could be easily found rocky cliffs made of Triassic granite which represents one of the dominant lithology of the northern part of the coast (the Hajodae in this case) and sand beaches which would likely be supplied with weathered materials from nearby rocky coasts or small rivers/streams. A typical gravel beach is shown in the picture taken in the southern part of the east coast. In the last picture (*lower left*) an exposed volcanic rock outcrop with columnar joints could be seen in Gangdong. Volcanic rocks and tuffaceous Tertiary strata are relatively common in the southern part of the coast, which implies the adjoining beaches could be supplied with coarse materials from nearby sources

from the weathered materials of the adjacent rocky shores or from sediments transported through small river/stream systems (Lee et al. 2013). The average length of sand beaches is about 2.7 km increasing northward and their beach-face slopes are generally steeper than those on the western and southern coasts of Korea (Fig. 16.15). Gravel (shingle) beaches can be found more frequently in the southern part of the east coast. This seems to be from the result of differential dominance of the unconsolidated Tertiary gravel-rich strata in the southern part. Recursive measurements of beach profiles, CCTV monitoring and remote-sensing data analysis have revealed that most of the beaches are governed mainly by weather, climate, sediment supply and artificial influences (Korea Hydrographic and Oceanographic Administration 2009, 2010; Park et al. 2009; Oh and Bang 2013). It has been reported that serious geomorphologic changes or erosions at some beaches might result from dike or breakwater construction (Ministry of Marine Affairs and Fisheries 2004; Yoo et al. 2008).

Fig. 16.14 A picture taken in Naksan of Gangwon-do in the eastern coast of Korea. A typical sand beach could clearly be seen from the north on the roof of a hotel where video cameras are installed by the government for monitoring the beach morphology change. Along the east coast, small-scale rivers/streams are commonly interrelated with sand beaches implying that beach-forming sands could easily be supplied by limited river/stream systems as well as by waves weathering adjacent headlands. Once fed by the river/stream channels sands are seemingly transported alongshore by wave-generated coastal current

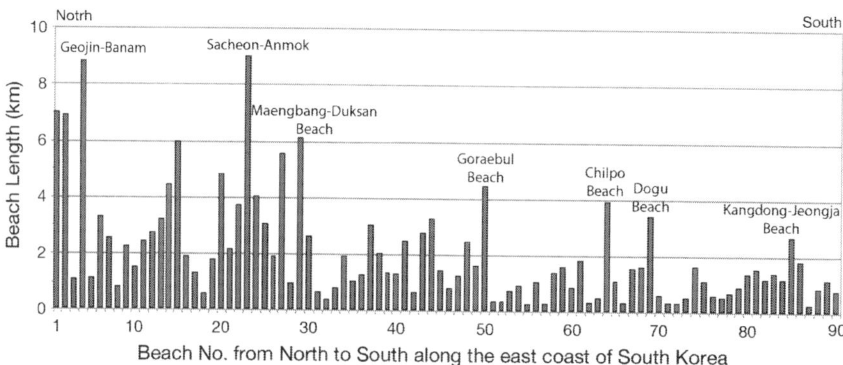

Fig. 16.15 Beach length statistics along the eastern coast of South Korea

At the southeastern end of the eastern coast of Korea several cases of rip currents have been reported, the characteristics of which were numerically simulated by a number of researchers (Fig. 16.16; Song and Bae 2011; Choi et al. 2012b, 2013; Yoon et al. 2012). Well-defined transverse (crescentic) sandbar systems were

Fig. 16.16 A picture showing a rip current formed in the middle part of the Haeundae Beach in Busan. Busan is located in the southern part of the east coast and famous for well-prepared resort areas. Every year during summer high season highly-attended observation and monitoring systems are operated for tourist safety due to unexpected rip current occurrence. It is notable in this picture that the people swimming in the upper shoreface could easily be dragged away from the shore by the rip current. The picture was taken in 2012 and granted by the courtesy of the Korea Meteorological Administration

recognized in recent nearshore bathymetric survey data and aerial photographs, implying active longshore sediment transports and bedform migrations induced mainly by oblique incident waves processes (Fig. 16.17).

16.2.3.2 Lagoons

Lagoons are the most distinct geomorphological features along the eastern coast of Korea. Forty-eight lagoons (64.85 km^2) out of fifty-seven natural lakes were recognized in the coastal zones by GIS analysis for topographic maps and satellite images (Lee et al. 2006). These lagoons have most likely formed after the sea level has transgressed relatively fast during the last deglaciation (Fig. 16.18; Pirazzoli 1991). Following the shoreline retreat, the seaward mouths of the submerged areas became blocked by spit systems or longshore sandbars around 4000 yr BP. Once protected from the sea most of the areas were turned into low-energy depositional environments and were eventually filled with fluvial sediments (Park and Kim 1981; Lee and Yu 2011; Go et al. 2013).

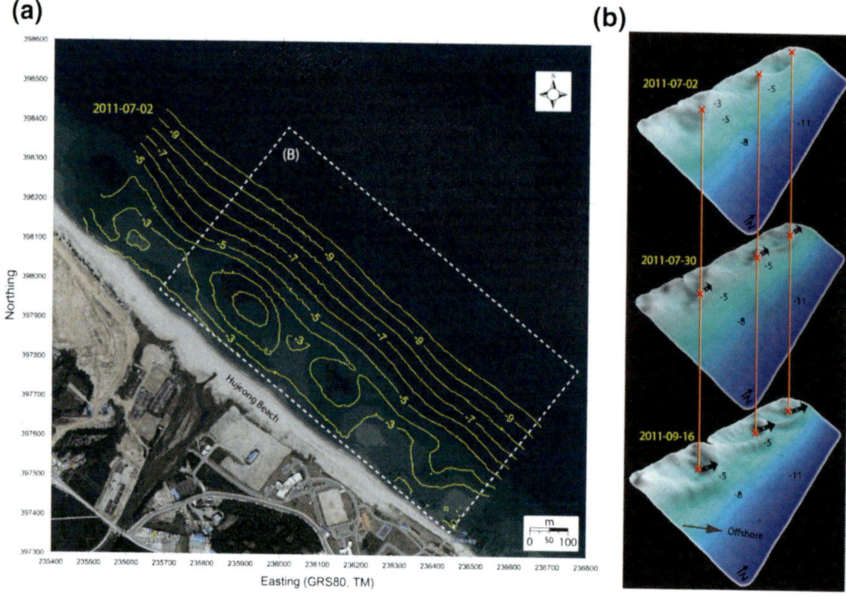

Fig. 16.17 A typical crescentic sand bar system along the eastern coast of Korea (Hujeong Beach, Uljin): **a** Nearshore bathymetric map (surveyed on July 2, 2011) overlapped with the aerial photograph taken on August 20, 2010 (elevation in meters); **b** a time series of the crescentic bar migration and deformation from July 2, 2011 to September 16, 2011

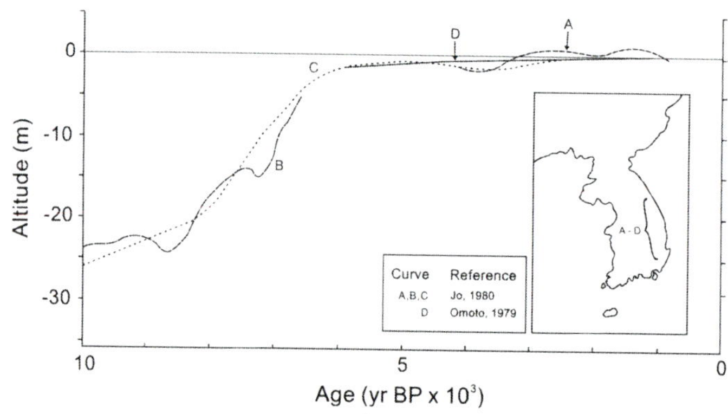

Fig. 16.18 Sea level change records reported from the eastern coast of the Korean Peninsula (Redrawn from Pirazzoli 1991). Despite of small variations the sea level in the East Sea (*A–D*) has transgressed 20–30 m during the Holocene and became stable since 6–7 kyr BP. In some areas shown in the curve *A* the sea level was presumed to be higher than at present

Some of them, however, still remain unfilled and are connected to the sea through very shallow inlets. For example, Cheongchoho Lake in Sokcho city (Gangwon-do) is only 1–2 m in depth and approximately 5 km in circumference (Yeon et al. 2007). The salinity range of the lagoon water is generally very wide due

Fig. 16.19 A picture showing a lagoon (Cheongchoho). The seaward mouth and inlet of the lagoon are maintained for ship's navigation. The surrounding area is well developed for a public park establishment

to fresh water input through small rivers/streams and rainfall (Park and Kim 1981; Yum et al. 2002; Ha et al. 2009; Heo et al. 2011). Sporadically high-salinity seawater intrudes when overwash by strong waves or freshwater flooding events breach the mouth bars or spits. Except for a few lagoons close to the northern end of South Korea (Yum et al. 2002; Park et al. 2007; Lee and Yu 2011), most of the lagoons along the east coast are strongly disturbed or polluted by human activities, and their seaward mouths are artificially maintained (Fig. 16.19; Park and Kim 1981).

16.2.3.3 Marine Terraces

Morphologic features of marine terraces have been studied frequently on the eastern coast of Korea, mainly for ground stability assessments for nuclear power plant construction (Choi et al. 2008). The basement block of the East Sea is inferred to have been uplifted at a slow rate during the Quaternary period because the Ulleung Basin is still in its compressional stage (Chough et al. 2000). Continuous uplift of the eastern coastal zones has displaced paleo-shorelines upwards so that several groups of marine terraces in 4–6 height levels are reported along the coast (Chun et al. 2005). Most of the terraces occur as a flat surface and are frequently used for agricultural fields or coastal roads. The height of those terraces generally ranges from 3 to 130 m above sea level (Fig. 16.20; Choi and Kim 2005; Kim et al. 2005, 2007). The terrace deposits are about 5–10 m in thickness and are dominated by well-round beach gravels in unconsolidated muddy or silty matrix. It is notable that the occurrence of marine terraces is more frequent in the southern part of the eastern coast than in the northern part, indicating possible differences in uplift

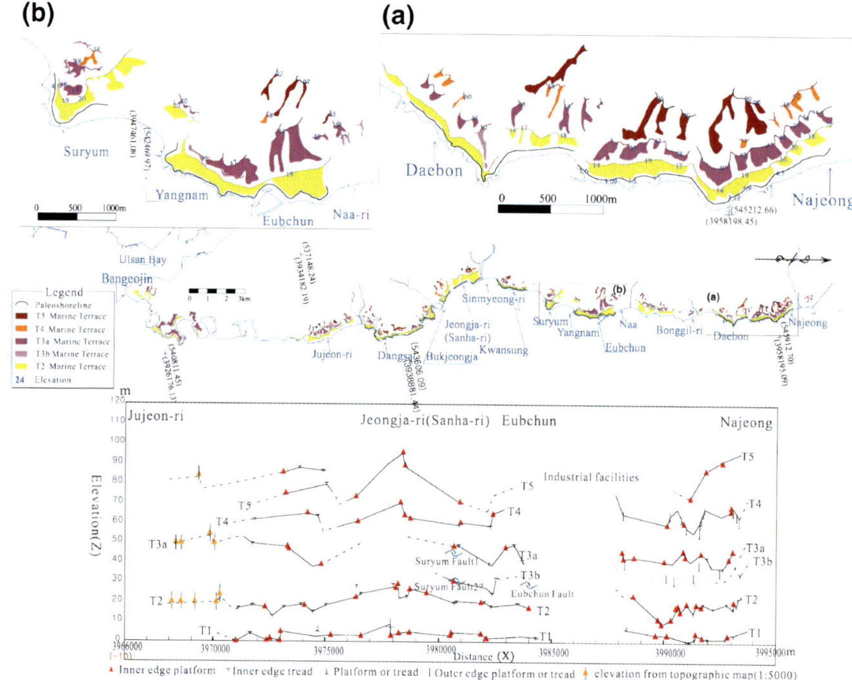

Fig. 16.20 Vertical and horizontal distribution of marine terraces along the southern part of the east coast (from Choi et al. 2008)

rate between the two parts (Choi et al. 2008). Based on various ages of the terrace deposits long-term models of relative sea level change or basement uplift were suggested for the Quaternary period by several scientists. According to their studies the uplift rate of the southern part of the east coast is less than 0.3 m/kyr (Kim et al. 2005, 2007, 2008; Choi et al. 2008, 2009).

16.2.3.4 Coastal Dunes

Thirty two coastal dune fields are located along the eastern coast of Korea (Ministry of Environment 2001). About 40 % of them are more than 2 km in length and around 18 % are naturally preserved due to their locations in the demilitarized zone or on military bases (Ministry of Environment 2001). The others generally have been removed or modified for construction of tourist facilities and coastal roads. The coastal dunes consist of medium sands. Most of the dunes develop in a series of long narrow trains classified as beach-foredune ridges (National Institute of Environmental Research 2009). Recent studies using optically stimulated luminescence dating and ground penetrating radar reveal that the beach-foredune ridges in the northern part of the eastern coast of Korea started to form at around 7000 years ago when the sea-level transgression slowed down, and experienced multiple stages of seaward progradation during Holocene epoch (Choi et al. 2010).

Acknowledgments This is partly supported by the project of 'International Ocean Discovery Program' funded by the Ministry of Oceans and Fisheries (Korea), and the project funded by the Korea Meteorological Industry Promotion Agency. The authors thank a reviewer for improving the manuscript.

References

Cha HJ, Choi MS, Lee CB et al (2007) Geochemistry of surface sediments in the southwestern East/Japan Sea. J Asian Earth Sci 29:685–697

Choi KH, Kim JW (2005) Detection of a paleoshore platform on a coastal terrace using the seismic refraction method. J Korea Geomorphol Assoc 12(3):89–97 (in Korean)

Choi SJ, Merritts DJ, Ota Y (2008) Elevations and ages of marine terraces and late Quaternary rock uplift in southeastern Korea. J Geophys Res 113(B10403):1–15

Choi JH, Kim JW, Murray AS et al (2009) OSL dating of marine terrace sediments on the south eastern coast of Korea with implications for Quaternary tectonics. Quatern Int 199:3–14

Choi KH, Kim JW, Choi JH (2010) Reconstruction of coastal progradation during the Holocene in Kangneung area along the east coast of Korea, based on OSL dating and GPR survey on beach-foredune ridges. Geophys. Res Abstracts 12 EGU2010-7603-1 EGU General Assembly 2010

Choi J, Lee KE, Lee HJ (2012a) Spatial and temporal changes in sea surface temperature, circulation and subpolar front of the East Sea (Japan Sea) during the last 130,000 years. Palaeogeogr Palaeocl 313–314:225–233

Choi J, Park WK, Bae JS et al (2012b) Numerical study on a dominant mechanism of rip current at Haeundae Beach: honeycomb pattern of waves. J Korean Soc Civil Eng 32(5):321–329 (in Korean)

Choi J, Shin CH, Yoon SB (2013) Numerical study on sea state parameters affecting rip current at Haeundae Beach: wave period, height, direction and tidal elevation. J Korea Water Resour Assoc 46(2):205–218 (in Korean)

Chough SK, Lee HJ (1987) Stability of sediments on the Ulleung Basin slope. Mar Geotechnol 7:123–132

Chough SK, Lee SH, Kim JW et al (1997) Chirp (2–7 kHz) echo characters in the Ulleung Basin. Geosci J 1:143–153

Chough SK, Lee HJ, Yoon SH (2000) Marine geology of Korean Seas. Elsevier, Amsterdam

Chun HY, Lim SB, Koh HJ et al (2005) Natural environments and geology of the east coast. Korea Institute of Geoscience and Mineral Resources, Yewon Publisher, Seoul (in Korean)

Crusius J, Pedersen TF, Calvert SE et al (1999) A 36 kyr geochemical record from the Sea of Japan of organic matter flux variations and changes in intermediate water oxygen concentrations. Paleoceanography 14:248–259

Fujine K, Tada R, Yamamoto M (2009) Paleotemperature response to monsoon activity in the Japan Sea during the last 160 kyr. Palaeogeogr Palaeocl 280:350–360

Go A, Yukiya T, Kaoru K (2013) Sedimentary environment of Hwajinpo using diatom analysis. J Korea Geomorphol Assoc 20(2):15–25 (in Korean)

Gorbarenko SA, Southon JR (2000) Detailed Japan Sea paleoceanography during the last 25 kyr: constraints from AMS dating and $\delta^{18}O$ of planktonic foraminifera. Palaeogeogr Palaeocl 156:177–193

Ha KH, Choi KH, Kim JW (2009) Geomorphological characteristics and salinity distribution of Yeongok estuary in Kangneung, Korea. J Korea Geomorphol Assoc 16(1):89–100 (in Korean)

Heo WM, Choi SG, Kwak SJ et al (2011) The study of water environment variations in Lake Hwajinpo. Korean J Limnol 44(1):9–21 (in Korean)

Hong GH, Kim SH, Chung CS et al (1997) ^{210}Pb-derived sediment accumulation rates in the southwestern East/Japan Sea (Sea of Japan). Geo-Mar Lett 17:126–132

Hyun S, Bahk JJ, Suk BC et al (2007) Alternative modes of Quaternary pelagic biosiliceous and carbonate sedimentation: a perspective from the East Sea (Japan Sea). Palaeogeogr Palaeocl 247:88–99

Khim BK, Bahk JJ, Hyun S et al (2007) Late Pleistocene dark laminated mud layers from the Korea Plateau, western East Sea/Japan Sea, and their paleoceanographic implications. Palaeogeogr Palaeocl 247:74–87

Khim BK, Ikehara K, Irino T (2012) Orbital- and millennial-scale paleoceanographic changes in the north-eastern Japan Basin, East Sea/Japan Sea during the late Quaternary. J Quat Sci 27:328–335

Kido Y, Minami I, Tada R et al (2007) Orbital-scale stratigraphy and high-resolution analysis of biogenic components and deep-water oxygenation condition in the Japan Sea during the last 640 kyr. Palaeogeogr Palaeocl 247:32–49

Kim JW, Choi JH, Choi JH et al (2005) The morphological characteristics and geochronological ages of coastal terraces of Heunghae region in northern Pohang city, Korea. J Korea Geomorphol Assoc 12(1):103–116 (in Korean)

Kim JW, Chang HW, Choi JH et al (2007) Optically stimulated luminescence dating on the marine terrace deposits of Hujeong-Jukbyeon region in Uljin. Korea. J Korea Geomorphol Assoc 14(1):15–27 (in Korean)

Kim JY, Oh KC, Yang DY et al (2008) Stratigraphy, chronology and implied uplift rate of coastal terraces in the southeastern part of Korea. Quatern Int 83:76–82

Korea Hydrographic and Oceanographic Administration (2009) Final report of the contracted project on coastal marine survey and character study in the Northern Gangwon and Southeastern Jeju: basic data and feature survey. MONO1201107811 T0033891, Registered Publication No. 11-1611234-000054-01, p 250 (in Korean)

Korea Hydrographic and Oceanographic Administration (2010) Coastal erosion monitoring. GOVP1201131296 627.5-11.12, Registered Publication No. 11-1611234-000121-01, p 205 (in Korean)

Lee KE (2007) Surface water changes recorded in late Quaternary marine sediments of the Ulleung Basin, East Sea (Japan Sea). Palaeogeogr Palaeocl 247:18–31

Lee SH, Chough SK (2001) High-resolution (2–7 kHz) acoustic and geometric characters of submarine creep deposits in the Korea Plateau, East Sea. Sedimentology 48:629–644

Lee JL, Kim IH (2006) A survey and analysis of swim zone width and beach scale factor for Gangwon beaches. Korean Soc Coast Marine Eng 18(3):241–250 (in Korean)

Lee SH, Yu KM (2011) Depositional environments of Holocene laminated layers in the core SJ99 collected from the Songjiho Lagoon on the eastern coast of Korea. J Geol Soc Korea 47(2):123–137 (in Korean)

Lee HJ, Chun SS, Yoon SH et al (1993) Slope stability and geotechnical properties of sediment of the southern margin of Ulleung Basin, East Sea (Sea of Japan). Mar Geol 110:31–45

Lee HJ, Chough SK, Yoon SH (1996) Slope-stability change from late Pleistocene to Holocene in the Ulleung Basin, East Sea (Japan Sea). Sediment Geol 104:39–51

Lee SH, Chough SK, Back GG et al (1999) Gradual downslope change in high-resolution acoustic characters and geometry of large-scale submarine debris lobes in Ulleung Basin, East Sea (Sea of Japan), Korea. Geo-Mar Lett 19:154–161

Lee SH, Chough SK, Back GG et al (2002) Chirp (2–7 kHz) echo characters of the South Korea Plateau: styles of mass movement and sediment gravity flow. Mar Geol 184:227–247

Lee SH, Bahk JJ, Chough SK (2004a) Late Quaternary sedimentation in the eastern continental margin of the Korean Peninsula. In: Clift P, Kuhnt W, Wang P et al (eds) Continent—ocean interactions within east Asian marginal seas. Geophysical monograph series, vol 149. American Geophysical Union, Washington, DC, pp 205–233

Lee SH, Bahk JJ, Chough SK et al (2004b) Late Quaternary sedimentation in the Ulleung Interplain Gap, East Sea (Korea). Mar Geol 206:225–248

Lee MB, Kim NS, Lee GR (2006) The distribution and geomorphic changes of natural lakes in east coast of Korea. J Korean Assoc Reg Geog 12(4):449–460 (in Korean)

Lee KE, Bahk JJ, Choi J (2008) Alkenone temperature estimates for the East Sea during the last 190,000 years. Org Geochem 39:741–753

Lee SH, Bahk JJ, Kim HJ et al (2010) Changes in the frequency, scale, and failing areas of latest Quaternary (<29.4 cal. ka B.P.) slope failures along the SW Ulleung Basin, East Sea (Japan Sea), inferred from depositional characters of densely dated turbiditic successions. Geo-Mar Lett 30:133–142

Lee JY, Yoon HS, Hong SS (2011) Geology of the Korean Peninsula based on the 1/1,000,000 Tectonic Map. STN Ltd., Seoul (in Korean)

Lee SH, Lee HJ, Park JY et al (2013) Possible origin of coastal sands and their long-term distribution along the high slope-gradient, wave-dominated eastern coast, Korea. Geosci J 17(2):163–172

Lee SH, Bahk JJ, Kim HJ et al (2014) Contrasting development of the latest Quaternary slope failures and mass-transport deposits in the Ulleung Basin, East Sea (Japan Sea). In: Krastel S, Behrmann JH, Völker D et al (eds) Submarine mass movements and their consequences (Adv Nat Technol Haz 37:403–412)

Lim D, Xu Z, Choi J et al (2011) Paleoceanographic changes in the Ulleung Basin, East (Japan) Sea, during the last 20,000 years: Evidence from variations in element composition of core sediments. Prog Oceanogr 88:101–115

Locat J, Lee HJ (2002) Submarine landslides: advances and challenges. Can Geotech J 39:193–212

Ministry of Environment (2001) Current status of Sand Dunes in Korea and Study on the Preservation and Management Plan. Ministry of Environment, Seoul (in Korean)

Ministry of Marine Affairs and Fisheries (2004) Coastal erosion monitoring system construction (I, II). GOVP1200512011. Ministry of Marine Affairs and Fisheries, Seoul (in Korean)

Nagashima K, Tada R, Matsui H et al (2007) Orbital- and millennial-scale variations in Asian dust transport path to the Japan Sea. Palaeogeogr Palaeocl 247:144–161

Nakajima T (2006) Hyperpycnites deposited 700 km away from river mouths in the central Japan Sea. J Sediment Res 76:60–73

Nakajima T, Kanai Y (2000) Sedimentary features of seismoturbidites triggered by the 1983 and older historical earthquakes in the eastern margin of the Japan Sea. Sediment Geol 135:1–19

Nakajima T, Satoh M (2001) The formation of large mudwaves by turbidity currents on the levees of the Toyama deep-sea channel, Japan Sea. Sedimentology 48:435–463

Nakajima T, Satoh M, Okamura Y (1998) Channel-levee complexes, terminal deep-sea fans and sediment wave fields associated with the Toyama Deep-Sea Channel system in the Japan Sea. Mar Geol 147:25–41

National Institute of Environmental Research (2009) Research on coastal landscape and the conservational strategy (2007~2009). NIER NO. 2009-52-1108, Publication No. 11-1480523-000214-12. National Institute of Environmental Research, Incheon (in Korean)

Niino H, Emery KO, Kim CM (1969) Organic carbon in sediments of Japan Sea. J Sediment Petrol 39:1390–1398

Oba T, Kato M, Kitazato H et al (1991) Paleoenvironmental changes in the Japan Sea during the last 85,000 years. Paleoceanography 6:499–518

Oh JK, Bang KY (2013) Variations of sediment textural parameters and topography around Gangneung harbor after the completion of harbor construction. J Korean Earth Sci Soc 34(2):120–135 (in Korean)

Otosaka S, Togawa O, Baba M et al (2004) Lithogenic flux in the Japan Sea measured with sediment traps. Mar Chem 91:143–163

Park BK, Kim WH (1981) The depositional environments of lagoons in the east coast of Korea. J Geol Soc Korea 17(4):241–249 (in Korean)

Park S, Choi J, Choi E et al (2007) The characteristics of fish community in the lagoon Hwajinpo, Korea. Korean J Limnol 40(3):449–458 (in Korean)

Park J, Jeong E, Lee S et al (2009) Morphologic change monitoring in an erosional coastal environment using a LiDAR system and remote sensing data: a case study at the Hujeong-ri Beach, East Coast of South Korea. American Geophysical Union, fall meeting 2009, AGU, Washington, DC

Pirazzoli PA (1991) World Atlas of Holocene sea-level changes. Elsevier Oceanography Series, vol 58. Elsevier, Amsterdam

Satake K (1985) The mechanism of the 1983 Japan Sea earthquake as inferred from long-period surface waves and tsunamis. Phys Earth Planet Inter 37:249–260

Song DS, Bae HK (2011) Observation and forecasting of rip current generation in Haeundae Beach, Korea Plan and Experiment. J Coast Res SI64:946–950

Tada R, Irino T, Koizumi I (1999) Land-ocean linkages over orbital and millennial timescales recorded in late Quaternary sediments of the Japan Sea. Paleoceanography 14:236–247

Tanioka K, Satake K, Ruff L (1995) Total analysis of the 1993 Hokkaido Nansei-oki earthquake using seismic wave, tsunami, and geodetic data. Geophys Res Lett 22:9–12

Ujiie H, Ichikura M (1973) Holocene to uppermost Pleistocene planktonic foraminifers in a piston core from off San-in District, Sea of Japan. Trans Proc Paleontol Soc Jpn 91:137–150

Yeon YJ, Kim YB, Kim JS et al (2007) Ocean Atlas of Korea, East Sea. Korea Hydrographic and Oceanographic Administration, Busan (in Korean)

Yoo HS, Kim KH, Joung EJ (2008) The analysis for the causes of beach erosion on Jeonchon-Najung Beach on the East Coast of Korea. J Korean Coast Ocean Eng 20(6):611–620 (in Korean)

Yoon SH, Chough SK (1995) Regional strike slip in the eastern continental margin of Korea and its tectonic implications for the evolution of Ulleung Basin, East Sea (Sea of Japan). Bull Geol Soc Am 107:83–97

Yoon SB, Kwon SJ, Bae JS et al (2012) Investigation of characteristics of rip current at Haeundae Beach based on observation analysis and numerical experiments. J Korean Civil Eng 32(4B):243–251 (in Korean)

Yum JG, Yu GM, Sampei Y et al (2002) Depositional environmental change during the last 400 years in the Hwajinpo lagoon on the eastern coast of Korea. J Geol Soc Korea 38(1):21–32 (in Korean)

Zou J, Shi X, Liu Y et al (2012) Reconstruction of environmental changes using a multi-proxy approach in the Ulleung Basin (Sea of Japan) over the last 48 ka. J Quat Sci 27:891–900

Chapter 17
Crustal Structure and Tectonic Evolution of the East Sea

Gwang Hoon Lee and Han-Joon Kim

Abstract The East Sea (Japan Sea) is a back-arc basin lying on the southeastern Amur plate, separated from the Philippine Sea, Eurasia, and Pacific-Okhotsk plates by a complex border. The opening of the East Sea was initiated in the Early Oligocene by rifting and extension in the Japan Basin, followed by seafloor spreading in the Late Oligocene. The opening of the Japan Basin was achieved by a counterclockwise rotation of the NE Japan Arc. The extension in the Japan Basin propagated southwestward toward the northern part of the eastern Korean margin, inducing NW-SE rifting in the South Korea Plateau. In contrast, the southern part of the Korean margin underwent E-W extension before the Early Miocene, probably due to the N-S trending subduction off the SW Japan Arc. Back-arc rifting at the Korean margin gave way to breakup at the base of the continental slope, after which the SW Japan Arc was separated southeastward with a clockwise rotation, opening the Ulleung Basin. Part of the Ulleung Basin is underlain by thicker-than-normal oceanic crust generated by seafloor spreading with hotter-than-normal mantle temperature. The plate reorganization in East Asia in the early Middle Miocene led to the East Sea closure. The anticlinal structures in the southwestern Ulleung Basin margin show NW-SE or N-S compression in the Middle Miocene-Early Pliocene and NE-SW or E-W compression since the Late Pliocene. The incipient subduction along the eastern margin of the East Sea also suggests E-W compression since the Late Pliocene. The E-W compression is probably due to the eastward movement of the Amur plate that began in the Pliocene.

G.H. Lee (✉)
Department of Energy Resources Engineering, Pukyong National University,
608-737 Busan, Republic of Korea
e-mail: gwanglee@pknu.ac.kr

H.-J. Kim
Korean Seas Geosystem Research Unit, Korea Institute of Ocean Science and Technology,
426-744 Ansan, Republic of Korea
e-mail: hanjkim@kiost.ac.kr

© Springer International Publishing Switzerland 2016
K.-I. Chang et al. (eds.), *Oceanography of the East Sea (Japan Sea)*,
DOI 10.1007/978-3-319-22720-7_17

Keywords Back-arc basin · Back-arc closure · Tectonics · Amur plate · Ulleung Basin · East Sea (Japan Sea)

17.1 Tectonic Setting

The East Sea is a back-arc basin located in the southeastern part of the Amur plate which includes the Korean Peninsula, SW Japan Arc, northeastern China, and Russian Far East (Fig. 17.1). The Amur plate is bounded on the north and west by the Eurasia plate, on the east and northeast by the Okhotsk plate, and on the south by the Philippine Sea plate. The Amur plate collides with the northwestward-moving Philippine Sea plate along the Nankai Trough where the young Shikoku Basin lithosphere is subducting beneath the SW Japan Arc. The Pacific plate is moving largely northwestward, subducting beneath the NE Japan Arc. The Amur plate has been moving eastward since the Pliocene (Tamaki and Honza 1985; Taira 2001; Lee et al. 2011). The boundary between the Amur and Okhotsk plates lies along the eastern margin of the East Sea and is characterized by thrusts and obduction of back-arc crust and sediments, suggesting incipient subduction of the Amur plate

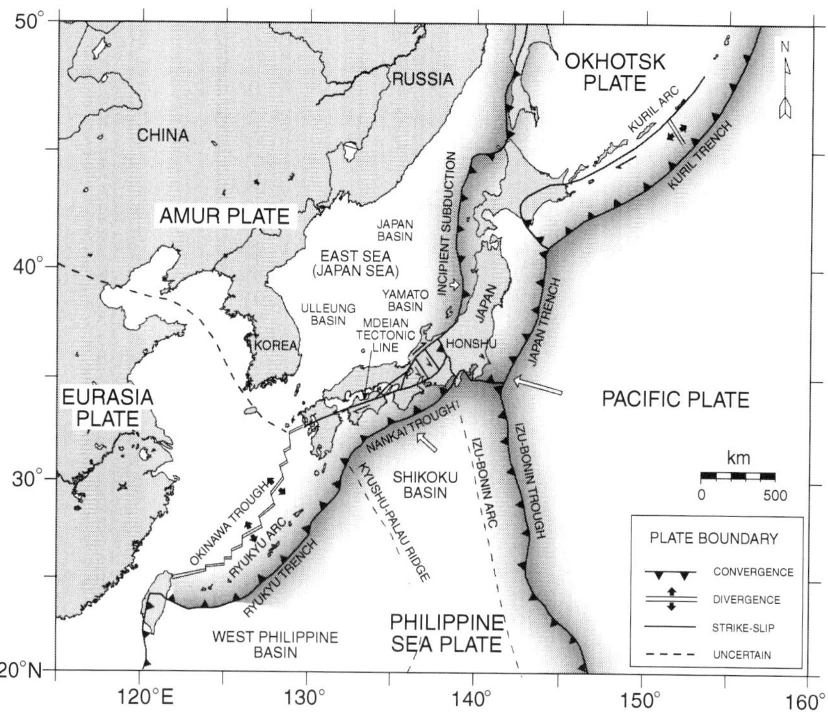

Fig. 17.1 Plate boundaries of the northeast Asia (from Taira 2001)

(Nakamura 1983; Tamaki and Honza 1985). The Baikal rift zone (not shown in Fig. 17.1) forms a boundary between the Amur plate and the Eurasia plate in the west (Zonenshain and Savostin 1981).

The East Sea consists of three deep basins (the Japan, Yamato, and Ulleung basins), separated by submerged continental remnants such as the Korea Plateau, Oki Bank, Yamato Rise and Kita-Yamato Ridge (Fig. 17.2). The opening of the East Sea was initiated by crustal thinning and rifting in the northeastern Japan Basin in the Early Oligocene, resulting from the subduction of the Pacific plate underneath the Eurasia plate (Tamaki et al. 1992). Rifting was followed by sea-floor spreading in the Late Oligocene. While seafloor spreading continued in the Japan Basin, the southern East Sea underwent extension, opening the Yamato and Ulleung basins. The major plate reorganization in East Asia in the early Middle Miocene (ca. 15 Ma) led to the East Sea back-arc closure that continues today (Sibuet et al. 2002; Sdrolias et al. 2004). Crustal shortening due to the back-arc closure resulted in regional deformation of the peripheral regions of the East Sea (Ingle 1992).

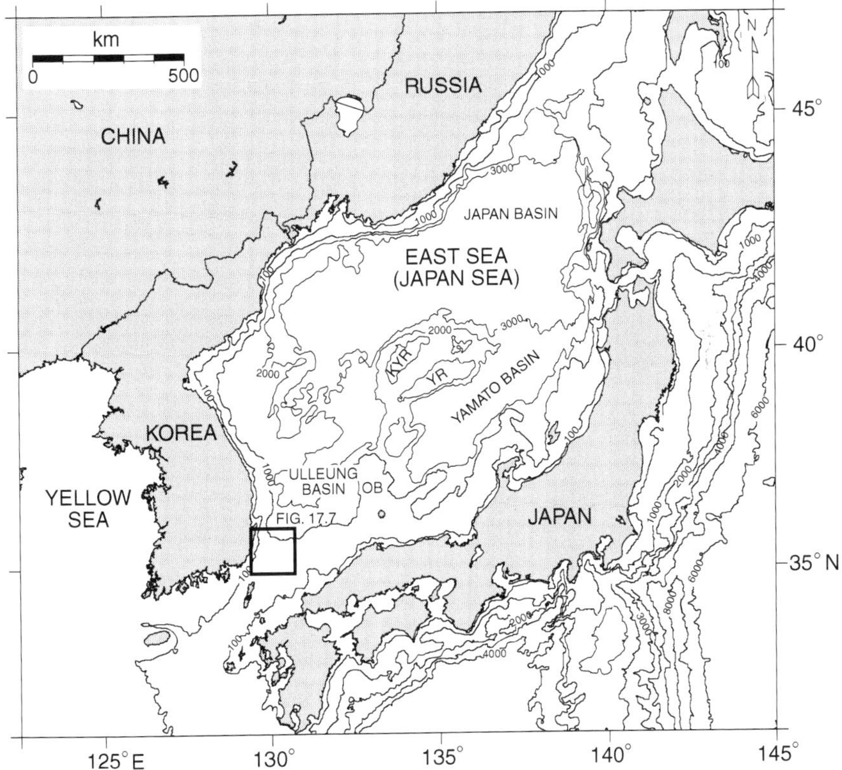

Fig. 17.2 Physiographic map of the East Sea. Box indicates the location of the map shown in Fig. 17.7. *OB* Oki Bank; *YR* Yamato Rise; *KYR* Kita-Yamato Ridge. Bathymetry in meters

17.2 Crustal Structure

Ludwig et al. (1975) used sonobuoys to determine the 1-D crustal seismic velocity distribution in the East Sea. They computed sedimentary and crustal structure down to the Moho discontinuity in the Japan and Yamato basins; however, they failed to estimate the Moho discontinuity depth in the Ulleung Basin because of thick sediments. Deep seismic experiments using ocean bottom seismometers (OBSs) revealed a wide range of compositions in the back-arc basin crust in the East Sea from typically oceanic to intermediate between oceanic and continental (Chung et al. 1990; Hirata et al. 1992; Kim et al. 1998). The crust in the eastern part of the Japan Basin is typical of oceanic crust in thickness (Hirata et al. 1992). Layer 2 has a thickness of less than 2 km and seismic velocities from 4.5 to 5.6 km/s; layer 3 is about 5 km thick with velocities from 6.6 to 6.7 km/s. Linear magnetic anomalies trending N40°E and N60°E in this area (Seama and Isezaki 1990) confirm the oceanic origin of the crust. In contrast, the crust in the Yamato Basin is 11–12 km thick (Chung et al. 1990); layer 2 is about 2 km thick with velocities from 5.1 to 5.4 km/s, and layer 3 is 10 km thick with velocities from 6.7 to 7.2 km/s. Shinohara et al. (1992) considered the crust underlying the Yamato Basin to be intermediate; they classified it as neither oceanic nor continental due to its anomalous thickness.

The small size of the Ulleung Basin (Fig. 17.2) may suggest that lithospheric extension in the basin did not evolve into full-fledged seafloor spreading (Lee et al. 1999). However, the crust underlying the basin plain, although thicker than typical oceanic crust, is essentially oceanic in character as it consists of laterally consistent layers with seismic velocities that are representative of oceanic crustal layers 2B, 2C, and 3 in velocity and gradient distribution (Figs. 17.3, 17.4 and 17.5; Kim et al. 1998; Lee et al. 1999; Kim et al. 2003). The uppermost layer 2A in the oceanic crust, consisting of volcanic extrusives, is only present on oceanic ridges near eruptive centers and shows a drastic increase in seismic velocity to more than 5 km/s once covered with a few hundred meters of sediments (Rohr 1994). Sediment cover plays an important role in increasing the seismic velocity in two ways. Firstly, it directly decreases the porosity of layer 2A. Secondly, it seals extrusive basalts, influencing the convective hydrothermal circulation that triggers mineralization in the abundant porosity in layer 2A (Purdy and Ewing 1986). Considering the thickness of sediment cover (Fig. 17.4) in the Ulleung Basin, it is appropriate to observe layer 2B velocities at the top of layer 2.

The most important feature of the crust in the Ulleung Basin is the presence of layer 2C, the transition between layer 2 and layer 3 in oceanic crust (Fig. 17.5; Kim et al. 1998). Earlier seismic studies in the Japan and Yamato basins (Ludwig et al. 1975; Hirata et al. 1992; Shinohara et al. 1992) did not report the presence of layer 2C. Layer 2C in the Ulleung Basin is characterized by a rapid increase in velocity from 5.7 to 6.3 km/s over a short depth interval of less than 1.5 km.

Velocities around 6 km/s are representative of the granitic basement rock of the Japanese mainland (Honshu) and continental margin on the East Sea side (Ludwig

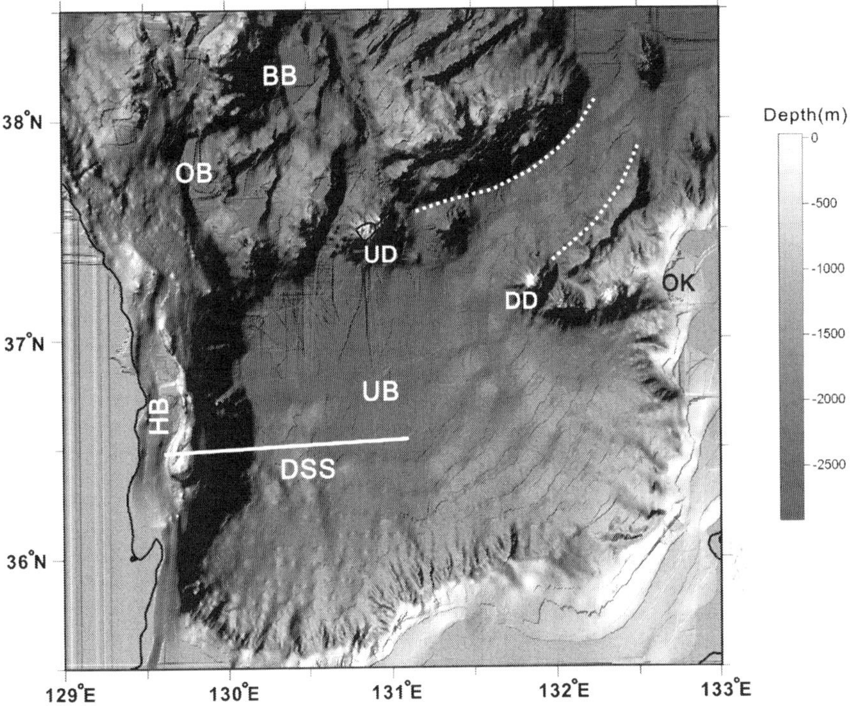

Fig. 17.3 Shaded bathymetry of the eastern Korean margin and the Ulleung Basin (Redrawn from Kim et al. 2007). DSS is the deep seismic sounding line of ocean bottom seismometers shown in Fig. 17.4. *UB* Ulleung Basin; *BB* Bandal Basin; *OB* Onnuri Basin; *HP* Hupo Basin; *UD* Ulleungdo; *DD* Dokdo; *OK* Oki Bank

et al. 1975). The thickness of the granitic basement in the SW Japan Arc is esti-mated to be more than 12 km (Hashizume and Matsui 1979). The Yamato Bank to the northeast of the Ulleung Basin bears significant resemblance to the Japan Arc in that it is underlain by a 10 km thick layer with 6 km/s velocity; therefore, it is interpreted as a stretched fragment of the continental crust of Honshu (Kurashimo et al. 1996).

If the crust in the Ulleung Basin is assumed to be extended continental crust of the Japan Arc, layer 2C in the Ulleung Basin would have to be the extended granitic rock underlying the Japan Arc. Then, an enormously large stretch fac-tor of about 10 would be required to explain the extension. Because the granitic rock in the brittle upper part of the crust cannot sustain such extension, Kim et al. (1998) interpreted that layer 2C in the Ulleung Basin did not originate from the southwestern Japan Arc crust. Furthermore, crustal velocity-depth profiles of the Ulleung Basin show that the velocity increases gradually from 5 to 7 km/s. This is a clear difference from the SW Japan Arc where a velocity disconti-nuity is observed at the boundary between the 6 km/s layer and the underlying

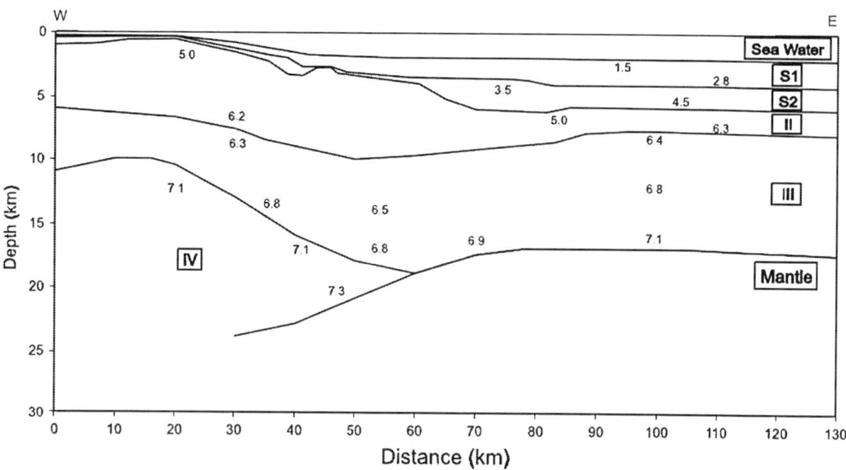

Fig. 17.4 Crustal velocity model from the continental shelf of the Korean Peninsula to the center of the Ulleung Basin (from Kim et al. 2003). *S1* and *S2* upper and lower sedimentary layers; *II* and *III* layers 2 and 3, respectively; *IV* abnormally high-velocity material interpreted as magmatic underplating. Velocities are in km/s. See Fig. 17.3 for location

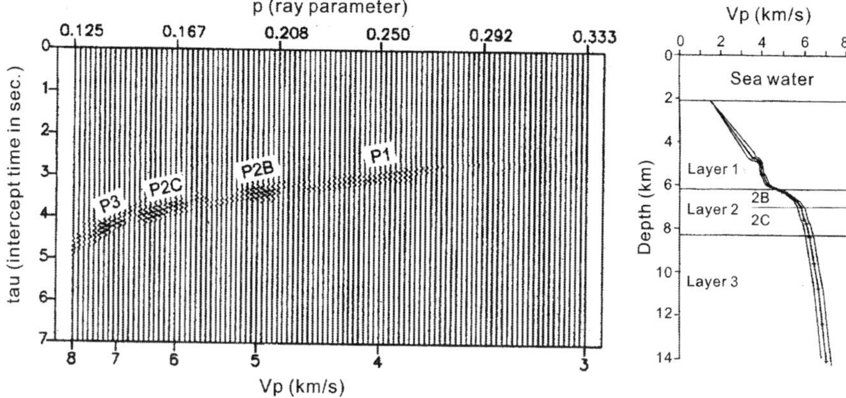

Fig. 17.5 *Left* the tau-p transform of OBS data recorded in the center of the Ulleung Basin (from Kim et al. 1998). *Right* the crustal velocity-depth profile derived from the tau-p transform (from Kim et al. 1998)

crust (Hashizume and Matsui 1979), typical of continental crust. It is known that thicker-than-normal oceanic crust forms when the mantle temperature is hotter than normal (Su et al. 1994). Therefore, the crust underlying the Ulleung Basin plain can be defined as oceanic crust, newly generated by seafloor spreading at the eastern Korean margin with hotter-than-normal mantle temperature. The hotter-than-normal mantle may be the result of asthenospheric upwelling induced above the subducting slab and rift-induced convection particularly in the narrow margin south of the Korea Plateau (Kim et al. 2007).

17.3 Tectonic Evolution

17.3.1 Opening

A variety of models has been proposed for the opening of the East Sea. One group of researchers inferred, based on paleomagnetic data, that the NE Japan Arc and SW Japan Arc underwent counterclockwise and clockwise rotations away from mainland Asia, respectively, since the Early Miocene (e.g. Otofuji and Matsuda 1987; Tosha and Hamano 1988). Their model is referred to as "fan-shaped opening." Another group proposed that the East Sea has opened as a mega-pull-apart basin since the Late Oligocene in association with the southward translation of the Japan Arc along large primary shear zones (e.g. Jolivet et al. 1994). This model is referred to as 'pull-apart opening'. Yoon and Chough (1995) interpreted the Ulleung and Hupo faults along the western margin of the Ulleung Basin as strike-slip faults that guided the early (Late Oligocene–Early Miocene) pull-apart opening of the Ulleung Basin. Lee et al. (1999) postulated that the Ulleung Basin is a small pull-apart basin within a much large pull-apart system, the East Sea.

In spite of numerous studies on the opening of the East Sea, the nature and spatial configuration of rifting at its continental margins, where the early history of rifting to breakup was recorded, were poorly understood until recently. Kim et al. (2007) identified rift basins such as the Onnuri and Bandal basins on the South Korea Plateau (Fig. 17.3). The continental shelf south of the Korea Plateau is also incised along its length by the Hupo Basin (Fig. 17.3). While the Onnuri and Bandal basins create a relatively broad and symmetric profile bounded by major border and minor antithetic faults, the Hupo Basin shows a narrow half-graben profile (Kim et al. 2007). The border fault zones on the exterior seaward side constitute elongated structural highs that are interpreted as uplifted rift flanks (Kim et al. 2007). Therefore, the rift basins at the Korean margin demonstrate a fundamental unit of rift architecture. Crustal rifting occurs in a remarkably similar manner between intracontinental, incipient ocean spreading settings and passive continental margins (Wright 1997). The presence of rift basins and their uplifted flanks at the Korean margin indicates that the continental crust here was subjected to extension by normal faulting. Two directions of extension for rifting are recognized at the Korean margin: E-W for the Onnuri and Hupo basins and NW-SE for the Bandal Basin. A distinct unconformity divides the sedimentary fill of the Onnuri and Bandal basins into syn- and post-rift units (Lee et al. 2003).

The bathymetry of the southwestern East Sea shows that the South Korea Plateau and the Oki Bank of the SW Japan Arc are a pair of conjugate continental margins (Figs. 17.2 and 17.3), suggesting that the SW Japan Arc was separated in a southeastward direction (Kim et al. 2007). The bathymetry of the East Sea also suggests that the Yamato Bank was separated southeastward from mainland Asia. Therefore, the possible positions of the structural highs of continental fragments including the Korea Plateau, the Oki Bank, and the Yamato Bank before back-arc rifting can be inferred by moving them backward principally in a northwestward

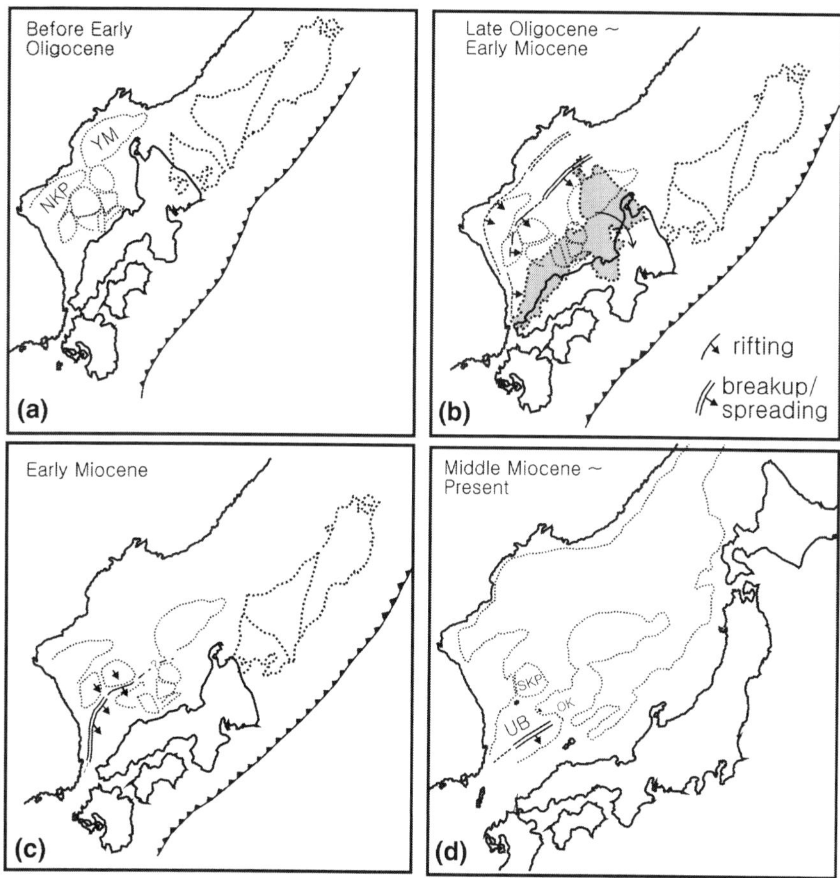

Fig. 17.6 **a** Positions of the Japan Arc and other tectonic units before back-arc rifting (Redrawn from Kim et al. 2007). **b** Back-arc rifting occurred along the eastern margin of Korea. While the Japan Basin is widened by seafloor spreading, a rift propagated southwestward through the Korea Plateau. **c** Breakup and initial seafloor spreading at the Korean margin. **d** Main opening phase of the Ulleung Basin and the present positions of tectonic units. The initial position of the NE Japan Arc was estimated from Otsuki (1992). *NKP* North Korea Plateau; *SKP* South Korea Plateau; *UB* Ulleung Basin; *YM* Yamato Bank; *OK* Oki Bank

direction (Fig. 17.6a; Kim et al. 2007). Data from Ocean Drilling Program drilling indicate that back-arc rifting in the East Sea commenced in the Late Oligocene in the northeastern part to form the Japan Basin (Tamaki et al. 1992). While back-arc spreading widened the Japan Basin, the eastern Korean margin experienced rifting in Late Oligocene to Early Miocene times. According to Kim et al. (2007), back-arc rifting at the Korean margin occurred in two stages. First, rifting propagated southwestward toward the Korean Peninsula during the opening of the Japan Basin but did not penetrate the strong lithosphere of the Korean Shield. The rifting then changed direction to the southwest to produce southeastward extension, creating the Bandal Basin (Figs. 17.3 and 17.6b). Then, the eastern Korean margin underwent extension in a southeastward direction orthogonal to the trench axis off the Japan Arc, responding to the subduction of the Pacific Plate. As a result, breakup occurred at the present slope base (Fig. 17.6c) and seafloor spreading continued to form the Ulleung Basin from the Early to Middle Miocene (Fig. 17.6c, d). Trench roll-back may have caused extension behind the arc and migration of spreading centers towards the trench. Considering that the southern portion of the eastern Korean margin is conjugate to the continental margin of the SW Japan Arc, a significant amount of clockwise rotation of the SW Japan Arc (Otofuji and Matsuda 1987) is likely to have occurred concurrently with the opening of the Ulleung Basin.

17.3.2 *Closure*

The East Sea back-arc closure began in the early Middle Miocene (Lee et al. 2001, 2011) as a result of the major plate reorganization in East Asia (Sibuet et al. 2002; Sdrolias et al. 2004). The regional plate reorganization, involving 13 plates, began in the middle Early Miocene (20 Ma) and culminated in the earliest Middle Miocene (ca. 15 Ma) (Sdrolias et al. 2004), resulting in a simple three-plate (Eurasian, Philippine, and Pacific) system (Sibuet et al. 2002). Coincident with the 15-Ma event is the acceleration in convergence of the Philippine Sea plate with the Amur plate along the Nankai Trough that led to the rapid subduction of young (<20 Ma) Shikoku Basin lithosphere under the SW Japan Arc (Sdrolias et al. 2004). The regional deformation of the peripheral regions of the East Sea (Ingle 1992), caused by the crustal shortening due to the back-arc closure, created a series of thrusts and anticlines. The deformation is particularly prominent along the southern margin of the East Sea (Itoh et al. 1997; Lee and Kim 2002; Lee et al. 2004, 2011).

The early Middle Miocene transition to a compressional tectonic regime was followed by Pliocene N-S trending inversion along the eastern margin of the East Sea off western Hokkaido and northeastern Honshu in Japan (Nakamura 1983; Tamaki and Honza 1985; Okamura et al. 1995; Taira 2001; Okui et al. 2008). This inversion, suggesting E-W compression, represents a nascent boundary of the Amur and Okhotsk plates (Fig. 17.1; Nakamura 1983; Kobayashi 1983; Tamaki and Honza 1985; Tokuyama et al. 1992). Various geological and geophysical studies, including GPS observations (e.g. Heki et al. 1999; Jin et al. 2007; DeMets et al. 2010), support the presence of the plate boundary or incipient subduction zone along the eastern margin of the East Sea. This young plate boundary runs from Sakhalin to the Itoigawa-Shizuoka Tectonic Line (Nakamura 1983; Seno et al. 1996) that divides the Japanese Island Arc into the NE Japan Arc and SW Japan Arc. However, because the E-W compression at this boundary cannot be accounted for by the subduction of the northwestward-moving Pacific and Philippine Sea plates, an eastward movement of the Amur plate has been suggested (Wei and Seno 1998; Heki 2007; Lee et al. 2011).

The recent study by Lee et al. (2011) reveals the detailed structural evolution of the southwestern margin of the Ulleung Basin (Fig. 17.7) since the onset of back-arc closure. The compressional deformation along the southwestern margin of the Ulleung Basin began in the earliest Middle Miocene, forming the NE-SW trending thrusts and anticlines that include the Dolgorae structures (Figs. 17.7, 17.8a, c). In the Late Miocene, the compressional deformation propagated or jumped to the Gorae structures (Figs. 17.7, 17.8a, b) in the southwest and then farther to the southwestern corner of the basin, creating an unnamed inversion structure (Fig. 17.7) in the Late Pliocene. The NE-SW trending Dolgorae structures indicate NW-SE or N-S compression, largely perpendicular to the direction of subduction of the Shikoku Basin lithosphere under the SW Japan Arc.

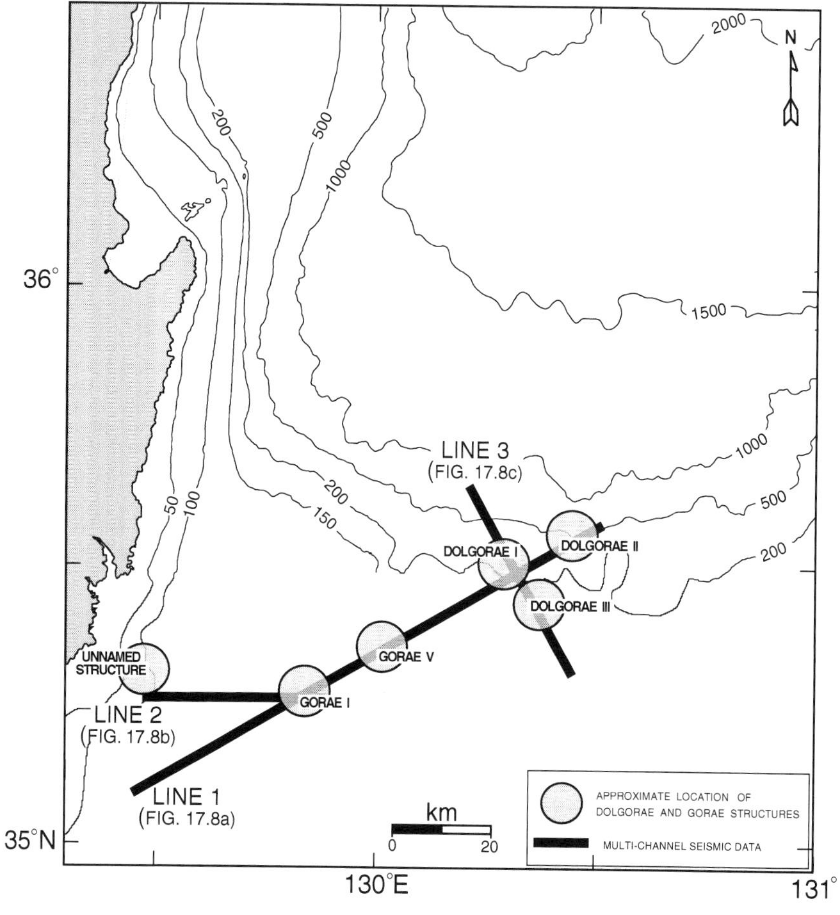

Fig. 17.7 Bathymetry of the southwestern Ulleung Basin and locations of seismic profiles shown in other figures and the Dolgorae *I*, *II* and *III*, Gorae *I* and *V*, and unnamed structures. Bathymetry in meters

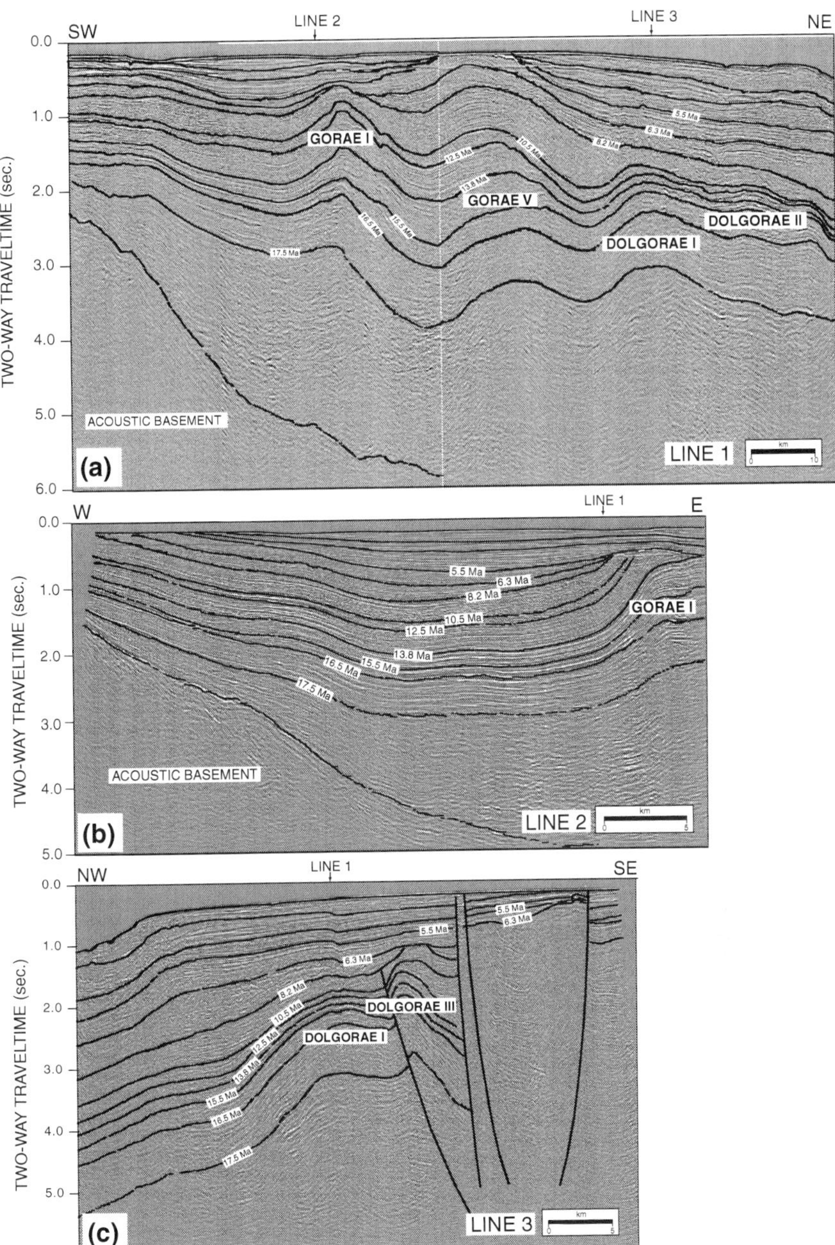

Fig. 17.8 Three regional seismic profiles from the southwestern Ulleung Basin margin. **a** *Line 1* traversing the shallow Gorae *I* and *V* structures and the deeper Dolgorae *I* and *II* structures. **b** *Line 2* traversing the shallow, unnamed structure in the southwestern corner of the Ullueng Basin and extending into the Gorae I structure. **c** *Line 3* traversing the Dolgorae *I* and *III* structures. See Fig. 17.7 for location

Much of the NE-SW trending Gorae I structure formed during the Late Miocene. Its orientation suggests NW-SE or N-S compression. On the other hand, the Gorae V structure trends NW-SE and extends orthogonally from the NE-SW trending thrust belt in the southeast, indicating NE-SW or E-W compression. Because the NW-SE or N-S compression still dominated in the area when the Gorae V structure began to form, the structure is likely to be the result of the superposition of the initial NW-SE or N-S compression and the ensuing, more significant NE-SW or E-W compression that began after 5.5 Ma. The youngest, unnamed inversion structure in the southwestern corner of the Ulleung Basin was also affected mostly by E-W compression.

The onset of compressional deformation along the southwestern margin of the Ulleung Basin at around 15 Ma coincides with the culmination of the regional plate reorganization in East Asia (Lee et al. 2011). The northeastward moving Shikoku Basin lithosphere also made full contact with the SW Japan Arc at around 15 Ma (Hibbard and Karig 1990; Sdrolias et al. 2004). The emplacement of felsic to intermediate magmas in the SW Japan Arc (12–15 Ma) (Kobayashi 1983) and the eruption of unusually high magnesium andesites in the area (13–15 Ma) (Furukawa and Tatsumi 1999) support the arrival of the Shikoku Basin at the SW Japan Arc in the Middle Miocene. The convergence of young, hot and buoyant oceanic lithosphere with continental lithosphere may induce low-angle or flat subduction, causing strong interplate coupling (Uyeda 1991; Gutscher 2001a, b) that leads to tectonic switching from extension to compression along the margin of the overriding plate (Gutscher et al. 2000; Collins 2002). The Nankai Trough is known for flat subduction with a shallow dip angle of about 10° (Jarrard 1986; Gutscher et al. 2000; Gutscher 2001a; Nakajima and Hasegawa 2007). The initiation of back-arc closure in the southern East Sea may be due to the tectonic switching from extension to compression with the arrival of the Shikoku Basin lithosphere at the SW Japan Arc at the culmination of the regional plate reorganization (Lee et al. 2011).

The abrupt change in the stress regime from NW-SE or N-S to NE-SW or E-W compression in the southwestern margin of the East Sea is probably related to a change in regional-scale plate geodynamics. The present-day stress field in East Asia, interpreted from focal mechanism solution of recent earthquakes, shows a strong eastward component (Fig. 17.9; Tamaki and Honza 1985; Xu et al. 1992; Zoback 1992; Ree et al. 2003; Park et al. 2006; Choi et al. 2012). Lee et al. (2011) postulated that the NE-SW or E-W compression that occurred after 5.5 Ma in the Gorae area and sometime in the Pliocene in the unnamed inversion structure in the southwestern corner of the Ulleung Basin is due to the eastward-moving Amur plate. The movement of the Amur plate may be related to far-field stresses imposed by the India-Eurasia collision (Park et al. 2006) or the reactivation of rifting in the Baikal zone (Taira 2001).

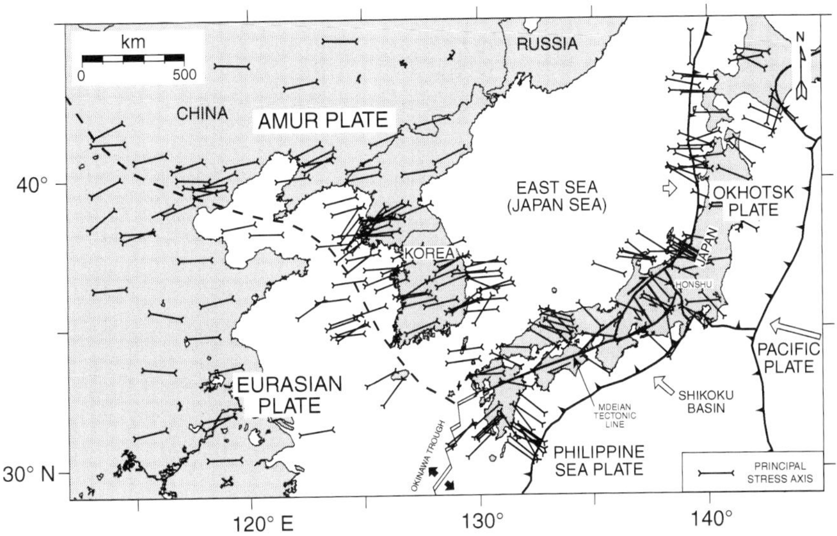

Fig. 17.9 Present-day stress field in East Asia, based on focal mechanism solution of recent earthquakes, showing a strong eastward component (Redrawn from Tamaki and Honza 1985; Xu et al. 1992; Zoback 1992; Ree et al. 2003; Park et al. 2006; Choi et al. 2012)

Acknowledgements Partial support for this work was provided by the project of 'International Ocean Discovery Program' funded by the Ministry of Oceans and Fisheries (Korea) and the project funded by the Center Atmospheric Sciences and Earthquake Research (Grant CATER 2012-8100). The authors thank a reviewer for improving the manuscript.

References

Choi H, Hong TK, He X et al (2012) Seismic evidence of reverse activation of a paleo-rifting system in the East Sea (Sea of Japan). Tectonophysics 572–573:123–133

Chung TW, Hirata N, Sato R (1990) Two-dimensional P and S wave velocity structure of the Yamato basin, the southwestern Japan Sea, from refraction data collected by ocean bottom seismographic array. J Phys Earth 38:99–146

Collins WJ (2002) Hot orogens, tectonic switching, and creation of continental crust. Geology 30:535–538

DeMets C, Gordan R, Argus D (2010) Geologically current plate motions. Geophys J Int 181:1–80

Furukawa Y, Tatsumi Y (1999) Melting of a subducting slab and production of high-Mg andesite magmas: unusual magmatism in SW Japan at 13-15 Ma. Geophys Res Lett 26:2271–2274

Gutscher MA (2001a) An Andean model of interplate coupling and strain partitioning applied to the flat subduction zone of SW Japan (Nankai Trough). Tectonophysics 333:95–109

Gutscher MA (2001b) Andean subduction styles and their effect on thermal structure and inter-polate coupling. J S Am Earth Sci 15:3–10

Gutscher MA, Sparkman W, Bijwaard H et al (2000) Geodynamics of flat subduction: seismicity and tomographic constraints from the Andean margin. Tectonics 19:814–833

Hashizume M, Matsui Y (1979) Crustal structure of southwestern Honshu, Japan, derived from explosion seismic waves. Geophys J Roy Astr S 58:181–189

Heki K (2007) Secular, transient, and seasonal crustal movements in Japan from a dense GPS array: Implications for plate dynamics in convergent boundaries. In: Dixon TH, Moore JC (eds) The seismogenic zone of subduction thrust faults. Columbia University Press, New York, pp 512–539

Heki K, Miyazaki S, Takahashi H et al (1999) The Amurian Plate motion and current plate kinematics in eastern Asia. J Geophys Res 104:29147–29155

Hibbard JP, Karig DE (1990) Alternative plate model for the early Miocene evolution of the southwest Japan margin. Geology 18:170–174

Hirata N, Karp BY, Yamaguchi T et al (1992) Oceanic crust in the Japan basin of the Japan Sea by 1990 Japan-USSR expedition. Geophys Res Lett 19:2027–2030

Ingle JC Jr (1992) Subsidence of the Japan Sea: stratigraphic evidence from ODP sites and onshore sections. In: Tamaki K, Suyehiro K, Allan J et al (eds) Proceedings of Ocean Drilling Program, scientific results, vol 127/128. Part 2, Ocean Drilling Program, College Station, Texas, pp 1197–1218

Itoh Y, Nakajima T, Takemura A (1997) Neogene deformation of the back-arc shelf of Southwest Japan and its impact on the palaeoenvironments of the Japan Sea. Tectonophysics 281:71–82

Jarrard RD (1986) Relations among subduction parameters. Rev Geophys 24:217–284

Jin S, Park P, Zhu W (2007) Micro-plate tectonics and kinematics in Northeast Asia inferred from a dense set of GPS observations. Earth Planet Sc Lett 257:486–496

Jolivet L, Tamaki K, Fournier M (1994) Japan Sea, opening history and mechanism; a synthesis. J Geophys Res 99:22237–22259

Kim HJ, Han SJ, Lee GH et al (1998) Seismic study of the Ulleung Basin crust and its implications for the opening of the East Sea (Japan Sea). Mar Geophys Res 20:219–237

Kim HJ, Jou HT, Cho HM et al (2003) Crustal structure of the continental margin of Korea in the East Sea (Japan Sea) from deep seismic sounding data: evidence for rifting affected by the hotter than normal mantle. Tectonophysics 364:25–42

Kim HJ, Lee GH, Jou HT et al (2007) Evolution of the eastern margin of Korea: Constraints on the opening of the East Sea (Japan Sea). Tectonophysics 436:37–55

Kobayashi K (1983) Cycles of subduction and Cenozoic Arc activity in the Northwestern Pacific margin. In: Hilde TWC, Uyeda S (eds) Geodynamics of the Western Pacific-Indonesian region. US Government Printing Office, Washington, pp 287–302

Kurashimo E, Shinohara M, Suyehiro K et al (1996) Seismic evidence for stretched continental crust in the Japan Sea. Geophys Res Lett 23:3067–3070

Lee GH, Kim B (2002) Infill history of the Ulleung Basin, East Sea (Sea of Japan) and implications on source rocks and hydrocarbons. Mar Petrol Geol 19:829–845

Lee GH, Kim HJ, Suh MC et al (1999) Crustal structure, volcanism, and opening mode of the Ulleung Basin, East Sea (Sea of Japan). Tectonophysics 308:503–525

Lee GH, Kim HJ, Han SJ et al (2001) Seismic stratigraphy of the deep Ulleung Basin in the East Sea (Japan Sea) back-arc basin. Mar Petrol Geol 8:615–634

Lee GH, Kim HJ, Jou HT et al (2003) Opal-A/opal-CT phase boundary inferred from bottom-simulating reflectors in the southern South Korea Plateau, East Sea (Sea of Japan). Geophys Res Lett 30:2238. doi:10.1029/2003GL018670

Lee GH, Kim B, Chang SJ et al (2004) Timing of trap formation in the southwestern margin of the Ulleung Basin, East Sea (Japan Sea) and implications for hydrocarbon accumulations. Geosci J 8:369–380

Lee GH, Yoon YH, Nam B et al (2011) Structural evolution of the southwestern margin of the Ulleung Basin, East Sea (Japan Sea) and tectonic implications. Tectonophysics 502:293–307

Ludwig WJ, Murauchi S, Houtz RE (1975) Sediments and structure of the Japan Sea. Geol Soc Am Bull 86:651–664

Nakajima J, Hasegawa A (2007) Tomographic evidence for the mantle upwelling beneath southwestern Japan and its implications for arc magmatism. Earth Planet Sc Lett 254:90–105

Nakamura K (1983) Possible nascent trench along the eastern Japan Sea as the convergent boundary between Eurasian and North American Plates. Bull Earthq Res Inst Univ Tokyo 58:711–722 (in Japanese)

Okamura Y, Watanabe M, Morijiri R et al (1995) Rifting and basin inversion in the eastern margin of the Japan Sea. Isl Arc 4:166–181

Okui A, Kaneko M, Nakanishi S et al (2008) An integrated approach to understanding the petroleum system of a frontier deep-water area, offshore Japan. Petrol Geosci 14:223–233

Otofuji Y, Matsuda T (1987) Amount of rotational motion of southwestern Japan: Fan shape opening of the southwestern part of the Japan Sea. Earth Planet Sc Lett 85:289–301

Otsuki K (1992) Oblique subduction, collision of microcontinents and subduction of oceanic ridge: their implications on Cretaceous tectonics of Japan. Isl Arc 1:51–63

Park Y, Ree JH, Yoo SH (2006) Fault slip analysis of Quaternary faults in southeastern Korea. Gondwana Res 9:118–125

Purdy GM, Ewing J (1986) Seismic structure of the oceanic crust. In: Vogt PR, Tucholke BE (eds) The geology of North America, vol M. Geological Society of America. The Western North Atlantic Region, pp 313–330

Ree JH, Lee YJ, Rhodes EJ et al (2003) Quaternary reactivation of Tertiary faults in southeastern Korean Peninsula and age constraint of faulting by optically stimulated luminescence dating. Isl Arc 12:1–12

Rohr KM (1994) Increase of seismic velocities in upper oceanic crust and hydrothermal circulation in the Juan de Fuca plate. Geophys Res Lett 21:2163–2166

Sdrolias M, Roest WR, Müller RD (2004) An expression of Philippine Sea plate rotation: the Parece Vela and Shikoku Basins. Tectonophysics 394:69–86

Seama N, Isezaki N (1990) Sea-floor magnetization in the eastern part of the Japan Basin and its tectonic implications. Tectonophysics 181:285–297

Seno T, Sakurai T, Stein S (1996) Can the Okhotsk plate be discriminated from the North American plate? J Geophys Res 101:11305–11315

Shinohara M, Hirata N, Nambu M et al (1992) Detailed crustal structure of northern Yamato basin. In: Tamaki K, Suyehiro K, Allan J et al (eds) Proceedings of ODP Scientific Results Scientific Results, Vol. 127/128, part 2. Ocean Drilling Program, College Station, Texas, pp 1075–1106

Sibuet JC, Hsu SK, Le Pichon X et al (2002) East Asia plate tectonics since 15 Ma: constraints from the Taiwan Region. Tectonophysics 344:103–134

Su W, Mutter CZ, Mutter JC et al (1994) Some theoretical predictions on the relationships among spreading rate, mantle temperature, and crustal thickness. J Geophys Res 99:3215–3227

Taira A (2001) Tectonic evolution of the Japanese Island Arc system. Annual Rev Planet Sci 29:109–134

Tamaki K, Honza E (1985) Incipient subduction and obduction along the eastern margin of Japan Sea. Tectonophysics 119:381–406

Tamaki K, Suyehiro K, Allan J et al (1992) Tectonic synthesis and implications of Japan Sea ODP drilling. In: Tamaki K, Suyehiro K, Allan J et al (eds) Proceedings of ODP Scientific Results Scientific Results, Vol 127/128, Part 2. Ocean Drilling Program, College Station, Texas, pp 1333–1348

Tokuyama H, Kuramoto S, Woon S et al (1992) Initiation of ophiolite emplacement: a modern example from the Okushiri Ridge, Northeast Japan Arc. Mar Geol 103:323–334

Tosha T, Hamano Y (1988) Paleomagnetism of Tertiary rocks from the Oga Peninsula and the rotation of Northeast Japan. Tectonics 7:653–662

Uyeda S (1991) The Japanese Isl Arc and the subduction process. Episodes 14:190–198

Wei D, Seno T (1998) Determination of the Amurian plate motion. In: Flower MFJ, Chung SL, Lo CH et al (eds) Mantle dynamic and plate interactions in East Asia. Geodynamics series, vol 27. American Geophysical Union, Washington, pp 337–346

Wright IC (1997) Morphology and evolution of the Remnant Colville and Active Kermadec Arc Ridge south of 33°30'S. Mar Geophys Res 19:177–193

Xu Z, Wang S, Huang Y et al (1992) Tectonic stress field of China inferred from a large number of small earthquakes. J Geophys Res 97:11867–11877

Yoon SH, Chough SK (1995) Regional strike slip in the eastern continental margin of Korea and its tectonic implications for the evolution of Ulleung Basin, East Sea (Sea of Japan). GSA Bulletin 107:83–97

Zoback MD (1992) First- and second-order patterns of stress in the lithosphere: the world stress map project. J Geophys Res 97:11703–11728

Zonenshain LP, Savostin LS (1981) Geodynamics of the Baikal rift zone and plate tectonics of Asia. Tectonophysics 76:1–45

Chapter 18
Stratigraphy

Seok Hoon Yoon

Abstract This section mainly focuses on the stratigraphy of the Ulleung Basin and its margins, where large amounts of data are available as Korea has committed substantial resources over the years. Although the deep basin is yet to be drilled, there has been extensive exploration, including drilling, elsewhere, particularly along the southwestern margin of the basin. We review the sedimentary sequences in terms of the seismic-stratigraphic framework; for basic concepts and technical terminology of seismic stratigraphy, refer to Mitchum et al. (Seismic stratigraphy—applications to hydrocarbon exploration. AAPG Memoir, pp. 117–133, 1977).

Keywords Seismic stratigraphy · Basin fillings · Depositional environment · Ulleung Basin · East Sea (Japan Sea)

18.1 Ulleung Basin

18.1.1 Acoustic Basement

The acoustic basement in the central part of the Ulleung Basin (Fig. 16.2) occurs mainly at a depth of 5 km below sea level (bsl) and progressively deepens southward to more than about 11 km bsl (Fig. 18.1). In seismic profiles, the top of the acoustic basement is generally identified by discontinuous, high-amplitude, parallel reflectors with a low relief and smooth surface (ca. 4.7 km/s in interval velocity). In some places, the basement has a mound configuration that shows lateral variation in amplitude, frequency, and continuity of reflection (Fig. 18.1). These

S.H. Yoon (✉)
Department of Earth and Marine Sciences, Jeju National University, Jeju 690-756, Republic of Korea
e-mail: shyoon@jejunu.ac.kr

© Springer International Publishing Switzerland 2016
K.-I. Chang et al. (eds.), *Oceanography of the East Sea (Japan Sea)*,
DOI 10.1007/978-3-319-22720-7_18

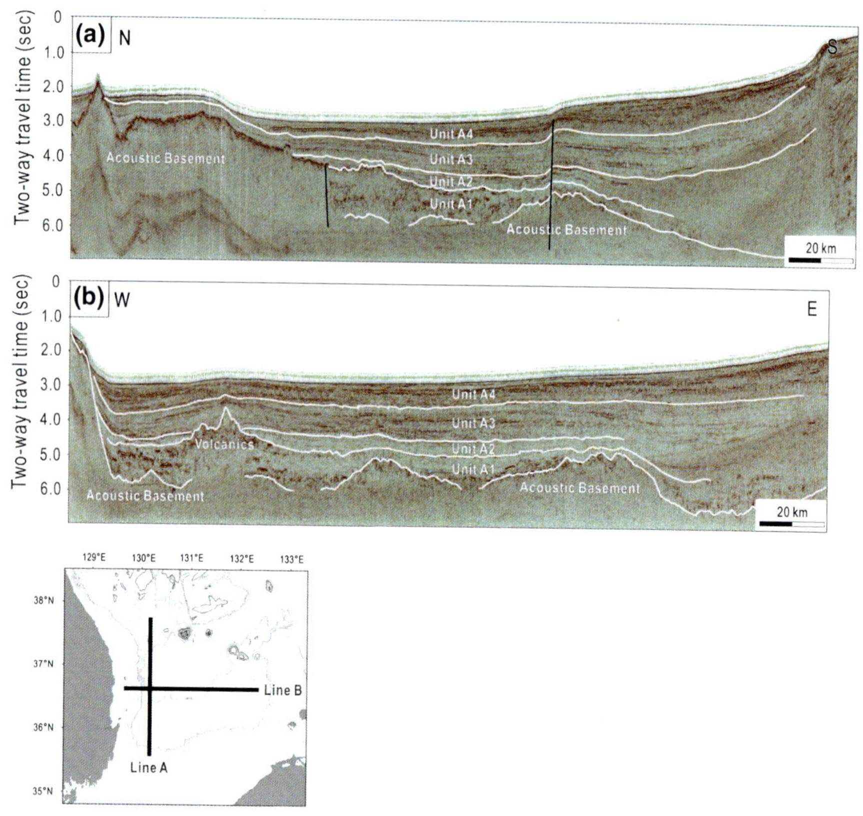

Fig. 18.1 Seismic profiles crossing the Ulleung Basin with stratigraphic interpretation

seismic reflection patterns most likely represent widespread volcanic materials, including sills and lava flows (Chough and Lee 1992). On the other hand, the acoustic basement in the northern part of the basin appears to be more or less ambiguous; the overlying sequence unit is highly reflective and shows discontinuous, variable-amplitude, parallel reflections that are interpreted as volcanic flows and sills interlayered with clastic sediments (Chough et al. 2000).

18.1.2 Sedimentary Sequences

Thickness distribution of the sedimentary section overlying the acoustic basement shows two depocenters separated by an intervening ENE-trending median high with thinner basin fill (Fig. 18.2; Lee et al. 2001). The basin fill is thickest (>11 km) in the southern depocenter near the southern margin of the basin and thins northward across the median high, exhibiting an overall wedge-like

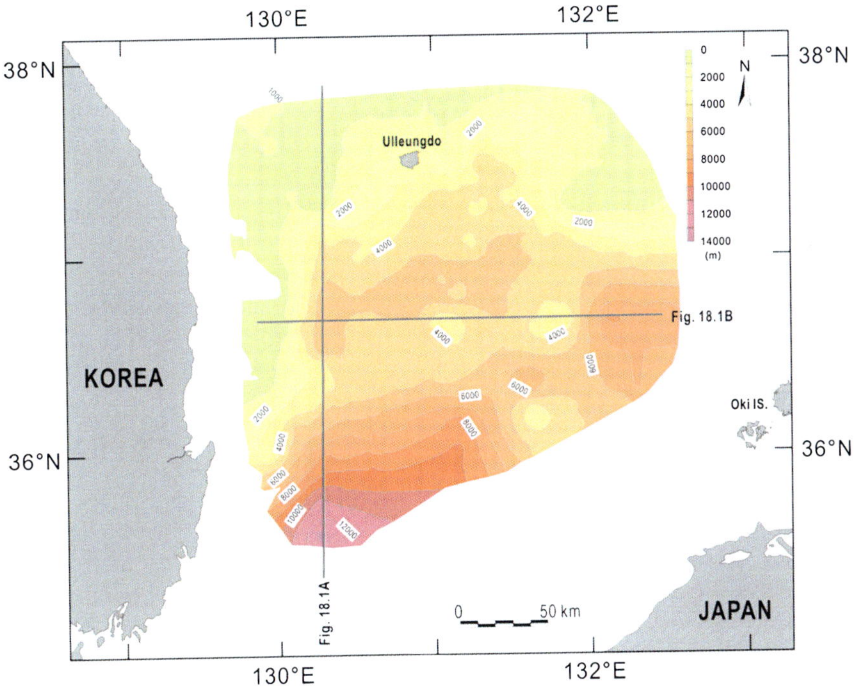

Fig. 18.2 Total sediment thickness between the top of the acoustic basement and the seafloor. Contour interval is 1000 m (from Lee et al. 2001)

geometry (Fig. 18.1). In the northern depocenter, the total sediment thickness is more than 5 km. The seismic stratigraphic framework in the Ulleung Basin proper was first presented by Lee (1992) and Chough and Lee (1992) based on the reflection configuration of multichannel seismic profiles provided by Korea National Oil Corporation. Lee et al. (2001) reinterpreted the same data for seismic stratigraphy. In this section, the sedimentary sequences are discussed based on the stratigraphic framework of Lee (1992) and Chough and Lee (1992), which consists of four seismic sequence units (here renamed A1, A2, A3 and A4 in ascending order). The geologic ages of the unit boundaries were estimated from correlation with the sequence boundaries defined in the southwestern basin margin where the ages are constrained by biostratigraphic data from some hydrocarbon exploratory wells (Lee 1994).

18.1.2.1 Unit A1

Unit A1 is the lowermost unit of the sequence. It is Early Miocene and ranges in thickness from 0.4 to 1.2 s in two-way travel time (twt) with an interval velocity of 3.6–4.8 km/s (Fig. 18.1; Chough et al. 2000). Acoustically, this unit is characterized by variable-amplitude, parallel to shingled reflections with poor lateral continuity in the northern part of the basin, and low-amplitude, parallel reflections in the southern part. Chough and Lee (1992) and Lee et al. (2001) interpreted the seismic facies in the north as volcanic flows and sills intercalating with sedimentary layers, and the facies in the south as bedded clastic or volcaniclastic debris. The transition zone between the acoustic basement and the overlying lowermost sedimentary sequence is also characterized by high-amplitude, discontinuous reflections, probably due to an abrupt contrast in acoustic impedance between the volcanic material and the sediments (Chough et al. 2000). In some places near the basement highs, Unit A1 exhibits onlap termination and/or internal convergence, suggesting that the basement highs existed prior to the deposition of the unit but continued to rise for some time after deposition began (Lee et al. 2001).

18.1.2.2 Unit A2

Unit A2, mainly comprising the Middle Miocene sediments, is relatively uniform in thickness, ranging from about 0.4 to 0.6 s twt in the central part of the basin. It progressively thickens southward, indicating a significant contribution of sediments from the southern margin of the Ulleung Basin (Fig. 18.1; Lee et al. 2001). In seismic reflection profiles, this unit is generally characterized by low-frequency, variable-amplitude, parallel reflections with poor-to-fair lateral continuity (Chough and Lee 1992). In the western margin of the basin, reflectors in the lower part of the unit show onlap onto the basement high. The unit also includes mound configurations and inclined reflectors downlapping on the lower unit boundary, which indicates a significant influx of volcaniclastics from the basin margins during the deposition of Unit A2 (Chough et al. 2000). In the northeastern part of the basin, Unit A2 is characterized by two distinct reflection configurations: low-amplitude, discontinuous parallel reflections in the lower part and high-amplitude, discontinuous parallel reflections in the upper part. The former is interpreted as massive sandstones or non-bedded volcaniclastics, and the latter as turbidites (Chough and Lee 1992).

18.1.2.3 Unit A3

Unit A3 is more than 2.0 s twt thick in the southern part of the basin, and progressively thins northward to less than about 0.5 s twt in the northern margin of the basin. Two zones of distinct seismic facies are recognized: low-to-moderate or variable amplitude, poor-to-low continuity reflections with hummocky and/ or structureless to chaotic zones in the southern part of the basin, and low-to-high amplitude, moderate-to-good continuity reflections in central part of the basin (Lee et al. 2001). The former is interpreted as stacks of mass-transport deposits formed by slides, debris flows and high-density turbidity currents derived mainly from the southern margin of the basin. As the mass flows approached the basin center (i.e. northward), they most likely transformed into low-density turbidity currents, resulting in widespread deposition of interbedded fine-grained turbidites and hemipelagites with reflection configurations of better continuity (Lee et al. 2001). During the later stage of Unit A3 deposition, the main sediment source region shifted from the southeast to the southwest, and the depocenter of mass-flow deposits extended farther basinward (Lee et al. 2001). In the upper part of Unit A3, normal faults are identified, which are probably related to volcanic intrusions that have disrupted and displaced the sediments (Lee et al. 2001). Chough and Lee (1992) correlated Unit A3 with the upper Miocene and lower Pliocene successions of Dolgorae-I well in the southern margin of the basin, whereas Lee et al. (2001) suggested the deposition in the late Middle Miocene to middle Late Miocene time.

18.1.2.4 Unit A4

Unit A4, the uppermost sequence, is characterized either by high-amplitude parallel reflections or by choppy reflections. In the central part of the basin, the unit is relatively uniform in thickness (0.6–0.8 s twt). The continuous parallel reflectors indicate interbedded hemipelagic and terrigenous sediments; the choppy reflections represent volcanic ash mixed with hemipelagic sediments (Chough and Lee 1992). Lee and Suk (1998) divided this sequence interval into five subunits, within which they distinguished five seismic facies on the basis of seismic reflection configuration and geometry. The lower two subunits are typically dominated by structureless-to-chaotic internal reflections without distinct external form which are interpreted as mass-flow deposits. Thickness variations and seismic facies distribution suggest that these sediments were derived mainly from the western and southern margins of the Ulleung Basin. In the upper three subunits, the occurrence of mass-flow deposits with structureless-to-chaotic reflections is limited to the base-of-slope area, suggesting a marked decrease in the generation of mass movements. The upper subunits are dominated by well-stratified, continuous reflectors that are interpreted as mixed deposits of turbidites and hemipelagites. Near the axis of the Ulleung Interplain Gap, low-relief undulating mounds suggest the occasional influence of bottom currents. The geological age of Unit A4 is still controversial. Chough and Lee (1992) suggested late Pliocene and Quaternary, whereas Lee et al. (2001) assigned middle to late Late Miocene ages to the lower part of the unit.

18.2 South Korea Plateau

18.2.1 Acoustic Basement

The basement topography of the South Korea Plateau (Fig. 16.2) appears different in its western and eastern parts. In the western South Korea Plateau, the acoustic basement is characterized by a highly indented topography at depths of

0.8–3.8 s twt bsl, and comprises a series of NNE–SSW trending, subparallel and elongate ridges and troughs controlled by normal faults, showing a typical graben-and-horst structure (Figs. 18.3 and 18.4; Kim et al. 2011). The bounding faults generally exhibit slightly curving fault planes, and vertical displacement reaches a few hundreds of meters. The basement highs (0.8–2.5 s twt) include seamounts and elongate ridges scattered in the plateau area. On the top of the basement underlying the troughs, various elevated topographic features such as hills, knolls and peaks rising generally less than 500 m can be identified. In the eastern South Korea Plateau, however, there is no significant topographic indentation except for the deep fault-controlled valley at the southern margin of the block (Kim et al. 2011). The fault lines generally trend ENE–WSW, which distinguishes them from

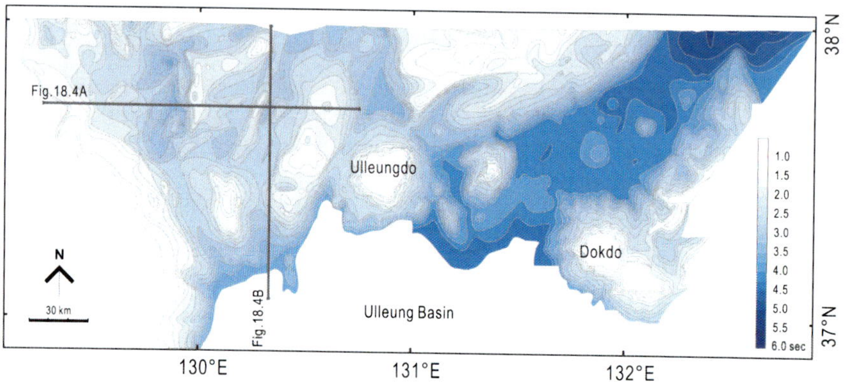

Fig. 18.3 Time-structure map of the top of the acoustic basement in the South Korea Plateau. Contours are in seconds (two-way travel time below sea level) (from Kim et al. 2011)

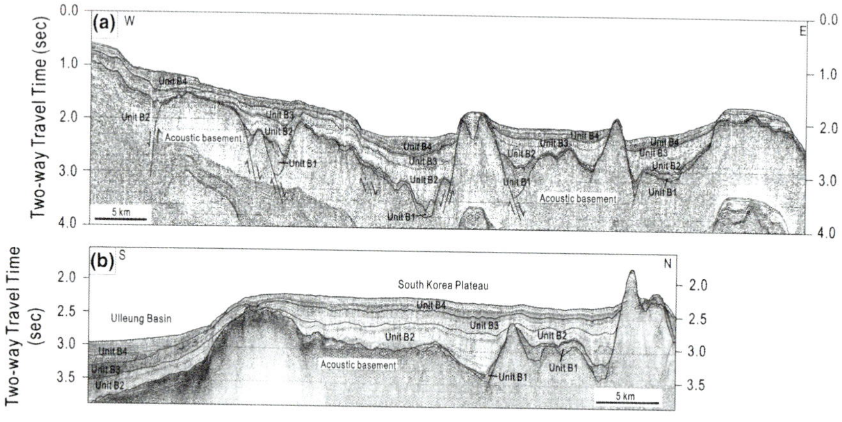

Fig. 18.4 Interpreted seismic profiles showing distribution of stratigraphic units in the South Korea Plateau (from Kwon et al. 2009). For locations, see Fig. 18.3

the NNE–SSW trending normal faults in the western part of the plateau. The depth of the acoustic basement in this region ranges from 0.8 to 3.5 s twt bsl.

18.2.2 Sedimentary Sequences

A seismic stratigraphic framework for the South Korea Plateau was first presented by KORDI (1999), which suggested three seismic sequence units with ages of Early Miocene to the present. Kwon (2005) presented a stratigraphic framework consisting of nine seismic units based on reflection character, coastal onlapping and erosional truncation. This work was recently revised by Kwon et al. (2009), giving a stratigraphic framework of four seismic sequence units (renamed Unit B1, B2, B3 and B4 in ascending order) (Fig. 18.4). In this section, the stratigraphy is described based on this new framework.

18.2.2.1 Unit B1

Unit B1, the lowermost seismic unit, is generally restricted to isolated basement lows. It has an overall wedge form with thickness varying from about 0.05 to more than 0.2 s twt, depending on the basement topography (Fig. 18.4). This unit is characterized by a wide range of reflection configurations, including short and tangled high-amplitude reflectors, contoured subparallel reflectors of low to moderate amplitude, and chaotic to transparent reflections. These features may suggest syn-rift localized deposition in isolated rift basins prior to marine incursion. Short and tangled high-amplitude reflectors are similar to the high-amplitude reflection package overlying the acoustic basement in the Yamato and Ulleung basins, which is interpreted as a volcanic sill/flow–sediment complex (Lee et al. 2001). Unit B1 is overlain by an erosional surface of high-amplitude reflections with moderate continuity. On the other hand, the fill strata show subparallel reflectors with low to moderate amplitudes; this most likely suggests more or less uniform deposition in a lacustrine environment with a fluvial input of sediments. The syn-rift sedimentary successions, consisting of volcanic rocks and tuff often associated with fan-delta and lacustrine sediments, crop out in northern Honshu in Japan (Sato and Amano 1991). Unit B1 is overlain by an erosional surface of high-amplitude reflections with good to moderate continuity. This suggests a long period of subaerial exposure prior to marine incursion.

18.2.2.2 Unit B2

Unit B2 occurs in the basement troughs. It has a maximum thickness of ca. 1.0 s twt. It is mostly characterized by stratified reflectors of low to intermediate amplitude, and partly by chaotic or hummocky to reflection-free or transparent configurations with low amplitudes and poor lateral continuity. The stratified reflectors commonly show onlap termination onto the basement highs, but in the deep troughs, individual strata drape over underlying small-scale irregularities. This indicates turbidites and hemipelagic sediments filling irregular topography. The chaotic or reflection-free strata appear as isolated sediment masses or parts of stratified strata of variable dimensions, and occasionally have hummocky or irregular upper surfaces. This indicates disorganized or remolded internal structures that formed by mass flows including slide/slump and debris flows. In the western part of the plateau, a sharp discordant interface is recognized within Unit B2; it approximately coincides with the boundary between the lower transparent or chaotic reflection and the upper stratified reflection. This interface represents an opal A/CT phase transition associated with chert diagenesis (Lee et al. 2003). The upper boundary of Unit B2 corresponds approximately to the early Late Miocene chronostratigraphic horizon identified in the Ulleung Basin (Lee et al. 2001).

18.2.2.3 Unit B3

Unit B3 has sheet-drape geometry with a relatively uniform thickness ranging from 0.1 to 0.3 s twt. It can be divided into two subunits based on seismic characteristics. The lower subunit is characterized by transparent to chaotic reflections with basal onlap-fill facies, while the upper subunit shows well-stratified reflectors with moderate to high amplitudes and good continuity (Fig. 18.4). The facies change from the lower transparent or chaotic reflections to the upper well-stratified reflections suggests a change from a depositional environment dominated by mass flows to one dominated by turbidity current/hemipelagic processes. Along the western margin of the plateau, sedimentary successions of Unit B2 and Unit B3 are locally compressed, giving rise to anticlinal folds and reverse faults. Some compressional structures seem to have developed in response to the reactivation of pre-existing extensional basement faults. Unit 3 corresponds approximately to the Late Miocene to early Early Pliocene sequence identified along the eastern Korean continental margin (Yoon and Chough 1995).

18.2.2.4 Unit B4

Unit B4 shows a sheet-draping reflector of fairly uniform thickness of 0.1–0.2 s twt (Fig. 18.4). The lower boundary of the unit is defined by a progressive onlap termination against the apex of the anticlinal folding. This unit comprises well-stratified continuous reflectors with high to moderate amplitude, which is suggestive of turbidites and hemipelagic sediments deposited in a deep marine environment (Lee and Suk 1998). Sediments of two drill cores (25 and 28 m long) that were recovered from the upper part of Unit B4 on the gentle slope comprise thin (<1 m) layers of bioturbated or laminated muds and massive sands (Kwon 2005). In some topographic lows, the upper part of the unit occurs as sediment fill with divergent patterns. On the other hand, along the gently sloping western margin of the plateau, sharp scars of large-scale slope failures are recognized, and blocky and lumpy masses of hyperbolic or chaotic reflectors are embedded downslope. The K/Ar-based ages of the tephra in the sediment cores indicate that the deposition of Unit B4 started prior to 2.7 Ma, probably in the middle Pliocene (Kwon 2005).

18.3 Southwestern Margin of the Ulleung Basin

The southwestern margin of the Ulleung Basin lies on the triple junction of a continent (the Korean Peninsula), a volcanic island arc (the Japanese Islands) and a deep-sea basin (the Ulleung Basin) (Fig. 16.1). An interaction among these tectonic provinces during the back-arc evolution of the East Sea is primarily responsible for the complex stress fields that caused a thick sedimentary succession to accumulate at this margin (Chough and Barg 1987; Park 1990). Since the early 1980s, due to high prospective hydrocarbon potential, the margin has been intensively probed with several exploratory wells (Dolgorae and Gorae series) and tens of thousands of kilometers of seismic profiling conducted by Korea National Oil Corporation. This abundant data provides an opportunity for insight into the more detailed stratigraphy of the margin.

18.3.1 Acoustic Basement

The southern part of the Ulleung Basin is occupied by a broad continental shelf and gentle slope (Fig. 16.2). Deep seismic profiling and gravity measurements delineate a deep, narrow, hinged depression deepening toward the east and northeast (Park 1992). The depression in the northeast is filled with a thick (ca. 8–10 km in thickness) succession of Early Miocene to Recent sediments (Barg 1986; Park 1992).

18.3.2 Sedimentary Sequences

Multi-channel seismic surveys and chronostratigraphic work on exploratory wells (Lee 1994) in the southwestern margin reveal that the present shelf and slope region is underlain by thick (>8 km in thickness) Tertiary sediments within a narrow, fan-shaped depression (Barg 1986; Park 1990, 1998). This Tertiary basin-margin fill has a recurrent progradational stacking pattern as a result of the N to NE advance of the shelf-slope system since the Early Miocene (Fig. 18.5). Yoon et al. (2002) identified ten seismic sequence units (here renamed Unit C1 to C10 in ascending order) from the Tertiary basin-margin fill. The unit boundaries are

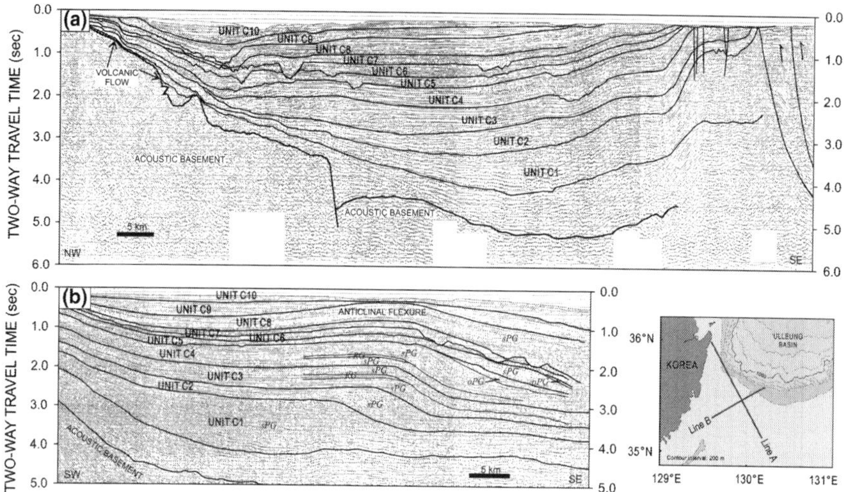

Fig. 18.5 Interpreted seismic profiles showing overall distribution of stratigraphic units in the southwestern margin of the Ulleung Basin (from Yoon et al. 2003)

defined in nearshore seismic profiles by the occurrence of erosional truncation and/or coastal onlap. The chronostratigraphic framework is primarily provided by biostratigraphic analysis of the Dolgorae-I well (Lee 1994), in which the age determination was mainly based on foraminiferal zonation (Table 18.1).

18.3.2.1 Unit C1 (Early Miocene?)

In the western part of the margin, Unit C1 appears to fill the basement lows, showing divergent or onlap fill configuration (Fig. 18.5). Toward the basin center, the sequence progressively thickens up to 1.5 s twt forming a monotonous ramp margin with a simple aggradational stacking pattern. Reflection characters are characterized by subparallel and discontinuous configurations with variable amplitude and moderate to high frequency. In the basin, this sequence consistently shows a sheet-drape geometry without any significant variation in reflection configuration. The age and depositional environment of this unit are still uncertain.

18.3.2.2 Units C2 and C3 (Early to Middle Miocene)

Units C2 and C3 are characterized by prograding clinoform geometry with a relatively high ratio of progradation to aggradation (Fig. 18.5). During the deposition of these sequences, the shelf break prograded more than 30 km northeastward and rose vertically by up to 0.7 s twt. In the nearshore profiles, the sequence boundaries are defined by onlap-bounding surfaces, and their upper boundaries

Table 18.1 Stratigraphic correlation of Tertiary sedimentary successions in the southwestern part of the Ulleung Basin

Age(Ma)		Lee (1994)		Park (1998)		Yoon et al. (2002)		Jung and Hwang, (2008)		
		Foraminiferal zonation	Environ-ment	Depositional sequence unit	Age (Ma)	Depositional sequence unit	Age (Ma)	Depositional sequence unit		Age (Ma)
Quaternary		*Globorotalia truncatulinoides*	Neritic	VIII		X	1.6	VI	VId / VIc / VIb	
Pliocene L		*Cassidulina laevigata*			3.8			V	VIa / Vb	3.0 / 3.8 / 4.2
Pliocene E				VII	5.3		5.5	IV	Va / IVc / IVb	
		Rare zone		VI	6.3	IX	6.3		IVa	5.5 / 6.3
Miocene Late		Barren zone	Coastal	V		VIII	8.2	III		8.2
					10.3	VII	10.5	II		10.5
Miocene Middle		*Martinottiella communis*		IV	11	VI	12.5	I		12.5
				III	12.5	V	13.8			
		Cyclammina japonica		II	13.8	IV	15.5			
			Bathyal	I	15.5 / 16.5	III	16.5			
						II	17.5			
Miocene Early						I				
						-----?-----				
Oligo. Late										

are marked by conspicuous irregular erosional truncation features. Individual strata are arranged in a well-developed sigmoid progradational stacking pattern. Reflection configurations of the topset (shelf) strata are characterized by moderate-amplitude reflections with moderate continuity. The foreset (slope) strata show discontinuous reflections with low to moderate amplitudes and poor continuity. In the basin area, Units C2 and C3 show a sheet-drape geometry without significant thickness variation. Reflections at the base of the sequence are generally concordant with the top of sequence I. In Dolgorae I-well, Units C2 and C3 consist dominantly of mudstone with intercalated sandstone which were deposited in a slope to basin setting (KIER 1982; Lee 1994).

18.3.2.3 Unit C4 (Middle Miocene)

The reflection configuration of Unit C4 is similar to those of the underlying units. The shelf strata recovered in the Gorae-I and V wells consist predominantly of sandstone and silty sandstone beds intercalated with claystone layers. This unit decreases in thickness toward the basin area, where it shows shale-dominant facies in the Dolgorae-I well. Unit C4 can be divided into three subunits. The bottom subunit is about 0.2 s twt thick and characterized by a strongly progradational stacking pattern; the shelf margin prograded more than 4 km during the deposition of this subunit. The middle subunit occurs as a thin (ca. 0.1 s twt) retrogressive parasequence set in which successively younger strata are deposited progressively landward, giving a retreat of the shelf margin of more than 2 km. The top subunit is characterized by a thick (ca. 0.3 s) progradation which is more aggradational than the bottom subunit.

18.3.2.4 Unit C5 (Middle Miocene)

Unit C5 is characterized by a strong progradation of more than 8 km. The lower boundary is correlated with a landward onlapping surface and lacks subaerial erosion; the upper boundary shows pervasive erosional truncation on the shelf. Reflection configurations of shelf and slope strata are not significantly different from those of the underlying sequences. In exploration wells, the shelf strata (Gorae-I and V) are dominated by sandstones and silty sandstones which are replaced by shale beds in the basin area (Dolgorae-I well).

18.3.2.5 Units C6 and C7 (Late Middle to Early Late Miocene)

Units C6 and C7 are characterized by rapid progradation of the shelf margin. The shelf-margin complex of these sequences prograded at least 18 km toward the northeast to north. The sequence boundaries are defined by frequent occurrences of erosional truncation and channel incision which can be traced over the entire shelf. Each unit consists of two parasequence sets. The lower parasequence sets are thin wedge-shaped bodies showing an oblique progradational stacking pattern, characteristic of the forced-regressive systems tract of Hunt and Tucker (1992). The progradational stacking pattern consists of a number of dipping parasequences terminating updip by toplap at or near the flat upper depositional surface, and downdip by downlap against the lower depositional surface. The upper parasequence sets are characterized by a sigmoid progradational stacking pattern with a minor component of aggradation. The shelf strata show moderate- to high-amplitude (sub)parallel reflections with moderate continuity, whereas the prograding slope strata have moderate to high amplitude reflections with poor continuity.

Near the Dolgorae thrust belt, Units C6 and C7 comprise gently dipping, wedge-shaped units filling the depression between the uplifted thrust belt and the axis of the frontal fold. Away from the thrust belt, these sequence units show a very high ratio of progradation to aggradation with unconformable sequence boundaries. However, the forced-regressive systems tracts with oblique progradational patterns are not distinct. Instead, thin lens- or wedge-shaped bodies of a lowstand fan are identified at the base of the slope. The rapid progradation may be attributed to frequent intercalation of slide/slump masses which show low-amplitude chaotic internal reflection. Individual mass-flow deposits are separated by highly reflective and continuous thin (<0.1 s twt) shelf deposits. Units C6 and C7 comprise siltstones with intercalation of claystone and thin sandstone layers, deposited most likely in a shallowing-upward environment from a bathyal to a coastal setting (KIER 1982).

18.3.2.6 Units C8, C9 and C10 (Late Miocene to Present)

Units C8, C9 and C10 in the basin center consist of four parasequence sets from bottom to top: oblique progradational (forced regressive), sigmoid progradational (lowstand), retrogradational (transgressive) and sigmoid progradational (highstand) stacking patterns. The sequence boundaries comprise downlap- or rarely onlap-bounding surfaces which include some erosional truncations or fluvial incisions. The shelf deposits with sheet-drape geometry are characterized by continuous and subparallel reflection configurations with higher amplitude and frequency. The sediment cores are predominantly sand layers alternating with thinner silt to clay layers; fossil assemblages suggest a progressive deepening-upward from coastal to outer neritic environments (KIER 1982; Lee 1994). The foresets near the shelf break and upper slope region predominantly exhibit very poorly organized reflections or reflection-free configurations. These reflection characteristics suggest frequent slope failures and subsequent sediment gravity flows (i.e. slump, slide, and debris flows) in an unstable steep slope setting.

18.4 Eastern Continental Margin of Korea

The eastern continental margin of Korea is the north-south trending tectonic transition zone between the eastern Asian continent and the East Sea back-arc basin (Yoon and Chough 1995). This transition zone is presently occupied by a narrow continental shelf and steep continental slope bounded on the east by the Ulleung Basin and the Korea Plateau. Yoon (1994) analyzed the sequence stratigraphy and geologic structures of this marginal area based on single-channel airgun profiles.

18.4.1 Acoustic Basement

Along the eastern Korean continental margin, the acoustic basement generally lies 0.2–3.0 s twt bsl (Fig. 18.6) and exhibits a faintly stratified or opaque reflection.

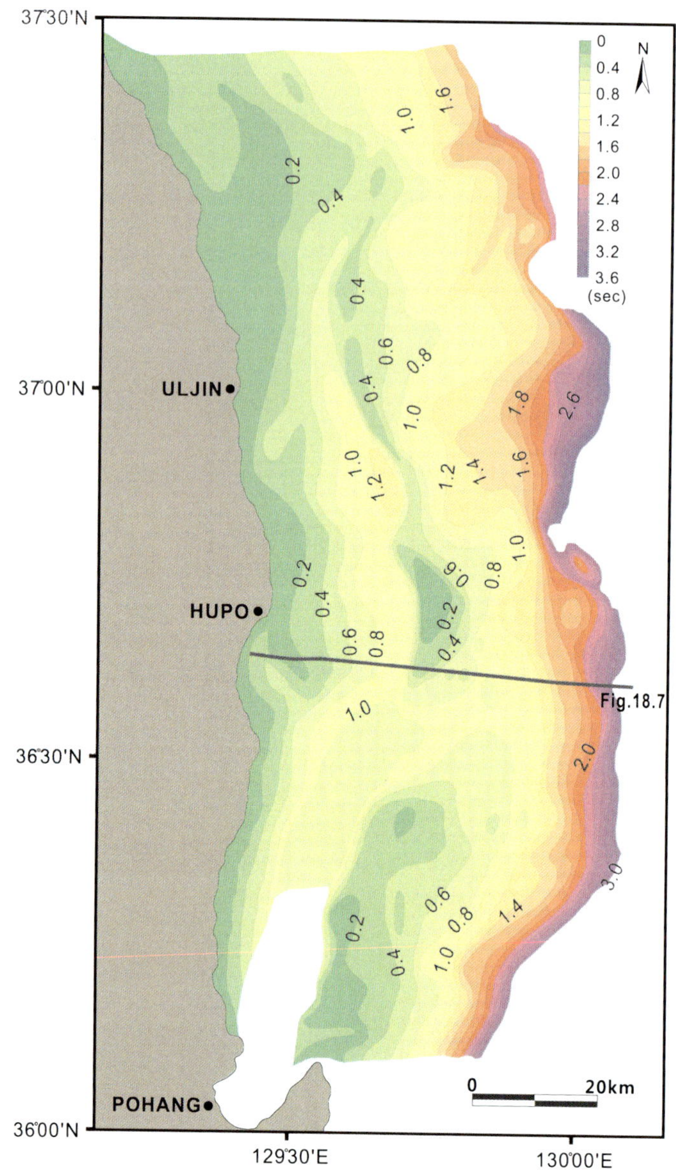

Fig. 18.6 Time-structure map of the acoustic basement in the eastern Korean continental margin. Contours are in seconds (two-way travel time below sea level) (from Yoon 1994)

It is most likely an extension of the Precambrian-Mesozoic metamorphic and igneous complexes that are exposed along the adjacent coastal region and were intruded by volcanics in the continental slope. The basement is characterized by a complex series of elongated ridges, troughs, and domes. The ridges and troughs generally trend either NE-SW or N-S (Fig. 18.6) and are bounded by basement

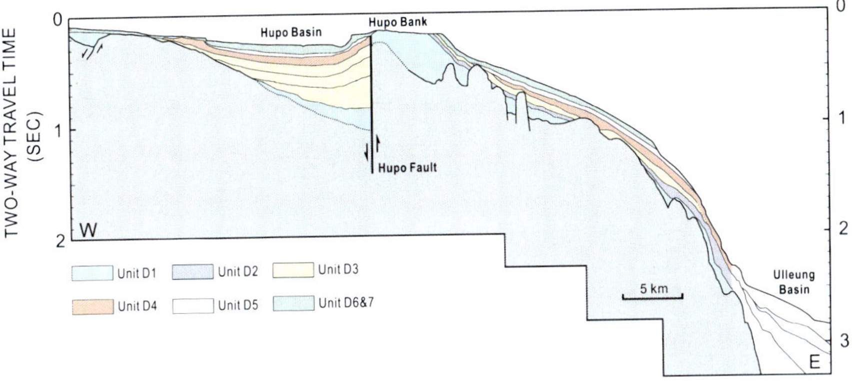

Fig. 18.7 Interpreted seismic profile crossing the eastern Korean continental margin. For location, see Fig. 18.6 (from Yoon 1994)

faults and basement/sequence-involved faults. The irregular topography of the basement delineates shelf- or slope-perched, full- or half-graben, including the Pohang, Hupo, and Mukho basins (Yoon 1994) where thick Neogene sediments accumulated (Fig. 18.7). Beneath the lower slope, the acoustic basement is much steeper (>30°) than that of the upper- to mid-slope, and forms a buried escarpment with relatively subdued surface morphology and small-scale hyperbolic reflectors (Fig. 18.7). Yoon et al. (1997) traced the basement escarpment along the entire western margin of the Ulleung Basin and suggested that this could be part of the Ulleung Fault.

18.4.2 Sedimentary Sequences

The eastern continental margin of Korea is covered with a sedimentary sequence ranging in thickness from a few tens of meters to 1200 m depending on basement topography (Schluterr and Chun 1974). Yoon (1994) defined the seismic stratigraphy of seven seismic sequence units (here renamed Unit D1 to D7) (Fig. 18.7). These sedimentary units are usually bounded by extensive erosional or nondepositional unconformities.

18.4.2.1 Unit D1 (Early to Middle Miocene)

The lowermost unit (Unit D1) is acoustically faintly- to well-stratified and deformed by reverse faults and folds. The unit largely occurs within basement lows and reaches to the coast only in the southern Youngduk Basin (Yoon 1994). More than 0.6 s twt of Unit D1 accumulated in intra-shelf and slope-perched basins (basement lows) and beneath the lower continental slope, whereas the shelf north of Hupo and basement highs in the slope region are barren of Unit D1. This unit can be correlated with sedimentary deposits exposed on land around Pohang city, based on spatial distribution and similarity of the internal sequence configuration (e.g. structural deformation) (Huntec Ltd. 1968). In the upper slope, the onlap-filling of Unit D1 strata above the rough-surfaced basement indicates that the unit consists dominantly of mass-flow deposits. These terrigenous sediments were funneled actively by fan-delta systems onshore and other paralic environments. However, transport farther downslope by the mass flows seems to have been significantly hindered by the mid-slope basement highs. On the mid-to-lower slope, Unit D1 is relatively thinner and the draping of the mid-to-lower slope deposit reflects an increased proportion of hemipelagic sediments. These indicate a significantly lowered accumulation rate of sediments from the shelf and the upper slope. This is attributed to basement highs in the mid-slope hindering sediment supply farther downslope and a steeper slope basement, which resulted in mass flows bypassing the area. Unit D1 appears to have been significantly deformed, especially in the fault-bounded basement lows, and shows various compressional structures. The upper boundary of Unit D1 is generally an even and highly-reflective surface below which reflectors terminate laterally, forming an erosional truncation. This unconformable boundary is correlated with the Late Miocene (6.5 Ma) unconformity of the Dolgorae wells (Yoon 1994). The widespread subaerial exposure of Unit D1 seems to have been controlled by eustatic sea-level fall coupled with tectonic uplift.

18.4.2.2 Unit D2 (Late Miocene)

Unit D2 occurs mainly as sheet- or wedge-form deposits in the slope region. It is absent on the shelf and on some basement highs on the slope. This unit reaches more than 0.3 s twt in thickness (average less than 0.1 s twt) in the lower slope. The unit is characterized by subparallel reflection configurations with variable amplitude and continuity. The reflections lap seaward against irregular acoustic basement or structure-controlled topographic highs of the underlying sequence unit. The onlap-fill character of the unit above the rough-surfaced basement indicates that it consists dominantly of mass-flow deposits. A downslope facies change occurs in the slope region from well-stratified reflections with high amplitude and continuity to faintly-stratified or transparent reflections with lower amplitude and continuity. This facies change is indicative of submarine slope failure (i.e. slide and slump) deposits, which have disorganized and thoroughly remolded internal structures. Such active redistribution of the sediments was interpreted to be a consequence of higher sediment input and tectonic instability.

18.4.2.3 Unit D3 (Late Miocene to Early Pliocene)

Unit D3 is a conformable set of strata characterized by subparallel reflections with variable amplitude and continuity. The lower unit boundary is generally an erosional truncation surface and largely coincides in timing with the structural deformation along the Hupo fault: Units D1 and D2 near the Hupo fault are significantly deformed and tilted, while the overlying Unit D3 is relatively undeformed. In the mid-to-lower slope, Unit D3 is further divided into two subunits, bounded by an onlap-bounding or downslope conformable surface of contrasting reflection characters. The lower subunit shows relatively continuous and higher-amplitude reflections and thins downslope, whereas the upper subunit has discontinuous and weak-amplitude reflections which grades downslope into higher amplitude, diffuse reflections. In the shelf and slope regions, the well-stratified reflection of the lower subunit indicates the alternation of hemipelagic mud layers and current-emplaced sand layers. On the other hand, the poorly-organized upper subunit probably reflects the sand-prone deposits with infrequent vertical occurrences of mud layer; the downslope facies change into higher-amplitude reflection suggests that increased mud interlayering is causing an acoustic impedance contrast.

Between 36° 20′ and 37°15′N, Unit D3 is thick (>0.4 s twt in maximum thickness) in a half-graben of the Hupo Basin and in the slope region off the Hupo Bank (Fig. 18.7). Unit D3 in the Hupo Basin is characterized by subparallel, continuous, high-amplitude reflections. It dominantly shows an onlap termination over the eastward-dipping (2.0–4.5°) erosional upper boundary of Unit D1 or the acoustic basement; individual reflectors are slightly (less than 1°) inclined eastward. Particularly within the lower subunit, the inclination of individual reflectors is steeper in the lower part of the unit. This differential tilting of individual reflectors suggests that the western block of the Hupo fault has progressively tilted toward the fault plane; hence, the lower subunit is interpreted as a syn-deformational sedimentary deposit. Along the eastern margin of the Hupo Basin, the unit consists of aggrading sediment aprons built out from the Hupo Bank (Fig. 18.7). Unit III in the slope is less than 0.35 s twt in thickness, which is much thinner than that within the Hupo Basin.

18.4.2.4 Units D4 and D5 (Early to Late Pliocene)

Units D4 and D5 consist of continuous, parallel and well-stratified reflectors, and are thick in the Hupo Basin and the slope region. In the shelf and upper slope regions, the base of Unit D4 is generally marked by an erosional unconformity, suggestive of widespread subaerial exposure during a lowstand of relative sea level. Off the coast between Hupo and Samchuck, Unit D4 is offset or significantly tilted near the Hupo fault indicating fault reactivation; the structural tilting is evinced by onlapping and interval thinning of Unit D4. In the Hupo Basin, Units D4 and D5 comprise subparallel, continuous reflections with high amplitude; the combined thickness of these two units reaches up to 0.2 s twt in the center of the basin and decreases progressively toward both northern and southern margins. Locally, prograding deltas near the present shelf break are delineated in the upper part of Unit D4, which grades basinward into a sheet-form deposit (Fig. 18.7). In the eastern margin, wedge-shaped sediment aprons are built out from the Hupo Bank and taper toward the center of the basin, showing a downlap termination. These sediment aprons show predominantly transparent or disorganized reflections indicating mass failures such as slumps, slides and debris flows. Frequent mass failures were likely a result of active readjustment of the Hupo fault and other deformation structures. In the slope region, Units D4 and D5 have various external forms including sheet-drape, wedge-shaped and lens-shaped geometries; their occurrence seems to depend on the deformation features of the underlying sequences because they fill the lows on the structural flanks. In the steep lower slope off the Hupo Bank, the well-stratified reflections of Units D4 and D5 abruptly pass into poorly-stratified and disorganized seismic reflections and sequence boundaries become more or less indistinct. This downslope seismic facies change indicates frequent slope failures or sediment bypass due to a higher slope gradient. Toward the basin plain of the Ulleung Basin, Units D4 and D5 grade into a poorly-organized sequence unit and an overlying acoustically well-stratified sequence unit, respectively.

18.4.2.5 Units D6 and D7 (Quaternary)

Units D6 and D7 occur dominantly in the Hupo Basin and slope region; they are thin below seismic resolution in the continental shelf. These units are characterized by well-stratified reflection with continuous and high-amplitude reflectors. The lower sequence boundaries are commonly characterized by widespread sub-aerial exposures, which are indicated by erosional truncation and frequent channel incision. Several folds are identified mainly within Unit D6; they are mostly restricted to the slope region bordering on the Korea Plateau, north of latitude 37° 10′N.

In the southern part of the eastern Korean continental margin, each unit is generally less than 0.08 s twt in thickness and pinches out landward, showing onlap termination at the base of the unit. In the Hupo Basin, each unit has a sheet-drape geometry that generally thickens eastward; the thickness of individual units is mostly less than 0.05 s twt. Near the eastern margin of the Hupo Basin, erosional channeling is also recognized at the top of these units. In the mid-slope region, Units D6 and D7 show lens-shaped geometries. Landward thinning is caused by onlap at the base of each unit, and the downslope thinning results from downlap around structural highs and thinning of individual strata on the steep slope. Stacking of individual strata within each unit shows dominantly simple draping with onlap-fill geometry.

In the northern continental margin, Unit D6 commonly forms a prograding clinoform in the landward margin of the unit which can be placed within the highstand system tract because it overlies downslope onlapping strata (i.e.

transgressive systems tract). Near the Hupo Fault, Units D6 and D7 progressively infill the structure-controlled depressions, showing onlap- or divergent-fill geometry.

In the steep lower slope between 36° 25′N and 37° 00′N, these units merge into a faintly-stratified or acoustically transparent deposit (less than 0.1 s twt thick) which is interpreted to have been emplaced by large-scale slope failures dislocating upslope sequences. At the base of the slope, Units D4 and D5 grade into the uppermost sedimentary unit of the Ulleung Basin that includes an acoustically well-stratified sequence.

References

Barg E (1986) Cenozoic geohistory of the southwestern margin of the Ulleung Basin, East Sea. Dissertation, Seoul National University

Chough SK, Barg E (1987) Tectonic history of Ulleung Basin margin, East Sea (Sea of Japan). Geology 15:45–48

Chough SK, Lee KE (1992) Multi-stage volcanism in the Ulleung back-arc basin, East Sea (Sea of Japan). Isl Arc 1:32–39

Chough SK, Lee HJ, Yoon SH (2000) Marine geology of Korean seas. Elsvier, Amsterdam

Hunt D, Tucker ME (1992) Stranded parasequences and the forced regressive wedge systems tract: deposition during base-level fall. Sediment Geol 81:1–9

Huntec Ltd. (1968) Report on the offshore geophysical survey in the Pohang area, Republic of Korea. UN ECAFE/CCOP Tech Bull 1:1–12

Jung CM, Hwang IG (2008) Tectono-stratigraphic evolution of the SW margin of Ulleung Basin since the late Middle Miocene. J Geol Soc Korea 44:241–258 (in Korean with English abstract)

KIER (1982) Petroleum resources potential in the continental shelf of Korea (Block 6 and Blocks 2, 4 and 5). KIER research report, Korea Institute of Energy and Resources, p 342 (in Korean)

Kim GB, Yoon SH, Chough SK et al (2011) Seismic reflection study of acoustic basement in the South Korea Plateau, the Ulleung Interplain Gap, and the northern Ulleung Basin: Volcano-tectonic implications for Tertiary back-arc evolution in the southern East Sea. Tectonophysics 504:43–56

KORDI (1999) Marine environment changes and basin evolution in the East Sea of Korea. Report MECBES-99, Korea Ocean Research and Development Institute, p 388 (in Korean)

Kwon YK (2005) I. Sequence stratigraphy of the Taebaek Group (Cambrian-Ordovician), Mideast Korea and II. Seismic stratigraphy of the Western South Korea Plateau, East Sea. Dissertation, Seoul National University

Kwon YK, Yoon SH, Chough SK (2009) Seismic stratigraphy of the western South Korea Plateau, East Sea: implications for tectonic history and sequence development during back-arc evolution. Geo-Mar Lett 29:181–189

Lee KE (1992) Geological structure of Ulleung back-arc basin, East Sea. Dissertaiton, Seoul National University

Lee HY (1994) Neogene foraminifera biostratigraphy of the southern margin of the Ulleung Basin, East Sea, Korea. Dissertation, Seoul National University

Lee GH, Suk BC (1998) Latest Neogene-Quaternary seismic stratigraphy of the Ulleung Basin, East Sea (Sea of Japan). Mar Geol 146:205–224

Lee GH, Kim HJ, Han SJ et al (2001) Seismic stratigraphy of the deep Ulleung Basin in the East Sea (Japan Sea) back-arc basin. Mar Petrol Geol 18:615–634

Lee GH, Kim HJ, Jou HT et al (2003) Opal-A/opal-CT phase boundary inferred from bottom-simulating reflectors in the southern South Korea Plateau, East Sea (Sea of Japan). Geophys Res Lett 30:1755–1758

Mitchum RM, Vail PR Jr, Sangree JB (1977). Stratigraphic interpretation of seismic reflection patterns in depositional sequences. In: Payton CE (ed) Seismic stratigraphy—applications to hydrocarbon exploration, vol 26. AAPG Memoir, pp 117–133

Park KS (1990) The seismic stratigraphy, structure and hydrocarbon potential of the Korea Strait. Dissertation, University of London

Park KS (1992) Geologic strucuture and seismic stratigraphy of the southern part of Ulleung basin. In: Chough SK (ed) Sedimentary basins in the Korean Peninsula and Adjacent seas. Korean Sedimentology Research Group Special Publication, Harnlimwon Publishers, Seoul, pp 40–59

Park SJ (1998) Stratal patterns in the southwestern margin of Ulleung back-arc basin: a sequence stratigraphic analysis. Dissertation, Seoul National University

Sato H, Amano K (1991) Relationship between tectonics, volcanism, sedimentation and basin development, Late Cenozoic, central part of Northern Honshu, Japan. Sediment Geol 74:323–343

Schluterr HV, Chun WC (1974) Seismic survey of the East Coast of Korea. UN ECAFE/CCOP Tech Bull 8:1–16

Yoon SH (1994) The eastern continental margin of Korea: seismic stratigraphy, geologic structure and tectonic evolution. Dissertation, Seoul National University

Yoon SH, Chough SK (1995) Regional strike-slip in the eastern continental margin of Korea and its tectonic implications for the evolution of Ulleung Basin, East Sea (Sea of Japan). Geol Soc Am Bull 107:83–97

Yoon SH, Park SJ, Chough SK (1997) Western boundary fault systems of Ulleung back-arc basin: further evidence of pull-apart opening. Geosci J 1:75–88

Yoon SH, Park SJ, Chough SK (2002) Evolution of sedimentary basin in the southwestern Ulleung Basin margin: sequence stratigraphy and geologic structures. Geosci J 6:149–159

Yoon SH, Chough SK, Park SJ (2003) Sequence model and its application to a Miocene shelf-slope system in the tectonically active Ulleung Basin, East Sea (Sea of Japan). Mar Petrol Geol 20:1089–1103

Index

© Springer International Publishing Switzerland 2016
K.-I. Chang et al. (eds.), *Oceanography of the East Sea (Japan Sea)*,
DOI 10.1007/978-3-319-22720-7